Shephe

Handbook of Brain Microcircuits

Handbook of Brain Microcircuits

Edited by

Gordon M. Shepherd, MD, DPhil

Department of Neurobiology
Yale University School of Medicine
New Haven, CT

Sten Grillner, MD

Nobel Institute for Neurophysiology
Department of Neuroscience
Karolinska Institutet
Stockholm, Sweden

OXFORD
UNIVERSITY PRESS
2010

OXFORD
UNIVERSITY PRESS

Oxford University Press, Inc., publishes works that further
Oxford University's objective of excellence
in research, scholarship, and education.

Oxford   New York
Auckland   Cape Town   Dar es Salaam   Hong Kong   Karachi
Kuala Lumpur   Madrid   Melbourne   Mexico City   Nairobi
New Delhi   Shanghai   Taipei   Toronto

With offices in
Argentina   Austria   Brazil   Chile   Czech Republic   France   Greece
Guatemala   Hungary   Italy   Japan   Poland   Portugal   Singapore
South Korea   Switzerland   Thailand   Turkey   Ukraine   Vietnam

Published by Oxford University Press, Inc.
198 Madison Avenue, New York, New York 10016

www.oup.com

Oxford is a registered trademark of Oxford University Press

Library of Congress Cataloging-in-Publication Data

Handbook of brain microcircuits / edited by Gordon M. Shepherd, Sten Grillner.
p. ; cm.
Includes bibliographical references and index.
ISBN 978-0-19-538988-3 (alk. paper)
1. Brain.   2. Neural circuitry.   3. Neurophysiology.   I. Shepherd, Gordon M., 1933–
    II. Grillner, Sten, 1941–
[DNLM: 1. Brain—physiology.   2. Nerve Net.   3. Neurons—physiology.   4. Synaptic Transmission.
    WL 300 H2355 2010]
QP376.H36 2010
612.8'2—dc22
2009053027

ISBN-13 9780195389883
9 8 7 6 5 4 3 2 1
Printed in China
on acid-free paper

Contents

Contributors

Yuri I. Arshavsky, PhD
Institute for Nonlinear Science
University of California-San Diego
San Diego, CA

Craig H. Bailey, PhD
Department of Neuroscience
Columbia University
New York Psychiatric Institute
Kavli Insitute for Brain Sciences
New York, NY

Cornelia I. Bargmann, PhD
Howard Hughes Medical Institute
The Rockefeller University
New York, NY

Edi Barkai, PhD
Department of Neurobiology &
 Ethology
University of Haifa
Mt. Carmel, Israel

Aaron Beyerlein
Department of Neuroscience
ARL-Division of Neurobiology
University of Arizona
Tucson, AZ

Gene D. Block, PhD
David Geffen School of Medicine
University of California-Los Angeles
Los Angeles, CA

J. Paul Bolam, PhD
MRC Anatomical Neuropharmacology
 Unit
Department of Pharmacology
University of Oxford
Oxford, England

György Buzsáki, MD, PhD
Center for Molecular and Behavioral
 Neuroscience
Rutgers University
Newark, NJ

Ronald L. Calabrese, PhD
Department of Biology
Emory University
Atlanta, GA

Nirupa Chaudhari, PhD
Department of Physiology & Biophysics
 and Program in Neuroscience
Miller School of Medicine, University of
 Miami
Miami, FL

Peter Dallos, PhD
Department of Neurobiology and
 Physiology
Northwestern University
Evanston, IL

Javier DeFelipe, PhD
Instituto Cajal
Madrid, Spain

Tatiana G. Deliagina, PhD
Department of Neuroscience
Karolinska Institutet
Stockholm, Sweden

Jonathan B. Demb, PhD
Department of Ophthalmology and
   Visual Sciences
The University of Michigan Kellogg
   Eye Center
Ann Arbor, MI

Kimberly J. Dougherty, PhD
Mammalian Locomotor Laboratory
Department of Neuroscience
Karolinska Institutet
Stockholm, Sweden

Rodney J. Douglas, PhD
Institute for Neuroinformatics
Zurich, Switzerland

John Dowling, PhD
Department of Molecular & Cell
   Biology
Harvard University
Cambridge, MA

A. El Manira, PhD
Department of Neuroscience
Karolinska Institutet
Stockholm, Sweden

Donald S. Faber, PhD
Department of Neuroscience
Albert Einstein College of Medicine
Bronx, NY

Jack L. Feldman, PhD
Department of Neurobiology
David Geffen School of Medicine
University of California Los Angles
Los Angeles, CA

Joseph R. Fetcho, PhD
Department of Neurobiology and
   Behavior
Cornell University
Ithaca, NY

Wolfgang Otto Friesen, PhD
Department of Biology
University of Virginia
Charlottesville, VA

Apostolos P. Georgopoulos, MD, PhD
University of Minnesota
Minneapolis, MN

Ann M. Graybiel, PhD
Department of Brain and Cognitive
   Sciences
McGovern Institute for Brain Research
Massachusetts Institute of Technology
Cambridge, MA

Sten Grillner, MD
Nobel Institute for Neurophysiology
Department of Neuroscience
Karolinska Institutet
Stockholm, Sweden

Robert D. Hawkins, PhD
Department of Neuroscience
Columbia University
New York Psychiatric Institute
New York, NY

John G. Hildebrand, PhD
Department of Neuroscience
ARL-Division of Neurobiology
University of Arizona
Tucson, AZ

Bryan M. Hooks, PhD
Janelia Farm Research Center
Howard Hughes Medical Institute
Ashburn, VA

Tadashi Isa, MD, PhD
Department of Developmental
    Physiology
National Institute for Physiological
    Sciences
Okazaki, Japan

Masao Ito, MD, PhD
Lab Memory & Learning
RIKEN Brain Science Institute
Wako, Japan

Luke R. Johnson, PhD
Center for the Study of Traumatic
    Stress (CSTS)
Department of Psychiatry and Program
    in Neuroscience
Uniformed Services University (USU)
Bethesda, MD

Edward G. Jones, MD, DPhil
Center for Neuroscience
University of California-Davis
Davis, CA

Eric R. Kandel, MD
Department of Neuroscience
Columbia University
New York Psychiatric Institute
Kavli Insitute for Brain Sciences
Howard Hughes Medical Insitute
New York, NY

Katsuyuki Kaneda, PhD
Department of Developmental
    Physiology
National Institute for Physiological
    Sciences
Okazaki, Japan

Paul S. Katz, PhD
The Neuroscience Institute
Georgia State University
Atlanta, GA

Ole Kiehn, MD, PhD
Mammalian Locomotor Laboratory
Department of Neuroscience
Karolinska Institutet
Stockholm, Sweden

Arlette Kolta, PhD
Faculté de Médecine Dentaire
Université de Montréal
Montréal, Canada

Henri Korn, MD
Department of Neuroscience
Institut Pasteur
Paris, France

William B. Kristan, Jr., PhD
Neurobiology Section
Division of Biological Sciences
University of California-San Diego
San Diego, CA

Simon B. Laughlin, PhD
Department of Zoology
University of Cambridge
Cambridge, England

Joseph E. LeDoux, PhD
Center for Neural Science
New York University
New York, NY

Hong Lei, PhD
Department of Neuroscience
ARL-Division of Neurobiology
University of Arizona
Tucson, AZ

Wenchang Li, PhD
School of Biology
University of St. Andrews
St. Andrews, Scotland

Rodolfo R. Llinás, MD, PhD
Department of Physiology and
 Neuroscience
NYU Medical School
New York, NY

James P. Lund, PhD
Faculty of Dentistry
McGill University
Montréal, Canada

Gabriella B. S. Lundkvist, PhD
Karolinska Institutet
Stockholm, Sweden

Eve Marder, PhD
Volen Center
Brandeis University
Waltham, MA

Henry Markram, PhD
Brain Mind Institute, EPFL
Lausanne, Switzerland

Kevan A. C. Martin, PhD
Institute for Neuroinformatics
Zurich, Switzerland

David A. McCormick, PhD
Department of Neurobiology
Yale University School of Medicine
New Haven, CT

Randolf Menzel, PhD
Institut fuer Neurobiologie
Freie Universitat
Berlin, Germany

Michele Migliore, PhD
Institute of Biophysics
National Research Council
Palermo, Italy

Edvard I. Moser, PhD
Norwegian University of Science and
 Technology
Medical-Technical Research Centre
 (NTNU)
Centre for the Biology of Memory
Trondheim, Norway

May-Britt Moser, PhD
Norwegian University of Science and
 Technology
Medical-Technical Research Centre
 (NTNU)
Centre for the Biology of Memory
Trondheim, Norway

Frédéric Nagy, PhD
INSERM
Neurocentre Magendie
Institut des Neurosciences de Bordeaux
Bordeaux, France

Donata Oertel, PhD
Department of Physiology
University of Wisconsin-Madison
 Medical School
Madison, WI

Lynne A. Oland, PhD
Department of Neuroscience
ARL-Division of Neurobiology
University of Arizona
Tucson, AZ

Grigori N. Orlovsky, PhD
Department of Neuroscience
Karolinska Institutet
Stockholm, Sweden

Jeffrey A. Riffell, PhD
Department of Neuroscience
ARL-Division of Neurobiology
University of Arizona
Tucson, AZ

Stephen D. Roper, PhD
Department of Physiology & Biophysics
    and Program in Neuroscience
Miller School of Medicine, University of
    Miami
Miami, FL

Edwin W. Rubel, PhD
Virginia Merrill Bloedel Hearing
    Research Center
Department of Otolaryngology–Head
    and Neck Surgery
University of Washington
Seattle, WA

Jürgen Rybak, PhD
Institut fuer Neurobiologie
Freie Universitat
Berlin, Germany

Jason Tait Sanchez, PhD
Virginia Merrill Bloedel Hearing
    Research Center
Department of Otolaryngology–Head
    and Neck Surgery
University of Washington
Seattle, WA

Gordon M. Shepherd, MD, DPhil
Department of Neurobiology
Yale University School of Medicine
New Haven, CT

Gordon M. G. Shepherd, MD, PhD
Department of Physiology
Northwestern University
Chicago, IL

S. Murray Sherman, PhD
Department of Neurobiology
The University of Chicago
Chicago, IL

Keith T. Sillar, PhD
School of Biology
University of St. Andrews
St. Andrews, Scotland

Peter Somogyi, FRS, FMedSci
MRC Anatomical Neuropharmacology
    Unit
Department of Pharmacology
University of Oxford
Oxford, England

Costas N. Stefanis, MD
University Mental Health Research
    Institute

Nicholas J. Strausfeld, PhD
Department of Neurobiology
Center for Insect Science
University of Arizona
Tucson, AZ

Karel Svoboda, PhD
Janelia Farm Research Center
Howard Hughes Medical Institute
Ashburn, VA

Joseph B. Travers, PhD
Neuroscience Graduate Studies
    Program
College of Dentistry
The Ohio State University
Columbus, OH

Susan P. Travers, PhD
Neuroscience Graduate Studies
    Program
College of Dentistry
The Ohio State University
Columbus, OH

Peter Wallén, PhD
Nobel Institute for Neurophysiology
Department of Neuroscience
Karolinska Institutet
Stockholm, Sweden

Xiao-Jing Wang, PhD
Department of Neurobiology
Kavli Institute for Neuroscience
Yale University School of Medicine
New Haven, CT

Yuan Wang, PhD
Virginia Merrill Bloedel Hearing
    Research Center
Department of Otolaryngology–Head
    and Neck Surgery
University of Washington
Seattle, WA

Frank S. Werblin, PhD
Department of Molecular & Cell
    Biology
Division of Neurobiology
University of California-Berkeley
Berkeley, CA

David C. Willhite, PhD
Department of Neurobiology
Yale University School of Medicine
New Haven, CT

Charles J. Wilson, PhD
Department of Biology
University of Texas-San Antonio
San Antonio, TX

Donald A. Wilson, PhD
Emotional Brain Institute
Nathan Kline Institute for Psychiatric
    Research
New York University School of
    Medicine
New York, NY

Menno P. Witter, PhD
Department of Anatomy
VU University Medical Center
Amsterdam, The Netherlands

Eric D. Young, PhD
Department of Neuroscience
Johns Hopkins University
Baltimore, MD

Pavel V. Zelenin, PhD
Department of Neuroscience
Karolinska Institutet
Stockholm, Sweden

Introduction

The term *brain microcircuit* is defined for the purposes of this handbook as the way that nerve cells (and associated cells such as glia) are organized to carry out specific operations within a region of the nervous system.

The concept dates back to the founding of cellular neuroscience by the great pioneer histologist Santiago Ramón y Cajal. His monumental textbook of 1911, based on the stain invented by Camillo Golgi, provided comprehensive coverage of cell types throughout the nervous system. He also began to address the underlying principles of how the nerve cells are organized to carry out functional operations in the different regions. The first diagrams of this nature showed the cellular organization of the retina and the olfactory bulb, which he used to make the fundamental point that in both regions the flow of activity went through the dendrites as well as the axons, to prove that dendrites play an essential role in nervous activity. These diagrams were also the first to align two regions side by side in order to carry out detailed comparisons between them and to begin to extract organizing principles that are common across regions. In these respects, Cajal both experimentally and conceptually was the founder of the approach that underlies this handbook.

An allied step was the synthesis by von Waldeyer in 1891 of the new evidence from Cajal and the other pioneering histologists in which he introduced the term *neuron* and the concept of the "neuron doctrine." Among the regions cited, the spinal cord was of special significance. Together with the physiological investigations of Charles Sherrington beginning in that same period, the spinal cord has played a central role in developing the concepts of the functional organization of a brain region. In the 1950s John Eccles and his collaborators introduced intracellular recording techniques to work out the specific interneuronal circuits involved in the reflex activity of the spinal cord (see Eccles, 1957), the first examples of the kinds of functional cellular connectivity represented in this book.

The impetus toward the present volume got another big boost from the introduction of electron microscopy in the 1950s. By the 1970s, it was possible to identify not only the main types of synaptic connections between specific cell types in selected brain regions but also to provide, in correlation with intracellular recordings, the first diagrams of the synaptic circuits. One term introduced for these was *local circuits* (Rakic, 1976). Another was *basic circuits* (Shepherd, 1974). A third was *microcircuit*, in analogy with the new technology in the computer industry of building complex processing units on single chips (Byrne et al., 1978; Shepherd, 1978). A final suggestion has been *canonical circuit*, similar to *basic circuit*, to represent the idea of generic types of circuits carrying out generic types of functional operations (Douglas and Martin, 1990).

Much of this early data was summarized in *The Synaptic Organization of the Brain*, first published in 1974, most recently in a fifth edition (2004). It provided in-depth chapters on 10 of the best-understood regions of the nervous system. Successive editions incorporated pharmacology, biochemistry and molecular biology of the synaptic interactions, as well as direct recordings from dendrites with patch recordings and computational simulation of the specific dendritic and microcircuit propetrties. In contemplating a new edition, it seemed that the time had come to extend the coverage to many more brain regions, while at the same time focusing on the circuits themselves.

During this period the concept of microcircuits was gaining widespread utility. This led to an extensive review by Grillner et al. (2005) of five regions of critical importance in the neural organization of behavior. During this time, a Dahlem conference was held under the chairs of Sten Grillner and Ann Graybiel, which provided a forum for deeper exploration of the microcircuit concept, resulting in the book *Microcircuits*, (Graybiel and Grillner, 2006) the first book dealing with the new concept. The recent literature in neuroscience contains increasing application of the term to local organization in many brain regions.

There has thus been a convergence of recent work to characterize synaptic organization under the rubric of brain microcircuits. The result is the present volume, embracing some 50 regions of the nervous system, described by over 80 leaders in their respective fields. Building on the foundation extending from Cajal to the present, our collective aim is severalfold. First, we wish to distil the current knowledge of synaptic and functional organization of each brain region so that the most basic aspects can be summarized in synthesizing circuit diagrams. Second, these diagrams represent specific types of operations, and in so doing function as canonical circuits, that is, circuits that can be identified as carrying out generic operations that are essential to what a region contributes to the neural basis of behavior. Finally, by gathering these circuits within one volume, we hope it will be possible to begin to identify the canonical operations across different regions that within each region are fine-tuned to the particular form of the information in that region and the output targets for the operations.

We welcome feedback from readers on the utility of the microcircuit approach to giving deeper insight into the underlying principles of brain organization and function, and in practical terms providing hypotheses that can be tested experimentally and computationally. The present selection of chapters represents most of the best understood brain regions at present in terms of their microcircuit organization. It is estimated that the mammalian brain comprises at least some 400 different brain regions, so there is much more work to be done. We welcome suggestions from readers for additional brain regions to be included in the projected expansion of the handbook in the future.

REFERENCES

Byrne JH, Castellucci VF, Kandel ER (1978) Contribution of individual mechanoreceptor sensory neurons to defensive gill-withdrawal reflex in Aplysia. *J Neurophysiol* 41: 418–431.

Cajal, S Ramon y (1911) Histologie du systeme nerveux de l'homme et des vertebres (Translation by L. Azoulay). Paris: Maloine, 2 vols). English translation: Histology of the Nervous System of Man and Vertebrates by S. Ramon y Cajal, (Transl. by N Swanson and L Swanson) New York: Oxford University Press, 1995.

Douglas R, Martin KAC (1991) A functional microcircuit for cat visual cortex. *J. Physiol* 440:735–769.

Eccles JC (1957) The Physiology of Nerve Cells. Baltimore: Johns Hopkins University Press.

Graybiel AM, Grillner S (2006) Microcircuits: The Interface between Neurons and Global Brain Function. Cambridge MA: MIT Press.

Rakic P (ed.) (1975) Local circuit neurons. *NRP Bull* 3:291–446.

Shepherd GM (1974) The Synaptic Organization of the Brain. New York: Oxford University Press. (5th ed., 2004).

Shepherd GM (1978) Microcircuits in the nervous system. *Sci Am* 238:93–103.

Silberberg G, Grillner S, LeBeau FE, Maex R, Markram H. (2005) Synaptic pathways in neural microcircuits. *Trends Neurosci* 28:541–551.

Waldeyer-Hartz HWG (1891) Uber einige neuere Forschungen im Gebiete der Anatomie des Centralnervensystems. *Deutsch med Wschr* 17: 1213–1218.

Part I
Vertebrates

Section 1

Neocortex

1

Neocortical Microcircuits

Javier DeFelipe and Edward G. Jones

From the earliest studies into the cerebral cortex, researchers have been trying to represent the components of the cortex and their possible connections in simplified schemes. The early wiring diagrams attempted to illustrate the flow of information from input to output. The introduction of bioinformatics into computational neuroscience, and computer simulations are now making it possible to learn more about the role of each element in the input–output circuit. The ultimate goal is to try to understand the functional implications of cortical organization using circuit diagrams more detailed than previously possible. This is a particularly difficult task when dealing with the neocortex not only because of its complexity but also because of the variations observed between different cortical areas and across species. In the present chapter, we first describe some general aspects of the main neuronal components of the neocortex and then reflect on how these components are interconnected in a general, basic microcircuit, followed by a consideration of the variations that arise in this basic microcircuit (specialized microcircuits).

NEURONAL COMPONENTS OF THE NEOCORTEX: GENERAL ASPECTS

The neurons of the neocortex can be grouped into three major classes: pyramidal cells, spiny nonpyramidal cells, and aspiny nonpyramidal cells. *Pyramidal cells* are located in all layers except layer I, and they are the most abundant cortical neurons (estimated at 70%–80% of the total population). They represent the vast majority of projection neurons, but their axons have many intracortical collaterals. Because the cells are glutamatergic neurons, these collaterals are a main source of excitatory synapses in the cortex.

In addition to size and layer location, this class of neurons is frequently subdivided according to projection site (Jones, 1984). *Spiny nonpyramidal cells* are, with some exceptions, short-axon cells (spiny interneurons) that are located in the middle layers of the neocortex (especially in layer IV). Various types of spiny nonpyramidal cell have been recognized on the basis of their dendritic morphology, axonal arbors, and laminar connections (Lund, 1984). The typical spiny nonpyramidal cell (the granule cell or spiny stellate cell of layer IV) is a glutamatergic neuron whose morphology although often star shaped can adopt a form approaching that of a pyramidal cell. There is relatively little information available about this class of cell. *Aspiny nonpyramidal neurons* (smooth interneurons) are short-axon cells that have few or no dendritic spines; they adopt extremely varied morphologies, especially when the distributions of their axons are revealed. Smooth interneurons are found in all layers. They represent the vast majority of short-axon cells and 15%–30% of the total population of cortical neurons; the percentage varies by species, with fewer found in rodents than in primates. Most smooth interneurons are GABAergic and are therefore the main source of inhibitory synapses in the cortex. The different types of smooth interneurons are characterized by their axonal ramifications, particular synaptic connectivity, and the expression of a variety of cotransmitters, neuroactive peptides, or calcium-binding proteins (Peters and Jones, 1984; Hendry et al., 1989; Somogyi et al., 1998). Classically, they have been given names reflective of their morphological character, especially their axonal ramifications, but because this tends to be species specific and because some forms may not be present in all species, we have adopted a more general approach to naming them.

There are two major morphological types of cortical synapse, asymmetric and symmetric, corresponding to Gray's type I and II synapses, respectively (Peters et al., 1991). These synapses are mainly distinguished by the thickness of the postsynaptic density, with asymmetric synapses displaying a prominent postsynaptic density and symmetric synapses a thinner and less prominent one. Quantitative electron microscopy has shown that there are considerably fewer symmetric synapses than asymmetric synapses in the neocortex. In a variety of species and cortical areas that have been examined, approximately 80%–90% of the synapses are asymmetric and 10%–20% are symmetric (reviewed in DeFelipe et al., 2002). The major sources of asymmetric synapses are the axons of spiny neurons (pyramidal and spiny nonpyramidal cells) and those of extrinsic axons entering the cortex from subcortical and other cortical sites. By contrast, the vast majority of symmetric synapses are of intrinsic origin and are formed by the axons of smooth interneurons. Since spiny neurons and the major cortical afferent systems are excitatory in function, and the majority of aspiny nonpyramidal cells are GABAergic, it can be assumed that axon terminals observed forming asymmetric synapses in the cortex are excitatory and that those forming symmetric synapses are inhibitory.

A Basic Microcircuit

Although there is little truly detailed information about the synaptic connectivity of the neocortex, in general connections appear to be governed by relatively straightforward principles. Extrinsic afferent fibers coming from the thalamus or from other cortical areas, ipsilaterally or contralaterally, establish synapses with the dendritic spines and to a lesser extent with the dendritic shafts of all spiny neurons within their layers of termination, as well as with the dendrites of many smooth interneurons. The terminals of the nonspecific afferent systems from the monoaminergic systems of the brainstem and the cholinergic systems of the basal forebrain have more varied terminations, often without identifiable synaptic contacts, on both distal and proximal dendrites of spiny and nonspiny neurons. The axon collaterals of *pyramidal* cells are a major source of intrinsic connections. They establish asymmetric synapses with the somata and dendrites of smooth interneurons (but not with their axon initial segments), as well as with the dendritic spines (mainly) and the dendritic shafts of other pyramidal cells (avoiding their somata, axon initial segments, and proximal dendrites). The first asymmetric synapses on pyramidal cells are located on the most proximal spines of the dendritic arbor some 10–50 µm from the soma. The axons of *smooth interneurons* establish synapses with the somata and dendrites of other smooth interneurons (and in some cases with the axon initial segment as well), and with all regions of the pyramidal cells (dendritic shafts, somata, axon initial segments, and to a lesser extent with dendritic spines). Although little is known about the connections of *spiny stellate cells*, their axons appear mainly to form asymmetric synapses with dendritic spines of other spiny neurons. Smooth interneurons generally show a preference for certain other cell classes as postsynaptic partners, and the axons of certain types may terminate exclusively on a particular region of the postsynaptic cell. However, this is by no means an absolute phenomenon, and the axons of many smooth interneurons can be found ending on all regions of both pyramidal and nonpyramidal cells.

An intrinsic circuit with the pyramidal neuron as its central element is shown in Figure 1.1. On the basis of the preferred postsynaptic region of the pyramidal neuron that their axons target, three general groups of smooth interneurons can be defined: *axo-dendritic cells* or cells whose axons form synapses primarily with dendrites (shafts and spines); *axo-somatodendritic cells* or cells whose axons form multiple synapses with both dendrites and the somata, but with a variable preference for somata or dendrites; and *axo-axonic or chandelier cells*, whose axons synapse only with the axon initial segment.

With the exception of chandelier cells that exclusively establish synapses with pyramidal cells, the other types of interneurons so far examined form synapses with both pyramidal and spiny nonpyramidal cells and with other smooth interneurons. The thalamic and cortico-cortical afferent systems

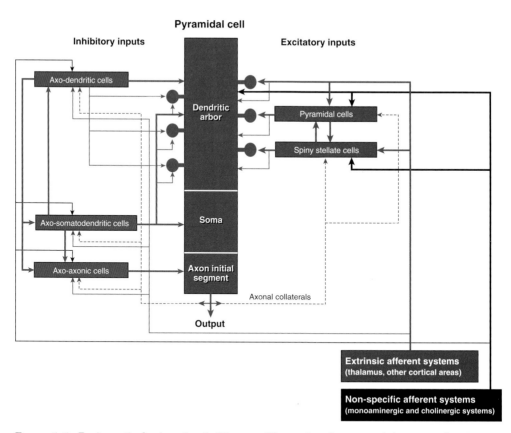

Figure 1–1. Basic cortical microcircuit. Diagram illustrating the synaptic inputs and main patterns of local connections of pyramidal cells. *Excitatory* inputs (mostly glutamate) only arrive at the dendritic arbor, and they originate from extrinsic afferent systems (coming from the thalamus or other cortical areas) or from the axon collaterals of spiny cells (pyramidal cells and spiny stellate cells). Most of these inputs terminate on dendritic spines. *Inhibitory* inputs, which mostly originate from GABAergic interneurons, terminate on the dendrites, soma, axon initial segment, and, to a lesser extent, on the dendritic spines of pyramidal cells. These interneurons receive inputs from extrinsic afferent systems, and they are interconnected among themselves, with the exception of chandelier cells that only form synapses with the axon initial segment of pyramidal cells. The terminals of the nonspecific afferent systems (coming from subcortical extrathalamic nuclei) have more varied terminations, often without identifiable synaptic contacts, on both distal and proximal dendrites. Red, glutamatergic cells; blue, GABAergic cells; black, subcortical extrathalamic afferents that use nonglutamatergic/non-GABA transmitters (e.g., serotonin, dopamine, and acetylcholine). (Modified from DeFelipe, 2002)

generally establish synapses with the dendritic spines of pyramidal and spiny stellate neurons and with the dendrites of smooth interneurons, avoiding the cell bodies of pyramidal neurons (reviewed in White, 1989). Some extrathalamic subcortical afferents may form multiple contacts (synapses or appositions) with the somata and proximal dendrites of certain smooth interneurons.

Variations of the Elemental Neocortical Microcircuit:
Specialized Microcircuits

Most research into the cerebral cortex is based on the idea that the organizational patterns of connectivity described in the foregoing are universal. However, as we will see, we cannot yet define an elemental microcircuit of the mammalian neocortex based on microanatomical/neurochemical characteristics alone. For example, the microcircuit shown in Figure 1.2 that is based on studies on the primate and cat neocortex may be not be identical in other species. A comprehensive account of the intrinsic circuitry of the cortex of these animals, seen from the perspective of the different cell classes and their relationships to one another seen from the perspective of thalamic and pyramidal cell collateral inputs, can be found in Douglas et al. (2004). The following sources of variation need to be taken into account.

The neocortex shows variations in both the vertical and horizontal dimensions. In the vertical dimension, the cortex is divided into layers and

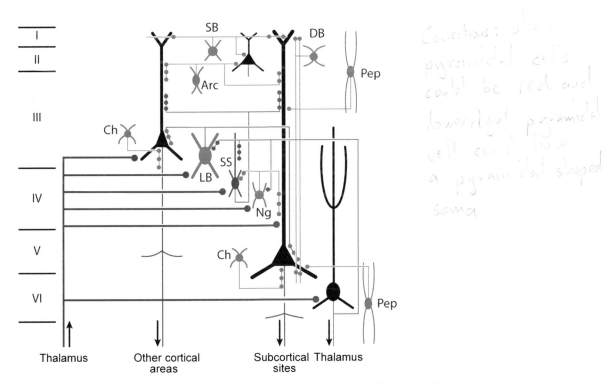

Figure 1–2. A schematic circuit based upon the known cortical cells upon which thalamic afferent fibers terminate in cats and monkeys. The GABAergic smooth interneurons (blue) are identified by the names that they have received in these species. Arc, neuron with arciform axon; Ch, chandelier cell; DB, double bouquet cell; LB, large basket cell; Ng, neurogliaform cell; Pep, peptidergic neuron. Excitatory neurons (red) include pyramidal cells of layers II–VI and the spiny stellate cells (SS) of layer IV. (Based on Jones, 2007)

sublayers with significant variation depending on species and cortical area. The classical division of the neocortex into six layers has many exceptions in the adult animal. Layer IV as seen in a Nissl stain, for example, is lost from the motor and premotor areas as they develop and these areas thus become agranular (Huntley and Jones, 1991). More than six layers are recognized in the primary visual cortex of adult primates, and certain of the numbered layers can be clearly divided or merged in other cortical areas of primates and nonprimates (DeFelipe et al., 2002). Nor is there any absolute rule as to the layer(s) in which extrinsic afferents terminate. For thalamic afferents, this may be layers III, IV, or both (Jones and Burton, 1976). And for corticocortical afferents, there are numerous patterns of laminar termination depending on the area (Felleman and Van Essen, 1991).

In the horizontal dimension, neurons can be organized into periodic groups of synaptically associated cells that receive common inputs from or project to a particular part of the brain. These patterns vary depending on the cortical area and species (Foote and Morrison, 1987; White, 1989; Lund et al., 1994). For a given cortical area, the distribution, density, and extent of the terminal axonal arborizations of the majority of the various afferent systems (thalamic and nonthalamic) need not be homogeneous. Therefore, neurons located in different parts of the same layer could receive different numbers and proportions of synapses from cortical afferents. Among the numerous variations in the disposition of neurons, one of the most striking is the barrel cortex of rodents in which the neuronal clusters called barrels in layer IV receive one class of thalamic afferents and few or no callosal afferents, while intervening regions of layer IV receive inputs from a second class of thalamic afferents and from callosal afferents (reviewed in Keller, 1995). However, many other cytoarchitectonic variations based on differential neuronal clustering are also known (reviewed in DeFelipe et al., 2002).

There is evidence that the number of neurons and the relative proportions of excitatory and inhibitory neurons may vary from species to species. For example, in the somatosensory cortex of a pygmy shrew *(Suncus etruscus)*, whose brain weight is only 0.062 g, the neocortex is approximately 400 μm thick and there are approximately 170,000 neurons/mm^3. By contrast, the brain of the mole *(Talpa europaea)* weighs 1.02 g, its neocortex is approximately 1200 μm thick, and there are only 40,000 neurons/mm^3. These densities are 7 and 1.6 times greater than in the human temporal cortex (24,186 neurons/mm^3), respectively (DeFelipe et al., 2002). In the rat, GABAergic cells represent no more than 15% of the total population of all the cortical areas (e.g., Beaulieu, 1993), whereas in the macaque monkey they constitute 20% in the visual cortex and up to 25% in other cortical areas (e.g., Hendry et al., 1987; Beaulieu et al., 1992).

The clear difference in the proportions of GABAergic neurons in the cortex of rodents and primates and the presence in the primate cortex of smooth interneurons with morphologies unlike those found in rodents, along with differences in the developmental origins of the GABAergic cortical interneurons in rodents and primates, has suggested that in the course of evolution of the primate cortex more GABA neurons and newer forms of GABA neurons may have appeared (DeFelipe et al., 2002; Jones, 2009). A striking example is the GABAergic double bouquet cell described originally by Santiago Ramón y Cajal as a prominent feature of the human cortex and later discovered in monkeys as well (Jones, 1975; Szentágothai, 1975; Valverde, 1978; Fig. 1.2) but not in rodents or lagomorphs and only variably present in carnivores (Ballesteros-Yáñez et al., 2005). The descending bundles of axon collaterals of the double bouquet cells form the basis of a dense series of microcolumns extending through layers II–VI of the primate cortex. They are the source of a large number of GABAergic synapses on small dendritic shafts and dendritic spines of pyramidal cells within a very narrow column of cortical tissue (DeFelipe et al., 1990). The radial fasciculi (vertical bundles of myelinated axons originating from aggregations of pyramidal cells) and the axonal columns of the double bouquet axons overlap on a one-to-one basis (Ballesteros-Yáñez et al., 2005). If the double bouquet cell is not represented in rodents, then clearly the microcircuit of the rodent cortex may lack an element of connectivity that is fundamental to that of the primate. All or almost all cortical GABAergic interneurons in rodents arise from the ganglionic eminence at the base of the developing cerebral hemisphere, while in monkeys and humans a significant number is born in the neuroepithelium of the lateral ventricular wall as well (reviewed in Jones, 2009). Conceivably, this latter source may be the origins of the additional numbers and additional forms of GABergic cortical interneurons in the primate cortex. With them may have come additional elements of intrinsic cortical circuitry.

There are important variations in the individual neurons contributing to the basic microcircuit shown in Figure 1.1, both in terms of morphology and neurochemical characteristics. For example, the basal dendritic arbor of pyramidal cells displays remarkable differences in terms of size, number of bifurcations, and spine density between comparable cortical areas of different species (e.g., Elston, 2003). Many other examples of atypical cells have now been reported. In the sensorimotor cortex of the cat (areas 3a and 4), the apical dendrites of layer V pyramidal neurons do not ascend straight toward the pial surface as pyramidal cells typically do. After a short vertical ascent, these dendrites bifurcate just above layer V and give rise to a large number of U-shaped secondary dendrite branches (Fleischhauer, 1974). This variation suggests that these pyramidal cells have a broader extent of surface available for the receipt of synapses than other pyramidal cells and thus any microcircuit of which they form a part will be different from that formed by pyramidal cells with less extensive dendritic arborizations. Some layer V pyramidal cells in the monkey and cat motor cortices, and in the cat somatosensory, parietal, and visual cortices, are almost devoid of dendritic spines (reviewed in DeFelipe and Fariñas, 1992). Variations in spine density imply variations in excitatory connections because each spine invariably receives an excitatory synapse in the cortex.

The neurochemical characteristics of cortical inhibitory interneurons can also vary across species. Although all appear to be GABAergic, their expression of neuropeptide cotransmitters and calcium-binding proteins can differ. Chandelier cells, for example, contain the calcium-binding protein parvalbumin and sometimes a second, calbindin, but not calretinin. In addition, these cells contain the neuropeptide corticotropin-releasing factor but not other neuropeptides such as cholecystokinin, somatostatin, neuropeptide Y, vasoactive intestinal polypeptide, or tachykinin (reviewed in Jones, 1993; DeFelipe, 1999). However, chandelier cells are chemically heterogeneous because the expression of these substances varies across species and across cortical areas and layers. For example, chandelier cell axon terminals containing corticotropin-releasing factor are found in the prefrontal and occipital cortex of the squirrel monkey and not in those areas of macaques (Lewis and Lund, 1990). In the prefrontal cortex, these terminals are located mainly in layer IV. Calbindin immunocytochemistry labels a small subpopulation of chandelier terminals mainly located in layers V and VI of the human neocortex, but not in other species like the macaque monkey where chandelier cell axon terminals are parvalbumin positive (DeFelipe, 1999). These and other variations in the neurochemical characteristics and laminar distributions of chandelier cell axon terminals imply that the circuits into which they enter are also different.

CONCLUSION

There are numerous common features of cortical organization that make it possible to define an elementary cortical microcircuit centered on the pyramidal cell that may be common to all neocortical areas and to all mammalian species (Fig. 1.1). However, there are numerous differences in laminar and sublaminar organization, in the nature and distributions of cortical afferents, and in the numbers, morphological variations, and neurochemical characteristics of the contributing cells across cortical areas and across species. These differences suggest the presence of specialized microcircuits that may be area and species specific and determined by evolutionary adaptations and functional demands (Fig. 1.2).

REFERENCES

Ballesteros-Yáñez I, Muñoz A, Contreras J, Gonzalez J, Rodriguez-Veiga E, DeFelipe J (2005) The double bouquet cell in the human cerebral cortex and a comparison with other mammals. *J Comp Neurol* 486:344–360.

Beaulieu C (1993) Numerical data on neocortical neurons in adult rat, with special reference to the GABA population. *Brain Res* 609:284–292.

Beaulieu C, Kisvarday Z, Somogyi P, Cynader M, Cowey A (1992) Quantitative distribution of GABA-immunopositive and -immunonegative neurons and synapses in the monkey striate cortex (area 17). *Cereb Cortex* 2:295–309.

DeFelipe J (1999) Chandelier cells and epilepsy. *Brain* 122:1807–1822.

DeFelipe J (2002) Cortical interneurons: from Cajal to 2001. *Prog Brain Res* 136:215–238.

DeFelipe J, Fariñas I (1992) The pyramidal neuron of the cerebral cortex: Morphological and chemical characteristics of the synaptic inputs. *Prog Neurobiol* 39:563–607.

DeFelipe J, Hendry SH, Hashikawa T, Molinari M, Jones EG (1990) A microcolumnar structure of monkey cerebral cortex revealed by immunocytochemical studies of double bouquet cell axons. *Neuroscience* 37:655–673.

DeFelipe J, Alonso-Nanclares L, Arellano JI (2002) Microstructure of the neocortex: comparative aspects. *J Neurocytol* 31:299–316.

Douglas R, Markram H, Martin K (2004) Neocortex. In: Shepherd GM, ed. *The Synaptic Organization of the Brain*, pp. 499–558. New York: Oxford University Press.

Elston GN (2003) Cortex, cognition and the cell: new insights into the pyramidal neuron and prefrontal function. *Cereb Cortex* 13:1124–1138.

Felleman DJ, Van Essen DC (1991) Distributed hierarchical processing in the primate cerebral cortex. *Cereb Cortex* 1:1–47.

Fleischhauer K (1974) On different patterns of dendritic bundling in the cerebral cortex of the cat. *Z Anat Entwickl-Gesch* 143:115–126.

Foote SL, Morrison JH (1987) Extrathalamic modulation of cortical function. *Ann Rev Neurosci* 10:67–95.

Hendry SHC, Schwark HD, Jones EG, Yan J (1987) Numbers and proportions of GABA-immunoreactive neurons in different areas of monkey cerebral cortex. *J Neurosci* 7:1503–1519.

Hendry SHC, Jones, EG, Emson PC, Lawson DEM, Heizmann CW, Streit P (1989) Two classes of cortical GABA neurons defined by differential calcium binding protein immunoreactivities. *Exp Brain Res* 76:467–472.

Huntley GW, Jones EG (1991) The emergence of architectonic field structure and areal boundaries in developing monkey sensorimotor cortex. *Neuroscience* 44:287–310.

Jones EG (1975) Varieties and distribution of non-pyramidal cells in the somatic sensory cortex of the squirrel monkey. *J Comp Neurol* 160:205–268.

Jones EG (1984) Laminar distribution of cortical efferent cells. In: Peters A and Jones EG, eds. *Cerebral Cortex,. Vol. 1. Cellular Components of the Cerebral Cortex*, pp. 521–553. New York: Plenum Press.

Jones EG (1993) GABAergic neurones and their role in cortical plasticity in primates. *Cereb Cortex* 3:361–372.

Jones EG (2007) *The Thalamus*. 2nd ed. Cambridge, England: Cambridge University Press.

Jones EG (2009) Origins of cortical interneurons: mouse vs monkey and human. *Cereb Cortex* 9(9):1953–1956.

Jones EG, Burton H (1976) Areal differences in the laminar distribution of thalamic afferents in cortical fields of the insular, parietal and temporal regions of primates. *J Comp Neurol* 168:197–247.

Keller A (1995) Synaptic organization of the barrel cortex. In: Jones EG and Diamond IT, eds. *Cerebral Cortex, Vol. 11. The Barrel Cortex of Rodents*, pp. 221–262. New York: Plenum Press.

Lewis DA, Lund JS (1990) Heterogeneity of chandelier neurons in monkey neocortex: corticotropin-releasing factor- and parvalbumin-immunoreactive populations. *J Comp Neurol* 293:599–615.

Lund J S (1984) Spiny stellate neurons. In: Peters A and Jones EG, eds. *Cerebral Cortex, Vol. 1. Cellular Components of the Cerebral Cortex*, pp. 255–308. New York: Plenum Press.

Lund JS, Yoshioka T, Levitt JB (1994) Substrates for interlaminar connections in area V1 of the macaque monkey cerebral cortex. In: Peters A and Rockland KS, eds. *Cerebral Cortex, Vol. 10. Primary Visual Vortex in Primates*, pp. 37–60. New York: Plenum Press.

Peters A, Jones EG (1984) Classifications of cortical neurons. In: Peters A and Jones EG, eds. *Cerebral Cortex, Vol. 1. Cellular Components of the Cerebral Cortex*, pp. 107–121. New York: Plenum Press.

Peters A, Palay SL, Webster H deF (1991) *The Fine Structure of the Nervous System: Neurons and Their Supporting Cells*. New York: Oxford University Press.

Somogyi P, Tamás G, Lujan R, Buhl EH (1998) Salient features of synaptic organisation in the cerebral cortex. *Brain Res Rev* 26:113–135.

Szentágothai J (1975) The "module-concept" in cerebral cortex architecture. *Brain Res* 95:475–496.

Valverde F (1978) The organization of area 18 in the monkey. A Golgi study. *Anat Embryol (Berl)* 154:305–334.

White EL (1989) Cortical circuits: synaptic organization of the cerebral cortex. Boston: Birkhäuser.

2

Canonical Cortical Circuits

Rodney J. Douglas and Kevan A. C. Martin

The observation that neural circuits of the neocortex are adapted to many different tasks raises deep questions of how they are organized and operate. Most theories of cortical computation propose that the cortex processes its information in a feedforward manner through a series of hierarchically organized stages and that each of these stages is dominated by the pattern of the input to the local cortical circuit. The most influential of these models of the local circuit is Hubel and Wiesel's (1962) proposal for the circuits that underlie simple and complex cells in the cat's primary visual cortex. Felleman and Van Essen (1991) extended the notion of a processing hierarchy in their comprehensive summary wiring diagram for the primate visual system. In these models of intra- and interareal cortical circuits, sensory information from the retina is passed through successive stages of cortical processing, each of which increases the feature selectivity of visual receptive fields. Thus, from the concentric center-surround receptive fields of the retina and dorsal lateral geniculate nucleus, simple cells are created, then complex cells from simple cells, and eventually the face cells, object-specific cells, and 3-D motion-specific cells of the high levels of the cortical processing hierarchy.

This serial processing schema is conceptually simple, which makes it very attractive for theorists (e.g., Riesenhuber and Poggio, 1999). More recent experimental and theoretical considerations of the cortical circuits, however, have suggested a rather different architecture: one in which local circuits of cortical neurons are connected in a series of nested positive and negative feedback loops, called "recurrent circuits" (Fig. 2.1; Douglas et al., 1989; Douglas and Martin, 2004, 2007). Excitatory neurons outnumber the inhibitory neurons by 5 to 1, so this ratio might be expected to create an unstable positive feedback. However, because the recurrent connections also exist

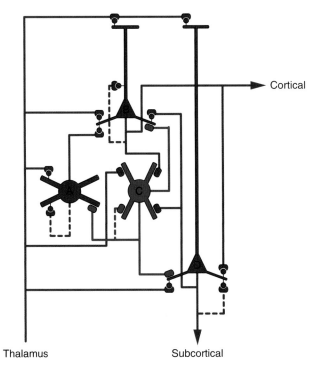

FIGURE 2–1. A canonical circuit for neocortex. Thalamic relay cells mainly form synapses in the middle layers of cortex, but they also form synapses with neurons in all six cortical layers, including the tufts of pyramidal cells in layer 1. In all layers the excitatory (red) and inhibitory (blue) neurons form recurrent connections with like cells within the same layer (dashed lines) and with other cell types (continuous lines). Layer 4 in some primary sensory cortical areas contain a specialist excitatory cell type, the spiny stellate cell (A), which projects to pyramidal cells and inhibitory cells in layer 4 and other layers. The superficial layer pyramidal cells (B) connect locally and project to other areas of cortex. Inhibitory neurons (C) are found in all layers (only one representative is shown here), and they constitute about 15% of the neurons in the neocortex. The deep layer pyramidal cells (D) also connect recurrently locally and project to subcortical nuclei in the thalamus, midbrain, and spinal cord.

between excitatory and inhibitory neurons, inhibition increases in proportion to excitation and the two opposing forces remain approximately in balance.

In the feedforward model, the thalamic input is strong, and it dominates the output of the neurons. In the recurrent model, however, input to the local circuits from the thalamus, or from other cortical areas, is thought to be relatively weak and the recurrent circuits either amplify or suppress this input (Douglas et al., 1989). The oldest and most notable example of "selective" amplification is the orientation preference of the neurons in the layer 4 of the cat's primary visual cortex. Although these neurons receive monosynaptic input from thalamic neurons that have nonoriented receptive fields, they can amplify the excitation generated by optimally oriented stimuli and suppress the thalamic excitation generated by nonoptimal stimuli. Thus, the goodness of "fit" of the input pattern to the "expectation" of the cortical circuits determines whether the input is amplified. These features of recurrent excitation and inhibition, amplification of weak inputs, and balanced excitation and inhibition, are fundamental attributes of the cortical circuits. To the extent they are features that appear in all cortical areas so far examined, they are defining characteristics of the proposed "canonical" circuit of neocortex (Fig. 2.1; Douglas et al., 1989, Douglas and Martin, 2004).

What is the experimental evidence that the thalamic input, which provides the cortex with its major input from the peripheral sense organs and from the basal ganglia, is relatively weak? The best evidence is from cat area 17, where anatomical and physiological studies indicate that the thalamus provides only a fraction (10%) of the total excitatory input to their main target neurons (Douglas et al., 1989; Binzegger et al., 2004; Da Costa and Martin 2009). The remaining excitatory synapses in layer 4 are contributed by other cortical neurons. Electrophysiological studies in slices of cat area 17 showed that while the amplitudes of the excitatory postsynaptic potentials (EPSPs) generated by putative thalamic axons were two-fold larger than those from local cortical neurons when stimulated at 1 Hz, they depressed with repeated stimulation (Stratford et al., 1996; Bannister et al., 2002). Thus, in vivo, where the spontaneous activity of thalamic afferents is relatively high, the amplitude of thalamocortical EPSPs may be considerably reduced by synaptic depression even before a stimulus arrives. In the rodent sensory cortices, the thalamic synapses are also outnumbered by the synapses arising from neighboring cortical neurons (White, 1989).

The evidence for recurrent connections between cortical neurons comes from a consideration of the distributions of cortical synapses. The most comprehensive analysis of the cortical circuit (Binzegger et al., 2004) indicates how much recurrent excitatory connections dominate within and between cortical layers. The intralaminar excitatory connections are most prominent in layers 2 and 3, where the pyramidal cells form most of their local excitatory synapses with each other, so much so that their recurrent connections involve one-fifth of all the excitatory synapses in area 17 (Binzegger et al., 2004). The consequence of this is that the recurrent connections between layer 2 and 3 pyramidal cells may predominate, whereas for other layers the interlaminar recurrent connections may have a greater role. For example, the spiny stellate neurons in layer 4 of cat visual cortex receive 40% of their excitatory synapses from pyramidal cells in layer 6 and only about 20% from their neighboring spiny stellate cells in layer 4.

The concept of serial processing within a cortical "column," introduced by Hubel and Wiesel in 1962, brought to attention the importance of the interlaminar connections. However, neurons live in a 3-D space and they can have extensive projections not just within a column, but laterally (Fig. 2.2). One of the most impressive examples of this is that of the superficial layer

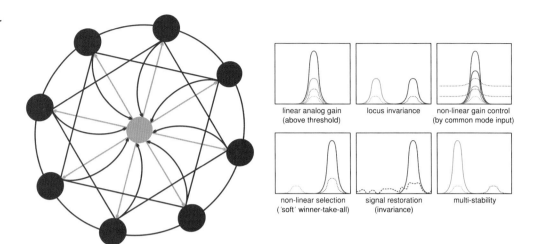

linear analog gain
(above threshold)

locus invariance

non-linear gain control
(by common mode input)

non-linear selection
('soft' winner-take-all)

signal restoration
(invariance)

multi-stability

FIGURE 2–2. Recurrent circuit formed by lateral connections and some of the basic computations it could perform. (*Left*) Single or pools of excitatory neurons (red filled circles) are recurrently connected (red curved lines) with their neighbors and with a pool of inhibitory cells (blue filled circle). A parameter (e.g., orientation preference) is mapped around the circle of excitatory neurons, so that nearest neighbors lie closer together in the parameter space (have similar preferences) and are more strongly connected than more distant neurons in the map. Cortical Daisies are represented here by bidirectional excitatory connections that skip nearest neighbors (straight red lines) and may connect to neurons with dissimilar functional preferences. (*Right*) Illustration of various computations such as linear analog gain, where above threshold, the network amplifies its hill-shaped input (stippled lines) with constant gain (output, solid lines). Locus invariance occurs when the gain remains the same across the map (provided that the connection weights are homogenous across the network). In gain modulation, the network gain is modulated by an additional constant input applied to all the excitatory neurons and superimposed on the hill-shaped input. The gain is least when no constant input is applied (input, orange stippled line; output, orange solid line) and largest for a large constant input (mauve lines). When two inputs of different amplitude are applied to the network, it selects the stronger one by a nonlinear selection or "winner-take-all" operation. Signal restoration restores the hill-shaped input, even when that input is embedded in noise. When separate inputs have the same amplitudes, multistability is the operation that selects only one input: which input is selected depends upon the initial conditions of the network at the time the input is applied.

pyramidal cells, which distribute their synaptic boutons in patches or clusters. If a small cluster of neurons is viewed from the surface of the cortex, their axons form patches of terminals that have the appearance of the petals of a flower. This structure, which we refer to as the cortical "Daisy" (Douglas and Martin, 2004), is found in the cortical areas of all nonrodent species studied so far. Our view is that these horizontal axon clusters are the means by which pyramidal cells collectively participate in a selection network (Fig. 2.2). The selection mechanism is a soft winner-take-all or soft MAX mechanism, which is an important element of many neuronal network models (Riesenhuber and Poggio, 1999; Maass, 2000; Yuille and Geiger, 2003). In this way, the superficial layer neurons would cooperate to explore all possible interpretations of input, and so select an interpretation consistent with their various subcortical inputs. However, these same pyramidal neurons not only participate in the local cortical circuit, but many of them also project outside their own cortical area to other cortical areas or subcortical structures and do this according to precise rules that govern the numbers and laminar origins of the pyramidal cells that form the interareal projections (Kennedy and Bullier, 1985; Barone et al., 2000). Thus, many of same neurons that form a Daisy in one area also provide input to Daisies in other cortical areas.

The inhibitory cells are recurrently connected with the spiny excitatory cells and with each other (Figs. 2.1 and 2.2). This arrangement is probably an early feature in the evolution of nervous systems, not just neocortex, and was originally revealed by Charles Sherrington in his studies of the spinal cord reflexes. For Sherrington, excitatory and inhibitory neurons were always in tandem and together they provided the algebra of the nervous system: "The net change which results there when the two areas are stimulated concurrently is an algebraic sum of the plus and minus effects producible separately by stimulating singly the two antagonistic nerves" (Sherrington, 1908). All cortical inhibitory neurons are GABAergic, and they conveniently divide according to which of three different calcium-binding proteins they contain (Douglas and Martin, 2004). The presence of these calcium-binding proteins correlates with the morphology of the different types of inhibitory cells and their specific connections with the spiny cells, which form 85% of their synaptic targets. The parvalbumin-containing cells, like chandelier and basket cells, which target the soma, proximal dendrites, and axon initial segment, seem well-positioned to control the output of the cell. The calbindin- or calretinin-containing cells, such as the double bouquet cells or Martinotti cells, form synapses with the more distal dendrites and thus are probably concerned with controlling the input to pyramidal cells, which are their major targets. The GABAergic cells may also colocalize polypetides such as somatostatin, vasointestinal polypeptide, or cholecystokinin. Interestingly, although the basket cells were so named because they formed a pericellular nest of terminals around the cell body of pyramidal cells, they actually form most of their synapses with the dendrites of their target excitatory cells (Douglas and Martin, 2004). Some evidence for the effectiveness of distal inhibition has come from studies of the apical dendritic tuft of large layer 5 pyramidal neurons in the somatosensory cortex of the rat. This neuron, which projects to subcortical structures of the thalamus and midbrain, possesses the longest apical dendrite of any neuron in the cortex. Its apical tuft is the source of a calcium spike that can be gated by a distal inhibitory input, probably from Martinotti cells (Murayama et al., 2009).

Recurrent inhibition has found its most universal incarnation in the "normalization" model of visual cortex (Carandini et al., 1997). This model was developed to correct for the deficiencies of models of simple cells and complex cells, in which the inputs were summed linearly and their output passed through a spiking threshold. Standard linear models with rectification do not explain many experimental observations, such as why the responses of all cortical cells saturate, or adapt, or are suppressed by masking stimuli. The modification is to add a recurrent inhibitory pathway in which the inhibition is proportional to the pooled activity of a large number of cortical cells and acts to divide the firing rate of each cell in the pool. Quite how this might be implemented mechanistically is not at all clear, but such normalization models do at least offer one means of correcting the deficiencies of the linear models.

The normalization model requires a collective computation of all the neurons in the circuit. However, each neuron forms 5000 or more synapses and the low firing rates of cortical neurons indicate that only a restricted subset of these 5000 can be active at one time. Thus, while the combinatorial possibilities of the 5000 or more inputs onto a single cortical neuron provide numbers that are more than astronomical, cortical neurons provide outputs that are highly robust and reliable in space and time, with the result that most of the time our perceptions and actions are well-matched to the environment. Here inhibition can play a key role in determining which few hundreds of neurons constitute the effective circuit at any moment, because the effective circuit is created only by those neurons that are above threshold. Although it seems likely that there is a degree of redundancy in the inputs, even then only a very restricted subset of outputs typically should occur. The number of different parameters represented in the output of a cortical neuron is likely to be tens, not hundreds or thousands. We have referred to this constraint in numbers and patterns of active neurons in a recurrently connected population as the "permitted set,". This set is the combination of active neurons whose effective weight matrix is stable and allows the network to converge to a steady state in a given context (Hahnloser et al., 2000; Douglas and Martin, 2007). This would require that perhaps only 10% of the synapse to be active. Thus, the nature of the recurrent activity and the size of the projective fields of clusters of neurons that share common inputs indicate that the computed output will be represented by the activity of less than 1000 neurons.

The organization of the neocortical circuits and the principles of their operation are still only very partially understood. However, each technical advance over the past century has reaffirmed that repeated patterns of structure and function are seen at every level, from molecule to cell to circuit, and that many of these patterns are common across cortical areas and species. In this context, the concept of a canonical circuit, like the concept of hierarchies of processing, offers a powerful unifying principle that links structural and functional levels of analysis across species and different areas of cortex.

ACKNOWLEDGMENTS

We thank Elizabetta Chicca and Nuno Da Costa for their creative graphics. The research was supported by EU Daisy grant FP6-2005-015803.

REFERENCES

Bannister NJ, Nelson JC, Jack JJ (2002) Excitatory inputs to spiny cells in layers 4 and 6 of cat striate cortex. *Philos Trans R Soc London B* 357:1793–1808.

Barone P, Batardiere A, Knoblauch K, Kennedy H (2000) Laminar distribution of neurons in extrastriate areas projecting to visual areas V1 and V4 correlates with the hierarchical rank and indicates the operation of a distance rule. *J Neurosci* 20:3263–3281.

Binzegger T, Douglas RJ, Martin KAC (2004) A quantitative map of the circuit of cat primary visual cortex. *J Neurosci* 24:8441–8453.

Carandini M, Heeger DJ, Movshon JA (1997) Linearity and normalization in simple cells of the macaque primary visual cortex. *J Neurosci* 17:8621–8644.

Da Costa N, Martin KAC (2009) The proportion of synapses formed by the axons of the lateral geniculate nucleus in layer 4 of area 17 of the cat. *J Comp Neurol* 143:101–108.

Douglas RJ, Martin KAC (2004) Neuronal circuits of the neocortex. *Ann Rev Neurosci* 27:419–451.

Douglas RJ, Martin KAC (2007) Mapping the matrix: the ways of neocortex. *Neuron* 56:226–238.

Douglas RJ, Martin KAC, Whitteridge D (1989) A canonical microcircuit for neocortex. *Neural Comput* 1:480–488.

Felleman DJ, Van Essen DC (1991) Distributed hierarchical processing in the primate cerebral cortex. *Cereb Cortex* 1:1–47.

Hahnloser R, Sarpeshkar R, Mahowald M, Douglas R, Seung S (2000) Digital selection and analogue amplification coexist in a cortex-inspired silicon circuit. *Nature* 405:947–951.

Hubel D, Wiesel T (1962) Receptive fields, binocular interaction and functional architecture in the cat's visual cortex. *J Physiol* 160:106–154.

Kennedy H, Bullier J (1985) A double-labeling investigation of the afferent connectivity to cortical areas V1 and V2 of the macaque monkey. *J Neurosci* 5:2815–2830.

Maass W (2000) On the computational power of winner-take-all. *Neural Comput* 12:2519–2535.

Murayama M, Pérez-Garci E, Nevian T, Bock T, Senn W, Larkum ME (2009) Dendritic encoding of sensory stimuli controlled by deep cortical interneurons. *Nature* 457:1137–1141.

Riesenhuber M, Poggio T (1999) Hierarchical models of object recognition in cortex. *Nat Neurosci* 2:1019–1025.

Sherrington C (1908) On the reciprocal innervation of antagonistic muscles. Thirteenth note. On the antagonism between reflex inhibition and reflex excitation, *Proc R Soc London, B* 80b:565–578 (reprinted *Folia Neuro-Biol* 1908, 1:365).

Stratford KJ, Tarczy-Hornoch K, Martin KAC, Bannister NJ, Jack JJ (1996) Excitatory synaptic inputs to spiny stellate cells in cat visual cortex. *Nature* 382:258–261.

White EL (1989) *Cortical Circuits: Synaptic Organization of the Cerebral Cortex. Structure, Function and Theory*. Boston: Birkhauser.

Yuille AL, Geiger D (2003) Winner-take-all networks. In: Arbib M, ed. *The Handbook of Brain Theory and Neural Networks*, pp. 1228–1231. Cambridge, MA: MIT Press.

3

Microcircuitry of the Neocortex

Henry Markram

The neocortex is a sheet of neurons organized in six layers, each receiving and projecting to specific brain areas depending on the neocortical region. This neuronal sheet, with some exceptions, displays very little horizontal anatomical segregation, but it is dynamically segregated into functional modules during stimulation. The neocortical column is one such functional module, which emerges with a diameter corresponding approximately to the expanse of the basal dendrites of thick tufted layer 5 pyramidal neurons (Markram, 2008). These columns are composed of minicolumns with a diameter of 20–30 mm, containing around 120 neurons.

The rat neocortical column is composed of 6–10,000 neurons interconnected by approximately 10 million local circuit synapses. The neocortical mircocircuitry is highly stereotypical in terms of layering, neuron types, synapse types, and interconnectivity patterns, across neocortical regions and mammalian species, with variations in neuron composition, detailed neuronal morphology, synaptic and spine densities, and functional properties appropriate for the specific function(s) of each neocortical region and in each species (DeFelipe et al., 2002; Silberberg et al., 2002; Thomson and Bannister, 2003; Douglas and Martin, 2004).

Around 86% of all synapses in the column are excitatory and 14% are inhibitory (DeFelipe et al., 1999). Roughly a third of the excitatory synapses are formed by the axons of neurons within that column, a third from neurons in neighboring columns, and a third from neurons in more distant brain regions (other cortical regions or the opposite hemisphere and subcortical brain regions). The precise distribution of connections can vary considerably between layers and columns depending on the region. Most of the inhibitory synapses arise from neurons within the same column, some from immediately neighboring columns, and a minority from more distant columns within the same neocortical region. Synaptic connections in the neocortex rarely consist of a single synapse; 3–12 synapses make up glutamatergic connections, and 5–30 make up GABAergic connections.

THE PRINCIPAL NEURONS

The principal neurons of the neocortex are excitatory pyramidal neurons receiving several thousand (2000–20,000) synaptic inputs and are found in layers II to VI (Fig. 3.1). Layer II/III pyramidal neurons are not easily divisible into separate morphological classes. Layer IV contains two main morphological types of pyramidal neurons, with classical and star pyramids. In primary sensory areas, layer IV additionally contains the glutamatergic spiny stellate cells, which are an important target population for thalamic innervation. Layer V contains two main morphological types of pyramidal neurons with the thin untufted pyramids that project to the opposite hemisphere, and the thick tufted pyramids that project subcortically. Layer VI has the greatest diversity of pyramidal morphologies, with at least four distinct types depending on the projection region (cortico-cortical, cortico-thalamic, cortico-callosal, and cortico-claustral).

Pyramidal neurons in general belonging to the same morphological class can be further divided into "projection subclasses" depending on the region(s) of the brain to which they project and/or from which they receive projections. The local arborization of a single pyramidal neuron can innervate 1%–30% of neighboring pyramidal neurons depending on the layer and the type of pyramidal target. Pyramidal interconnectivity within the dimensions of a minicolumn generally decreases from supragranular to infragranular layers (layer II/III, around 30%; layer IV, around 20%; layer V, around 10%; layer VI, around 1%). There is strong connectivity between layers, with at least one main directional tendency from layer IV to II/IIII and from II/III down to the infragranular layers (Thomson and Bannister, 2003).

THE INTERNEURONS

Interneurons receive only a few hundred (200–1000) synapses (because of their relatively more simple dendritic arborization) from within a column, neighboring columns, and from distant brain regions and are found in all six layers of the neocortex (Fig. 3.1). Interneurons are generally considered local circuit neurons because they mostly innervate neurons within the dimensions of a neocortical column.

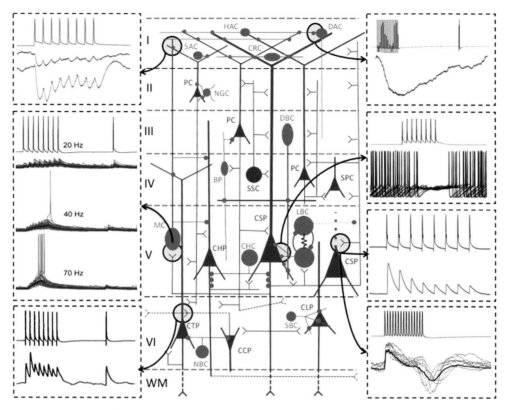

Figure 3–1. Simplified schematic representation of the neocortical microcircuitry. Red indicates excitatory neurons, dendrites, and axons; blue indicates inhibitory neurons, dendrites, and axons. Inhibitory synapses are marked in blue dots, and excitatory synapses are marked in red forks. From the top left and down, the insert illustrates a synaptic response from an MC onto a PC, a PC onto an MC, and a CCP onto a CTP. From top right and down, the inserts illustrate synaptic responses from an HAC on a VAC, an LBC on a PC, a PC response on a PC, and a disynaptic PC response on a PC via an MC. In all cases the presynaptic action potentials are above and the postsynaptic responses are below. Layers are indicated in roman numerals. Axons projecting beyond the neocortical dimensions are indicated by dotted lines. For the PCs, axons are thin lines relative to the dendrites; for the inhibitory neurons, only axons are schematized. Black arrows from grayed background circles indicate the synaptic locations for the inserted illustrated responses. BP, bipolar cell; CCP, cortico-cortical pyramid; CHC, chandelier cell; CHP, cortico-hemispheric pyramid; CLP, cortico-claustral pyramid; CRC, Cajal-Retzius cell; CSP, cortico-spinal pyramidal; CTP, cortico-thalamic pyramid; DBC, double bouquet cell; HAC, horizontal axon cell; LBC, large basket cell; MC, Martinotti cell; NBC, nest basket cell; NGC, neurogliaform cell; PC, pyramidal cell; SBC, small basket cell; SPC, star pyramidal cell; SSC, spiny stellate cell; DAC, descending axon cell; SAC, short axon cell; WM, white matter.

There are four major morphological types of interneurons in layer 1 (Cajal-Retzius, small axon cell, horizonal axon cell and descending axon cell) and nine in layers 2–6 (large basket, nest basket, small basket, bitufted, bipolar, neurogliaform, Martinotti, Double bouqet, and chandelier cells) (DeFelipe, 2002; Markram et al., 2004; Ascoli et al., 2008) (Fig. 3.1). Large basket cells with long horizontal axonal branches are the major source of longer distance inhibition across columns. Together with the horizontally projecting layer I cells, Martinotti cells also project to neighboring columns within layer I, where their axonal arborization fans out beyond the dimensions of a column (Wang et al., 2004). Each anatomical type of neuron can express up to 8 of 15 major types of electrical behaviors, giving rise to as many as 200 morpho-electrical types of interneurons in a neocortical column when also considering layer differences (Markram et al., 2004). Electrical diversity in neocortical neurons is achieved by expressing and distributing around 10% of the 200 main ion channels that are expressed in the neocortex.

Interneurons can be further subclassified according to molecular expression patterns, yielding an even greater potential diversity of morpho-electrical-molecular cell types. Interneurons are, for example, often classified according to their expression of different calcium-binding proteins (e.g., calbindin, calretinin, and parvalbumin) and a spectrum of neuropeptides, but in most cases expression of these proteins is found in more than one morpho-electrical type of neuron (Markram et al., 2004; Toledo-Rodriguez et al., 2005).

Target Selectivity

There are numerous examples of target selectivity in the neocortex (Thomson et al., 2002) with the strictest form displayed by chandelier cells, which target only pyramidal neurons and mainly on their axon initial segments, while completely avoiding other interneurons (Somogyi et al., 1998). The mechanism employed to avoid forming synapses on all other cells and other domains is not known. Interconnectivity between interneurons seems to be higher for immediate neighboring interneurons of the same type and connections and often also involves electrical synapses. However, while some types of interneurons, such as large basket cells, are highly interconnected, others, for example, double bouquet cells, are much less interconnected, if at all. Target selectivity is also evident among glutamatergic connections. For example, the thick tufted pyramidal neuron innervates around 10% of other thick tufted pyramidal neurons in the same layer (Markram, 1997), while they hardly innervate the thin untufted pyramidal neurons that lie within the same neuropil (Le Be et al., 2007). The thin untufted pyramidal neurons are also only sparsely interconnected (around 1%), much lower than the thick tufted pyramidal neurons (around 10%).

MULTISYNAPSE CONNECTIONS AND DOMAIN TARGETING

Each anatomical type of interneuron innervates its target cells by distributing multiple synapses in a characteristic manner onto selected domains of the neuron (axon initial segments [AISs], somata, proximal and distal dendritic shafts and spines, dendritic tufts). This domain targeting is easily observed onto pyramidal neurons because of their stereotypical morphology. Numerous mechanisms for domain targeting have been proposed, but how this is achieved in the neocortex is still a mystery.

Glutamatergic neurons employ 3–12 synapses to innervate interneurons (Wang et al., 1999; Gupta et al., 2000). These synapses typically only form onto a small fraction of dendrites, and they tend to cluster their innervation, which contrasts with the highly distributed manner in which glutamatergic synapses innervate excitatory cells. Most synapses are formed on dendrites, but importantly, glutamatergic synapses can also be formed on the cell bodies of interneurons, contrasting with the lack of excitatory synapses on pyramidal somata. The precise domain of a neuron targeted ("domain targeting") by a given class of pyramidal neuron differs between interneuron classes, but the mechanism for such differential targeting is not known. Glutamatergic synaptic transmission at connections onto interneurons utilizes different AMPA receptor subunits and some classes utilize fewer NMDA receptors.

Inhibitory neurons generally form a much larger number of synapses onto their target cells (up to 30 synapses/connection) than excitatory neurons (Gupta et al., 2000). Axo-dendritic inhibitory synapses are typically highly distributed across the dendritic surface of target cells and are mainly formed onto dendritic shafts. Whenever formed onto spines, they provide an additional input (mainly "displaced" toward the spine neck region) to the excitatory synapse, which always impinges onto the spine head in mature circuits.

In addition to fast $GABA_A$ receptor–mediated inhibition, slow inhibitory synaptic responses have been recorded in neocortical neurons. These slow responses, mediated by metabotropic $GABA_B$ receptors, have mainly been detected after strong extracellular stimulation (i.e., activation of several presynaptic interneurons or by repetitive activation of one or a few inputs). More recently, single neurogliaform cells and some layer I interneurons have been shown to be capable of activating $GABA_B$ receptors with the GABA release resulting from a single action potential (Tamas et al., 2003).

HETEROGENEITY OF SYNAPTIC DYNAMICS

Neocortical synaptic connections can display one of six types of short-term plasticity ("synaptic dynamics") depending on the ratio of the time constants of synaptic depression and facilitation (F1, F>>D; F2, D>>F; F3, D==F), which yields three main classes that are each further divisible by a low or high probability of release (Wang et al., 2006). The specific type of synaptic dynamic deployed between any two neurons is genetically determined, developmentally expressed, relatively independent of the mammalian species, and cannot be "switched" by synaptic plasticity. The axon of a neocortical neuron, and even sequential boutons on the same axon collateral, can deliver synapses that exhibit quite different dynamic properties depending on the postsynaptic target (Markram et al., 1998).

The type of synaptic dynamics expressed between pyramidal and interneurons is highly predictable from the morpho-electrical nature of both pre- and postsynaptic neurons (Gupta et al., 2000). This strongly suggests that a combinatorial identity match drives diversity in the mapping of synaptic dynamics between neurons of a neocortical column. The synaptic type is less reliably predicted for connections between interneurons, suggesting that additional factors probably determine the identity match and hence the mapping of synaptic dynamics. With around 200 morpho-electrical types of interneurons in a column, the diversity in the mapping of the six types of dynamic synapses is enormous.

Interpyramidal glutamatergic synapses more typically display synaptic depression (F2 type). During development, a fast time constant of facilitation emerges, but the time constant is still shorter than that governing depression (Reyes and Sakmann, 1999). The interpyramidal synapses in higher neocortical regions, such as the prefrontal cortex, display all six types of synaptic dynamics. Glutamatergic synaptic connections display both dynamic properties dependent upon the type of postsynaptic neuron as well as differential synaptic dynamics within a class of synaptic dynamics onto a population of the same type of target neuron (Markram et al., 1998; Wang et al., 1999). Deploying synapses with different dynamics onto different target neurons enables differential activation of target neurons within a layer, across layers, across columns, and in more distant brain regions.

GABAergic synapses display all six types of synaptic dynamics, but F1 and F3 types are more common, giving an overall impression of more synaptic facilitation at GABAergic synapses than at interpyramidal synapses. GABAergic synaptic connections formed by each type of interneuron also express different synaptic dynamics depending on the class of postsynaptic neuron as with the glutamatergic synapses, but they display a striking contrast in that all synaptic connections onto a population of the same type of target neurons express perfect homogeneity of synaptic dynamics. This homogeneous mapping of synaptic dynamics onto a homogeneous population of neurons is called "GABA grouping" (Gupta et al., 2000). GABA grouping could allow each interneuron to impose the same synchronization effect on a population of neurons of a given type and a different synchronization effect on populations containing different types of neurons. The uniqueness of dynamic synapses formed by different interneurons could additionally allow each class of interneuron to apply a spectrum of unique synchronization effects onto its various target populations.

MICROCIRCUIT, SYNAPTIC, AND META PLASTICITY

The circuit formed by the interconnectivity between neurons in the neocortical microcircuit is largely shaped by evolutionary and genetic factors (inferred from consistency of these properties within and across species), but it can be structurally and functionally altered by various forms of plasticity. Long-term microcircuit plasticity (LTMP) is a form of plasticity that reorganizes the structure of the circuit as it drives neurons to disconnect from some neurons and connect with others within a time scale of hours (Le Be and Markram, 2006). Spike timing-dependent plasticity (STDP) determines the magnitude and direction of change at existing synaptic connections depending on millisecond precision in the arrival time of a presynaptic input and the response of the postsynaptic neuron (Markram et al., 1997). The nature of the change can be in the form of a change in synaptic strength (number of synapses per connection, and receptors and/or receptor efficacy at individual synapses) and/or a change in probability of release, time constants of depression, and/or facilitation. Redistribution of synaptic efficacy (RSE) refers to redistributing the existing synaptic strength temporally across a train of presynaptic action potentials (Markram and Tsodyks, 1996), which is caused by changing probability of release, and time constants of depression and facilitation (Markram et al., 1998). STDP therefore determines the driving force for change and RSE describes the nature of the change. Neuromodulation, via acetylcholine, for example, may gate plasticity by modulating NMDA receptor efficacy (Markram and Segal, 1990) and downstream pathways to allow metaplastic control of STDP and RSE and hence allow neocortical columns to adapt to their stimulus environment in the context of behavior.

ALTERATIONS IN DISEASE

The neocortical column microcircuitry is the elementary foundation for emergent properties of the neocortex. The interconnectivity (neurons targeted, fraction of neurons targeted, number of synapses used to target a specific type of neuron) seems to be highly preserved features on which the microcircuitry is based. Pathological alterations in these features can result in profound disorders of higher brain function (Dierssen et al., 2003; Alonso-Nanclares et al., 2005; Markram et al., 2007; Knafo et al., 2009). In a rat model of Autism, glutamatergic fibers hyperconnect onto excitatory and inhibitory targets (Rinaldi et al., 2007s), over express NMDA receptors and display hyperplasticity (Rinaldi et al., 2007b). The microcircuit as a whole becomes more reactive to sensory stimulation. Hyperfunctionality of the neocortical microcircuitry has been proposed to result in an Intense World Syndrome (Markram et al., 2007).

REFERENCES

Alonso-Nanclares L, Garbelli R, Sola RG, Pastor J, Tassi L, Spreafico R, DeFelipe J (2005) Microanatomy of the dysplastic neocortex from epileptic patients. *Brain* 128(Pt 1): 158–173.

Ascoli GA, Alonso-Nanclares L, Anderson SA, Barrionuevo G, Benavides-Piccione R, Burkhalter A, et al. (2008) Petilla terminology: nomenclature of features of GABAergic interneurons of the cerebral cortex. *Nat Rev Neurosci* 9(7):557–568.

DeFelipe J (2002) Cortical interneurons: from Cajal to 2001. *Prog Brain Res* 136:215–238.

DeFelipe J, Alonso-Nanclares L, Arellano JI (2002) Microstructure of the neocortex: comparative aspects. *J Neurocytol* 31(3–5):299–316.

DeFelipe J, Marco P, Busturia I, Merchan-Perez A (1999) Estimation of the number of synapses in the cerebral cortex: methodological considerations. *Cereb Cortex* 9(7):722–732.

Dierssen M, Benavides-Piccione R, Martinez-Cue C, Estivill X, Florez J, Elston GN, DeFelipe J (2003) Alterations of neocortical pyramidal cell phenotype in the Ts65Dn mouse model of Down syndrome: effects of environmental enrichment. *Cereb Cortex* 13(7):758–764.

Douglas RJ, Martin KA (2004) Neuronal circuits of the neocortex. *Annu Rev Neurosci* 27:419–451.

Gupta A, Wang Y, Markram H (2000) Organizing principles for a diversity of GABAergic interneurons and synapses in the neocortex. *Science* 287(5451):273–278.

Knafo S, Alonso-Nanclares L, Gonzalez-Soriano J, Merino-Serrais P, Fernaud-Espinosa I, Ferrer I, DeFelipe J (2009) Widespread changes in dendritic spines in a model of Alzheimer's disease. *Cereb Cortex* 19(3):586–592.

Le Be JV, Markram H (2006) Spontaneous and evoked synaptic rewiring in the neonatal neocortex. *Proc Natl Acad Sci USA* 103(35):13214–13219.

Le Be JV, Silberberg G, Wang Y, Markram H (2007) Morphological, electrophysiological, and synaptic properties of corticocallosal pyramidal cells in the neonatal rat neocortex. *Cereb Cortex* 17(9):2204–2213.

Markram H (1997) A network of tufted layer 5 pyramidal neurons. *Cereb Cortex* 7(6): 523–533.

Markram H (2008) Fixing the location and dimensions of functional neocortical columns. *Hfsp J* 2(3):132–135.

Markram H, Segal M (1990) Long-lasting facilitation of excitatory postsynaptic potentials in the rat hippocampus by acetylcholine. *J Physiol* 427:381–393.

Markram H, Tsodyks M (1996) Redistribution of synaptic efficacy between neocortical pyramidal neurons. *Nature* 382(6594):807–810.

Markram H, Lubke J, Frotscher M, Sakmann B (1997) Regulation of synaptic efficacy by coincidence of postsynaptic APs and EPSPs. *Science* 275(5297):213–215.

Markram H, Pikus D, Gupta A, Tsodyks M (1998) Potential for multiple mechanisms, phenomena and algorithms for synaptic plasticity at single synapses. *Neuropharmacology* 37(4–5):489–500.

Markram H, Wang Y, Tsodyks M (1998) Differential signaling via the same axon of neocortical pyramidal neurons. *Proc Natl Acad Sci USA* 95(9):5323–5328.

Markram H, Toledo-Rodriguez M, Wang Y, Gupta A, Silberberg G, Wu C (2004) Interneurons of the neocortical inhibitory system. *Nat Rev Neurosci* 5(10):793–807.

Markram H, Rinaldi T, Markram K (2007) The intense world syndrome—an alternative hypothesis for autism. *Front Neurosci* 1(1):77–96.

Reyes A, Sakmann B (1999) Developmental switch in the short-term modification of unitary EPSPs evoked in layer 2/3 and layer 5 pyramidal neurons of rat neocortex. *J Neurosci* 19(10):3827–3835.

Rinaldi T, Silberberg G, Markram H (2007a) Hyperconnectivity of Local Neocortical Microcircuitry Induced by Prenatal Exposure to Valproic Acid. *Cereb Cortex*.

Rinaldi T, Kulangara K, Antoniello K, Markram H (2007b) Elevated NMDA receptor levels and enhanced postsynaptic long-term potentiation induced by prenatal exposure to valproic acid. *Proceedings of the National Academy of Sciences* 104:13501–13506.

Silberberg G, Gupta A, Markram H (2002) Stereotypy in neocortical microcircuits. *Trends Neurosci* 25(5):227–230.

Somogyi P, Tamas G, Lujan R, Buhl EH (1998) Salient features of synaptic organisation in the cerebral cortex. *Brain Res Brain Res Rev* 26(2–3):113–135.

Tamas G, Lorincz A, Simon A, Szabadics J (2003) Identified sources and targets of slow inhibition in the neocortex. *Science* 299(5614):1902–1905.

Thomson AM, Bannister AP (2003) Interlaminar connections in the neocortex. *Cereb Cortex* 13(1):5–14.

Thomson AM, Bannister AP, Mercer A, Morris OT (2002) Target and temporal pattern selection at neocortical synapses. *Philos Trans R Soc Lond B Biol Sci* 357(1428): 1781–1791.

Toledo-Rodriguez M, Goodman P, Illic M, Wu C, Markram H (2005) Neuropeptide and calcium-binding protein gene expression profiles predict neuronal anatomical type in the juvenile rat. *J Physiol* 567(Pt 2):401–413.

Wang Y, Gupta A, Markram H (1999) Anatomical and functional differentiation of glutamatergic synaptic innervation in the neocortex. *J Physiol Paris* 93(4):305–317.

Wang Y, Toledo-Rodriguez M, Gupta A, Wu C, Silberberg G, Luo J, Markram H (2004) Anatomical, physiological and molecular properties of Martinotti cells in the somatosensory cortex of the juvenile rat. *J Physiol* 561(Pt 1):65–90.

Wang Y, Markram H, Goodman PH, Berger TK, Ma J, Goldman-Rakic PS (2006) Heterogeneity in the pyramidal network of the medial prefrontal cortex. *Nat Neurosci* 9(4):534–542.

4

Barrel Cortex

Karel Svoboda, Bryan M. Hooks, and
Gordon M. G. Shepherd

Rodents move their whiskers (vibrissae) to explore textures, identify objects, and measure distances to navigational landmarks (Diamond et al., 2008). Whisker-based somatosensory perception depends on the whisker representation area of the somatosensory cortex (posteromedial barrel subfield or "barrel cortex"), which contains a prominent map of the 34 large facial whiskers (Figs. 4.1A–B) (Woolsey and van der Loos, 1970). The barrel cortex derives its name from clusters of cells (barrels) and thalamocortical terminals in layer (L) 4 (Figs. 4.1B–C). Between barrels are septa.

Over the last decade the barrel cortex has emerged as a major model system for the analysis of the structure, function, and experience-dependent plasticity of neocortical microcircuits (Petersen, 2007; Diamond et al., 2008). Circuit studies of the barrel cortex are about equally divided between rats and mice. Indications are that their cortical layers, cell types, and the intralaminar connectivity are similar, with one exception: in the rat, barrels and septa are associated with different thalamocortical, local cortical (Shepherd and Svoboda, 2005), and cortico-cortical circuits (Alloway, 2008). In the mouse barrel cortex, septa are small and cell poor (Woolsey and van der Loos, 1970), and the distinction between barrel and septum circuits is unclear (Bureau et al., 2006). Here we combine conclusions gained from experiments in rats and mice, but all quantitative data pertain to the mouse barrel cortex.

The L4 barrels are landmarks that define functional columns spanning all cortical layers (Fig. 4.1C). Mouse barrel columns, with a mean diameter of

31

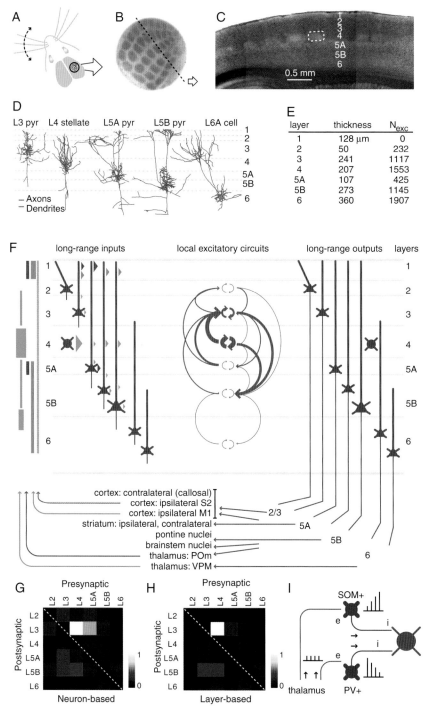

FIGURE 4–1. Barrel cortex microcircuits. (*A*) The left whiskers map onto the right barrel cortex (circle). (*B*) Image of cytochrome oxidase-stained barrels (corresponding to the circle in *A*). (*C*) Image of barrels and cortical layers in a living slice. Dashed line marks one of the L4 barrels. (*D*) Reconstructed barrel cortex neurons (rat) (adapted from Shepherd et al., 2005).

Figure 4–1. continued
(*E*) Layer thickness and number of excitatory neurons (N_{exc}) per column (300 µm diameter) and layer (mouse). (*F*) Wiring diagram depicting major long-range input pathways, local excitatory pathways, and long-range output pathways. The arrows depicting local excitatory pathways are based on the laminar connectivity matrix (*G*) and data from channelrhodopsin-2-assisted circuit mapping (Petreanu et al., 2007; Petreanu et al., 2009) and pair recording (Lefort et al., 2009). (*G–H*) Laminar connectivity matrices, for interlaminar excitatory pathways, measured using laser scanning photostimulation (e.g., Bureau et al., 2006). Intralaminar data (along dashed line) were omitted. (*G*) Neuron-based connectivity matrix. Pixels represent the average strength of connections between individual neurons in each layer. (*H*) Layer-based connectivity matrix. Pixels represent the total strength of connections between layers. The values were derived from the neuron-based connectivity matrix by multiplying by the number of postsynaptic and presynaptic neurons (*E*). (*I*) Feedforward inhibitory microcircuits. Inhibitory interneurons expressing somatostatin (SOM) show facilitating responses to thalamic inputs, and those expressing parvalbumin (PV) show depressing responses.

300 micrometers and spanning the entire thickness of the cortex (1.35 millimeters), contain approximately 10,000 neurons per column (Lefort et al., 2009). Neurons in each barrel are excited best, and with short latencies, by stimulation of a particular whisker (the principal whisker), and more weakly, and with longer latencies, by neighboring whiskers (the surround whiskers) (Petersen, 2007). Excitation then spreads to layers above and below the barrel (barrel-related column) and to neighboring columns.

Cortical Layers and Their Neurons

The spread of excitation through the barrel cortex in response to deflection of a whisker is understood in terms of the major connections between the cell types residing in different cortical layers (Bureau et al., 2006; Lubke and Feldmeyer, 2007; Petersen, 2007; Lefort et al., 2009) (Fig. 4.1C). These connections have been mapped in great detail with electrophysiological circuit mapping methods in brain slices (Thomson and Lamy, 2007). Here we focus on quantitative results obtained with these physiological methods, supplemented by information from standard anatomical methods.

Barrel cortex layers can be visualized using histochemical stains or bright field microscopy in brain slices (Bureau et al., 2006; Lefort et al., 2009) (Fig. 4.1C). Each layer corresponds to a characteristic set of cell types and connections to other cortical and subcortical targets. Six layers are easily distinguishable and in common use (layers 1, 2/3, 4, 5A, 5B, 6, numbered from the pia downward). Although L2 and L3 are not clearly demarcated by cytoarchitecture, we subdivide L2/3 based on the distinct circuits made by cells in a thin (~50 micrometers) layer of superficial pyramidal cells (L2) compared to deeper neurons (L3) (Bureau et al., 2006). Future circuit studies will likely demand further subdivision of other layers.

Approximately 85% of barrel cortex neurons are spiny and glutamatergic (Lefort et al., 2009). The diversity of glutamatergic neurons corresponds neatly to cortical layers (Fig. 4.1D). L1 is distinguished by the absence of excitatory neurons. L2 contains mainly small pyramidal neurons with short or horizontal apical dendrites and extensive basal dendrites (Bureau et al., 2006). L3 has classical pyramidal neurons with a prominent apical dendrite ascending into L1 and extensive basal dendrites centered on the soma. L4 contains small stellate cells. In mice these cells are arranged in ring-like clusters (the barrel walls), each of which surrounds a cell poor cylinder (the barrel hollow). Stellate cell dendrites point into the barrel hollow and respect barrel boundaries. L5A contains "thin-tufted" pyramidal neurons with extensive basal dendrites and a relatively thin apical dendrite with a small tuft in L1. In addition to thin-tufted cells, L5B contains "thick-tufted" cells with a thick apical dendrite and a large tuft in L1. L6 contains a heterogeneous collection of pyramidal neurons. Most of these cells have short apical dendrites terminating in the deep or middle layers, with only a small fraction reaching L1. The density of glutamatergic neurons varies across layers, ranging from $5 \times 10^4/mm^3$ in L5A to $12 \times 10^4/mm^3$ in L4 (Lefort et al., 2009) (Fig. 4.1E). The density of glutamatergic synapses is relatively homogeneous across layers, ranging from $2 \times 10^9/mm^3$ in L5A to $3 \times 10^9/mm^3$ in L4 (DeFelipe et al., 1997).

Approximately 15% of barrel cortex neurons are aspiny, or sparsely spiny, and GABAergic (Lefort et al., 2009). These interneurons have diverse morphologies and firing patterns and express a variety of protein markers (Thomson and Lamy, 2007). A detailed discussion of the classification of GABAergic interneurons is beyond the scope of this review. Two major classes stand out: parvalbumin-expressing (PV+), fast-spiking, soma-targeting interneurons account for approximately 50% of the total interneuron population. Somatostatin-expressing (SOM+), regular-spiking, dendrite-targeting cells make up about half of the remainder. The density of GABAergic interneurons in different layers ranges from $5 \times 10^3/mm^3$ in L6 to $9 \times 10^3/mm^3$ (Lefort et al., 2009) in layers 2–6; L1 contains a low density of interneurons ($2 \times 10^3/mm^3$), most of which express calretinin.

INPUTS

Inputs from the whiskers ascend through the trigeminal ganglion and the trigeminal nucleus of the brainstem to the thalamus and cortex in multiple parallel pathways (Pierret et al., 2000; Bureau et al., 2006; Yu et al., 2006; Petersen, 2007) (Fig. 4.1F). In the lemniscal pathway, a single whisker excites approximately 300 neurons in one barreloid of the ventral posterior medial nucleus (VPM) of the thalamus. These in turn project to a single barrel column in the cortex. The VPM provides the major source of ascending excitation to the barrel cortex, accounting for 20% of excitatory synapses on stellate cells in L4 and a smaller fraction (likely 10-fold less) on pyramidal cells in other layers. VPM projections also directly excite PV+ and SOM+ interneurons in L4 (Cruikshank et al., 2007; Kapfer et al., 2007) and L5 (Tan et al., 2008).

In the paralemniscal pathway, whiskers excite the medial subdivision of the posterior nucleus (POm). The POm provides a parallel source of topographically diffuse thalamic input into the barrel cortex. POm axons arborize in L5A and in L1. They make synapses with L5A neurons and more weakly with L3 neurons, but they avoid L4 and L5B neurons (Petreanu et al., 2009). Weaker projections from other thalamic nuclei have also been described.

Barrel cortex receives prominent inputs from other cortical areas, especially primary vibrissal motor cortex (vM1) and the secondary somatosensory cortex (S2). Neuromodulatory inputs from multiple sources provide serotonergic (dorsal raphe), noradrenergic (locus ceruleus), dopaminergic (ventral tegmental area), and cholinergic (nucleus basalis) innervation.

EXCITATORY INTERLAMINAR AND INTERCOLUMNAR CONNECTIONS

The local excitatory circuits (Fig. 4.1F) can be summarized by a connectivity matrix (Figs. 4.1G, H), showing the strengths of the excitatory connections between layers. Within layers, nearby neurons are densely interconnected (Lefort et al., 2009). Across layers, the dominant projections are (in order of decreasing strength) (Figs. 4.1F, H): L4→L3, L2/3→L5, L4→L5, L3→L2, L5A→L2. All layers except L4 receive substantial interlaminar cortical input; interlaminar input to L6 is weak.

In supragranular and infragranular layers, excitatory neurons project to neighboring barrel columns (Bernardo et al., 1990). This local connectivity is more limited than the highly organized, long-range horizontal connectivity seen in cat and monkey visual cortex.

GABAERGIC CIRCUITS

The circuits involving GABAergic interneurons are less well understood compared to excitatory neurons. Reconstructions of the axonal arbors of interneurons show that some have strictly intralaminar and intracolumnar projections, whereas others project to layers above or below, or even to neighboring barrel columns (Helmstaedter et al., 2008). One remarkable class of SOM+ interneurons, the Martinotti cell, is distinguished by projections to L1. A prominent class of SOM+ cells projects from L5 to inhibit L4. An interesting class of PV+ interneurons feeds back from L3 to inhibit L4. Although such interlaminar inhibition is likely to be important, we currently know little about the prevalence of these structural motifs or their function.

In some cases the dynamics of circuit modules involving excitatory neurons and multiple types of interneurons have been worked out in some detail. For example (Fig. 4.1H), feedforward excitation from VPM excites L4 stellate cells and, more strongly, L5 PV+ cells (Cruikshank et al., 2007; Kapfer et al., 2007). The PV+ cells inhibit stellate cells, thereby temporally sharpening the effect of ascending excitation. However, because VPM to PV+ cell synapses depress in a use-dependent manner, maintained VPM activity diminishes the influence of PV+ cells. Synapses from VPM to SOM+ positive interneurons instead facilitate. Thus, for low-frequency thalamocortical activity, PV+ cell inhibition will dominate, whereas for prolonged trains SOM+ inhibition will take over. Similar circuits operate in L5 (Tan et al., 2008).

OUTPUTS

Pyramidal neurons in the infragranular layers provide the major output to subcortical and other cortical targets (Veinante et al., 2000; Hattox and Nelson, 2007). In addition, pyramidal neurons in supragranular layers contribute to cortico-cortical connections. Cells in layers 1 and 4 do not project out of the barrel cortex.

Thick-tufted L5B neurons project to POm, to pontine nuclei, and to other targets in the brainstem. Neurons in L6 project to VPM. Thin-tufted L5A neurons and L3 cells send output to several cortical areas, including prominent projections to vM1, S2, and perirhinal cortex. A weaker projection targets lateral parts of the contralateral somatosensory cortex via the corpus callosum. A strong projection from L5A descends into the dorsolateral striatum. Several studies suggest that the outputs to separate targets originate from largely separate cell populations even within a cortical layer (Hattox and Nelson, 2007).

FUNCTIONAL IMPLICATIONS

Information about touch to a single whisker is conveyed by the lemniscal pathway to stellate cells in one L4 barrel. The only other excitatory input received by stellate cells comes from other stellate cells in the same barrel, which are presumably excited by deflection of the same whisker (Fig. 4.1F). The L4 barrel therefore relays peripheral signals to large numbers of cortical targets, mainly in L3. The recurrent connectivity in L4 may serve to amplify small, but temporally synchronous inputs. Inhibitory neurons in L3 and L5 with projections to L4 are poised to gate peripheral input within L4.

The lemniscal signal flows through L4 to the rest of the barrel column, where it interacts with other inputs, including signals representing whisker movement (e.g., POm), motor planning (e.g., vM1), and behavioral context (e.g., neuromodulatory systems). The convergence of touch signals and whisker-position signals in the barrel cortex is likely required to decode object location and object identity (Curtis and Kleinfeld, 2009).

The major excitatory connections of the thalamocortical and intracortical circuits described above are sufficient to explain the structure and dynamics of barrel cortex receptive fields as measured in anesthetized rodents. However, more needs to be discovered about the precise relationships between action potentials in defined neurons within the barrel cortex and active somatosensation in awake rodents.

REFERENCES

Alloway KD (2008) Information processing streams in rodent barrel cortex: The differential functions of barrel and septal circuits. *Cerebral Cortex* 18:979–989.

Bernardo KL, McCasland JS, Woolsey TA, Strominger RN (1990) Local intra- and interlaminar connections in mouse barrel cortex. *J Comp Neurol* 291:231–255.

Bureau I, von Saint Paul F, Svoboda K (2006) Interdigitated paralemniscal and lemniscal pathways in the mouse barrel cortex. *PLoS Biol* 4:e382.

Cruikshank SJ, Lewis TJ, Connors BW (2007) Synaptic basis for intense thalamocortical activation of feedforward inhibitory cells in neocortex. *Nat Neurosci* 10:462–468.

Curtis JC, Kleinfeld D (2009) Phase-to-rate transformations encode touch in cortical neurons of a scanning sensorimotor system. *Nat Neurosci* 12:492–501.

DeFelipe J, Marco P, Fairen A, Jones EG (1997) Inhibitory synaptogenesis in mouse somatosensory cortex. *Cereb Cortex* 7:619–634.

Diamond ME, von Heimendahl M, Knutsen PM, Kleinfeld D, Ahissar E (2008) "Where" and "what" in the whisker sensorimotor system. *Nat Rev Neurosci* 9:601–612.

Hattox AM, Nelson SB (2007) Layer V neurons in mouse cortex projecting to different targets have distinct physiological properties. *J Neurophysiol* 98:3330–3340.

Helmstaedter M, Sakmann B, Feldmeyer D (2008) Neuronal correlates of local, lateral, and translaminar inhibition with reference to cortical columns. *Cereb Cortex* 19(4):926–937.

Kapfer C, Glickfeld LL, Atallah BV, Scanziani M (2007) Supralinear increase of recurrent inhibition during sparse activity in the somatosensory cortex. *Nat Neurosci* 10:743–753.

Lefort S, Tomm C, Floyd Sarria JC, Petersen CC (2009) The excitatory neuronal network of the C2 barrel column in mouse primary somatosensory cortex. *Neuron* 61:301–316.

Lubke J, Feldmeyer D (2007) Excitatory signal flow and connectivity in a cortical column: focus on barrel cortex. *Brain Struct Funct* 212:3–17.

Petersen CC (2007) The functional organization of the barrel cortex. *Neuron* 56:339–355.

Petreanu L, Huber D, Sobczyk A, Svoboda K (2007) Channelrhodopsin-2-assisted circuit mapping of long-range callosal projections. *Nat Neurosci* 10:663–668.

Petreanu L, Mao T, Sternson SM, Svoboda K (2009) The subcellular organization of neocortical excitatory connections. *Nature* 457:1142–1145.

Pierret T, Lavallee P, Deschenes M (2000) Parallel streams for the relay of vibrissal information through thalamic barreloids. *J Neurosci* 20:7455–7462.

Shepherd GMG, Svoboda K (2005) Laminar and columnar organization of ascending excitatory projections to layer 2/3 pyramidal neurons in rat barrel cortex. *J Neurosci* 25:5670.

Shepherd GMG, Stepanyants A, Bureau I, Chklovskii DB, Svoboda K (2005) Geometric and functional organization of cortical circuits. *Nature Neuroscience* 8:782–790.

Tan Z, Hu H, Huang ZJ, Agmon A (2008) Robust but delayed thalamocortical activation of dendritic-targeting inhibitory interneurons. *Proc Natl Acad Sci USA* 105:2187–2192.

Thomson AM, Lamy C (2007) Functional maps of neocortical local circuitry. *Front Neurosci* 1:19–42.

Veinante P, Lavallee P, Deschenes M (2000) Corticothalamic projections from layer 5 of the vibrissal barrel cortex in the rat. *J Comp Neurol* 424:197–204.

Woolsey TA, van der Loos H (1970) The structural organization of layer IV in the somatosensory region (S1) of mouse cerebral cortex. *Brain Res* 17:205–242.

Yu C, Derdikman D, Haidarliu S, Ahissar E (2006) Parallel thalamic pathways for whisking and touch signals in the rat. *PLoS Biol* 4:e124.

5

The Motor Cortical Circuit

Apostolos P. Georgopoulos and Costas N. Stefanis

The core motor cortical circuit (cMCC) consists of a cell column ("minicolumn"), perpendicular to the cortical surface, ~30 µm in width (Georgopoulos et al., 2007). Each minicolumn contains pyramidal (~72%) and nonpyramidal (~28%) cells (Sloper et al., 1979). The greater MCC (gMCC) comprises minicolumns surrounding the core and receiving dense horizontal (i.e., tangential to the surface) projections from the core. These projections form a cylinder of ~500 µm diameter, centered on the core minicolumn (Gatter and Powell, 1978) (Figs. 5.1A and B). This cylinder contains ~278 minicolumns, given a minicolumn tangential area of $\pi R^2 = 3.141 \times \sim 15^2 = \sim 707$ µm^2, and a cylinder tangential area of $3.141 \times 250^2 = 196{,}312$ µm^2. Afferent fibers to, and efferent fibers from, the MCC are arranged in parallel to the minicolumns. Inputs external to the MCC (from the thalamus, contralateral hemisphere, and ipsilateral hemisphere) are excitatory and terminate mostly on dendritic spines of pyramidal cells and dendritic shafts and somata of nonpyramidal cells. Finally, the MCC also receives extensive monoaminergic innervation from all monoamine systems (dopamine, norepinephrine, serotonin, and acetylcholine) the functional role of which is not well understood.

SPATIAL ASPECTS

In the arm area of the motor cortex, MCC neurons are tuned to the same direction of movement (Georgopoulos et al., 2007). Preferred MCC directions are repeatedly mapped with a spatial repetition periodicity of ~240 µm (Georgopoulos et al., 2007). Remarkably, this periodicity closely corresponds to the radius of the gMCC (Fig. 5.1C), as defined above based on degeneration studies. This finding indicates that a given cMCC with preferred

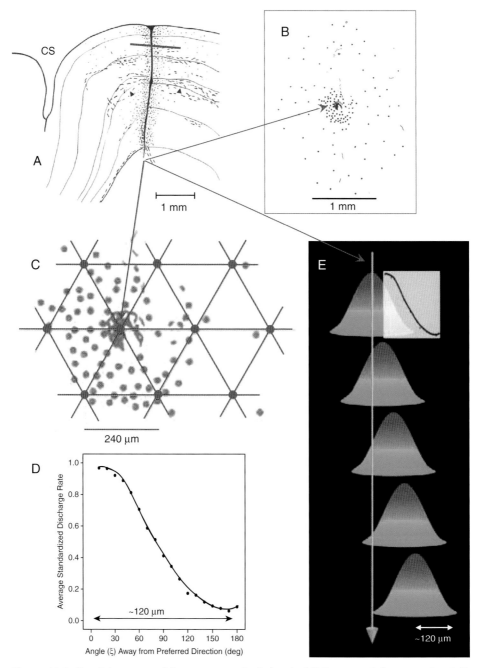

FIGURE 5–1. Spatial aspects of the motor cortical circuit. (*A*) Pattern of degeneration following insertion of a microelectrode in monkey motor cortex. Dots indicate degeneration of terminals (adapted from Fig. 1 of Gatter and Powell, 1978). (*B*) Distribution of terminal degeneration in a tangential section of the motor cortex (adapted from Fig. 9 of Gatter and Powell, 1978). (*C*) The greater motor cortical circuit (area of dense terminal degeneration from *B*, blown up to scale) is demarcated by minicolumns (cMCC) with the same preferred directions, placed at the corners

FIGURE 5–1. continued

of a regular hexagon (adapted from Georgopoulos et al., 2007). The remainder of the lattice is filled with minicolumns of spatially orderly varying preferred directions (see Figs. 6 and 7 of Georgopoulos et al., 2007). (*D*) Standardized, average directional tuning in the arm area of the motor cortex (adapted from Fig. 2 of Georgopoulos and Stefanis, 2007). (*E*) Observed average standardized directional tuning of *D* above partially superimposed on a model of spatial tuning field (adapted from Fig. 5 of Georgopoulos and Stefanis, 2007). The arrow indicates the fixed position of a hypothetical microelectrode recording from a cell for which the preferred direction coincides with the direction at the center of the field. The displaced directional tuning field indicates activation of other preferred directions farther and farther away from that of the top. It is hypothesized that the directional tuning curve of the recorded cell reflects the progressively decreasing influence of the spatially moving away directional tuning field, visualized, for example, as the length of the line crossing the directional tuning field at a particular location.

direction *C* interacts intensely with the other ~278 minicolumns within the gMCC demarcated by cMCCs of the same preferred direction *C*, roughly placed at the corners of a regular hexagon (Fig. 5.1C).

The MCC directional tuning (Fig. 5.1D) arises from orderly interactions among neighboring cMCCs (Georgopoulos and Stefanis, 2007; Merchant et al., 2008). This is illustrated in Figure 5.1E, where the observed directional tuning curve (insert) has been placed next to a spatial tuning profile (Georgopoulos and Stefanis, 2007). A systematic increase in the angle between the preferred direction *C* of a *cMCC* and the preferred direction *C'* of cMCCs at distances farther and farther away from *C* (Georgopoulos et al., 2007). This local shaping of the gMCC arises from the orderly excitatory and inhibitory interactions among neighboring cMCCs within a given gMCC, as follows.

TEMPORAL ASPECTS

The driving force for the local shaping of the gMCC (Georgopoulos and Stefanis, 2007) comes from the recurrent collaterals of pyramidal cell axons and their spatially orderly excitatory and inhibitory effects mediated through MCC interneurons (Stefanis and Jasper 1964a, 1964b; Eccles 1966; Stefanis, 1969). Figure 5.2A illustrates these recurrent effects in a diagrammatic form (Eccles, 1966). Figure 5.2B illustrates antidromically elicited postsynaptic potentials recorded intracellularly from pyramidal tract cells in the motor cortex (*a*, excitatory postsynaptic potential; *b* and *c*, inhibitory postsynaptic potentials, IPSP). The key player here is the *recurrent inhibition* (Stefanis and Jasper 1964a, 1964b; see Eccles, 1966 for a review). Recurrent inhibition is most likely polysynaptic and can last for up to 200 ms (Eccles, 1966), as evidenced by (*1*) the duration of antidromically elicited IPSPs and (*2*) the duration of depression of cell response following conditioning antidromic stimuli (Stefanis and Jasper, 1964a, 1964b).

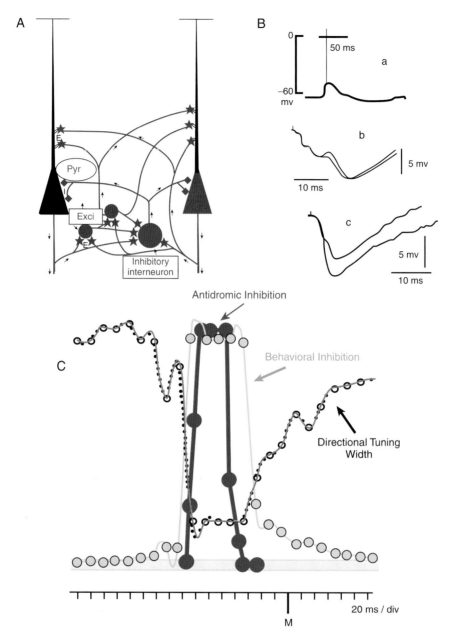

Figure 5–2. Recurrent and behavioral inhibition. (*A*) Schematic diagram of local motor cortical circuitry (adapted from Eccles, 1964). Synaptic actions are color coded; all pyramidal cell actions (in blue) are excitatory. (*B*) Examples of postsynaptic potentials recorded intracellularly from pyramidal tract neurons in cat motor cortex in response to antidromic stimulation (see Stefanis and Jasper, 1964a; Stefanis and Jasper, 1964b for experimental details). (*a*) Excitatory PSP followed by a shallow IPSP (adapted from Fig. 3 of Stefanis and Jasper, 1964a); (*b*) Graded, summed IPSPs (adapted from Fig. 5 of Stefanis and Jasper, 1964a); (*c*) Graded IPSPs in response to graded stimulus intensity (adapted from Fig. 5 of Stefanis and Jasper, 1964a). (*C*) Time course of recurrent and behavioral inhibition. Gray and turqoise lines denote time-varying change in tuning width and in behavioral inhibitory drive, respectively (adapted from Fig. 8 of Merchant et al., 2008, where experimental details can be found). Thick blue denotes the time course of an instance of recurrent inhibition (adapted from Fig. 3 of Stefanis and Jasper, 1964b; the solid line in that figure was inverted and rescaled in time to the behavioral inhibition curve). IPSP, inhibitory postsynaptic potential; M, movement onset.

BEHAVIORAL ASPECTS

Motor cortical cells in the arm area are directionally tuned (Georgopoulos et al., 1982), such that a cell discharges the highest for a movement of the arm in a particular direction, the cell's *preferred direction*, whereas the discharge rate decreases progressively with movements farther and farther away from this preferred direction. A recent study (Merchant et al., 2008) examined in detail the role of putative inhibitory mechanisms underlying the directional tuning, as follows. Cells were recorded during free, reaching movements of the arm toward eight targets in 3-D space, and their directional tuning calculated. Next, a combination of measurements was used to classify recorded cells into three categories: putative pyramidal cell 1 (PP1), putative pyramidal cell 2 (PP2), and putative interneurons (PI). Then, a detailed analysis was carried out, for every cell, on two concurrently estimated parameters, namely (1) the time-varying tuning width (i.e., sharpness of tuning), and (2) the time-varying putative inhibitory drive ("behavioral inhibition"). It was found that the two were significantly and positively correlated, such that the tuning width decreased as the inhibitory drive increased (see Fig. 7 in Merchant et al., 2008). An example is illustrated in Fig. 5.2C (gray and turqoise lines). Interestingly, this association was observed only for the PP1 cell. It was concluded that a timely inhibitory drive plays a major role in "sculpting" pyramidal cell discharge during the preparation and execution of movement. It was further hypothesized that this inhibitory drive may arise from recurrent pyramidal cell collaterals.

Indeed, a close examination of the time courses of the recurrent and behavioral inhibition supports this hypothesis. An example is shown in Figure 5.2C, where the time course of an instance of recurrent inhibition (Stefanis and Jasper, 1964b) is superimposed on an instance of behavioral inhibition (Merchant et al., 2008). Of course, more important than a single illustration is the quantitative picture. First, we deal with the duration of inhibition. With respect to recurrent inhibition, IPSPs typically lasted from 100–150 ms (Stefanis and Jasper, 1964a) and recurrent inhibition typically lasted from 25 to 120 ms (Stefanis and Jasper, 1964b). Behavioral inhibition lasted from 60 to 225 ms (Figs. 8 and 9 in Merchant et al., 2008). These values are very similar, in support of the hypothesis that recurrent inhibition underlies behavioral inhibition. In general, the latter tended to be a little longer than the former, and this can reasonably be accounted for by temporally staggered recurrent inhibitory effects arising from staggered pyramidal cell recruitment (Georgopoulos et al., 1982). The second quantitative aspect refers to the relative frequency of observed occurrence of recurrent and behavioral inhibition. Recurrent inhibitory effects were observed in 48 of 172 tested neurons (27.9%) (Stefanis and Jasper, 1964b). In the behavioral inhibition study (Merchant et al., 2008), 1206 cells were identified as pyramidal cells (928 as PP1 and 278 as PP2, see above). Behavioral inhibition was observed only in directionally tuned PP1 neurons (N = 366). This yields a prevalence of behavioral inhibition of 366/1206 = 30.3%. These two percentages are very close. Thus, both lines of evidence (i.e., duration and frequency of occurrence) speak for a key role of recurrent inhibition in shaping the directional tuning.

SPATIAL, TEMPORAL, AND BEHAVIORAL INTEGRATION

We reviewed earlier the basic layout of the motor cortical circuit as well as spatial, temporal, and behavioral aspects of it, with an emphasis on recurrent inhibition. The MCC operates as an integrated network at the gMCC level the output of which is an orderly tuning function. This tuning function can refer to the direction of arm movement in space in the arm area (Georgopoulos et al., 1982, 2007; Georgopoulos and Stefanis, 2007) or to a combination of finger movements in the hand area of the motor cortex (Georgopoulos et al., 1999). It is likely that a suitable tuning function pertains in different parts of the motor cortex, depending on the relevant parameters dictated by the somatotopic arrangement; for example, tuning of motor cortical cells with respect to the direction of tongue protrusion has been described in the orofacial area of the motor cortex (Murray and Sessle, 1992).

In real time, external, synchronous, excitatory, converging inputs to the relevant part of the motor cortex would initiate *gMCC* activation and pyramidal cell discharge; within a short time (a few tens of milliseconds), recurrent excitatory and inhibitory actions would ensure shaping the local motor cortical landscape, enhancing activity at its center (by boosting excitation) and gradually reducing activity at its periphery (by recurrent inhibition), that is, enhancing the *motor contrast* (Stefanis and Jasper, 1964b; Georgopoulos and Stefanis, 2007). External inputs might also preshape the field by suitable feed-forward activation of inhibitory interneurons (Eccles, 1966). The graded output of the tuning field is then transmitted downstream to spinal and other subcortical areas as well as to other cortical areas. It is remarkable that directional tuning has been observed in practically all motor areas that have been investigated, including the premotor cortex, parietal cortex, and cerebellum. This indicates a formal correspondence in the coding of movement direction in space across motor areas.

ACKNOWLEDGMENTS

This work was supported by the United States Department of Veterans Affairs and the American Legion Brain Sciences Chair.

References

Eccles JC (1966) Cerebral synaptic mechanisms. In: Eccles JC, ed. *Brain and Conscious Experience*, pp. 24–58. New York: Springer.

Gatter KC, Powell TPS (1978) The intrinsic connections of the cortex of area 4 of the monkey. *Brain* 101:513–541.

Georgopoulos AP, Stefanis CN (2007) Local shaping of function in the motor cortex: Motor contrast, directional tuning. *Brain Res Rev* 55:383–389.

Georgopoulos AP, Kalaska JF, Caminiti R, Massey JT (1982) On the relations between the direction of two-dimensional arm movements and cell discharge in primate motor cortex. *J Neurosci* 2:1527–1537.

Georgopoulos AP, Pellizzer G, Poliakov AV, Schieber MH (1999) Neural coding of finger and wrist movements. *J Comput Neurosci* 6:279–288.

Georgopoulos AP, Merchant H, Naselaris N, Amirikian B (2007) Mapping of the preferred direction in the motor cortex. *Proc Natl Acad Sci USA* 104:11068–11072.

Merchant H, Naselaris T, Georgopoulos AP (2008) Dynamic sculpting of directional tuning in the primate motor cortex during three-dimensional reaching. *J Neurosci* 28:9164–9172.

Murray GM, Sessle BJ (1992) Functional properties of single neurons in the face primary motor cortex of the primate. III. Relations with different directions of trained tongue protrusion. *J Neurophysiol* 67:775–785.

Sloper JJ, Hiorns RW, Powell TPS (1979) A qualitative and quantitative electron microscopic study of the neurons in primate motor and somatic sensory cortices. *Phil Trans Royal Soc London B* 285:141–171.

Stefanis C (1969) Interneuronal mechanisms in the cortex. In: Brazier MAB, ed. *The Interneuron*, pp. 497–526. Berkeley: University of California Press.

Stefanis C, Jasper H (1964a) Intracellular microelectrode studies of antidromic responses in cortical pyramidal tract neurons. *J Neurophysiol* 27:828–854.

Stefanis C, Jasper H (1964b) Recurrent collateral inhibition in pyramidal tract neurons. *J Neurophysiol* 27:855–877.

6

Prefrontal Cortex

Xiao-Jing Wang

The frontal lobe, the most anterior part of the neocortex, is conventionally defined by its afferent pathways from the mediodorsal thalamus. It subdivides into agranular areas (which lack a granular layer 4) and granular areas (which have a layer 4) (Wise, 2009). The agranular frontal areas are shared by all mammals and include parts of the orbitofrontal cortex (OFC) and anterior cingulate cortex (ACC). The granular frontal areas, collectively called the prefrontal cortex (PFC), include the dorsolateral prefrontal cortex (DLPFC), ventral prefrontal cortex, frontal pole cortex, dorsal and medial prefrontal areas, and rostral orbitofrontal cortex. The granular frontal cortex is present in the primates but not rodents. Its volume is about 6700 mm^3 in the chimpanzee and 34,800 mm^3 in the human; the corresponding surface is 52.7% and 83.2% of the frontal lobe, or 11.3% and 28.5% of the entire neocortex, respectively (Elston, 2007).

The PFC plays a central role in a wide range of cognitive functions, such as working memory, decision making, planning, self-control, and problem solving (Miller and Cohen, 2001; Fuster, 2008). The functional versatility of the PFC is due in part to its extensive input–output connections with the rest of the brain. A recent study examined afferent connections into 25 cytoarchitectonically defined frontal areas in macaque monkey, using neuroinformatics analysis of anatomical connectivity data (Averbeck and Seo, 2008). It was found that inputs from 68 sensory, motor, and limbic areas reach each of 25 frontal areas either directly or through a single intermediate step (on average) within the frontal network. The frontal network is highly interconnected, with each area sending output to about nine other frontal areas with an intermediate or strong connection. At the same time, it is a heterogeneous structure: 25 frontal areas are hierarchically organized into five clusters, each defined by a unique set of inputs. Inputs to each cluster from the frontal network arc dominated by the other areas within the same cluster. The extrinsic inputs to each cluster are roughly characterized as follows: *(1)* The ventrolateral group receives inputs from ventral visual and auditory areas, *(2)* the dorsolateral group receives inputs from dorsal visual and auditory areas, *(3)* the caudoorbital group receives chemosensory (gustatory and olfactory) and interoceptive inputs, *(4)* the dorsomedial group is defined by its motor inputs, and *(5)* the ventromedial group is defined by its limbic inputs (hippocampus and amygdala). The PFC projects to many posterior cortical areas (but not the primary visual cortex, V1), as well as to the thalamus, basal ganglia, hippocampus, amygdala, and superior colliculus.

Despite the diversity in the degree of identifiable laminae across frontal areas, there is a simple organization rule for the projections between two frontal areas. Namely, when frontal areas are classified based on the number and definition of its cortical layers (level 1, lowest; level 5, highest), projection neurons from a lower level area originate mostly in the deep layers (5–6), and their axons terminate predominantly in the upper layers of a higher level area. Conversely, projection neurons from a higher level cortex are located mostly in the upper layers (2–3), and their axons terminate predominantly in the deep layers of a lower level cortex (Barbas et al., 2002).

MICROCIRCUITRY

Local circuitry within a PFC area shares the general layout with other neocortical areas but also displays marked differences. Notably, in macaque monkey and human, the basal dendrites of layer 3 pyramidal neurons have up to 10 times more spines, the site of excitatory synapses, in PFC than in the primary visual cortex (V1). This is not just because cells are larger but also spine density is higher in PFC: the basal dendritic arbors are more widespread in PFC pyramidal cells, and the spine density (the number of spines per unit of dendritic length) is four times greater, compared to V1. There is a progressive increase in pyramidal cells' synaptic integration along the processing hierarchy of the visual system, from V1, V2, V4, TEO, TE to PFC. Furthermore, pyramidal cells in DLPFC are larger and have more branched dendrites and more spines than those in the premotor cortex, which in turn are more spinous and display larger and more branched dendrites than in the primary motor cortex (Elston, 2007). Therefore, along the sensory-association-motor axis, prefrontal pyramidal neurons are empowered with the greatest capability of integrating synaptic inputs. If dendritic trees are composed of relatively independent compartments, large and highly branched dendrites of prefrontal pyramidal cells would enable them to differentially process and gate information flows from different sensory, motor, and limbic areas in a way and to an extent unlike any other cortical area.

Many computational purposes can be served by this capability, such as combining sensory cues, reward signals, and task rules in decision making. Equally importantly, this means that intrinsic PFC microcircuitry is endowed with strong excitatory recurrent connections. Indeed, a large fraction of excitatory synapses onto pyramidal cells originate from the local circuit. In the cat V1, ~20% of all excitatory synapses are horizontal synaptic connections between pyramidal cells in layer 2/3. Assuming that this holds true across species and cortical areas, combined with the fact that pyramidal cells have more than 10-fold more excitatory synapses in PFC than in V1, it is expected that pyramidal cells in PFC layer 2/3 are endowed with severalfold stronger interconnections than in V1. Furthermore, the patterns of these intrinsic connections are also unique in the PFC. In the superficial layers 2/3 of sensory cortical areas of macaque monkey, axonal collaterals from pyramidal cells form patches of terminals, with an average width of 230 μm and patch center-to-center distance of 430 μm. The patchy connections link neurons that are separated at long distances but display similar stimulus selectivity. Horizontal axonal projections also form patch-like patterns in motor cortex. By contrast, in the PFC, horizontal connections form strip-like patterns, rather than patches. The strip length is more than 1mm, the width of strips is about 270 μm, and the strip center-to-center distance averages 530 μm. The PFC in primates thus exhibits unique, strip-like, intrinsic connections (Lund et al., 1993).

REVERBERATORY EXCITATION

Strong lateral connections between pyramidal cells may be key to understanding the PFC circuitry and functions. The most studied process that depends on PFC is working memory, the brain's ability to actively hold and manipulate information in the absence of direct sensory stimulation. In monkey experiments, when a subject is required to actively hold information about a sensory cue (e.g., a visual object, a vibrotactile stimulus frequency, or a spatial location) across a short delay, neurons in the PFC display stimulus-selective persistent activity during the delay period (Fuster, 2008). This mnemonic activity must be internally maintained in order to subserve working memory, a candidate mechanism underlying persistent activity is recurrent synaptic excitation that is sufficiently strong to sustain cross-talk among pyramidal neurons (Goldman-Rakic, 1995; Wang, 2006). Computational models have shown that, in a canonical cortical circuit, self-sustained persistent activity emerges when the amount of recurrent connections exceeds a threshold level (Wang, 2006). Thus, the PFC may have a similar intrinsic organization as sensory areas, but quantitative differences (e.g., in the strength of interconnections) may be sufficient to give rise to qualitatively different behaviors (the emergence of persistent activity).

Interestingly, biophysically based circuit modeling predicted that recurrent excitation in a working memory circuit should not only be strong but also slow relative synaptic inhibition in order to ensure network stability. More recent work showed that slow excitatory reverberation also provides a candidate circuit mechanism for gradually integrating information over time in decision making. A candidate substrate to implement slow excitation is the NMDA receptor–mediated synaptic transmission at local synapses. This possibility has been tested in *in vitro* physiological studies where the fast AMPA receptor (with a time constant of ~2 ms) and slow NMDA receptor (time constant ~50–100 ms) mediated components of synaptic currents were measured between pairs of connected pyramidal neurons. It was observed that, in adult rodents, pyramidal cells express more the NR2B NMDA subunits in the medial frontal area than in V1. As a result, the NMDA receptor–mediated currents at local synapses between pyramidal cells exhibit a two-fold longer decay time-constant and temporally summate a train of stimuli more effectively, in the frontal cortex compared to those in the primary visual cortex. Moreover, dopamine modulation greatly affects PFC functions, and dopamine D1/D5 receptors selectively enhance the NMDA receptor–mediated excitatory postsynaptic current. Finally, in behaving monkeys performing a working memory task, iontophoresis of drugs that blocked the NMDA receptors suppressed delay-period persistent activity of PFC, in support of an important role of the NMDA receptors in PFC processes.

Other slow positive feedback mechanisms have also been identified in the PFC. In particular, excitatory synapses between layer 5 pyramidal cells exhibit a stronger propensity for short-term facilitation (time constant of several hundred milliseconds) in the PFC than in V1 of young rodents, which could enhance recurrent excitation in an activity-dependent manner. There is also evidence that, in rodent layer 5 pyramidal cells of the medial frontal cortex, a calcium-dependent inward current induces a slow afterdeporalization (time constant of a few seconds). Prolonged depolarization of pyramidal neurons leads to further mutual excitation, providing another slow positive feedback through the interplay between synaptic dynamics and intrinsic ion channels in single cells (Wang, 2006). A commonality of these diverse types of cellular and synaptic positive feedback mechanisms is that they are slow, operating on the timescale of many tens of milliseconds to seconds. This is in support of the prediction from computational models that slow reverberatory dynamics represent a characteristic feature of PFC microcircuits, and it is well suited for underlying working memory and decision-making computations.

SYNAPTIC INHIBITION AND ITS BALANCE WITH EXCITATION

A general principle of cortical organization is a delicate balance between excitation and inhibition. Insofar as PFC local circuits are empowered by strong synaptic excitation, they should also be endowed with specialized inhibitory circuitry. Traditionally, fast-spiking, perisomatic targeting basket cells have been the focus of studies of synaptic inhibition. However, in the cortex, there is a wide diversity of GABAergic interneurons, with regard to their morphology, electrophysiology, chemical markers, synaptic connections, and short-term plasticity molecular characteristics. Three largely nonoverlapping subclasses of inhibitory cells can be identified according to the expression of calcium-binding proteins parvalbumin (PV), calbintin (CB), or calretinin (CR). Interestingly, in macaque monkey, the distributions of PV, CB, and CR interneurons appear to be quite different in PFC compared to V1. In primary visual cortex, PV-containing interneurons (including fast-spiking basket cells) are prevalent (~75%), whereas the other two CB- and CR-containing interneuron types constitute of about 10% each of the total GABAergic neural population. By contrast, in the PFC, the proportions are about 24% (PV), 24% (CB), and 45% (CR), respectively. Thus, the non-PV-containing interneurons are predominant in the macaque monkey PFC. Curiously, this does not hold true in the rat frontal cortex, where PV-containing interneurons constitute 43%–61% of all GABAergic cells. GABAergic neurons represent a larger proportion of all neurons in monkey cortex (~25% in the medial PFC) than in rat frontal cortex (16%). This difference may reflect a differential increase of the absolute number of non-PV interneurons in monkeys. Therefore, primate PFC circuits are characterized by an increased proportion of GABAergic cells relative to rodents and a predominance of non-PV interneurons, unlike early sensory areas.

A circuit model suggests how these different interneuron types may work together in the PFC (Fig. 6.1). This model incorporates three subtypes of interneurons classified according to their synaptic targets and their prevalent interconnections. First, PV interneurons project widely and preferentially target the perisomatic region of pyramidal neurons, thereby controlling the spike output of principal cells and sculpt the tuning of the network activity pattern. Second, CB interneurons act locally, within a cortical column. They predominantly target dendritic sites of pyramidal neurons, hence controlling the inputs onto principal cells. Third, CR interneurons also act locally and project preferentially to CB interneurons. Note that the three interneuron types in the model should be more appropriately interpreted according to their synaptic targets, rather than calcium-binding protein expressions. For example, PV cells display a variety of axonal arbors, among which the large basket cells are likely candidates for the widely projecting perisoma-targeting cells. Similarly, CB interneurons show a high degree of heterogeneity, but some of them (such as double bouquet cells or Martinotti cells) are known to act locally and preferentially target dendritic spines and shafts of pyramidal cells. Finally, although many CR interneurons do project to pyramidal cells, anatomical studies show that a subset of CR cells avoids pyramidal cells, at least in the same cortical layer and preferentially targets CB interneurons. It is also possible that axonal innervations of a CR cell project onto pyramidal

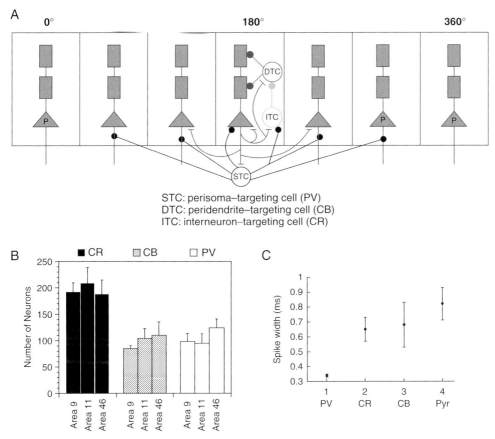

FIGURE 6–1. (*A*) A spatial working memory model with three subclasses of GABAergic interneurons. Pyramidal (P) neurons are arranged according to their preferred cues (a directional angle, from 0 to 360 degrees). There are localized recurrent excitatory connections and broad inhibitory projections from perisoma-targeting (parvalbumin-containing, PV) fast-spiking neurons to P cells. Within a column, calbindin-containing (CB) interneurons target the dendrites of P neurons, whereas calretinin-containing (CR) interneurons preferentially project to CB cells. Excitation of a group of pyramidal cells recruits locally CR neurons, which sends enhanced inhibition to CB neurons, leading to dendritic disinhibition of the same pyramidal cells. (Adapted from Wang et al., 2004 with permission) (*B*) Proportional distribution of PV-, CB-, and CR-expressing GABAergic cells in three subregions of the monkey prefrontal cortex. (Reproduced with permission from Condé et al., 1994) (*C*) Half-peak spike width for pyramidal neurons and three types of interneurons of macaque monkey prefrontal cortex. (Reproduced with permission from Zaitsev et al., 2005 and Povysheva et al., 2006)

cells in a different cortical layer while selectively targeting inhibitory neurons in the same layer.

Figure 6.2a–b shows a computer simulation of a biophysically detailed implementation of this circuit model for a spatial working memory task. When pyramidal cells in a column are excited by a transient extrinsic input,

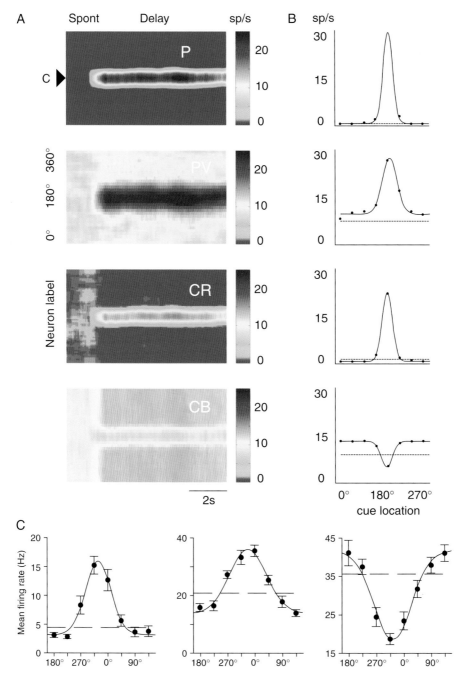

FIGURE 6–2. Computer simulation of a spatial working memory model schematically shown in Figure 6.1a, and comparison between the model and recorded PFC neuronal tuning curves. (A) Spatiotemporal activity patterns for the pyramidal cells and the three (PV, CB, and CR) inhibitory neuron populations during the cue and delay periods. Instantaneous firing rates are color coded. (B) Observed neuronal tuning curves (solid lines) during the delay period in the model simulations. Eight different cue positions are used. Dashed lines, spontaneous firing

FIGURE 6–2. continued
rate during the resting state. (*C*) Three kinds of recorded tuning curves in monkey dorsolateral prefrontal cortex during a spatial working task, with the same conventions as in (*B*). Solid line, the best Gaussian fit; dotted line, average firing rate during the last second of fixation. Note that the putative fast-spiking PV cell (*center*) has a higher spontaneous firing rate and wider tuning than the regular-spiking putative pyramidal cell (*left*), similar to what is found in the network simulations (*B*). An example of the inverted tuning curve is shown (*right*), with a high baseline activity (dashed horizontal line), strong reduced delay period activity for some cues, and slightly increased delay period activity for other cues. There are about 5% of recorded neurons showing these properties, which the model predicts to be putative CB interneurons that preferentially target pyramidal dendrites. Consistent with slice physiology (Fig. 6.1c), the spike width is the shortest for putative PV cells, the longest for putative pyramidal cells, and intermediate between the two for putative CB interneurons. (Reproduced from Wang et al., 2004)

they excite each other through interconnections. At the same time, activated CR interneurons suppress CB interneurons within the same column, leading to reduced inhibition (disinhibition) of the dendrites of the same pyramidal cells. The concerted action of recurrent excitation and CR interneuron-mediated disinhibition generates self-sustained persistent activity in these neurons, and the network activity pattern is shaped by synaptic inhibition from PV interneurons. Moreover, CB interneurons in other columns might be driven to enhance their firing activity; therefore, pyramidal cells in the rest of the network would become less sensitive to external inputs, ensuring that working memory storage is not vulnerable to behaviorally irrelevant distracters. A prediction of this model is that a small fraction of (putative CB) PFC neurons recorded from behaving monkey should show inverted tuning of mnemonic delay period activity, that is, a reduced firing relative to spontaneous activity selectively for some sensory cues. This prediction was confirmed in data analysis from a monkey spatial working memory task (Fig. 6.2c). Roughly 5% of recorded neurons in that experiment showed behavior that was predicted by the model for dendrite-targeting CB interneurons, consistent with the crude estimate of ~6% CB-containing interneurons (~24% of GABAergic cells, which in turn represent ~25% of all neurons).

Hence, different interneuron cell types show both division of labor and cooperation in subserving mnemonic PFC functions: stimulus selectivity, memory storage, and resistance against distracters. They also contribute differentially to the temporal dynamics, such as synchronous oscillations, during working memory (Wang, 2006). The inhibitory circuitry across different cortical layers may also show some specialized features in the PFC. For instance, a recent model suggests that connections from layer 5 to layer 2/3 excitatory and inhibitory neurons, and those from layer 6 to layer 4 interneurons, ought to be stronger in the frontal eye field than in V1 (Heinzle, 2007). Intriguingly, non-PV interneuron types may be differently involved in projections between different prefrontal subregions. Indeed, there is anatomical evidence that, in the monkey PFC area 9, inputs from the neighboring area 46 and from anterior cingulated cortex area 32 similarly innervate excitatory neurons. However, GABAergic neuron targets are different: inputs from area 46 prevalently terminate onto CR-containing interneurons, while those from ACC predominantly terminate onto CB-containing interneurons (Medalla and Barbas, 2009). According to the model of Figure 6.1a, these findings imply that inputs from DLPFC area 46 serve to disinhibit pyramidal cells and boost their activity, while inputs from ACC effectively serve to inhibit dendrites of pyramidal cells, presumably contributing to gating inputs and resisting distraction, as the PFC actively maintains internal representations of sensory information, task rule, and so on. Note that in view of the high degree of heterogeneity among distinct areas in the PFC, it is likely that the inhibitory circuitry is also heterogeneous, adaptively tailored to each area's functional demands.

NEUROMODULATION

The PFC is a prominent target of afferents from the dopamine, norepinephrine, serotonin, and acetylcholine systems. Dopamine D1/D5 receptors are particularly concentrated in PFC and are prevalently located in the spines; thus, they are in a privileged position to modulate synaptic inputs. There is also evidence that both D1 and D2 dopamine receptors are expressed in GABAergic interneurons. Iontophoresis studies using behaving monkeys showed that mnemonic PFC neural activity in a delayed response task exhibits an inverted U-shaped dependence on the level of D1 receptor agonist concentration: working memory is impaired with either too little or too much dopamine D1 activity. Norepinephrine modulation of the PFC activity during working memory is also characterized by an inverted U-shaped influence curve. At the present, little is known about how neuromodulators differentially target distinct subclasses of interneurons. Notable is the anatomical evidence that serotonin 5-HT$_{2A}$ receptors are preferentially expressed in pyramidal cells and perisomatic targeting interneurons, whereas 5-HT$_3$ receptors are mostly expressed in calbintin-containing dendrite-targeting interneurons and calretinin-containing interneurons. The functional implications of this specialization remain to be understood, in the context of recurrent network dynamics underlying cognitive functions.

SUMMARY

The PFC circuits are characterized by several features. First, their input–output connections with the rest of the brain are extraordinarily extensive.

Pyramidal neurons in PFC are greatly more spinous than in V1, and thus they have a very large capacity for synaptic integration. Second, PFC areas are endowed with strong intrinsic excitatory and inhibitory connections that are sufficient to generate persistent activity underlying working memory and competitive neurodynamics for decision making. A general principle is that excitatory feedback mechanisms underlying reverberation should be slow, in order to ensure network stability and to best serve such cognitive functions as gradual time integration of information in decision making. Third, excitation and inhibition are balanced dynamically. In the PFC, the synaptic inhibitory circuit is predominated by GABAergic cell subclasses that are not fast-spiking PV-containing interneurons. The increased abundance of other interneuron types (compared to sensory areas), which target pyramidal dendrites or regulate inhibitory circuit itself, may reflect the functional demand of selectively gating input pathways into the PFC in accordance with the behavioral context and goals.

REFERENCES

Averbeck BB, Seo M (2008) The statistical neuroanatomy of frontal networks in the macaque. *PLoS Comput Biol* 4:e1000050.

Barbas, H, Ghashighaei HT, Rempel-Clower N, Xiao D (2002) Anatomical basis of functional specialization in prefrontal cortices in humans. In: Grafman J, ed. *Handbook of Neuropsychology*, Vol. 7, 2nd ed., pp. 1–27. New York: Elsevier.

Condé F, Lund JS, Jacobowitz DM, Baimbridge KG, Lewis DA (1994) Local circuit neurons immunoreactive for calretinin, calbindin D-28k or parvalbumin in monkey prefrontal cortex: distribution and morphology. *J Comp Neurol* 341:95–116.

Elston GN (2007) Specialization of the neocortical pyramidal cell during primate evolution. In: Kaas J and Preuss TM, eds. *Evolution of Nervous Systems: A Comprehensive Reference*, pp. 191–242. New York: Elsevier.

Fuster J (2008) *The Prefrontal Cortex*. 4th ed. New York: Academic Press.

Goldman-Rakic PS (1995) Cellular basis of working memory. *Neuron* 14:477–485.

Heinzle J, Hepp K, Martin KA (2007) A microcircuit model of the frontal eye fields. *J Neurosci* 27:9341–9353.

Lund JS, Yoshioka T, Levitt JB (1993) Comparison of intrinsic connectivity in different areas of macaque monkey cerebral cortex. *Cereb Cortex* 3:148–162.

Medalla M, Barbas H (2009) Synapses with inhibitory neurons differentiate anterior cingulate from dorsolateral prefrontal pathways associated with cognitive control. *Neuron* 61:609–620.

Miller EK, Cohen JD (2001) An integrative theory of prefrontal cortex function. *Ann Rev Neurosci* 24:167–202.

Povysheva NV, Gonzalez-Burgos G, Zaitsev AV, Kröner S, Barrionuevo G, Lewis DA, Krimer LS (2006) Properties of excitatory synaptic responses in fast-spiking interneurons and pyramidal cells from monkey and rat prefrontal cortex. *Cereb Cortex* 16: 541–552.

Wang X-J (2006) A microcircuit model of prefrontal functions: ying and yang of reverberatory neurodynamics in cognition. In: Risberg J and Grafman J, eds. *The Prefrontal Lobes: Development, Function and Pathology*, pp. 92–127. Cambridge, England: Cambridge University Press.

Wang X-J, Tegner J, Constantinidis C, Goldman-Rakic PS (2004) Division of labor among distinct subtypes of inhibitory neurons in a microcircuit of working memory. *Proc Natl Acad Sci (USA)* 101:1368–1373.

Wise SP (2009) Forward frontal fields: phylogeny and fundamental function. *Trends Neurosci* 31:599–608.

Zaitsev AV, Gonzalez-Burgos G, Povysheva NV, Kröner S, Lewis DA, Krimer LS (2005) Localization of calcium-binding proteins in physiologically and morphologically characterized interneurons of monkey dorsolateral prefrontal cortex. *Cereb Cortex* 15:1178–1186.

Section 2
Thalamus

7

The Thalamus

Edward G. Jones

The thalamus of mammals is composed of three fundamental entities, distinguished by different developmental histories and connections (reviewed in Jones, 2007). The *epithalamus*, consisting of the habenular and paraventricular nuclei, is mainly connected with the hypothalamus and will not be considered here. The *dorsal thalamus* is the large cell mass, divided into multiple nuclei, through which information from the sense organs, the motor systems, and other intrinsic brain sources is relayed to the cerebral cortex and basal ganglia. Its connections with the cortex are bidirectional: thalamocortical and corticothalamic. The *ventral thalamus* covers the anterior, lateral, and ventral aspects of the dorsal thalamus and is composed of the ventral lateral geniculate nucleus, reticular nucleus, zona incerta, and field of Forel. Only the reticular nucleus will be considered here on account of its intimate connections with the underlying dorsal thalamus. All data from which the following description is derived can be found in these reviews: (Steriade et al. 1990); (McCormick 1992); (Steriade et al. 1993); (McCormick and Bal 1997); (Steriade et al. 1997); (Sherman and Guillery 2001); and (Jones 2007).

The nuclei of the dorsal thalamus and their intrinsic circuitry appear to be constructed on a common theme (Figs. 7.1 and 7.2), although most information comes from the principal sensory relay nuclei, the ventral posterior, dorsal lateral geniculate, and medial geniculate. Three cellular elements lie at the heart of thalamic circuitry: *relay neurons* that project their axons to the cerebral cortex (*thalamocortical fibers*), *intrinsic GABAergic interneurons* located within the relay nuclei, and *extrinsic GABAergic neurons* located in the reticular nucleus that send their axons into the dorsal thalamus. The principal axonal elements of the circuitry consist of *afferent fibers* entering the dorsal

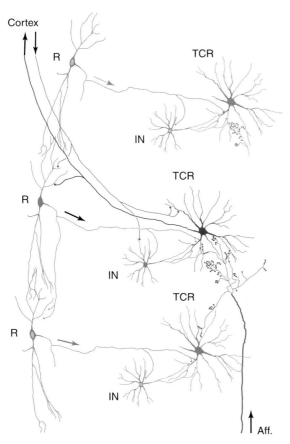

FIGURE 7–1. Schematic figure showing the basic circuitry of the thalamus, made up of connections between afferent fibers (Aff.), thalamocortical relay cells (TCR), intrinsic interneurons (IN), reticular nucleus cells (R), and the cerebral cortex. (Based on Steriade et al.,1997)

thalamus from subcortical sources and commonly forming well-known fiber tracts such as the medial lemniscus and optic tract, *corticothalamic fibers* returning to the dorsal thalamic nucleus or nuclei from which their parent cortical area(s) received thalamocortical input, and the *reticulothalamic axons* from the reticular nucleus. As thalamocortical and corticothalamic fibers traverse the reticular nucleus en route to cortex or thalamus, respectively, they each give off *collaterals* to the reticular nucleus. These collaterals form the principal afferent drive to the reticular nucleus. Exerting a modulatory influence over the excitability of the thalamus, by means of their diffusely distributed *nonspecific afferent fibers*, are the cholinergic, serotoninergic, and noradrenergic systems of the brainstem. Dopaminergic input to the thalamus is very small or absent.

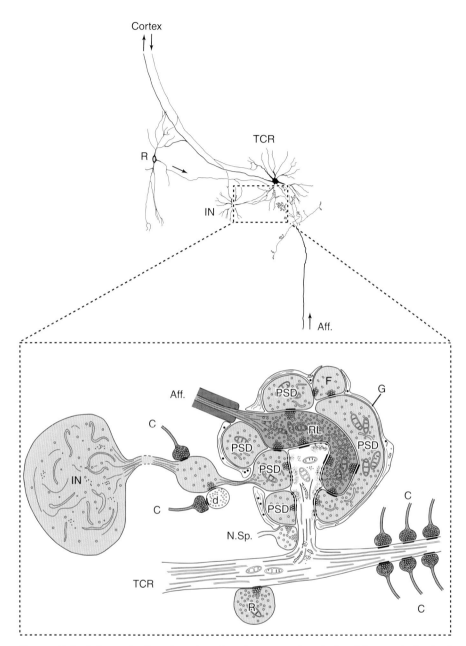

Figure 7–2. Schematic drawing showing the synaptic relationships typical of relay cells in dorsal thalamic nuclei in which intrinsic interneurons are present. Proximal dendrites or dendritic protrusions of relay neurons (TCR) receive terminals (RL) of subcortical afferent fibers (Aff.) and the presynaptic dendritic terminals (PSD) of interneurons (IN) in complex glial (G) ensheathed synaptic aggregations. Presynaptic dendrites of interneurons are postsynaptic to the RL terminals and also to one another and can form triadic arrangements in which a RL terminal ending on the relay cell dendrite ends on a PSD which also ends on the same part of the relay cell dendrite. R indicates terminals of reticulothalamic axons, located mainly on relay cell dendrites. Terminals labeled F are predominantly the terminals of reticulothalamic axons, but an unknown number may originate from axons of interneurons when present. Corticothalamic terminals (C) end in large numbers on the relay cell dendrites distal to the synaptic aggregations and to a lesser extent on the parent dendrites of the PSDs. Terminals (N.Sp.) of cholinergic afferents from the brainstem end close to the principal afferent terminals. (Based on Jones, 2007)

RELAY NEURONS

With the exception of the dorsal lateral geniculate nucleus of carnivores, thalamic relay neurons show a remarkable uniformity of morphology and dendritic architecture across nuclei and across species. Typically, a relay cell is moderately large with a soma c. ~400 μm^2 in area and radially oriented dendrites. The primary dendrites are thick and their secondary branches commonly devolve into a tuft of four or more tertiary branches, which give the cells a characteristically bushy appearance. Found mainly on the primary and secondary dendrites are dendritic appendages of variable size, number, and complexity; they are considerably larger than the dendritic spines of pyramidal neurons in the cerebral cortex or of Purkinje cells in the cerebellar cortex, and their internal structure is quite unlike that of a dendritic spine. The axon that the relay cells give off to the cerebral cortex is thick for a central axon (1.5–3 μm in diameter).

The bushy thalamic relay cell, with its symmetrical dendritic field and a greater or lesser number of dendritic appendages, represents the fundamental relay cell type, expressed in all dorsal thalamic nuclei of all species, generally with only minor variations. The dorsal lateral geniculate nucleus of the cat and certain other carnivores is an exception. Here, although the relay neurons can be seen as conforming in basic essentials to the standard bushy cell type, there is a di- or trimorphism that distinguishes relay neurons innervated specifically by different classes of retinal ganglion cell. Cells referred to as Y cells because they receive input from the Y class of retinal ganglion cells have large somata (250–800 μm^2 in area) and radially symmetrical dendritic fields. The dendrites are thin, lack significant numbers of appendages, and can cross borders between laminae of the nucleus. The axon measures 1.5 to >3 μm in diameter.

X cells, named for their input from retinal ganglion cells that possess the capacity of resolving high degrees of spatial contrast, have small or medium-sized somata (50–450 μm^2 in area) and elongated or oval dendritic fields that have the long axis oriented across the geniculate lamina in which the cell lies; the dendrites do not cross the borders between laminae. The dendrites are relatively thin and bear numerous appendages. These are associated with a somewhat more complex synaptic organization than in Y cells. The axon measures 1–1.5 μm in diameter.

W cells have small somata, similar in size to those of X cells, and thin, varicose, relatively long dendrites forming an elongated field horizontally extended within the borders of the lamina in which the cell lies. The axon is thin and, although myelinated, measures no more than 1 μm in diameter.

Generally, the passive cable properties of relay neurons are relatively uniform, reflecting the lack of morphological variation; but in the feline dorsal lateral geniculate nucleus, the different dendritic configurations are associated with differences in passive cable properties: X cells have higher specific membrane resistances and longer membrane time constants than Y cells, although the two cell types have similar electrotonic lengths and dendritic-to somatic conductances.

Intrinsic Interneurons

Intrinsic interneurons are found in all dorsal thalamic nuclei of many species and account for ~25% of the neuronal population. But in rodents, lagomorphs, bats, and marsupials they are essentially absent from all nuclei except the dorsal lateral geniculate nucleus. In these animals inhibition is provided solely by the input from the reticular nucleus. In the dorsal lateral geniculate nucleus of rodents, approximately 20% of the neurons are GABAergic, while in other nuclei, only a very rare GABA cell is found. In other mammals, the density of GABA cells is relatively high in all nuclei, the cells accounting for not less than 30% of the volume of the nucleus and in some cases (e.g., the ventral lateral posterior nucleus) as much as 45%.

The GABAergic intrinsic interneurons are much smaller than the relay cells with somal areas approximately half the size of those of relay neurons. These small intrinsic interneurons have relatively few (3–4) dendrites, from which emerge many thin, lengthy, and often branched appendages bearing knob-like dilations. Electron microscopy shows that the knob-like structures are vesicle-filled terminals that end in symmetrical synapses on dendrites of relay cells and on one another. The dendrodendritic synapses thus formed represent the principal type of synaptic connection made by intrinsic interneurons. In general, intrinsic interneurons lack an axon, although axons have been described on an occasional interneuron in the dorsal lateral geniculate nucleus of the cat. Here, they may represent displaced neurons of the perigeniculate nucleus, a part of the reticular nucleus.

In animals such as rodents, most of whose thalamic nuclei lack intrinsic interneurons, disynaptic inhibition from the reticular nucleus is the only source of inhibition occurring in a relay nucleus during afferent driving. Intrinsic interneurons, even when present, could be relatively ineffectual in causing major hyperpolarization changes in relay cells during afferent driving because the presynaptic dendritic terminals may be isolated electrically from the soma of the parent cell. The attenuated nature of their dendrites may isolate inputs to parts of the dendrites so that localized membrane changes are not registered at the soma or throughout the dendritic tree. Most afferent inhibition, even in nuclei containing intrinsic interneurons, would thus be attributable to the inputs from the reticular nucleus.

Neurons of the Reticular Nucleus

The neurons of the reticular nucleus are very different from relay cells of the dorsal thalamus. All are GABAergic with large, fusiform, or triangular somata and long dendrites mainly disposed parallel to the surface of the underlying dorsal thalamus and at right angles to the fibers entering and leaving the dorsal thalamus. A typical reticular nucleus cell possesses two or more polar dendrites that branch once or twice, giving the cell a bitufted dendritic field, where the nucleus is relatively thin. In thicker parts of the nucleus, the cells can have a more radial dendritic field. The dendrites of neighboring cells overlap extensively and span the full thickness of the nucleus. The dendrites are spine free. In some species, such as cats, the tertiary dendrites possess knob-like processes that prove, on electron microscopic examination, to be presynaptic dendritic terminals. They may not be present in other species. Reticular nucleus cells have thin axons that, after emitting two or more intra-nuclear collaterals, enter the dorsal thalamus, where they ramify widely although with a certain degree of topographic specificity.

Afferent Fibers

Afferent fibers entering and terminating within a thalamic relay nucleus from subcortical sites have a uniform morphology (Jones, 2007). The parent axons are thick (2–5 µm in diameter) and give off preterminal and terminal branches over a relatively compact territory 100–200 µm by 500–1000 µm in extent before devolving into a spray of unmyelinated terminal branches studded with large boutons ~1–5 µm in diameter, many of them forming grape-like clusters of 5–30 boutons embracing the proximal dendrites of relay cells.

The majority of corticothalamic fibers arise from modified pyramidal cells of layer VI of the cerebral cortex; they are thin, relatively straight axons approximately 1 µm in diameter with short side branches ending in single, small boutons. These thin corticothalamic fibers outnumber the subcortical afferents by as much as 10 to 1. Their terminals stud the secondary and tertiary dendrites of relay cells. Other corticothalamic fibers, fewer in number and arising from layer V pyramidal cells, are thicker and end in grape-like clusters of larger boutons not unlike those of the subcortical afferents. Their numbers may vary from nucleus to nucleus.

The axons entering the dorsal thalamus from the reticular nucleus are thin and branch extensively throughout the thalamus, ending in spray-like terminal branches studded with boutons of relatively large size that make symmetrical synapses, primarily on the dendrites of relay neurons.

Other afferent axons entering the nuclei of the dorsal thalamus belong to the cholinergic, noradrenergic, serotoninergic, histaminergic, and peptidergic pathways that arise from a variety of brainstem, hypothalamic, and basal forebrain sites and form the so-called nonspecific pathways to the thalamus. They are distributed diffusely throughout the dorsal and ventral thalamus.

MICROCIRCUITRY OF A RELAY NUCLEUS

A central feature of all dorsal thalamic nuclei is the presence of large aggregations of synaptic terminals centered on those derived from the principal subcortical afferent fiber system and clustered around the proximal dendrites of the relay cells (Fig. 7.2). These synaptic aggregations are sometimes referred to as glomeruli or synaptic islands. They are especially prevalent upon X-type neurons of the feline dorsal lateral geniculate nucleus but are found to a varying extent on all relay neurons.

In each synaptic aggregation there are three major elements: the dendrite of a relay neuron or one or more of its dendritic appendages and two presynaptic components. The first presynaptic element is made up of one or more of the large axon terminals derived from the principal subcortical afferent fiber system to the nucleus, but some cholinergic terminals derived from axons originating in the brainstem may also be found in this position. The dominant terminal is usually large, up to 3 µm in diameter and 5–7 µm in length with a rather dense cytoplasm that is packed with synaptic vesicles. These terminals are glutamatergic and make multiple, asymmetrical, synaptic contacts with the dendritic elements of the aggregation and with dendrites belonging to the thalamic interneurons, when these are present. The second presynaptic component of the synaptic aggregations is formed by the presynaptic dendritic terminals of interneurons, when present. They are GABAergic and resemble axon terminals in containing synaptic vesicles. They make symmetrical synaptic contacts on the relay cell dendrite in the aggregations and on one another. They are themselves postsynaptic, at asymmetrical synapses, to the subcortical afferent terminal(s). In this way, serial, reciprocal, and triadic synaptic arrangements can be found. A triad is formed when a subcortical afferent terminal ends on the dendrite of a relay neuron and on the presynaptic dendritic terminal of an interneuron that, in close proximity, is itself presynaptic to the relay cell dendrite. At the perimeter of a synaptic aggregation and distributed along the length of the relay cell dendrites, other GABAergic axon terminals can be found. These are mainly derived from axons of the reticular nucleus. Eighty percent of the reticulothalamic terminals are on the dendrites of relay neurons, no more than 20% being on the dendrites of intrinsic interneurons when present.

Synaptic aggregations are absent or rare on the second- and third-order dendrites of relay cells. Instead, these more peripheral dendrites are studded with asymmetrical, glutamatergic synapses formed by large numbers of small axon terminals, 1–1.5 µm in diameter mainly derived from corticothalamic axons. A smaller number may be derived from the nonspecific afferent systems.

Quantification of the distribution of synapses derived from subcortical afferents, cortical afferents, interneurons, and reticular nucleus axons on the soma-dendritic surfaces of physiologically characterized relay cells has

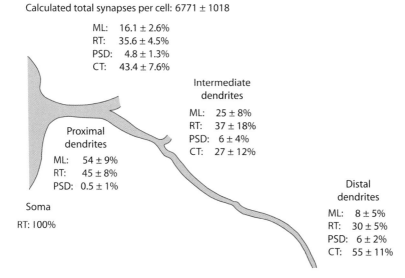

Calculated total synapses per cell: 6771 ± 1018

ML:　16.1 ± 2.6%
RT:　35.6 ± 4.5%
PSD:　4.8 ± 1.3%
CT:　43.4 ± 7.6%

Intermediate
dendrites

ML:　25 ± 8%
RT:　37 ± 18%
PSD:　6 ± 4%
CT:　27 ± 12%

Proximal
dendrites
ML:　54 ± 9%
RT:　45 ± 8%
PSD:　0.5 ± 1%

Soma
RT: 100%

Distal
dendrites

ML:　8 ± 5%
RT:　30 ± 5%
PSD:　6 ± 2%
CT:　55 ± 11%

FIGURE 7–3. The total number of synapses, their relative proportions, and their distributions on intracellular labeled cells in the ventral posterior nucleus of cats. CT, corticothalamic terminals; ML, medial lemniscal terminals; PSD, presynaptic dendrites of interneuron; RT, reticulothalamic terminals. (Based on Liu et al., 1995a; Liu et al., 1995b)

revealed that most relay neurons in the cat receive 4000 to 8000 synapses with a density of 0.6–1 synapses per micrometer (Fig. 7.3). On X and Y cells examined in the lateral geniculate nucleus, retinal fiber terminals accounted for 15%–20%, inhibitory terminals for 25%–30%, and corticothalamic terminals for 40%–50%; presynaptic dendritic of interneurons were included in the inhibitory class. On cells examined in the ventral posterior nucleus 12%–29% of the terminals were medial lemniscal, 29%–44% were inhibitory and mainly derived from the reticular nucleus axons, 23%–53% were corticothalamic, and 2%–7% were presynaptic dendritic terminals.

The proximo-distal distribution of synaptic types on relay neurons, reading outward from the soma, is as follows (Fig. 7.3): On the soma 100% of the terminals are inhibitory, although they are very few in number; on proximal dendrites subcortical afferent terminals form 43%–62% of the total synaptic contacts, reticulothalamic terminals form 38%–54%, presynaptic dendritic terminals form less than 2%, and there are no corticothalamic terminals. On second-order dendrites, terminals of subcortical afferents form 16%–35% of the total, reticulothalamic terminals remain relatively constant at 38%–50%, presynaptic dendritic terminals form 8%–9%, and corticothalamic terminals form 9%–42% of the contacts. On third-order dendrites, subcortical afferent terminals account for 4%–13% of the total synaptic contacts, reticulothalamic terminals continue to form 25%–40%, presynaptic dendritic terminals form 1% or less, and corticothalamic terminals form 46%–66%.

Overall, the density of corticothalamic terminals is remarkable high, forming more than 80% of all synapses on a relay cell in comparison with about 2% for subcortical afferent terminals and ~15% for reticulothalamic terminals.

The reticulothalamic axons are also targeted primarily at the relay cells. It was found that 82% of the reticular nucleus terminals made synaptic contact with the dendrites of relay neurons, 8.5% with dendrites or presynaptic dendrites of interneurons and 9.3% with somata of interneurons or relay cells. These reticulothalamic terminals account for the majority of GABAergic synapses in a relay nucleus.

Synaptic Chemistry

The excitatory response of a thalamic relay cell to natural or artificial stimulation of afferent fibers is determined by the action of the released excitatory amino acid transmitter upon NMDA and non-NMDA receptors. The excitatory postsynaptic potentials (EPSPs) induced by afferent stimulation are strongly dependent on non-NMDA receptors, but a sustained element of the response depends on NMDA receptors. GluR3 and GluR4 are the principal AMPA receptor subunits expressed in the dorsal thalamus. Kainate receptors play little part in the responses of thalamic cells to stimulation of subcortical or cortical afferents. Metabotropic glutamate receptors play a less important role in subcortical activation of relay cells than they do in corticothalamic activation. The effects of mGluR antagonists upon the responses of relay cells to peripheral activation may depend upon mGluR effects on thalamic interneurons. Localization on presynaptic processes away from the points of synaptic contact may explain why mGluRs usually need high-frequency stimulation to become activated and to inhibit transmitter release.

Corticothalamic synapses are associated with NMDA, AMPA, and metabotropic glutamate receptors, each contributing a characteristic form to the EPSPs elicited by corticothalamic stimulation. The metabotropic receptor, mGluR1α, in particular has been associated with the terminations of corticothalamic fibers on distal dendrites of relay cells. The distal dendrites on which the corticothalamic terminals are concentrated are where high-threshold P/Q type calcium channels are concentrated and when a relay cell is relatively depolarized, corticothalamic stimulation will elicit high-frequency (~40 Hz) oscillations of neuronal discharge.

Reticulothalamic synapses are associated with both $GABA_A$ and $GABA_B$ receptors. The $GABA_B$ receptor–mediated inhibitory postsynaptic potentials (IPSPs) engendered in relay cells under the influence of the reticular nucleus are larger and activate much more slowly than the $GABA_A$ receptor–mediated IPSPs because of the kinetics of the metabotropic response, and they become more prominent with prolonged bursting of reticular nucleus cells.

When relay cells are hyperpolarized and the low-threshold T-type calcium channels (located mainly on the proximal dendrites of relay cells) are deinactivated, corticothalamic stimulation and the excitation of reticular nucleus cells that it produces will induce large IPSPs in the relay cells; because of the deinactivation of the T channels, these IPSPs are followed by burst discharges of the relay cells and low-frequency oscillations.

The Reticular Nucleus

All axons passing bidirectionally between the dorsal thalamus and cerebral cortex and between the dorsal thalamus and basal ganglia pass through the reticular nucleus. En route thalamocortical and corticothalamic fibers give off collateral branches whose terminals represent the principal excitatory drive to the reticular nucleus cells.

In the elementary circuit diagram of the thalamus (Figs. 7.1 and 7.2), subcortical afferent fibers arriving at a dorsal thalamic nucleus excite relay neurons and the collaterals of their thalamocortical axons, in turn, excite reticular nucleus cells. The axons of the reticular nucleus cells then provide an inhibitory feedback to the relay neurons. In the descending direction, corticothalamic fibers projecting to the relay neurons of the dorsal thalamus, on passing through the reticular nucleus, give off collaterals that excite the reticular nucleus cells whose axons in this case provide a feedforward inhibition to the relay cells. The reciprocal circuit produced by the collateral innervation of the reticular nucleus is of profound importance in setting up coherent activity of large ensembles of neurons in the thalamocortical network.

Quantification of the synaptology of reticular nucleus neurons in the rat revealed the following distribution of synapses of various types (Fig. 7.4): on proximal dendrites approximately 50% of the synapses were derived from corticothalamic collaterals; 30%–40% were derived from thalamocortical collaterals and 0%–25% were GABAergic. On second-order dendrites, 60%–65% of the terminals were from corticothalamic collaterals, 20% from thalamocortical collaterals, and 15% were GABAergic. Overall, corticothalamic terminals account for nearly 70% of the synapses that these cells receive. Other, less common synaptic types, derived from the nonspecific afferent systems of the thalamus and made up of serotoninergic, noradrenergic, cholinergic, histaminergic, and a variety of peptidergic types, have not been quantified but qualitative descriptions place them as illustrated in Figure 7.4.

The terminals of the collaterals of thalamocortical fibers in the reticular nucleus are large; filled with synaptic vesicles; and end in large, multisegmented, asymmetrical contacts on somata, and on proximal and intermediate dendrites. The EPSPs can readily be evoked at disynaptic latencies by electrical stimulation of the medial lemniscus, optic tract, or cerebellar nuclei, and monosynaptically by action potentials in individual dorsal thalamic neurons.

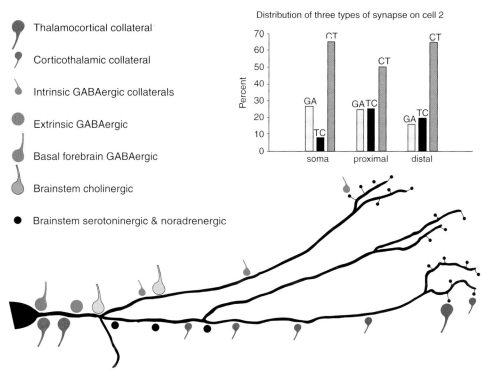

Thalamocortical collateral

Corticothalamic collateral

Intrinsic GABAergic collaterals

Extrinsic GABAergic

Basal forebrain GABAergic

Brainstem cholinergic

Brainstem serotoninergic & noradrenergic

FIGURE 7–4. Schematic map of the distribution of synapses made by terminals of various types and derived from various sources, on the soma-dendritic membrane of a reticular nucleus cell. (*Inset*) Percentages of the three main types of synapse on the dendrites of an intracellular injected reticular nucleus cell of a rat. CT, terminals derived from collaterals of corticothalamic fibers; GA, GABA immunoreactive terminals representing mainly collateral terminals of reticular nucleus axons; TC, terminals derived from collaterals of thalamocortical axons. (Based on Liu and Jones, 1999)

The collateral terminals of corticothalamic fibers arising from layer VI cells of the cerebral cortex are small and similar to those of the same corticothalamic fibers in the dorsal thalamus. The single small postsynaptic density associated with corticothalamic synapses reflects the presence of a single vesicle release site. The larger, although less frequent synapses derived from collaterals of thalamocortical fibers are distinguished by the presence of large, perforated postsynaptic densities, indicative of multiple vesicle release sites. The differences between the two types of collateral synapse in the reticular nucleus are reflected in the variability, amplitudes, and rise times of unitary excitatory postsynaptic currents induced in these cells by stimulation of the two sets of collaterals.

Amplitudes of excitatory post synaptic currents (EPSCs) elicited in reticular nucleus neurons by minimal stimulation of corticothalamic fibers are approximately 2–4 times larger than those induced in relay neurons of the dorsal thalamus and quantal size is 2–6 times greater. These differences are associated with 3–7 times more GluR4 receptor subunits at corticothalamic synapses on reticular nucleus neurons. GluR3 subunits are found in approximately equal numbers at the postsynaptic densities of corticothalamic synapses on the two kinds of cell. At synapses of thalamocortical collaterals in the reticular nucleus, GluR3 subunits outnumber GluR4 subunits (Golshani et al., 2001). Larger conductances prevail at the thalamocortical collateral synapses on account of larger overall numbers of subunits and the presence of multiple release sites.

Inhibitory GABAergic terminals in the reticular nucleus are formed mainly by the intranuclear collaterals of reticular nucleus axons. They end on the reticular nucleus cells in symmetrical contacts on dendrites of varying sizes, and on presynaptic dendrites when these are present. In certain species, other inhibitory contacts between the cells are formed by presynaptic dendrites.

Excitation of reticular nucleus cells by collateral thalamocortical or corticothalamic inputs is followed at disynaptic latency by a hyperpolarization, which reflects the inhibitory connections between the reticular cells. This disynaptic inhibition is mediated primarily by $GABA_A$ receptors. The collateral branches of the axons of other reticular nucleus cells may be the main source of this inhibition. Gap junctions between reticular nucleus cells may serve to synchronize activity of reticular nucleus cells, especially if $GABA_A$ receptor function is reduced.

Like relay neurons, reticular nucleus cells possess the low-threshold calcium current, I_T, and both burst and tonic modes of action potential generation can be induced by changes in membrane potential. Because the threshold for inactivation or deinactivation of the low-threshold calcium channels, at about −60 mV, lies much closer to the resting membrane potential in reticular nucleus cells than in relay cells, reticular nucleus cells are always poised close to the threshold for either tonic or burst firing. The concentration of the channels that underlie I_T in the dendrites of reticular nucleus cells, as opposed to their largely somal location in relay cells, may also make the low-threshold calcium current unusually voltage dependent at the somata of reticular nucleus cells (and confer on the cells the capacity to fire in long bursts of action potentials).

NONSPECIFIC AFFERENTS

The dorsal thalamic nuclei and the reticular nucleus are innervated by nonspecific afferent fiber systems, most of which ascend from the brainstem reticular formation. Cholinergic, noradrenergic, and serotoninergic fibers predominate but glutamatergic, histaminergic, and peptidergic fibers can be found as well. The monoamine fibers end in small terminals that only rarely (5%–10%) possess the distinct membrane specializations typical of synapses, although the terminals make close membrane appositions that probably represent transmitter release sites on proximal dendrites or somata of relay cells and on presynaptic dendrites of interneurons. There are also membrane appositions in relation to GABA immunoreactive terminals, most of which are derived from reticulothalamic axons. These findings suggest the probable release of transmitter on both relay neurons and interneurons and on terminals of reticular nucleus axons.

The cholinergic and noradrenergic inputs can disinhibit relay cells and block long-lasting rhythmic hyperpolarizations during brain activation, mostly by their actions on the reticular nucleus. Acetylcholine has a pronounced excitatory effect produced by a fast, nicotinic receptor–based depolarization that is followed by a slower, longer-lasting, depolarization due to muscarinic receptor activation and a reduction in the potassium leak current, I_{KL}. This switches the relay cells from bursting to tonic, single-spike firing. By contrast, acetylcholine leads to inhibition of interneurons and reticular nucleus neurons via effects on muscarinic receptors and an increase in the potassium current, I_{KG}. The resulting disinhibition of relay cells effectively reinforces the excitation of relay neurons by acetyl choline.

Noradrenaline excites relay cells and reticular nucleus cells by acting on α_1 adrenoreceptors, leading also to a reduction in I_{KL} and tonic, single-spike firing. Noradrenaline has little or no effect on intrinsic interneurons.

Serotonin causes weak hyperpolarization-dependent inhibition of relay cells, with suppression of spontaneous discharges caused by an increase in potassium conductance and mediated by 5-HT$_{1A}$ receptors. Acting via 5HT$_2$, receptors, serotonin inhibits the calcium-activated potassium current responsible for the slow after-hyperpolarization that succeeds a spike discharge. Serotonin leads to prolonged excitation of reticular nucleus cells due to 5-HT$_7$ receptor activation and a reduction in I_{KL}, leading to tonic, single-spike activity. This causes inhibition of relay cells. Serotonin has a depolarizing effect on interneurons that reinforces the inhibition of relay neurons.

Histamine or adenosine, nitric oxide (co-released from cholinergic synapses), and various neuropeptides have relatively modest effects on relay neurons, usually by altering resting membrane potential, and have no effect on interneurons. Histamine acts through H$_1$ receptors to decrease I_{KL} and through H$_2$ receptors to change the voltage dependency of I_H and helps, therefore, to switch the firing of thalamic neurons from bursting to tonic.

NETWORK ACTIVITIES IN THE THALAMOCORTICAL SYSTEM

The thalamocortical pathway and its reciprocal corticothalamic pathway, with the embedded connections between the reticular nucleus and the relay cells, form an integrated network (Fig. 7.5). Coherent activity of the cells in this network is an emergent property that is manifested in rhythmic

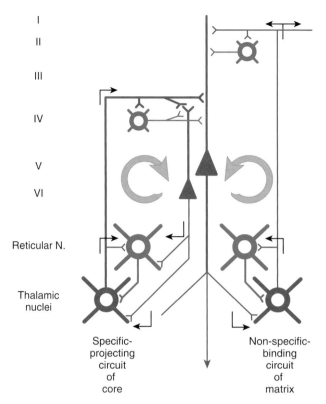

I

II

III

IV

V

VI

Reticular N.

Thalamic
nuclei

Specific-
projecting
circuit
of
core

Non-specific-
binding
circuit
of
matrix

FIGURE 7–5. Coincidence detection circuit in the cerebral cortex, formed by differential laminar terminations of axons arising from middle layer-projecting (core) and superficially projecting (matrix) neurons in the dorsal thalamus. High-frequency oscillations in the core and matrix cells are integrated over the dendritic trees of pyramidal neurons, and oscillatory activity promoted in the cortical cells is further promoted by feedback to the initiating thalamic cells by layer VI corticothalamic neurons. Widespread extent of matrix cell terminations in the cortex and of layer V corticothalamic axons in the thalamus spread synchronous activity across large parts of the cortex and thalamus. (From Jones, 2007, based on Llinás and Paré, 1997, and Jones, 2001)

oscillations of the electroencephalogram. This emergent property is dependent upon the two modes of spike generation in relay cells, which is dependent in the living animal upon the inputs from the nonspecific afferent systems, on the feedback and feedforward inhibition of the reticular nucleus, and on the re-entrant corticothalamic excitatory pathway.

During wakefulness, directed attention and higher cognitive performance, relay neurons, primarily under the influence of the cholinergic inputs, are relatively depolarized and fire tonically in response to peripherally or centrally generated afferent activity. This is manifested in high-frequency oscillations in the network. By contrast, during drowsy inattentiveness and slow-wave sleep, the neurons, in the absence of the nonspecific brainstem influences, trend toward hyperpolarization and to burst firing as the result of deinactivation of the low threshold Ca^{2+} channels. In this state the influence of the reticular nucleus is at its strongest; its bursts lead to repetitive hyper-polarizations of the relay cells that serve to entrain the relay cells in a low-frequency oscillation that is reinforced both by the repetitive excitation of the reticular nucleus cells and by the collateral inputs from the bursting relay cells via their thalamocortical fibers and by the feedback from layer VI neurons of the cerebral cortex. The effect is to set up a low-frequency oscillation throughout the entire thalamocorticothalamic network that is characteristic of the sleeping state.

In the conscious, attentive state, relay neurons are relatively depolarized under the influence of the brainstem systems, their T channels are inactivated, and they now fire tonically. In this state the inhibitory input from the reticular nucleus is at its least effective, serving only to sharpen the contrast between signals relayed by subcortical afferents and noise, an important factor, nevertheless, in making the relay cells faithful transmitters of inputs from sensory receptors to the cerebral cortex. The corticothalamic system in the awake state is more effective in exciting the relay cells, high-threshold Ca^{2+} channels are activated, a ~40 Hz membrane oscillation occurs, and when engaged by corticothalamic synaptic activity, the tonic firing of the relay cells becomes entrained at ~40 Hz, an oscillation that is communicated to the whole thalamo-cortico-thalamic network.

The somewhat divergent thalamic terminations of layer VI corticothalamic axons are important in spreading activity across the thalamus and cortex by recruiting other relay cells into an assembly, including cells that project focally or diffusely upon the cortex. At the cortical level, relay cell oscillation will be conveyed via monosynaptic inputs to middle layers, that is, to pyramidal cells of layers III–VI, and to superficial layers, that is, to the apical dendritic sprays of these cells. This forms a coincidence detection circuit that further reinforces synchrony in the assembly. Eventually, the whole dorsal thalamus and cerebral cortex will be engaged in synchronous high-frequency oscillations that are the hallmark of consciousness.

REFERENCES

Golshani P, Liu X-B, Jones EG (2001) Differences in quantal amplitude reflect GluR4-subunit number at corticothalamic synapses on two populations of thalamic neurons. *Proc Natl Acad Sci USA* 98:4172–4177.

Jones EG (2001) The thalamic matrix and thalamocortical synchrony. *Trends Neurosci* 24:595–601.

Jones EG (2007) *The Thalamus*, 2nd ed., Cambridge, England: Cambridge University Press.

Liu X-B, Jones EG (1999) Predominance of corticothalamic synaptic inputs to thalamic reticular nucleus neurons in rats. *J Comp Neurol* 414:67–79.

Liu X-B, Honda CN, Jones EG (1995a) Distribution of four types of synapse on physi-ologically identified relay neurons in the ventral posterior thalamic nucleus of the cat. *J Comp Neurol* 352:69–91.

Liu X-B, Warren RA, Jones EG (1995b) Synaptic distribution of afferents from reticular nucleus in ventroposterior nucleus of cat thalamus. *J Comp Neurol* 352:187–202.

Llinás R, Paré D (1997) Coherent oscillations in specific and non-specific thalamocortical networks and their role in cognition. In: Steriade M, Jones EG, and McCormick DA, eds. *Thalamus, Vol. 2, Experimental and Functional Aspects*, pp. 501–516. Amsterdam: Elsevier.

McCormick DA (1992) Neurotransmitter actions in the thalamus and cerebral cortex and their role in modulation of thalamocortical activity. *Progr Neurobiol* 39:103–113.

McCormick DA, Bal T (1997) Sleep and arousal: thalamocortical mechanisms. *Annu Rev Neurosci* 20:185–215.

Sherman SM, Guillery RW (2001) *Exploring the Thalamus*. San Diego, CA: Academic Press.

Steriade M, Jones EG, Llinás RR (1990) *The Thalamus as a Neuronal Oscillator*. New York: Wiley.

Steriade M, McCormick DA, Sejnowski TJ (1993) Thalamocortical oscillations in the sleep-ing and aroused brain. *Science* 262:679–685.

Steriade M, Jones EG, McCormick DA (1997) *Thalamus*, Vols. 1–2. Amsterdam: Elsevier.

8

The Lateral Geniculate Nucleus

S. Murray Sherman

The circuitry of the thalamus is among the most thoroughly studied and best understood exemplars of functional connectivity in the brain (for details, see Sherman and Guillery, 2006; Jones, 2007). Here, we shall focus on the A laminae of the cat's lateral geniculate nucleus (LGN), which represents the relay of retinal input to cortex, because this has proven to be an excellent model for thalamus. There are two major payoffs for understanding this circuit: the basic plan revealed by LGN circuitry seems to be applied throughout thalamus, with some modifications, and so this provides general insights into overall thalamic functioning; and circuit principles first appreciated in the LGN may apply to other brain circuits.

BASIC CELL TYPES

As shown in Figure 8.1A and B, the basic circuit in LGN is comprised of three main cell types, with one of these having two distinct subtypes. The *relay cell* receives direct input from the retina and projects to visual cortex. It is a classical excitatory neuron that uses glutamate as its neurotransmitter. In the A laminae of the cat's LGN, there are two relay cell classes, X and Y, and these represent subtle differences in circuitry. These are recipient, respectively, of input from distinct retinal ganglion cell classes also known as X and Y, and thus the relay cells are incorporated into two parallel streams of information from retina to cortex (Sherman, 1982).

The interneuron is a local, GABAergic, inhibitory cell that resides in the A laminae among relay cells. With some exceptions, the relay cell to interneuron ratio throughout thalamus and in all mammalian species is roughly 3 to 1 (Sherman and Guillery, 2006; Jones, 2007). The interneuron is an unusual cell,

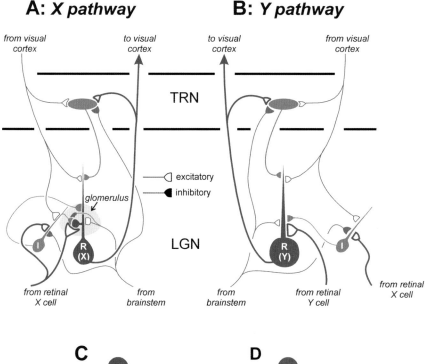

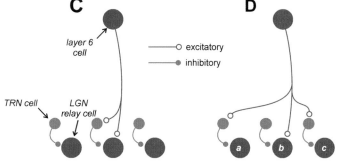

FIGURE 8–1. Overview of circuitry of LGN. (*A* and *B*) Detailed circuitry for X and Y relay cells of the LGN of the cat. (Redrawn from Sherman and Guillery, 2004). (*C* and *D*) Two possible patterns among others for corticogeniculate projection. (*C*) shows excitation and feedforward inhibition. (*D*) shows a more complicated pattern whereby a cortical axon can excite some relay cells directly (e.g., cell *b*) and inhibit others indirectly (e.g., cells *a* and *c*). I, interneuron; LGN, lateral geniculate nucleus; R, LGN relay cell; TRN, thalamic reticular nucleus. (Redrawn from Sherman and Guillery, 2004)

because while it has a conventional axon producing synaptic outputs, most of its synaptic efferents derive from its distal dendrites (Sherman, 2004). Furthermore, these dendritic terminals are both presynaptic to relay cells and postsynaptic to retinal or brainstem inputs (see also the section "Triads and Glomeruli") and are thus the only synaptic terminal type in thalamus with a postsynaptic status. One suggestion for the interneuron's function is that the axonal output is controlled conventionally by proximal inputs that determine the cell's firing, but that the inputs onto the dendritic terminals are so far electronically from the soma that they have little effect on the axonal output (Sherman, 2004). In this sense, the interneuron can multiplex by having separate input/output circuits operating through the axonal and dendritic terminals. As shown in Figure 8.1A, the retinal input to interneurons that determines its receptive field properties and axonal output is from axons of the X type (Sherman and Friedlander, 1988).

Finally, the cell located in the thalamic reticular nucleus (TRN),[1] a shell of neurons adjacent to the thalamus and through which all thalamocortical and corticothalamic axons pass, is another local, GABAergic, inhibitory cell.

Circuitry

General Circuit Features

Figure 8.1A and B also shows the major inputs to the relay cells. In addition to the retinal input, which represents the information relayed to cortex, there are a number of other inputs. These include inhibitory inputs from interneurons and TRN cells, a feedback, glutamatergic input from visual cortex, and assorted inputs from scattered cells in the brainstem. This last group represents mostly cholinergic inputs, but there are also inputs from serotonergic, noradrenergic, and histaminergic cells in the brainstem (for further details, see Sherman and Guillery, 2006; Jones, 2007).

Figure 8.2A shows a more detailed view of how these inputs innervate relay cells. Note that the different input types innervate different parts of the dendritic arbor (reviewed in Sherman and Guillery, 2006; Jones, 2007). Thus, retinal, brainstem, and interneuronal inputs innervate proximal dendrites, while cortical and TRN inputs innervate distal dendrites. Generally, it is thought the more distal the input, the less effective it is due to properties of electrotonic transmission, but this assumes passive cable properties of the relay cell dendrites, and this is one issue for which sufficient relevant information is unavailable. Thus, the significance of the differential distribution of synaptic inputs onto relay cell dendritic arbors remains to be fully determined. One difference between X and Y cells is the relationship of triadic inputs in glomeruli seen in X but not Y cells (Sherman, 2004; Sherman and Guillery, 2006); triads and glomeruli are considered more fully in the section "Triads and Glomeruli." Also note that interneuron axons, whose output is dominated by retinal X input, inhibit both X and Y relay cells, so at the level of LGN, there is some inhibitory mixture of these pathways.

Figure 8.2A also represents each input type in roughly proportional numbers. Each relay cell receives approximately 5000 synaptic inputs (Sherman and Guillery, 2006). Of these, about 5% are retinal in origin, and most of the rest are roughly equally divided among cortical, brainstem, and local

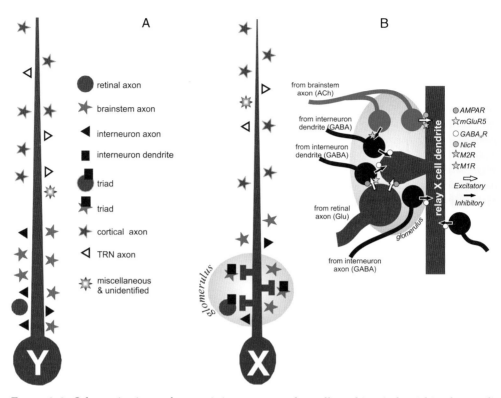

FIGURE 8–2. Schematic views of synaptic inputs onto relay cells and in triads within glomeruli. (*A*) Inputs onto schematic, reduced dendrite of an X and Y cell. Synaptic types are shown in relative numbers and locations. The main difference between X and Y cells is that the former has most retinal input filtered through triads in glomeruli, while the latter has a simpler pattern of retinal input. The triadic inputs and glomeruli typically occur on dendritic appendages of X cells. (Redrawn from Sherman and Guillery, 2004). (*B*) Triads and glomerulus. Shown are the various synaptic contacts (arrows), whether they are inhibitory or excitatory, and the related postsynaptic receptors. The "classical" triad includes the lower interneuron dendritic terminal and involves the retinal terminal. Another type of triad includes the upper interneuron dendritic terminal and also involves the brainstem terminals. For simplicity, the NMDA receptor on the relay cell postsynaptic to the retinal input has been left off. ACh, acetylcholine; AMPAR, (RS)-α-amino-3-hydroxy-5-methyl-4-isoxazolepropionic acid receptor; GABA, γ-aminobutyric acid; GABA$_A$R, type A receptor for GABA; Glu, glutamate; M1R and M2R, two types of muscarinic receptor; mGluR5, type 5 metabotropic glutamate receptor; NicR, nicotinic receptor; TRN, thalamic reticular nucleus. (Redrawn from Sherman, 2004)

GABAergic sources (Sherman and Guillery, 2006). Finally, roughly 5% cannot be identified as one of these major types.

Drivers and Modulators

At first glance the above ratios of different inputs to relay cells seem quite surprising, because the major information to be relayed is retinal, and yet this comprises only 5% of the synaptic input. Although small in number anatomically, retinal input is nonetheless quite powerful in driving relay cells, and so we refer to this as the *driver* input (Sherman and Guillery, 1998, 2006). If the retinal driver input represents the main information to be relayed, what of the other nonretinal inputs? These have been lumped together as *modulators*, because their main role seems to be one of modulating retinogeniculate transmission.

Driver (retinal) and modulator (nonretinal) inputs can be distinguished on a number of criteria (for a complete list and other details, see Sherman and Guillery, 1998, 2006), but the main ones are as follows:

- Driver inputs have large, powerful synapses, while modulator inputs are small and weak.
- Driver synapses have a high probability of release and produce large excitatory postsynaptic potentials (EPSPs) with paired-pulse depression, while modulator synapses generally have a low probability of release and produce small EPSPs (or inhibitory postsynaptic potentials [IPSPs]) with paired pulse facilitation.
- Driver synapses activate only ionotropic receptors (iGluRs; mostly AMPA but also NMDA), while modulator synapses in addition activate metabotropic receptors (i.e., metabotropic glutamate receptors, mGluRs, for cortical input, $GABA_B$ receptors for interneuron and TRN input, muscarinic receptors for brainstem input, etc.; for more information on metabotropic receptors, see Kandel et al., 2000).

Modulation can take many forms, including affecting the gain of retinogeniculate transmission, altering relay cell excitability, and controlling a number of voltage- and time-gated ionic conductances, such as I_T, I_A, and I_h (Jahnsen and Llinás, 1984; McCormick, 2004; Sherman and Guillery, 2006). I_T, a Ca^{2+} current, is particularly interesting, because it determines in which of two firing modes, burst or tonic, relay cells respond to retinal input, and this has important consequences for the relay of information (Sherman, 2001). If a relay cell is depolarized sufficiently (in amplitude and time), I_T is inactivated, and the cell responds in tonic mode; if instead the cell is sufficiently hyperpolarized, inactivation of I_T is removed, and the next effective excitatory input will activate I_T, leading to a burst of action potentials in the relay cell. The activation of metabotropic receptors is particularly important here, because they produce prolonged EPSPs or IPSPs, lasting hundreds of milliseconds to several seconds, and thus these produce membrane potential changes sufficient in amplitude and time to control the inactivation state of I_T and other such conductances. Ionotropic receptor activation typically produces postsynaptic potentials that are too brief to have a major effect on the inactivation state of these conductances.

This division of inputs to relay cells into drivers and modulators seems to be a general principle of thalamus, and identifying the driver input to a thalamic nucleus identifies the information to be relayed. The key point is that inputs to relay cells do not act equally as some sort of anatomical democracy. A study of most circuits laid out in textbooks will reveal that they are based on anatomical numbers almost exclusively. If one were just to consider numbers as the important variable, one might conclude that the LGN relays information mainly from brainstem cholinergic inputs, since these produce ~30% of synapses onto relay cells, while the small number of retinal inputs represents an obscure, unimportant input. An open question is the extent to which this driver versus modulator division of inputs to neurons extends to other areas of the brain, such as cortex (Lee and Sherman, 2008).

Effects of Extrinsic Modulatory Input

The two major extrinsic sources of modulatory input arrive from the brainstem and visual cortex.

Brainstem Input

The brainstem input, as noted earlier, is mostly cholinergic. A glance at Figure 8.1A shows an unusual feature of this input: different branches of the same brainstem axon excite relay cells and inhibit the inhibitory GABAergic cells (Sherman and Guillery, 2006). This remarkable trick is managed due to the different postsynaptic receptors involved. Relay cells respond to the cholinergic input with a depolarizing nicotinic receptor as well as one type of muscarinic receptor (M1), activation of which closes a leak K^+ channel, resulting in further depolarization. In contrast, interneurons and TRN cells respond mainly with another type of muscarinic receptor (M2) that leads to the opening of K^+ channels, resulting in a hyperpolarization. The net result is that increased activity in these brainstem axons leads to a direct depolarization of relay cells and indirect depolarization due to inhibition of GABAergic inputs to these cells. Thus, brainstem activation makes relay cells more responsive and less bursty (because the depolarization inactivates I_T). Indeed, as animals pass from sleep through drowsiness to vigilance, these cholinergic brainstem cells become more active, and LGN cells, in turn, become more active and less bursty (Datta and Siwek, 2002).

   Less is known about the other modulatory neurotransmitter systems, such as serotonergic, noradrenergic, and histaminergic inputs, but their overall effects seem similar to those of the cholinergic inputs (McCormick, 2004; Sherman and Guillery, 2006).

Cortical Input

The cortical input, which emanates from layer 6 cells, is glutamatergic. Its overall effect on relay cells is difficult to predict and depends on the details of circuitry, details that remain mostly obscure. That is, different branches of the same axon innervate relay cells and the local GABAergic cells, exciting all. Thus, from Figure 8.1A, it appears that the effect of this input is to directly excite and indirectly inhibit relay cells, but this may be an oversimplification.

As noted, the actual effects depend on circuit details, and two variants among others are illustrated in Figure 8.1C and D. Figure 8.1C shows the conventional view, which is a feedback inhibitory circuit. Since activation of the corticothalamic axons in this arrangement will provide a somewhat balanced direct depolarizing and indirect hyperpolarizing response in the relay cell, at first glance this might seem to be a fairly useless circuit. However, as Chance et al. (2002) have shown, increasing a fairly balanced inhibitory and excitatory input to a cell reduces its excitability, or in this case, activation of the corticothalamic axon reduces the gain of retinogeniculate transmission, a very effective modulatory function. This is achieved without a major change in the relay cell's membrane potential, partly by increasing synaptic conductance, which reduces neuronal input resistance, and partly by the increase in synaptic noise. Figure 8.1D shows something else altogether. In this circuit, activation of the corticothalamic axon directly excites some relay cells (e.g., cell *b*), thereby promoting tonic firing, while it indirectly inhibits others (e.g., cells *a* and *c*), promoting burst firing. There is some indirect evidence for such a circuit (Tsumoto et al., 1978).

Obviously, we must have a much better understanding of the details of corticothalamic circuitry before we can really understand its function. One key to this understanding is an appreciation that there may be no one function, but rather, many, and that multiple variations in the circuit such as those shown in Figure 8.1C and D, and other possible variants not considered here, may participate in the corticothalamic feedback.

Triads and Glomeruli

General Structure

Triads and glomeruli are ubiquitous features of thalamus, related to interneurons and found in most nuclei and species.[2] This is shown schematically in Figure 8.2B. A triad is a synaptic configuration comprised of three elements. The most common form involves a single retinal terminal that contacts both a dendritic terminal of an interneuron and a relay X cell, with the dendritic terminal contacting the same X cell (Sherman, 2004). The three synapses involved are retinal to dendritic terminal, retinal to relay cell, and dendritic terminal to relay cell. A variant of this involves a cholinergic brainstem axon that functionally replaces the retinal terminal: the brainstem axon contacts the interneuronal dendritic terminal and a relay X cell axon, via different brainstem terminals, with the dendritic terminal contacting the same relay cell.

All of these triadic contacts (plus some other simpler synapses involving axonal inputs onto relay X cells, mostly from interneurons) are contained within a glomerulus, which is thus a site of complex synaptic interaction involving inputs to X cells. Y cells are generally devoid of triadic inputs and glomeruli, so this appears to be a common variant in thalamic circuitry. What makes the glomerulus further distinct is the fact that the entire synaptic structure is contained within a single glial sheath (Szentágothai, 1963; Sherman and Guillery, 2006). Generally, each individual synapse in the brain is surrounded by a glial sheath, the function of which is obscure but is thought to play some role in synaptic regulation and neurotransmitter uptake (Bacci et al., 1999). Whatever that role for individual synapses may be, it appears to be missing in glomeruli because the individual synapses are naked. This has led to a number of hypotheses, one of which is that neurotransmitters released in the glomeruli are not limited to their immediately adjacent targets but may spill over to affect other processes as well. Whatever its functional significance, the glomerulus is a prominent component of LGN circuitry, and it seems likely it plays an important role in modulating retinogeniculate transmission.

Triadic Synaptic Properties: Retinal Inputs

One key to understanding the triad is appreciating the properties of the component synapses. We can start with a consideration of the "classical" triad involving retinal input and ask how it affects retinogeniculate transmission. At first glance, it seems organized in a feedforward inhibitory manner, with a direct monosynaptic EPSP in the relay cell followed by a disynaptic IPSP, perhaps organized to curtail prolonged excitatory input or provide gain control of retinogeniculate transmission much like the circuit of Figure 8.1C.

However, a look at the postsynaptic receptors involved suggests another, more interesting function. Note that the retinal-to-relay cell synapse activates only iGluRs, whereas the relay cell-to-dendritic terminal activates both iGluRs and mGluRs (Cox and Sherman, 2000; Sherman, 2004; Govindaiah and Cox, 2006). Activation of iGluRs typically occurs even at low rates of afferent activity, and so one would expect that at low retinal firing rates a simple feedforward inhibitory circuit would be activated. Activation of mGluRs usually requires higher rates of afferent activity, and so the prediction is that, as the retinal input fires at higher levels, extra inhibition is brought to bear via activation of the mGluRs. Furthermore, this extra inhibition evoked by higher retinal activity would be long-lasting due to the prolonged effects of activation of mGluRs; estimates indicate an effect that would outlast retinal activity by several seconds (Govindaiah and Cox, 2006).

This overall effect, including its time course, seems an ideal neuronal substrate for the function of contrast adaptation (Sclar et al., 1989; Demb, 2002; Solomon et al., 2004). This is an important property of vision, namely, the ability to adjust overall contrast sensitivity to the dynamic range of the visual stimuli, decreasing contrast sensitivity during epochs of high contrast, and vice versa. Evidence exists that retinal, LGN, and cortical circuitry all contribute to this (Sclar et al., 1989; Demb, 2002; Solomon et al., 2004). In general, retinal firing rates increase monotonically with increasing contrast in the stimulus. Thus, as increased contrast raises the firing of retinal inputs past a level sufficient to activate mGluRs on the interneuron dendritic terminals, extra inhibition of the relay cell kicks in, making the cell less sensitive, and this would outlast the increased period of contrast and elevated retinal firing by several seconds, all of which is precisely what occurs with contrast adaption. Note, however, that this property should be limited to the X system, since LGN Y cells lack triadic inputs. This, however, remains a hypothesis for the X system that has yet to be tested.

Triadic Synaptic Properties: Brainstem Cholinergic Inputs

The other sort of triad involving brainstem cholinergic inputs (see Fig. 8.2A) seems easier to understand (Cox and Sherman, 2000; Sherman, 2004). The terminal contacting the relay X cell activates M1 (metabotropic) and nicotinic (ionotropic) receptors, both producing excitation. The terminal contacting the interneuron dendritic terminal, in contrast, activates M2 (metabotropic) receptors, thereby inhibiting the terminal. Thus, in this circuit, just like that described in Figure 8.1A, the cholinergic brainstem input directly excites and indirectly disinhibits the relay X cell.

Concluding Remarks

As noted, LGN circuitry reflects that seen throughout thalamus, with some variations between species and nuclei. Thus, an appreciation of this circuitry helps us to understand the function of thalamus more generally. If we consider the role of the LGN in the visual system from the perspective of information processing, it appears to have a rather unique function. We can understand information processing at one level by determining how each stage in visual processing enhances and elaborates receptive field properties as one ascends the synaptic hierarchies (Van Essen and Maunsell, 1983; Hubel and Wiesel, 1998). Thus, as one passes within retina from receptors through interneurons to ganglion cells, at each stage receptive fields become more elaborate. The same is true as one ascends the hierarchy from LGN to and through the various levels of cortical processing. One clear exception to this pattern is the retinogeniculate synapse, because there seems little receptive field elaboration here. That is, the basic center-surround receptive field of the ganglion cell is seen also in the LGN relay cell, with only minor changes.

This means either that the retinogeniculate synaptic level has little real function (and the LGN was often in the past seen as an uninteresting, machine-like relay), which on the face of it seems absurd, or that this synapse has a unique role in visual processing. That role is not to further elaborate receptive field properties but rather to control the flow of retinal information to cortex. This control is accomplished via modulatory inputs that affect retinogeniculate transmission. One can see this control in a number of different forms, from obvious to fairly subtle. For instance, a glance at Figure 8.1A reveals that, if the local GABAergic (interneuron and TRN) cells are sufficiently active, relay cells will be so inhibited as to fail to relay any retinal information, and in this case, the thalamic gate is shut; conversely, silencing of the local GABAergic cells would open the gate. More subtle examples have been discussed earlier and include more continuously variable gain control of retinogeniculate transmission and control of burst versus tonic response modes. Many other modulatory functions are likely.

Behaviorally, control of information transfer might be related to arousal and attentional mechanisms. Indeed, LGN as well as other thalamic nuclei have been implicated in such behavioral phenomena (LaBerge, 2002; Kastner et al., 2004; McAlonan et al., 2006, 2008). This may well be the main role of thalamus, including LGN. All information reaching cortex must pass through thalamus, and as far as we know, all cortical regions receive a thalamic input. Thus, thalamus appears to play a key role in the flow of information to cortex, and this flow is related to behavioral states such as wakefulness and selective attention. This overview of LGN circuit properties is meant to provide some insights into how this function is achieved. While much is known, clearly this remains a ripe research area so that we can improve our knowledge of these thalamic relay functions.

1. This structure in the cat is actually named the perigeniculate nucleus, but it appears that this is indeed part of the TRN.
2. Exceptions seem to be rats and mice, which have interneurons in their LGN, but few if any are found in other thalamic nuclei (Arcelli et al., 1997). Because triads and glomeruli seem related to interneuronal dendritic terminals, these structures are also rare in these animals outside of the LGN. Other mammals so far studied, including other rodents, generally have interneurons, triads, and glomeruli throughout thalamus.

REFERENCES

Arcelli P, Frassoni C, Regondi MC, De Biasi S, Spreafico R (1997) GABAergic neurons in mammalian thalamus: a marker of thalamic complexity? *Brain Res Bull* 42:27–37.
Bacci A, Verderio C, Pravettoni E, Matteoli M (1999) The role of glial cells in synaptic function. *Philos Trans R Soc Lond B Biol Sci* 354:403–409.

Chance FS, Abbott LF, Reyes A (2002) Gain modulation from background synaptic input. *Neuron* 35:773–782.

Cox CL, Sherman SM (2000) Control of dendritic outputs of inhibitory interneurons in the lateral geniculate nucleus. *Neuron* 27:597–610.

Datta S, Siwek DF (2002) Single cell activity patterns of pedunculopontine tegmentum neurons across the sleep-wake cycle in the freely moving rats. *J Neurosci Res* 70: 611–621.

Demb JB (2002) Multiple mechanisms for contrast adaptation in the retina. *Neuron* 36: 781–783.

Govindaiah G, Cox CL (2006) Metabotropic glutamate receptors differentially regulate GABAergic inhibition in thalamus. *J Neurosci* 26:13443–13453.

Hubel DH, Wiesel TN (1998) Early exploration of the visual cortex. *Neuron* 20:401–412.

Jahnsen H, Llinás R (1984) Electrophysiological properties of guinea-pig thalamic neurones: an *in vitro* study. *J Physiol (Lond)* 349:205–226.

Jones EG (2007) *The Thalamus*, 2nd ed. Cambridge, England: Cambridge University Press.

Kandel ER, Schwartz JH, Jessell TM (2000) *Principles of Neural Science*. New York: McGraw Hill.

Kastner S, O'Connor DH, Fukui MM, Fehd HM, Herwig U, Pinsk MA (2004) Functional imaging of the human lateral geniculate nucleus and pulvinar. *J Neurophysiol* 91: 438–448.

LaBerge D (2002) Attentional control: brief and prolonged. *Psychol Res* 66:220–233.

Lee CC, Sherman SM (2008) Synaptic properties of thalamic and intracortical inputs to layer 4 of the first- and higher-order cortical areas in the auditory and somatosensory systems. *J Neurophysiol* 100:317–326.

McAlonan K, Cavanaugh J, Wurtz RH (2006) Attentional modulation of thalamic reticular neurons. *J Neurosci* 26:4444–4450.

McAlonan K, Cavanaugh J, Wurtz RH (2008) Guarding the gateway to cortex with attention in visual thalamus. *Nature* 456:391–394.

McCormick DA (2004) Membrane properties and neurotransmitter actions. In: Shepherd GM, ed. *The Synaptic Organization of the Brain*, 3rd ed., pp. 39–77. New York: Oxford University Press.

Sclar G, Lennie P, DePriest DD (1989) Contrast adaptation in striate cortex of macaque. *Vision Res* 29:747–755.

Sherman SM (1982) Parallel pathways in the cat's geniculocortical system: W-, X-, and Y-cells. In. Morrison AR and Strick PL (eds.), *Changing Concepts of the Nervous System*, pp. 337–359. Academic Press, Inc.

Sherman SM (2001) Tonic and burst firing: dual modes of thalamocortical relay. *Trends in Neurosci* 24:122–126.

Sherman SM (2004) Interneurons and triadic circuitry of the thalamus. *Trends in Neurosci* 27:670–675.

Sherman SM, Friedlander MJ (1988) Identification of X versus Y properties for interneurons in the A-laminae of the cat's lateral geniculate nucleus. *Exp Brain Res* 73: 384–392.

Sherman SM, Guillery RW (1998) On the actions that one nerve cell can have on another: distinguishing "drivers" from "modulators". *Proc Natl Acad Sci USA* 95:7121–7126.

Sherman SM, Guillery RW (2004) Thalamus. In: Shepherd GM, ed. *The Synaptic Organization of the Brain*, 3rd ed., pp 311–359. New York: Oxford University Press.

Sherman SM, Guillery RW (2006) *Exploring the Thalamus and Its Role in Cortical Function*. Cambridge, MA: MIT Press.

Solomon SG, Peirce JW, Dhruv NT, Lennie P (2004) Profound contrast adaptation early in the visual pathway. *Neuron* 42:155–162.

Szentágothai J (1963) The structure of the synapse in the lateral geniculate nucleus. *Acta Anat* 55:166–185.

Tsumoto T, Creutzfeldt OD, Legendy CR (1978) Functional organization of the cortifugal system from visual cortex to lateral geniculate nucleus in the cat. *Exp Brain Res* 32:345–364.

Van Essen DC, Maunsell JHR (1983) Hierarchical organization and functional streams in the visual cortex. *Trends in Neurosci* 6:370–375.

9

Thalamocortical Networks

David A. McCormick

The thalamus and cerebral cortex are intimately linked together through strong topographical connections, not only from the thalamus to the cortex but also from the cortex back to the thalamus. As in many parts of the brain, the basic circuit of thalamocortical connectivity is relatively simple, although intracortical and corticocortical connectivity provides a high level of complexity. One of the basic operations of the thalamocortical network is the generation of rhythmic oscillations, which are now relatively well understood. In the normal brain, these thalamocortical oscillations typically occur during sleep, although their pathological counterparts may appear as seizures during sleep or waking. Unfortunately, the normal function of reciprocal thalamocortical connectivity during the waking state is still unknown. Even so, focused research is yielding insights into the properties of each of the cellular and synaptic components of these networks and how they interact to perform circuit wide operations.

THALAMIC CIRCUIT

Cellular Properties

Thalamic relay neurons as well as cells of the thalamic reticular nucleus (nRt) possess two distinct firing modes. At depolarized membrane potentials, these cells generate trains of action potentials one at a time, depending upon the length of depolarization. Hyperpolarization of the membrane potential below approximately −65 mV, followed by depolarization, can result in the generation of a low threshold Ca^{2+} spike, which itself can be large enough to activate a high frequency burst of action potentials (Fig. 9.1E). In thalamic

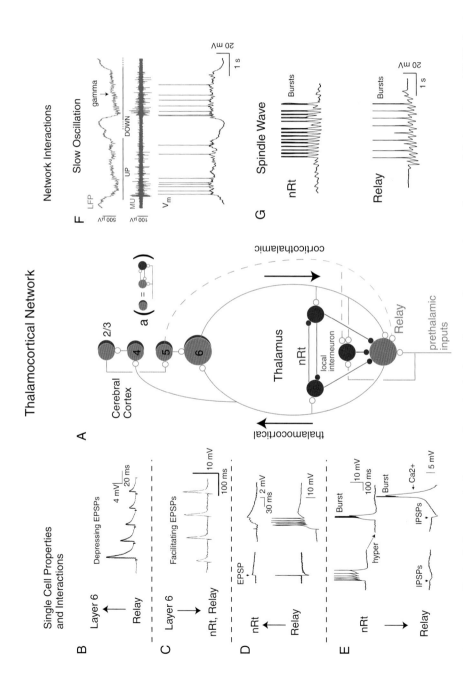

FIGURE 9–1. Basic thalamocortical connections, intrinsic and synaptic properties, and network interactions. (*A*) Thalamic relay cells and local thalamic GABAergic interneurons receive excitatory synaptic inputs from prethalamic inputs (e.g., retinal ganglion cells for the LGNd). Thalamic relay neurons project to layers 4 and 6 of the cerebral cortex. As these excitatory fibers pass through the thalamic reticular nucleus (nRt), they

synapse onto these GABAergic neurons. nRt neurons not only synapse on one another but also send a strong inhibitory feedback to thalamic relay cells. Layer 6 of the cortex projects strongly to the thalamus, sending their excitatory synaptic connections to not only relay neurons in a topographically precise manner but also onto local GABAergic inhibitory neurons as well as the appropriate region of the nRt. Some thalamic nuclei receive an excitatory input from layer 5 of the cortex. Within the cerebral cortex, there is a complex interconnectivity of different layers. Some of the major connections that are typically exemplified include the projection from layer 4 to layers 2/3, the projection from layer 2/3 to layer 5, and the projection from layer 5 to layer 6. All of these connections involve a direct excitatory connection that activates both feedforward and feedback inhibition (Aa; as indicated within the parentheses). (B) Excitatory connections between the thalamus and cerebral cortex typically exhibit depression with repetitive activation, while those between layer 6 and the thalamus are strongly facilitating (C). Synaptic connections from layer 5 to the thalamus are typically of the depressing type (e.g., B). (D) Relay cells monosynpatically excite thalamic reticular neurons, and a burst of spikes in the relay cell can cause a burst of EPSPs and an intrinsic low threshold Ca^{2+} spike mediated burst in the nRt neuron. (E) Tonic firing in the nRt neuron typically results in a tonic hyperpolarization of the relay cell, while burst firing in the nRt neuron (which occurs at a more hyperpolarized membrane potential) results in a large IPSP in relay cells, which can result in the generation of a rebound low-threshold Ca^{2+} spike and burst of action potentials. This cyclical burst-firing mediated interaction of nRt neurons and thalamic relay cells can generate spindle waves during slow-wave sleep (G). (F) Within the cerebral cortex, the activation of balanced recurrent excitatory and inhibitory networks can generate "Up" and "Down" states in which the Up states are characterized by the strong activation of gamma frequency (30–80 Hz) oscillations (upper right traces), which depend upon cyclical interactions between excitatory and inhibitory cells for their generation. Slow oscillations, which typically occur during slow-wave sleep, can also activate spindle waves in the thalamus, resulting in a "K-complex" of the sleep electroencephalogram. The waking, attentive state is associated with tonic firing and gamma-frequency oscillations, in similarity to the Up state of the slow oscillation. EPSPs, excitatory postsynaptic potentials; IPSPs, inhibitory postsynaptic potentials. (Recordings adapted from Bal et al., 1995a; Bal et al., 1995b; Kim et al., 1997; von Krosigk et al., 1999; Haider et al., 2006) relay neurons, a hyperpolarization that is long and large enough to result in a rebound low-threshold Ca^{2+} spike can be generated through the occurrence of a burst of action potentials in the GABAergic neurons of nRt, which subsequently activates $GABA_A$ receptors on the thalamic relay cell, and an inhibitory postsynaptic potential (IPSP) that is 50–100 msec in duration (Fig. 9.1E). The ability of an inhibitory potential to cause the thalamic cell to generate a burst of action potentials results in a rather unusual functional dynamic in which inhibition no longer reduces the probability of a cell from initiating action potentials, but actually increases it, if the membrane potential of the recipient relay neuron is in the appropriate range (e.g., around −65 mV).

Likewise, hyperpolarization of nRt neurons can also result in the cell switching from the single spike to burst firing mode. This hyperpolarization may be achieved through the removal of excitatory influences, such as excitatory postsynaptic potential (EPSP) barrages or neuromodulatory substances (McCormick and Bal, 1997). Once the cell is hyperpolarized and in the burst firing mode, the arrival of EPSP barrages from the thalamic relay neuron can result in the initiation of a low-threshold Ca^{2+} spike mediated burst of action potentials in these GABAergic cells, and subsequently a pronounced inhibition of thalamic relay neurons (Fig. 9.1E). If both the nRt and thalamic relay neurons are at the proper hyperpolarized membrane potentials, such as occurs during slow-wave sleep, then both will tend to generate bursts of action potentials and interact to rhythmically generate spindle waves (Fig. 9.1G).

Synaptic Properties

The thalamus contains excitatory relay cells, which project to the cortex and to the nRt (there are very few intrathalamic collaterals of thalamic relay neurons to other relay cells or local interneurons). Relay neurons and local interneurons receive excitatory, glutamatergic inputs from prethalamic inputs (e.g., retina, inferior colliculus, cerebellum, spinal cord, etc.). Activation of these excitatory inputs results in large EPSPs that exhibit synaptic depression with repetitive stimulation (similar to Fig. 9.1B). These prethalamic pathways can have a dominant, driving influence on thalamic activity owing to the innervation of the thalamic relay cell by many synapses from a small number of prethalamic axons, resulting in synchronized release and thus a large EPSP through precise temporal summation.

Layer 6 of the cerebral cortex sends a numerically impressive glutamatergic input back to the thalamus, where it synapses on all known cell types, including nRt neurons, local GABAergic interneurons, and thalamic relay neurons. Each thalamic cell is innervated by many corticothalamic fibers, with each giving rise to only a very small EPSP. Corticothalamic EPSPs from layer 6, however, exhibit strong facilitation with repetitive stimulation (Fig. 9.1C). Interestingly, repetitive activation of the layer 6 corticothalamic pathway can result in a slow depolarization of thalamic relay neurons through the closure of a resting K^+ conductance resulting from activation of glutamate metabotropic receptors (McCormick and von Krosigk, 1992).

Some thalamic nuclei also receive an input from cortical layer 5 pyramidal cells (Jones, 2007). This input acts more similar to the "prethalamic" inputs than a layer 6 excitatory pathway. Layer 5 cortical inputs to the thalamus activate large EPSPs that display synaptic depression with repetitive activation. It has been speculated that layer 5 may provide a "driving" influence to some thalamic nuclei, similar to that of the retina for the LGNd or inferior colliculus for the medial geniculate nucleus (Sherman and Guillery, 2006). In this way, some regions of the cortex may be acting as the driving or main informational input to selected thalamic nuclei, which then send this information on (following intrathalamic processing) to other cortical areas.

Thalamic relay cells strongly excite, through non-NMDA and NMDA receptors, nRt neurons, particularly if the thalamic relay neuron generates a burst of action potentials (Fig. 9.1D). The reaction of the nRt neuron to barrages of EPSPs arriving from thalamic relay cells (or deep-layer cortical neurons) depends upon the membrane potential of the nRt cell. At depolarized levels, the nRt neuron is in the single-spike firing mode and depolarizing waves of synaptic activity will modulate the tonic firing rate. However, once the nRt neuron hyperpolarizes to resting membrane potential or below, the cell may respond to excitatory synaptic barrages with a low-threshold Ca^{2+} spike and high-frequency burst of action potentials (Fig. 9.1E). These high-frequency bursts of action potentials in nRt neurons are essential to the generation of low-frequency rhythms such as spindle waves (Fig. 9.1G). Depolarizing GABAergic inhibitory potentials are not yet known to actually result in the generation of low-threshold Ca^{2+} spikes in nRt neurons under normal circumstances.

Thalamic reticular neurons densely innervate relay cells, where they strongly activate Cl^- conductances through $GABA_A$ receptors and, with prolonged and vigorous activity, they activate increases in membrane K^+ conductance through $GABA_B$ receptors (Kim et al., 1997; McCormick and Bal, 1997). Interestingly, nRt neurons also inhibit one another through the activation of $GABA_A$ receptors, thus spreading inhibition laterally within the sheet of the nRt (Bal et al., 1995a; McCormick and Bal, 1997; Huguenard and McCormick, 2007). These inhibitory interactions are important because they control thalamocortical interactions and rhythomogenesis in waking, sleep, and epilepsy (see 'Network Interactions' section).

Network Interactions

Much is known about how thalamic neurons interact to generate a prominent rhythmic activity of slow-wave sleep: sleep spindles. Much less is known about how thalamic neurons interact to generate the patterns of activity of the waking, attentive state. Although sleep spindles are generated within the thalamus, they are strongly modulated and synchronized by the interaction of the thalamus and the cerebral cortex (see 'Thalamocortical Interactions' section). Within the thalamus, sleep spindles are generated as hypothesized by Andersen and Andersson more than four decades ago (Andersen and Andersson, 1968): as a cyclical interaction between excitatory thalamic relay cells and inhibitory thalamic GABAergic neurons (which we now know reside within the nRt) (Fig. 9.1G). Burst firing in relay neurons initiates barrages of EPSPs and low-threshold Ca^{2+} spike-mediated bursts of action potentials in nRt neurons. The subsequent generation of bursts of action potentials in these GABAergic neurons results in the generation of large $GABA_A$ receptor–mediated IPSPs in thalamic relay neurons, some of which generate rebound low-threshold Ca^{2+} spike-mediated bursts, which subsequently initiates the next cycle of the oscillation. The time to complete one cycle of this activity (nRt-relay-nRt) is about 100–150 msec, and thus the oscillation occurs at about 6–10 Hz (Bal et al., 1995a,1995b; McCormick and Bal, 1997). Spindle waves got their name from their resemblance to a spindle of thread, owing to their propensity to wax and wane over a 1–2 second period. The growth, or "waxing," of the spindle wave results from the recruitment of more and more cells into the oscillation. The waning of spindle waves results from the gradual loss of the ability of thalamic neurons to generate burst firing owing to depolarization from the activation of the h-current, a mixed cation current that is sensitive to the intracellular levels of cAMP (which is sensitive to the levels of intracellular Ca^{2+} owing to Ca^{2+}-sensitive adenylate cyclases) (Luthi and McCormick, 1998). Thus, spindle waves appear and reappear with marked regularity during slow-wave sleep, as the cyclical interaction between the nRt and relay neurons waxes and wanes over a slow oscillatory periodicity of once every few seconds.

Importantly, during the generation of the intrathalamic oscillations, nRt neurons inhibit one another through $GABA_A$ receptors, dampening the activity of these neurons to a slight degree. The loss of this intra-nRt inhibition can result in large, synchronized oscillations at approximately 3 Hz, in similarity to the spike-and-wave discharges associated with absence epileptic seizures (Bal et al., 1995a, 1995b; McCormick and Bal, 1997).

CORTICAL CIRCUIT

Cellular Properties

There are two broad classes of cells in the cerebral cortex: excitatory and inhibitory, releasing glutamate and GABA, respectively. Within these two large groups, there are many subgroups, the number of which depends upon the classification scheme. Clearly every cell in the brain is unique in some aspect, meaning that one could create as many categories as there are neurons

(Ascoli et al., 2008). Useful categorization schemes will likely result in up to 100 different subtypes of GABAergic interneurons in the cerebral cortex, and perhaps dozens up to 100 different subtypes of excitatory cells. Electrophysiologically, the vast majority of cortical neurons generate some variation of single (also known as tonic) spiking. The depolarization of cortical cells with the intracellular injection of current results in trains of action potentials, the duration of which does not outlast the depolarization. This is true of both excitatory pyramidal cells and many subtypes of inhibitory GABAergic interneurons (Nowak et al., 2003). Notable exceptions to this generality are burst-generating neurons. Two subsets of pyramidal cells known as chattering neurons and intrinsic bursting neurons can generate, through intrinsic ionic mechanisms, bursts of action potentials when depolarized. Within these two broad electrophysiological classes (single spiking and bursting), there are many different subtle electrophysiological differences between different types of cortical neurons. The functional properties of these differences are largely unknown and are too numerous to cover here.

Synaptic Properties

The cerebral cortex consists of a large sheet of highly interconnected networks with the vast majority of long-range connections being formed by excitatory pyramidal cells, and local connections within and between layers being comprised of robust, but precise, recurrent excitatory and inhibitory networks. The physiological properties of these synaptic connections are varied and synapse dependent. Some connections, for example, exhibit depression with repetitive activation, while others exhibit facilitation (e.g., Figs. 9.1B and C) (Thomson and Deuchars, 1997). As with the projection from layer 6 to the thalamus, local connections from layer 6 to layer 4 exhibit strong synaptic facilitation, the function of which is not yet clear. One important operating principle of the cortex that has recently become clear is that there is a striking and important balance (proportionality) between the massive recurrent excitatory and inhibitory networks that characterize local cortical circuits (Haider et al., 2006). This balance between recurrent excitation and inhibition arises in large part owing to the recurring theme of feedback and feedforward inhibition, which is characteristic of nearly all intracortical connections (Fig. 9.1Aa).

Network Interactions

The highly interconnected nature of the cerebral cortex allows it to generate spontaneous rhythmic activities, both normal (e.g., slow oscillation) and pathological (e.g., epileptic seizures) (McCormick and Contreras, 2001; Steriade, 2006; Huguenard and McCormick, 2007). Perhaps the most well-studied rhythmic activity of the cortex is that of the slow oscillation between Up and

Down states, which normally occurs during slow-wave sleep (Fig. 9.1F). The slow oscillation is generated through the activation of local recurrent excitatory pathways, which cause a re-entrant excitation of all types of cortical neurons (Hasenstaub et al., 2005; Haider et al., 2006). Since inhibitory GABAergic neurons are highly interconnected within the local network, these cells are also strongly activated, and the degree to which they discharge depends upon the degree to which the local network is active. Thus, recurrent excitation and inhibition are both dependent upon the level of local network activity, and this gives rise to a balancing or proportionality of excitatory and inhibitory synaptic activity in pyramidal neurons. Interestingly, the Up state, which is similar in many regards to the waking cortex (Steriade et al., 2001), exhibits a strong activation of gamma frequency (30–80 Hz) oscillations (Fig. 9.1F). Intracellular recordings in cortical pyramidal cells during the Up state indicate that the main determinant of these oscillations in the activation of rhythmic IPSPs at 30–80 Hz, supporting the hypothesis that the gamma oscillation is generated through a cyclical interaction between excitatory and inhibitory neurons in cortical structures (Traub et al., 2004; Hasenstaub et al., 2005).

Thalamocortical Interactions

As we have seen so far, both the cerebral cortex and thalamus can generate autonomous activities when isolated. In vivo, these two structures are intimately interconnected and function as part of a greater whole. Some relatively simple interactions are now well understood. One of these is the generation of "K-complexes" during slow-wave sleep. K-complexes are associated with the initiation of a relatively large sharp event in the electroencephalogram (EEG) typically followed by the generation of a spindle wave. It now appears that these events are generated through the activation of an Up state within the cerebral cortex. This bombards the thalamus with a sudden barrage of EPSPs from layer 6 (and 5), resulting in the activation of nRt neurons. The large inhibition in thalamocortical relay cells will result in a rebound low-threshold Ca^{2+} spike in a subset of these cells, which then activates the next cycle of the spindle wave. Thus, the initiation of the Up state, which results in a large slow wave in the EEG, is followed by a spindle wave, as generated by the cyclical interaction of nRt and thalamic relay cells (Amzica and Steriade, 1998). The highly divergent and convergent nature of intracortical connectivity serves to synchronize and rapidly spread thalamocortical rhythms (Contreras et al., 1996; Contreras and Steriade, 1996).

Pathological interactions between the thalamus and cerebral cortex have also been examined in vivo and in vitro, in the form of epileptic seizures (Steriade, 2006; Huguenard and McCormick, 2007). The cortex and thalamus are interconnected in a reciprocal excitatory loop that is controlled by intracortical and intrathalamic GABAergic neurons (Fig. 9.1). If the cortex is abnormally activated (such as during or preceding a seizure resulting from a defect in the precise balance of recurrent excitation and inhibition), the large descending excitatory influence can have a strong activating effect on thalamic GABAergic cells (both nRt and local interneurons). The activation of these neurons can result in large IPSPs, mediated by both $GABA_A$ and $GABA_B$ receptors, in relay neurons, and subsequently the activation of large rebound bursts of action potentials (Blumenfeld and McCormick, 2000; McCormick and Contreras, 2001). At least in theory, this strong thalamic activation could then reactivate both the nRt as well as the cerebral cortex, resulting in another round of intense discharge in both of these structures. The cyclical pathological interaction of the thalamus and cortex in this manner may form the basis of at least some forms of absence epileptic seizures (Huguenard and McCormick, 2007).

Neuromodulatory Influences and the Control of Thalamocortical State

The state of thalamocortical networks changes markedly with the state of the animal and determines the degree to which information gains access to, and is processed by, the cerebral cortex. The largest state change that occurs in the normal brain is that from slow-wave sleep to waking. During this state change, thalamic relay and nRt neurons depolarize and shift their firing mode away from burst generation to tonic firing. This appears to be achieved in large part through the action of ascending modulatory neurotransmitters from the brainstem, including acetylcholine, norepinephrine, and serotonin, as well as the release of histamine from specialized neurons in the hypothalamus, and glutamate from descending corticothalamic projections (McCormick and Bal, 1997). Although the postsynaptic actions of these neuromodulators are complex and vary according to cell type, some prominent actions include depolarization of thalamic relay and nRt cells through the closure of "leak" K^+ currents; inhibition of local GABAergic interneurons through the activation of K^+ currents; and increases in neuronal excitability in the cerebral cortex by reduction of various K^+ currents that dampen neuronal discharge (McCormick and Bal, 1997). Through these complex interactions, the ascending systems from the brainstem and hypothalamus, as well as the activity of the thalamocortical circuit itself, determines the state of the network, waking the sleeping brain or allowing it to fall once again into restful reprise.

Concluding Remarks

Nearly all information to the cerebral cortex must pass through the thalamus. Far from being a mere relay station, the thalamus and cortex are intimately

and reciprocally interconnected. While the function of this high interconnectivity during the waking state is largely unknown, it is known that the thalamocortical networks are responsible for generating the rhythmic network
activities of sleep and epilepsy. These synchronized oscillations depend upon
the interactions of thalamic relay cells, reticular neurons, local GABAergic
neurons, and the complex interactions of the cortical sheet. Unraveling the
functional properties of thalamocortical actions in behavior will be a complex
and difficult task, owing to the high degree of interconnectivity, and the
precision of this connectivity, of the cerebral cortex and its descending projections to the thalamus.

REFERENCES

Amzica F, Steriade M (1998) Cellular substrates and laminar profile of sleep K-complex.
    Neuroscience 82:671–686.
Andersen P, Andersson SA (1968) *Physiological Basis of the Alpha Rhythm*. New York:
    Appleton-Century-Crofts.
Ascoli GA, Alonso-Nanclares L, Anderson SA, Barrionuevo G, Benavides-Piccione R,
    Burkhalter A, Buzsaki G, Cauli B, Defelipe J, Fairen A, et al. (2008) Petilla terminol-
    ogy: nomenclature of features of GABAergic interneurons of the cerebral cortex. *Nat
    Rev Neurosci* 9:557–568.
Bal T, von Krosigk M, McCormick DA (1995a) Role of the ferret perigeniculate nucleus in
    the generation of synchronized oscillations in vitro. *J Physiol* 483(Pt 3):665–685.
Bal T, von Krosigk M, McCormick DA (1995b) Synaptic and membrane mechanisms
    underlying synchronized oscillations in the ferret lateral geniculate nucleus in vitro.
    J Physiol 483(Pt 3):641–663.
Blumenfeld H, McCormick DA (2000) Corticothalamic inputs control the pattern of activ-
    ity generated in thalamocortical networks. *J Neurosci* 20:5153–5162.
Contreras D, Steriade M (1996) Spindle oscillation in cats: the role of corticothalamic feed-
    back in a thalamically generated rhythm. *J Physiol* 490(Pt 1):159–179.
Contreras D, Destexhe A, Sejnowski TJ, Steriade M (1996) Control of spatiotemporal
    coherence of a thalamic oscillation by corticothalamic feedback. *Science* 274:771–774.
Haider B, Duque A, Hasenstaub AR, McCormick DA (2006) Neocortical network activity
    in vivo is generated through a dynamic balance of excitation and inhibition. *J Neurosci*
    26:4535–4545.
Hasenstaub A, Shu Y, Haider B, Kraushaar U, Duque A, McCormick DA (2005) Inhibitory
    postsynaptic potentials carry synchronized frequency information in active cortical
    networks. *Neuron* 47:423–435.
Huguenard JR, McCormick DA (2007) Thalamic synchrony and dynamic regulation of
    global forebrain oscillations. *Trends Neurosci* 30:350–356.
Jones EG (2007) *The Thalamus*, 2nd ed. Cambridge, England: Cambridge University Press.
Kim U, Sanchez-Vives MV, McCormick DA (1997) Functional dynamics of GABAergic
    inhibition in the thalamus. *Science* 278:130–134.
Luthi A, McCormick DA (1998) Periodicity of thalamic synchronized oscillations: the role
    of Ca2+-mediated upregulation of Ih. *Neuron* 20:553–563.
McCormick DA, von Krosigk M (1992) Corticothalamic activation modulates thalamic
    firing through glutamate "metabotropic" receptors. *Proc Natl Acad Sci USA* 89:2774–
    2778.

McCormick DA, Bal, T (1997) Sleep and arousal: thalamocortical mechanisms. *Annu Rev Neurosci* 20:185–215.

McCormick DA, Contreras D (2001) On the cellular and network bases of epileptic seizures. *Annu Rev Physiol* 63:815–846.

Nowak LG, Azouz R, Sanchez-Vives MV, Gray CM, McCormick DA (2003) Electrophysiological classes of cat primary visual cortical neurons in vivo as revealed by quantitative analyses. *J Neurophysiol* 89:1541–1566.

Sherman SM, Guillery RW (2006) *Exploring the Thalamus and Its Role in Cortical Function.* Cambridge, MA: MIT Press.

Steriade M (2006) Grouping of brain rhythms in corticothalamic systems. *Neuroscience* 137:1087–1106.

Steriade M, Timofeev I, Grenier F (2001) Natural waking and sleep states: a view from inside neocortical neurons. *J Neurophysiol* 85:1969–1985.

Thomson AM, Deuchars J (1997) Synaptic interactions in neocortical local circuits: dual intracellular recordings in vitro. *Cereb Cortex* 7:510–522.

Traub RD, Bibbig A, LeBeau FE, Buhl EH, Whittington MA (2004) Cellular mechanisms of neuronal population oscillations in the hippocampus in vitro. *Annu Rev Neurosci* 27:247–278.

von Krosigk M, Monckton JE, Reiner PB, McCormick DA (1999) Dynamic properties of corticothalamic excitatory postsynaptic potentials and thalamic reticular inhibitory postsynaptic potentials in thalamocortical neurons of the guinea-pig dorsal lateral geniculate nucleus. *Neuroscience* 91:7–20.

Section 3

Circadian System

10

The Suprachiasmatic Nucleus

Gabriella B. S. Lundkvist and Gene D. Block

Diurnal variations in physiology and behavior are ubiquitous in higher organisms. Although some rhythms are driven directly by environmental cycles of light or temperature, most are generated by internal timers, commonly referred to as "biological clocks," that serve to create a rich internal temporal environment synchronized by geophysical cycles, stabilized against ambient temperature fluctuations, and shaped by seasonal as well as daily factors (e.g., day length). In mammals, including humans, these circadian (near 24-hr) properties are controlled by a central timer formed by a distinct regional network in the anterior hypothalamus close to the optic chiasm, the bilateral suprachiasmatic nucleus (SCN). Rodents with their SCN lesioned fail to exhibit diurnal variation in behavior. Impressively, rhythmicity can be restored by suprachiasmatic transplants (Ralph et al., 1990). The mechanism generating rhythmicity is contained within individual neurons; however, many of the properties of the circadian timing system derive from cellular interactions within the SCN. These microcircuits give rise to a functional clock capable of (a) receiving synchronizing signals about environmental cycles; (b) maintaining a coherent, stable period and phase; and (c) driving or synchronizing circadian rhythms in other brain regions.

THE SUPRACHIASMATIC NUCLEUS IS A HETEROGENEOUS TISSUE

The SCN is a heterogeneous tissue consisting of several cell types with regional specialization. Each nucleus contains 10,000–12,000 small neurons, tightly coupled to each other through synaptic connections and gap junctions. GABA is a dominant neurotransmitter within the entire nucleus and appears to play an interregional role in linking circuits of dorsal and ventral neurons (Albus et al., 2005). The ventral part of the SCN, which is the first region to receive synchronizing signals, receives monosynaptic retinal input mediated by glutamate and pituitary adenylate cyclase-activating polypeptide (PACAP). Ventral neurons contain the neuropeptides vasoactive intestinal peptide (VIP) and gastrin-releasing peptide (GRP), which have been implied in synchronization of neuronal circadian activity within the SCN (Aton et al., 2005; Maywood et al., 2006). The dorsal SCN receives projections from the ventral SCN and typically expresses vasopressin (VP) but also somatostatin (Fig. 10.1).

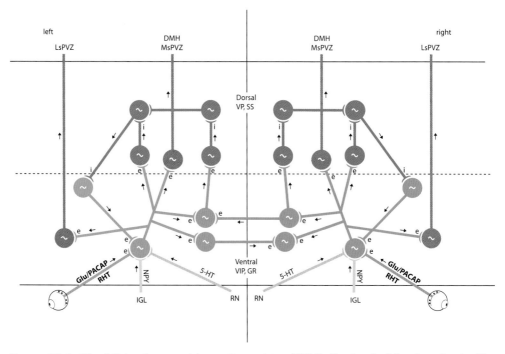

FIGURE 10–1. The bilateral suprachiasmatic nucleus (SCN) "brain clock" microcircuit. The suprachiasmatic neurons are self-sustained 24-hr molecular oscillators (~) that become synchronized to each other in the synaptic network. The ventral regions contain vasointestinal peptide (VIP) and gastrin-releasing peptide (GR) and receive monosynaptic light input from melanopsin-containing ganglion cells in the retina via glutamate (Glu) and PACAP signaling. The ventral regions also receive neuropeptide Y (NPY) input from the intergeniculate leaflet (IGL) and serotonergic (5-HT) input from the raphe nucleus (RN). The main intrinsic SCN transmitter is gamma amino-butyric acid (GABA), which is excitatory in the ventral region. The ventral SCN sends excitatory afferents to the dorsal SCN, which contains vasopressin (VP) and somatostatin (SS). GABA is inhibitory in the dorsal region and sends a small portion of inhibitory afferents to the ventral region. Thus, the ventral SCN phase shifts quickly in response to light, whereas the dorsal phase shifts more slowly. The dorsal regions send output efferents mainly to the dorsomedial hypothalamus (DMH) and the medial subparaventricular zone (MsPVZ). The ventral regions send efferents primarily to the lateral subparaventricular zone (LsPVZ). The two SCN nuclei send afferents to each other.

RHYTHM GENERATION

Rhythm generation is primarily the property of individual SCN neurons. Thus, unlike motor circuits (e.g., lamprey swimming) the basic rhythm is cell autonomous. This was demonstrated by recording rhythms in electrical activity from cultured SCN neurons plated at a low density, a condition under which individual neurons drift out of phase with one another, expressing their own intrinsic periodicities (Welsh et al., 1995). These cell-autonomous rhythms are generated by autoregulatory negative feedback loops, in which protein products of several "clock genes," including *Period* (*Per 1, 2,* and *3*) and *Chryptochrome* (*Cry 1* and *2*), inhibit their own transcription. The proteins ultimately become degraded, initiating a new cycle of clock gene transcription (Ko and Takahashi, 2006).

ROLES OF THE SUPRACHIASMATIC NUCLEUS MICROCIRCUITRY

Although rhythms are generated in single neurons, cell circuitry appears to be required to perform at least four chief functions.

Cell Circuitry Keeps the Rhythm Sustained as It Provides Excitatory Drive Required to Keep Neurons Oscillating

Per1 rhythmicity can be tracked in individual neurons in SCN brain slices from transgenic mice in which luciferase serves as a reporter for *Per1* activity. When excitability in SCN slices is blocked with the sodium channel blocker tetrodotoxin (TTX), *Per1* rhythmicity damps in individual neurons (Yamaguchi et al., 2003). The result indicates a close linkage between molecular rhythmicity and electrical signaling, suggesting that electrical activity may be required for both interoscillator synchronization and for maintaining cell-autonomous oscillations. Mice lacking VIP or the $VPAC_2$ receptor for VIP show poor behavioral, electrical, and molecular rhythmicity, perhaps because the SCN neurons are slightly hyperpolarized. Consistent with this view, if the SCN neurons from these mice are excited by depolarization (increased $[K^+]$), cellular rhythms can be restored (Maywood et al., 2006). The mechanism underlying degraded rhythmicity in VIP-deficient mice is not known, but it may involve a reduced calcium flux. In organotypically cultured SCN slices, the *Per* rhythm disappears in hyperpolarizing culture medium (decreased $[K^+]$) and in medium with low concentrations of Ca^{2+}, demonstrating that Ca^{2+} is essential for rhythmic expression of *Per1* and PER2 (Lundkvist et al., 2005). The results further suggest that membrane potential and a transmembrane Ca^{2+} flux in the SCN circuit may be critical for rhythm generation.

Cell Circuitry Synchronizes and "Averages the Period" of Multiple
Single-Cell Oscillators

Suprachiasmatic nucleus neurons plated at low cell density express a wide
range of circadian periods in electrical and molecular activity (Welsh et al.,
1995). In the tissue, in live animals, and even in high-density dispersals, the
periods are significantly more coherent. The synchronization process that
leads to coherence among SCN oscillatory neurons is not entirely clear; however, synchronizing candidates have been suggested. It has been demonstrated that electrical activity is crucial to maintain synchrony among clock
neurons, as TTX desynchronizes *Per1:luciferase* rhythms in individual SCN
neurons (Yamaguchi et al., 2003). Furthermore, in a recent study certain clock
genes were deleted that resulted in abolished circadian rhythmicity in single
dispersed neurons, but not in coupled neurons in tissue (Liu et al., 2007),
demonstrating the important role of synaptic and electrical contact between
SCN neurons in order to maintain circadian rhythmicity. Although other
synchronizing factors have been suggested (GABA, gap junctions), most
attention has been given to the synchronizing agent VIP, which is released
rhythmically in SCN slices. VIP is required for synchronization of electrical
and molecular rhythms in SCN neurons and in circadian behavior (Aton
et al., 2005; Maywood et al., 2006). Moreover, GRP enhances and synchronizes the *Per1* molecular rhythm (Maywood et al., 2006). Thus, peptide release
in the SCN appears to be necessary for neuronal synchrony.

Cell Circuitry Provides Pathways for Synchronization of Circadian
Oscillators by Light

Light is the primary environmental synchronizing signal for the SCN.
In several mammalian species, photic information is conveyed principally
to the ventral region of the SCN by glutamatergic input via the retino-
hypothalamic tract (RHT). PACAP is colocalized with glutamate in RHT
terminals and is most likely also involved in light synchronization. The
RHT consists of a group of axons within the optic nerve that originate from
a restricted group of retinal ganglion cells and project directly to the ventral
SCN. Photic input also reaches the ventral SCN indirectly via neuropeptide Y
(NPY) in the geniculo-hypothalamic tract (GHT) and the intergeniculate
leaflet (IGL) of the lateral geniculate nucleus (LGN). Pathways containing
5-hydroxytryptamine (5-HT) from the median raphe nucleus provide the
other major nonphotic signaling pathway to the SCN, which similar to NPY
phase shifts the clock in the absence of light and also modulates photic
signaling. For photic entrainment of the SCN, a subpopulation of melanopsin-
containing retinal ganglion neurons projecting to the SCN plays an important role in the circadian phototransduction cascade (Morin and Allen,
2006).

Cell Circuitry Encodes the Effects of Photoperiod through Phase Differences among Suprachiasmatic Nucleus Neurons

The SCN response to photoperiod is a circuit property. Recent work has demonstrated that the SCN exhibits a high level of plasticity in order to adjust to altered length of the day. This plasticity appears to occur on the network level, as the phase relationship between the neuronal rhythms changes in short days versus long days. Moreover, the SCN contains different cell populations that are differentially responsive at subjective dusk and dawn, respectively. One population peaks in activity closer to dusk, whereas the other peaks closer to dawn. During long days, these populations are therefore together active at a longer time span than during short days, thus providing a microcircuit "code" for reporting day length (Vansteensel et al., 2007). Interestingly, a high level of synchronization appears to enhance the clock's capacity to phase shift (vanderLeest et al., 2009).

Regional Specializations in Cell Types and Connectivity Lead to Complex Suprachiasmatic Nucleus Behavior

In addition to the cell–cell circuit interactions, the SCN tissue itself exhibits high temporal complexity. As a response to light, the ventral part shifts quickly, whereas the dorsal part, which receives information from the ventral SCN, shifts more slowly (Shigeyoshi et al., 1997; Albus et al., 2005; Nakamura et al., 2005). This asymmetry becomes more pronounced when the light/dark schedule is advanced instead of delayed. The complexity is increased further by the fact that the different clock genes within the ventral and dorsal regions shift at different rates in response to light; *Per1* and *Per2* shift rapidly, but *Cry 1* shifts slowly. Finally, *Per1* and *Per2* show different behavior after phase advances as compared to phase delays.

The functional significance of this complex asymmetry is not clear and the findings are therefore difficult to summarize in context of function. However, taken together, the studies that have been performed on this topic thus far suggest that the ventral, light-receiving portion of the SCN shifts immediately via glutamatergic signals (at least in terms of electrical output) and transmits excitatory GABAergic signals to the dorsal region, leading to a subsequent phase shift of the dorsal region. There also appears to be some feedback between the dorsal and ventral regions evidenced by the fact that the ventral region "overshoots" the final shift during phase advances and appears to be retarded back to the appropriate shift by the dorsal region. The coupling between the dorsal and ventral regions (Fig. 10.1) appears asymmetric in that the ventral region exerts a larger phase-shifting effect on the dorsal region, which could be explained by the fact that GABA is inhibitory in the dorsal region (Albus et al., 2005).

REFERENCES

Albus H, Vansteensel MJ, Michel S, Block GD, Meijer JH (2005) A GABAergic mechanism is necessary for coupling dissociable ventral and dorsal regional oscillators within the circadian clock. *Curr Biol* 15(10):886–893.

Aton SJ, Colwell CS, Harmar AJ, Waschek J, Herzog ED (2005) Vasoactive intestinal polypeptide mediates circadian rhythmicity and synchrony in mammalian clock neurons. *Nat Neurosci* 8(4):476–483.

Liu AC, Welsh D, Ko C, Tran H, Zhang E, Priest A, Buhr E, Singer O, Meeker K, Verma I (2007) Intercellular coupling confers robustness against mutations in the SCN circadian clock network. *Cell* 129(3):605–616.

Lundkvist GB, Kwak Y, Davis EK, Tei H, Block GD (2005) A calcium flux is required for circadian rhythm generation in mammalian pacemaker neurons. *J Neurosci* 25(33):7682–7686.

Ko CH, Takahashi JS (2006) Molecular components of the mammalian circadian clock. *Hum Mol Genet* 15(Spec No 2):R271–277.

Maywood ES, Akhilesh BR, Wong GKY, O'Neill JS, O'Brien JA, McMahon DG, Harmar AJ, Okamura H, Hastings MH (2006) Synchronization and maintenance of timekeeping in suprachiasmatic circadian clock cells by neuropeptidergic signaling. *Curr Biol* 16(6):599–605.

Morin LP, Allen CN (2006) The circadian visual system. *Brain Res Rev* 51(1): 1–60.

Nakamura W, Yamazaki S, Takasu NN, Mishima K, Block GD (2005) Differential response of Period 1 expression within the suprachiasmatic nucleus. *J Neurosci* 25(23): 5481–5487.

Ralph MR, Foster RG, Davis FC, Menaker M (1990) Transplanted suprachiasmatic nucleus determines circadian period. *Science* 247(4945):975–978.

Shigeyoshi Y, Taguchi K, Yamamoto S, Takekida S, Yan L, Tei H, Moriya T, Shibata S, Loros JJ, Dunlap JC, Okamura H (1997) Light-induced resetting of a mammalian circadian clock is associated with rapid induction of the mPer1 transcript. *Cell* 91(7): 1043–1053.

vanderLeest HT, Rohling JH, Michel S, Meijer JH (2009) Phase shifting capacity of the circadian pacemaker determined by the SCN neuronal network organization. *PLoS ONE* 4(3):e4976.

vanderLeest HT, Houben T, Michel S, Deboer T, Albus H, Vansteensel MJ, Block GD, Meijer JH (2007) Seasonal encoding by the circadian pacemaker of the SCN. *Curr Biol* 17(5):468–73.

Welsh DK, Logothetis DE, Meister M, Reppert SM (1995) Individual neurons dissociated from rat suprachiasmatic nucleus express independently phased circadian firing rhythms. *Neuron* 14(4):697–706.

Yamaguchi S, Isejima H, Matsuo T, Okura R, Yagita K, Kobayashi M, Okamura H (2003) Synchronization of cellular clocks in the suprachiasmatic nucleus. *Science* 302(5649):1408–1412.

Section 4

Basal Ganglia

11

Microcircuits of the Striatum

J. Paul Bolam

The striatum (or caudate-putamen, or caudate nucleus and putamen in those species in which they are divided by the internal capsule) is the major division of the basal ganglia, a group of structures involved in a variety of processes, including movement and cognitive and mnemonic functions. The striatum consists of a population of principal neurons, the medium-sized densely spiny neurons (MSNs; up to 97% of all neurons depending on species), which are the projection neurons of the striatum, several populations of GABAergic interneurons, and a population of cholinergic interneurons. The principal afferents of the striatum are glutamatergic, are derived from the cortex and thalamus, and mainly innervate the spines of MSNs. The essential computation performed by the striatum is the selection of which MSNs will fire, the consequence of which is altered firing of basal ganglia output neurons and hence the selection of the basal ganglia–associated behavior.

The essential aspects of the microcircuitry of the striatum are summarized as follows (also see Figs. 11.1– 11.3):

1. Under resting conditions the large population of MSNs is in a quiescent relatively hyperpolarized state, and selected MSNs or groups of MSNs are activated by afferent excitatory input/drive originating in the cortex and thalamus (Fig. 11.1).
2. The response of MSNs to excitatory drive from the cortex and thalamus is modified by the local collaterals of the MSNs (feedback inhibition) and the activity of GABAergic interneurons (feedforward inhibition; Fig. 11.2).
3. Short- and long-term plasticity of the excitatory transmission is under the control of modulatory inputs to the striatum, principally the dopaminergic nigrostriatal projection, but also possibly serotonergic and histaminergic afferents, and by the activity of cholinergic interneurons (Fig. 11.3).

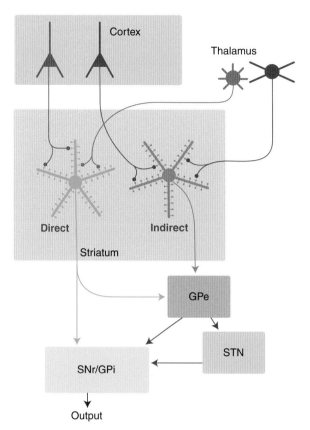

Figure 11–1. Microcircuit of the striatum. The majority of neurons in the striatum are the GABAergic output neurons, the medium-size densely spiny neurons (MSNs). MSNs are of two main types; on the left is an MSN of the direct pathway. These neurons *directly* innervate the output nuclei of the basal ganglia, the internal segment of the globus pallidus (GPi), and the substantia nigra pars reticulata (SNr), and they also send a collateral to the external segment of the globus pallidus (GPe). On the right is an MSN of the indirect pathway. These neurons *indirectly* innervate the output nuclei of the basal ganglia by innervating the GPe and then the output nuclei, and by innervation of the subthalamic nucleus (STN). Under resting conditions the MSNs are quiescent. They are activated by their main afferents, the excitatory, glutamatergic input originating in the cortex (CTX) and thalamus (THAL). Sufficient excitatory input to their 10–15,000 spines raises them to a depolarized "up-state" from which action potentials can be driven that then influence the activity of the output neurons of the basal ganglia via the direct and indirect pathways. The expression of basal ganglia function is thus a consequence of which MSNs are "selected," ultimately leading to the selection of the behavior. It is unknown as yet, whether individual MSNs receive input from both the cortex and the thalamus or whether individual cortical or thalamic axons innervate both direct and indirect pathways neurons. Red indicates glutamatergic structures; blue indicates GABAergic structures.

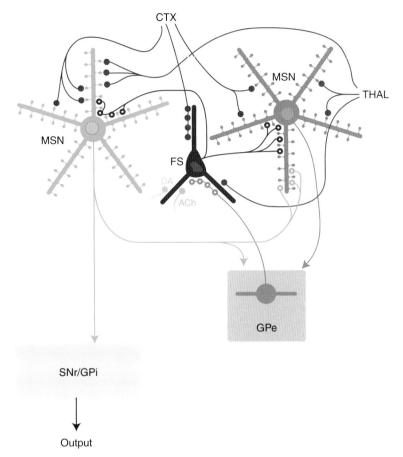

FIGURE 11–2. GABAergic inputs sculpt the response of medium-size densely spiny neurons (MSNs) to cortical and thalamic afferents. MSNs receive both feedback and feedforward GABAergic inhibition. Feedback inhibition is subserved by the axons of MSNs, which, in addition to giving rise to the output of the striatum, also give rise to local axon collaterals. The main synaptic target of the local collaterals is the dendrites (mainly distal) of other MSNs (only a connection between a direct pathway MSN and an indirect pathway MSN is illustrated here). In comparison to the interneurons, the response to MSN collateral activation is relatively weak when measured at the level of the soma. Feedback inhibition may thus involve dendritic processing rather than controlling the output of MSNs. Feedforward inhibition is mediated by GABAergic interneurons of which there are at least three types. Only one is illustrated here, the parvalbumin-positive fast-spiking GABAergic interneuron (FS). These neurons receive both cortical and thalamic input and in turn provide a powerful feedforward inhibition in the proximal regions of MSNs, which can prevent or delay the initiation of action potentials. The FS interneurons, as well as those that express NOS, also receive a prominent input to their cell bodies and proximal dendrites from the tonically active GABAergic neurons of the external globus pallidus (GPe). Thus, by virtue of their large axonal arbor, GABAergic interneurons are in a position to select which populations of MSNs will fire; they in turn are under the control of neurons of the GPe. Since the major input to neurons of the GPe are MSNs, the network involving the "indirect pathway" is critically involved in the selection of individual of groups of MSNs. GABAergic interneurons also receive input from dopaminergic terminals (DA, yellow) and cholinergic terminals (ACh, green). It is unknown as yet whether individual GABAergic interneurons receive input from both the cortex and the thalamus or whether the same cortical or thalamic axons innervate both MSNs and their connected GABAergic interneurons. Red indicates glutamatergic structures; blue indicates GABAergic structures.

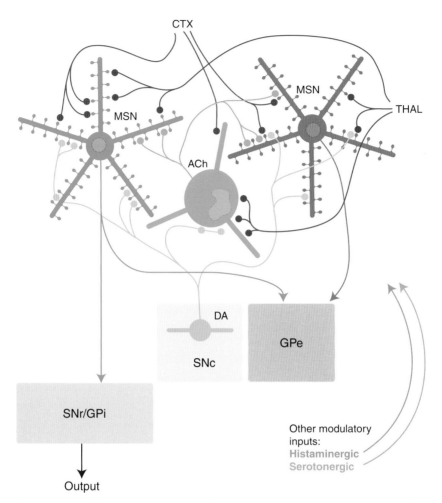

FIGURE 11–3. Modulatory control of the activity of medium-size densely spiny neurons (MSNs). The dopaminergic nigrostriatal projection (DA, yellow) provides a dense innervation of the striatum. Synaptically released dopamine, or dopamine spill-over from the synapse, modulates the responses of MSNs to the excitatory input at both pre- and postsynaptic sites. The net effect of dopamine is the long- or short-term enhancement or attenuation of gluta-matergic transmission. Cholinergic neurons (ACh, green) also underlie a modulatory control of MSNs. Released acetylcholine acting upon both muscarinic and nicotine receptors modu-lates the release of glutamate presynaptically and the responsiveness of MSNs to other affer-ents by a variety of mechanisms. Cholinergic interneurons, like other classes of interneurons in the striatum, are a site of interaction of neuromodulators within the striatum. The striatum also receives "modulatory input" from serotonergic-containing afferents originating in the dorsal raphé and histaminergic afferents from the tuberomammillary nucleus. Little is known of their connections and functional properties. It is unknown as yet whether individual cholinergic interneurons receive input from both the cortex and the thalamus or whether the same corti-cal or thalamic axons innervate both MSNs and their connected cholinergic interneurons. Red indicates glutamatergic structures; blue indicates GABAergic structures.

MEDIUM-SIZED DENSELY SPINY NEURONS AND THEIR EXCITATORY AFFERENTS

Medium-sized densely spiny neurons give rise to several dendrites that are initially spine free but then become densely laden with spines after the first bifurcation. They utilize GABA as their major neurotransmitter and are subdivided into two major subpopulations on the basis of their projection region, pattern of axonal collateralization, and their neurochemical content. One subpopulation gives rise to the "direct pathway," that is, they preferentially project *directly* to the output nuclei of the basal ganglia, the internal segment of the globus pallidus (GPi or entopeduncular nucleus in some species), and the substantia nigra pars reticulata (SNr) (but also sending a collateral to the external segment of the globus pallidus; GPe). They selectively express (among other molecular markers) the neuropeptides substance P and dynorphin, and the D1 subtype of dopamine receptor. The second subpopulation gives rise to the "indirect pathway"; they exclusively project to the GPe, which in turn projects to the output nuclei directly and via the subthalamic nucleus, that is, they thus *indirectly* innervate the output nuclei. They selectively express (among other molecular markers) enkephalin and the D2 subtype of dopamine receptors.

Under resting conditions, most neurons in the striatum (with the exception of cholinergic interneurons) are quiescent. The principal "driver" inputs to the striatum, and the basal ganglia in general, are the excitatory glutamatergic afferents from the cortex and the thalamus (mainly the intralaminar thalamic nuclei). The massive excitatory input from the cortex is derived from virtually the whole cortical mantle and provides motor, cognitive, and limbic information. Two main classes of pyramidal neurons give rise to the projection: pyramidal tract neurons (lower layer V) that give off collaterals within the ipsilateral striatum and pyramidal neurons that bilaterally innervate the cortex and striatum only (layer III and upper layer V). The main synaptic targets of the corticostriatal projection are the spines of both direct and indirect pathway MSNs. The two types of neurons seem to show selectivity for the different classes of MSNs (Lei et al., 2004). Corticostriatal terminals also make synaptic contact with the dendritic shafts of interneurons. Individual corticostriatal axons are likely to give rise to only a small number of synapses with an individual MSN, which, since MSNs possess 10,000–15,000 spines, implies a massive convergence of cortical input (Wilson, 2007).

The excitatory input from the thalamus is of the same order of magnitude as the cortical input when considering the number of synapses within the striatum (Lacey et al., 2005; Raju et al., 2008). The thalamostriatal projection mainly originates in the intralaminar nuclei, providing information about the external world and the state of arousal and wakefulness, however, many other thalamic nuclei also innervate the striatum. The main synaptic targets are the spines of both direct and indirect pathway MSNs, but they also innervate the dendritic shafts of MSNs and interneurons. Thalamostriatal neurons are of *at least* two markedly distinct types with respect to their somatodendritic morphology, their firing characteristics, and their synaptic targets within the striatum (Lacey et al., 2007). Those originating in the central lateral nucleus have a bushy dendritic arbor typical of thalamic neurons, give rise to low-threshold Ca^{2+} spike bursts, and innervate exclusively the spines of MSNs (presumably giving rise to both the direct and indirect pathways). In contrast, those in the parafascicular nucleus have long, infrequently branching dendrites, have a lower firing frequency, are less bursty, and mainly innervate dendritic shafts of striatal interneurons although the spines of MSNs are also targets. The precise distribution of postsynaptic targets is highly variable among individual parafascicular neurons. As with the cortical input, it is likely that there is a massive convergence of thalamic afferents onto individual MSNs. It is also likely that there is convergence of both cortical and thalamic inputs at the level of individual MSNs.

Both of these excitatory afferents are highly topographically organized. Corresponding functional territories of cortex and thalamus innervate similar regions of the striatum; thus, they impart the functionality upon the striatum and the basal ganglia in general. However, it should be noted that there are many sites and mechanisms in the striatum and other regions of the basal ganglia that underlie the integration of functionally diverse information.

Principles of Operation of the Microcircuits of the Striatum

Alterations in the firing of the output neurons of the basal ganglia (GABAergic, tonically active neurons in the SNr and GPi) underlie the expression of basal ganglia function. The principal players in the control of their activity are MSNs, either through their direct innervation of the output neurons or their indirect innervation of the output neurons through the "indirect pathway," or rather, the complex networks underlying the "indirect pathway." The role of the microcircuits of the striatum is the selection of individual, or groups, of MSNs to fire action potentials and thereby influence the activity of the output neurons of the basal ganglia and hence the selection, learning, and reinforcing of the appropriate behavior. Under resting conditions, MSNs are held at a relatively hyperpolarized and quiescent state as a consequence of the profile of ion channels they express. The excitatory drive to the MSNs carried by the glutamatergic corticostriatal and thalamostriatal pathways leads to depolarization of MSNs, bringing them to a so-called up-state. It is in this state that additional excitatory inputs, an alteration in the strengths of the synapses, or an alteration in the balance of excitatory and inhibitory inputs, leads to the firing of action potentials. Which individual MSNs or ensembles of MSNs fire at a given moment of time will depend on which afferent fibers are active, the pattern of innervation of individual MSNs by individual afferent axons, and the degree of convergence at the level of individual MSNs, the feedback and feedforward inhibition (via their local collaterals and via interneurons) and modulation of the excitatory transmission. According to the classical "direct/indirect pathway model," the firing of the selected population of MSNs will lead to the inhibition of a selected group of basal ganglia output neurons, which will then lead to the selection of the basal ganglia behavior. Simultaneous activation of MSNs of the indirect pathway is considered to temporally and/or spatially inhibit/attenuate or focus the basal ganglia behavior. However, the role of the "indirect" pathway is likely to be more complex and involve the selection process at every level of the basal ganglia.

SCULPTING OF THE RESPONSE OF MEDIUM-SIZED DENSELY SPINY NEURONS BY GABA TRANSMISSION

Feedback Inhibition

In addition to providing the output of the striatum, MSNs give rise to local axon collaterals, the main synaptic targets of which are the dendritic shafts of other MSNs. Direct pathway MSNs innervate other direct pathway neurons as do indirect pathway MSNs. Similarly, MSNs of the two pathways are synaptically interconnected. Synapses are located principally on the dendritic shafts in the more distal regions, that is, the spiny regions of the dendrites. The responses of spiking in a single presynaptic spiny neuron on postsynaptic MSNs are weak when measured at the level of the soma (particularly compared to the response to GABA interneurons; see Feedforward Inhibition) and do not significantly affect spike timing or generation (Tepper et al., 2008). This presumably relates to the location of the synapses on distal dendrites. Feedback inhibition may thus be involved in dendritic processing; however, sufficient convergent and simultaneous activation of MSN collaterals may influence the firing of the target MSNs (Wilson, 2007).

Feedforward Inhibition

The striatum also contains at least three populations of GABAergic interneurons, which account for ~3%–10% of striatal neurons depending on species: (1) Fast-spiking, parvalbumin-positive interneurons (FS-PV); (2) those that express nitric oxide synthase (NOS), neuropeptide Y, and somatostatin; and (3) those that express calretinin. The connections of GABAergic interneurons are summarized in Figure 11.2; most of the data are derived from the FS-PV neurons and to some extent the NOS-positive neurons. The GABAergic interneurons mainly innervate the dendritic shafts of MSNs and, at least in the case of FS-PV GABAergic interneurons, mainly in their proximal regions, including the perikaryon. They receive prominent input from terminals forming asymmetrical synapses derived from the cortex and thalamus. Unlike the feedback inhibition, feedforward inhibition is powerful and widespread. Spiking in a single interneuron leads to profound inhibitory postsynaptic potentials (IPSPs) in MSNs that are capable of blocking the generation of spikes in a large number of postsynaptic MSNs (Tepper et al., 2008). Although it is not known whether GABAergic interneurons receive the same cortical and thalamic input as do their target MSNs, it is clear that fast-spiking interneurons do provide a feedforward inhibition *in vivo* (Mallet et al., 2005). One of the roles of FS-PV interneurons may be to synchronize the activity of large populations of MSNs because their activation can lead to a brief delay in the firing of MSNs and their extensive axonal arborization (about 5000 synapses) means they are in contact with many (possibly hundreds) of MSNs (Koos and Tepper, 1999).

The FS-PV GABAergic interneurons, as well as the NOS-expressing GABAergic interneurons, receive a prominent innervation of their cell bodies and proximal dendrites from tonically active GABAergic neurons of the GPe (Bevan et al., 1998), which will presumably shunt excitatory cortical and thalamic drive. Firing of GABAergic interneurons will presumably only occur with the release or altered pattern of this inhibitory control. This will occur when MSNs of the indirect pathway are activated. Thus, GABAergic interneurons are involved in the spatiotemporal selection of which individual and ensembles of MSNs fire; in turn, they are under the control of the activity of neurons of the GPe and ultimately the activity of MSNs that give rise to the indirect pathway.

MODULATORY CONTROL OF THE ACTIVITY OF MEDIUM-SIZED DENSELY SPINY NEURONS

Modulatory Afferents

The activity of MSNs and excitatory afferents are under the *modulatory* control (short- and long-term plasticity) of several afferents to the striatum, including the histaminergic input from the tuberomammillary nucleus (providing information on wakefulness), serotonergic input originating in the dorsal raphé (providing information on basic physiological functions such as sleep, arousal, and satiety, as well as mood and emotion), and dopaminergic input from the substantia nigra pars compacta (SNc; providing information on motivation, reward, and stimulus salience). The density of the histaminergic and serotonergic innervation is relatively low compared to the dopaminergic innervation, and both seem to give rise to only few synapses; however, there little is known about their connections.

The dopaminergic innervation of the dorsal striatum originates in the SNc and a group of dopamine neurons located more ventrally in the SNr. Those dopaminergic neurons of the ventral tegmental area provide the main dopaminergic innervation of the ventral striatum and the prefrontal cortex. Dopamine neurons are heterogeneous with respect to their activity during reward-related activity (Bromberg-Martin and Hikosaka, 2009; Brown et al., 2009), but clearly their activities relate to reward, reward prediction error, and salience of an event and, as such, play a role in reinforcement-based learning.

Individual dopamine neurons provide a phenomenal innervation of the dorsal striatum. In the rat, the average total length of the axon of an individual dopamine neuron in the striatum is in the region of 47 cm and the arborization can occupy up to 5.7% of the volume of the striatum (Matsuda et al., 2009). This implies a large degree of convergence but similarly a large degree of divergence. Based on the numbers of neurons in the SNc and the striatum, and the known synaptic organization of the dopaminergic nigrostriatal projection, it is estimated that an individual dopaminergic neuron gives rise to between 170,00 and 408,000 synapses in the dorsal striatum (Moss and Bolam, 2010). The synapses are small and symmetrical (Gray's type 2) and are formed with dendritic spines, shafts, and perikarya of MSNs and interneurons. About 20% of those spines of MSNs that receive input from the *cortex* are apposed by a dopamine axon and about half make synaptic contact. Similarly, about 20% of those spines of MSNs that receive input from the *thalamus* are apposed by a dopamine axon and about half make synaptic contact. The association of the dopaminergic axons and synapses with spines and dendrites of MSNs, however, is unlikely to be a selective or targeted phenomenon as, when corrected for size, all structures in the striatum have an equal probability of being apposed by or in synaptic contact with dopaminergic axons (Moss and Bolam, 2008). Furthermore, the dopaminergic innervation is so dense that every point within the striatum is within 1 μm of a dopaminergic synapse. Since it has been proposed that synaptically released dopamine can diffuse up to 2–8 μm in a sufficient concentration to stimulate both high- and low-affinity dopamine receptors (Rice and Cragg, 2008), then every structure in the striatum is under the influence of dopamine.

Dopamine released at synapses located on structures postsynaptic to the excitatory cortical or thalamic synapses (and presumably dopamine that has diffused from synapses) modulates the response of MSNs to the excitatory input. Modulation of excitatory transmission can also occur at the level of the presynaptic terminal. There are many forms of dopamine-dependent plasticity or modulation of excitatory input to MSNs, including long-term potentiation, long-term depression, and changes in excitability and interactions at the level of receptors, signaling pathways, and gene regulation. Many factors influence the dopamine-dependent plasticity, including spike timing, the subclass of dopamine receptors involved, as well as the activation of NMDA receptors, the release of endocannabinoids, and the action of cholinergic interneurons (Wickens, 2009). The net effect is the long- or short-term enhancement or attenuation of glutamatergic transmission and the selection of which neurons reach or are prevented from reaching firing threshold, thus playing a role in reinforcement learning.

Cholinergic Interneurons

In addition to the GABAergic interneurons, the striatum also contains a population of large cholinergic interneurons (giant aspiny neurons) that give rise to a massive axonal arbor (the striatum contains the highest density of cholinergic markers in the brain). They innervate MSNs (in a similar manner to the dopaminergic innervation) and receive major inputs from the parafascicular nucleus of the thalamus (and presumably other thalamic nuclei), the cortex (seemingly mostly in their distal regions), and MSNs of both the direct and indirect pathway. They are considered to be the so-called tonically active neurons (TANs), whose activity is intimately related to the activity of dopamine neurons; a pause in their firing during reward-related paradigms is associated with the burst in activity in dopamine neurons and is considered to code the salience of an event. However, as with dopamine neurons the functional properties of cholinergic neurons are heterogeneous (Apicella et al., 2009). Released acetylcholine acting upon both muscarinic and nicotine receptors modulates the release of glutamate presynaptically and the responsiveness of MSNs to other afferents by a variety of mechanisms (Pisani et al., 2007). The effects of changes in the firing of cholinergic neurons, as with the dopamine input, thus plays a role in the selection of which individual MSNs or groups of MSNs are raised to firing threshold. It should be noted also that cholinergic interneurons, like other classes of interneurons in the striatum, are a site of interaction of neuromodulators within the striatum (Koos and Tepper, 2002).

ACKNOWLEDGMENTS

The author's own work described in this paper was supported by the Medical Research Council, the European Community, and The Parkinson's Disease Society (United Kingdom). The author would like to thank Ben Micklem for the preparation of the figures.

REFERENCES

Apicella P, Deffains M, Ravel S, Legallet E (2009) Tonically active neurons in the striatum differentiate between delivery and omission of expected reward in a probabilistic task context. *Eur J Neurosci* 30:512–526.
Bevan MD, Booth PAC, Eaton SA, Bolam JP (1998) Selective innervation of neostriatal interneurons by a subclass of neuron in the globus pallidus of the rat. *J Neurosci* 18: 9438–9452.

Bromberg-Martin ES, Hikosaka O (2009) Midbrain dopamine neurons signal preference for advance information about upcoming rewards. *Neuron* 63:119–126.

Brown MT, Henny P, Bolam JP, Magill PJ (2009) Activity of neurochemically heterogeneous dopaminergic neurons in the substantia nigra during spontaneous and driven changes in brain state. *J Neurosci* 29:2915–2925.

Koos T, Tepper JM (1999) Inhibitory control of neostriatal projection neurons by GABAergic interneurons. *Nature Neurosci* 2:467–472.

Koos T, Tepper JM (2002) Dual cholinergic control of fast-spiking interneurons in the neostriatum. *J Neurosci* 22:529–535.

Lacey CJ, Boyes J, Gerlach O, Chen L, Magill PJ, Bolam JP (2005) GABA(B) receptors at glutamatergic synapses in the rat striatum. *Neuroscience* 136:1083–1095.

Lacey CJ, Bolam JP, Magill PJ (2007) Novel and distinct operational principles of intralaminar thalamic neurons and their striatal projections. *J Neurosci* 27:4374–4384.

Lei W, Jiao Y, Del Mar N, Reiner A (2004) Evidence for differential cortical input to direct pathway versus indirect pathway striatal projection neurons in rats. *J Neurosci* 24:8289–8299.

Mallet N, Le Moine C, Charpier S, Gonon F (2005) Feedforward inhibition of projection neurons by fast-spiking GABA interneurons in the rat striatum in vivo. *J Neurosci* 25:3857–3869.

Matsuda W, Furuta T, Nakamura KC, Hioki H, Fujiyama F, Arai R, Kaneko T (2009) Single nigrostriatal dopaminergic neurons form widely spread and highly dense axonal arborizations in the neostriatum. *J Neurosci* 29:444–453.

Moss J, Bolam JP (2008) A dopaminergic axon lattice in the striatum and its relationship with cortical and thalamic terminals. *J Neurosci* 28:11221–11230.

Moss J, Bolam JP (2010) The relationship between dopaminergic axons and glutamatergic synapses in the striatum: structural considerations. In: *Dopamine Handbook*. Edited by Iversen LL, Iversen SD, Dunnett SB and Björklund A pp. 49–59. Oxford University Press.

Pisani A, Bernardi G, Ding J, Surmeier DJ (2007) Re-emergence of striatal cholinergic interneurons in movement disorders. *Trends Neurosci* 30:545–553.

Raju DV, Ahern TH, Shah DJ, Wright TM, Standaert DG, Hall RA, Smith Y (2008) Differential synaptic plasticity of the corticostriatal and thalamostriatal systems in an MPTP-treated monkey model of parkinsonism. *Eur J Neurosci* 27:1647–1658.

Rice ME, Cragg SJ (2008) Dopamine spillover after quantal release: rethinking dopamine transmission in the nigrostriatal pathway. *Brain Res Rev* 58:303–313.

Tepper JM, Wilson CJ, Koos T (2008) Feedforward and feedback inhibition in neostriatal GABAergic spiny neurons. *Brain Res Rev* 58:272–281.

Wickens JR (2009) Synaptic plasticity in the basal ganglia. *Behav Brain Res* 199:119–128.

Wilson CJ (2007) GABAergic inhibition in the neostriatum. *Prog Brain Res* 160:91–110.

12

Templates for Neural Dynamics in the Striatum: Striosomes and Matrisomes

Ann M. Graybiel

The striatum appears to be a relatively simple forebrain region, when compared to the overlying neocortex with its horizontal layers and vertical columns. In fact, however, the striatum in mammals has a sophisticated architecture of its own. This large subcortical region is now suspected of having a major influence on how the neocortex carries out its own functions—even functions related to human language (Graybiel, 2008; Enard et al., 2009). Furthermore, abnormalities in the striatum are increasingly being discovered in human disorders affecting cognitive as well as motor functions. It is, as a consequence, increasingly difficult to see the neocortex as a "higher structure" and the striatum as a "lower structure" in ranking their influence on behavior. In this chapter, I suggest that the functional architecture of the striatum provides a physical template for dynamic plasticity in striatal networks. I then propose that this dynamic plasticity may be key to understanding how the basal ganglia influence the neocortex as well as downstream action systems of the brain: first, by promoting adaptive behavioral flexibility, and second, by allowing the forebrain to create, as a result, chunked cognitive and motor action patterns.

THE STRIATUM HAS A MODULAR NEUROCHEMICAL ORGANIZATION: STRIOSOMES

The first clue to the modularity of the striatum came from the early application of methods for studying the distributions of neurotransmitter-related substances in the brain. For example, the striatum stands out sharply among forebrain regions as having very strong expression of markers for cholinergic and dopaminergic transmission, and these turn out not to be uniform. Nor are the distributions of other neurotransmitters, neuromodulators, receptors, and even second messenger molecules uniform. Nearly every one of these has a striking distribution, in which widely dispersed, relatively small zones stand out as being enriched or impoverished in the substance being analyzed. We called these zones "striosomes" (striatal bodies) to distinguish them from the much larger extrastriosomal striatal tissue, which we called the "matrix" (Graybiel and Ragsdale, 1978; Graybiel, 1990). Striosomal organization of striatal transmitter systems is present from quite early in embryonic development. In some instances, there are dramatic shifts in striosome/matrix expression patterns across development (e.g., for substances related to dopaminergic and cholinergic transmission), while for other substances the compartmental distributions are relatively stable.

The Striatum Has a Modular Input-Output Organization: Striosomes and Matrisomes

The modular organization of the striatum extends beyond a simple striosome/matrix organization, because the inputs to the matrix also are largely organized into modules. For example, the somatosensory cortex projects to the striatum in such a way that regions representing particular parts of the body map project to widely dispersed patchy regions in the matrix. Moreover, somatotopically matched parts of the somatosensory and motor cortex send overlapping patchy inputs to the striatum. This suggests that information about a given functional domain is collected into dispersed, but organized, convergence zones. This organization holds also for striatal inputs from the oculomotor cortex. It further has been shown that the dispersed matrisomes receiving inputs from corresponding somatosensory and motor cortex themselves can send convergent outputs to the basal ganglia output nuclei (Fig. 12.1A). Thus, there is a divergent-reconvergent architecture for the cortex-to-striatum-to-output nuclei of the basal ganglia (Flaherty and Graybiel, 1994). The entire map of striatal input-output organization has not been established. However, there is enough evidence to suggest highly principled connectivity between most areas of the neocortex and the striatum (Ragsdale and Graybiel, 1990).

Striosomes and Matrisomes as Templates for Plasticity

The divergent-reconvergent architecture of matrisomes seems ideal for allowing new combinations of inputs and outputs to be flexibly coordinated. For example, somatosensory input matrisomes receive inputs related to a given part of the body from corresponding parts of several somatosensory

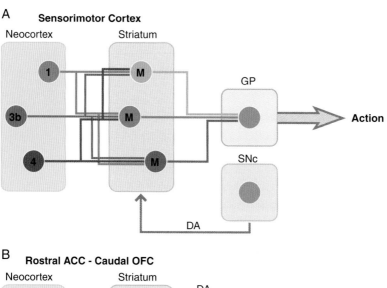

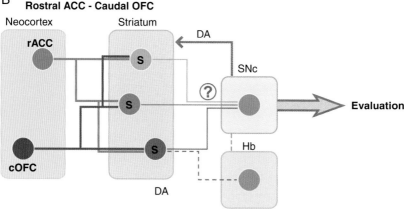

FIGURE 12–1. Schematic diagram of cortico-basal ganglia circuits leading through the striatum. (*A*) Sensorimotor circuits channeled through sensorimotor matrisomes. (*B*) Limbic-affective circuits channeled through striosomes. 1, 3b, 4, areas 1 and 3b of somatosensory cortex and area 4 of motor cortex; cOFC, caudal orbitofrontal cortex; DA, dopamine; GP, globus pallidus; Hb, habenula; M, matrisome; rACC, rostral (pregenual) anterior cingulate cortex; S, striosome; SNc, substantia nigra pars compacta.

areas and from the motor cortex. By having these different inputs converge, the modularity increases the chance for spatially local activation of striatal neurons. This convergence allows timing to become critical: when the inputs are convergent, they could differentially activate the matrisomes if the inputs were temporally correlated. This idea is supported by evidence that when somatosensory or motor cortex are stimulated to provide briefly simultaneous activation of cortical sites with matching somatotopy, striatal neurons in the corresponding matrisomes are stimulated to express early response genes (Parthasarathy and Graybiel, 1997). Thus matrisomes provide a way for information from different functionally related cortical regions to be recombined flexibly, rapidly, and selectively.

The striosomal system might be a comparable system to allow dynamic modulation of high-order circuits related to emotion, motivation, and evaluation (Fig. 12.1B). For example, striosomes in the anterior striatum receive converging inputs from the anterior cingulate cortex and the caudal orbitofrontal cortex, regions important for neural monitoring and control of motivation and emotion (Eblen and Graybiel, 1995). Moreover, neurons in striosomes project to the immediate vicinity of the dopamine-containing neurons of the substantia nigra, and they can also probably influence nigral activity via the lateral habenula (Rajakumar et al., 1993). This pattern suggests that the striosomes could have a powerful influence over what is learned through the action of basal ganglia-based circuits related to reinforcement-based learning (Eblen and Graybiel, 1995; Graybiel, 2008). Striosomes could also be related to the registration of salient stimuli and the reactions made to them, regardless of whether learning occurs. Far more needs to be learned about the connections of these striatal modules, and their physiology is still not known. Indirect evidence, however, some from physiology, supports the idea that striosomes might be related to processing salient stimuli, either rewarding or aversive (Aosaki et al., 1995; White and Hiroi, 1998; Blazquez et al., 2002).

LOCAL AND DISTRIBUTED PROCESSING BY STRIOSOMES AND MATRISOMES

Striosomes and matrisomes are three-dimensionally extended labyrinthine structures spread out widely within the striatum, not local patches or spheres (Graybiel and Ragsdale, 1978; Mikula et al., 2009). Moreover, the striosomes have distributions suggesting that they could provide nearly complete coverage of the part of the striatum in which they lie (Graybiel, 1984). This means that they could serve to coordinate, in space and time, the activity of many striatal neurons within their resident regions. This idea suggests that these modules could be important not only for local coordination of striatal processing—within individual striosomes or matrisomes—but also for coordination of more distributed striatal domains.

STRIATAL COMPARTMENTS AND STATE-LEVEL DYNAMIC PROCESSING

Coherent local field potential (LFP) activity is present in the striatum, even though only a small percentage of the single units in the striatum show coherent spike activity. For example, in monkeys performing oculomotor tasks, the LFP activity in the striatum exhibits high levels of beta-band coherence.

However, localized task-activated regions can exit from this general coherence in relation to task demands. These localized regions could represent matrisomes with task-related activity (Courtemanche et al., 2003). If so, the compartmentalization of the striatum would be important for organizing patterns of synchrony there, and thus for organizing global patterns of information flow.

A puzzle still to be addressed is how such broad synchrony of LFP activity, present even across regions that are parts of functionally distinct cortico-basal ganglia loops, is related to the distinctly different spike activity patterns that can be recorded in these different regions of the striatum. One idea is that the LFP coherence acts as a thresholding device, and that the pop-out of task-related activity represents supra-threshold activity (Courtemanche et al., 2003). The interneurons of the striatum are likely critical in setting these patterns of coherent activity and in organizing such dynamic filtering; all appear to have compartmentalized distributions, with most being more concentrated in the matrix or near striosome-matrix borders (Aosaki et al., 1995). The striosome and matrix compartments could thus be important for setting temporal patterns of activity in the striatum.

STRIOSOMES AND REPETITIVE BEHAVIORS

The transmitter-related specializations of striosomes relative to the surrounding matrix are impressive, suggesting that striosomes are likely to have particular functional effects due to their molecular expression patterns. The functional significance of this neurochemical specialization of the striosomes is still not understood. However, it is known that striosomes and matrix respond differently to treatments with dopamine receptor agonists. These differential responses likely have important functional implications. For example, the ratio of activity in striosomes and matrix in some parts of the striatum is highly correlated with the level of repetitive behavior induced by drugs such as amphetamine (Saka et al., 2004). The levels of levodopa-induced dyskinesias in models of parkinsonism are also correlated with differential striosome/matrix gene expression (Crittenden et al., 2009). One possibility raised by these findings is that differential activity in striosomes produces differential reinforcement-related signals leading to repetitive behaviors. Striosomes could be critical in influencing the balance between positive and negative reinforcement contexts and expectations (Graybiel, 2008). This suggestion is especially interesting given the differentially strong inputs to part of the striosomal system from the pregenual anterior cingulate cortex and the posterior orbitofrontal cortex. There is a pressing need to identify striosomes in electrophysiological recordings, and work in our laboratory is focused on this goal.

Striosomes and Human Neurologic and Neuropsychiatric Disorders

We and colleagues have identified differential effects on striosomes in primate and rodent models of addiction, parkinsonism, repetitive movement and thought disorders, and dopa-responsive dystonia. Moreover, in postmortem human brains, striosomes have been found to be differentially affected in subgroups of Huntington disease patients who in life suffered mood disorders as major symptoms (Tippett et al., 2007), and in patients with X-linked dystonia-parkinsonism (Goto et al., 2005). These findings raise the possibility that the compartmental organization of the striatum will prove to be important for understanding the etiology of some neurologic and neuropsychiatric disorders.

References

Aosaki T, Kimura M, Graybiel AM (1995) Temporal and spatial characteristics of tonically active neurons of the primate's striatum. *J Neurophysiol* 73:1234–1252.

Blazquez P, Fujii N, Kojima J, Graybiel AM (2002) A network representation of response probability in the striatum. *Neuron* 33:973–982.

Courtemanche R, Fujii N, Graybiel A (2003) Synchronous, focally modulated ß-band oscillations characterize local field potential activity in the striatum of awake behaving monkeys. *J Neurosci* 23:11741–11752.

Crittenden JR, Cantuti-Castelvetri I, Saka E, Keller-McGandy CE, Hernandez LF, Kett LR, Young AB, Standaert D, Graybiel AM (2009) Dysregulation of CalDAG-GEFI and CalDAG-GEFII predicts the severity of motor side-effects induced by antiparkinsonian therapy. *Proc Natl Acad Sci USA* 106:2892–2896.

Eblen F, Graybiel AM (1995) Highly restricted origin of prefrontal cortical inputs to striosomes in the macaque monkey. *J Neurosci* 15:5999–6013.

Enard W, Gehre S, Hammerschmidt K, Hölter SM, Blass T, Somel M, et al. (2009) A humanized version of Foxp2 affects cortico-basal ganglia circuits in mice. *Cell* 137:961–971.

Flaherty AW, Graybiel AM (1994) Input-output organization of the sensorimotor striatum in the squirrel monkey. *J Neurosci* 14:599–610.

Goto S, Lee LV, Munoz EL, Tooyama I, Tamiya G, Makino S, Ando S, Dantes MB, Yamada K, Matsumoto S, Shimazu H, Kuratsu J, Hirano A, Kaji R (2005) Functional anatomy of the basal ganglia in X-linked recessive dystonia-parkinsonism. *Ann Neurol* 58:7–17.

Graybiel AM (1984) Correspondence between the dopamine islands and striosomes of the mammalian striatum. *Neuroscience* 13:1157–1187.

Graybiel AM (1990) Neurotransmitters and neuromodulators in the basal ganglia. *Trends Neurosci* 13:244–254.

Graybiel AM (2008) Habits, rituals and the evaluative brain. *Annu Rev Neurosci* 31: 359–387.

Graybiel AM, Ragsdale CW, Jr. (1978) Histochemically distinct compartments in the striatum of human, monkey, and cat demonstrated by acetylthiocholinesterase staining. *Proc Natl Acad Sci USA* 75:5723–5726.

Mikula S, Parrish SK, Trimmer JS, Jones EG (2009) Complete 3D visualization of primate striosomes by KChIP1 immunostaining. *J Comp Neurol* 514:507–517.

Parthasarathy HB, Graybiel AM (1997) Cortically driven immediate-early gene expression reflects modular influence of sensorimotor cortex on identified striatal neurons in the squirrel monkey. *J Neurosci* 17:2477–2491.

Ragsdale CW, Jr., Graybiel AM (1990) A simple ordering of neocortical areas established by the compartmental organization of their striatal projections. *Proc Natl Acad Sci USA* 87:6196–6199.

Rajakumar N, Elisevich K, Flumerfelt BA (1993) Compartmental origin of the striato-entopeduncular projection in the rat. *J Comp Neurol* 331:286–296.

Saka E, Goodrich C, Harlan P, Madras BK, Graybiel AM (2004) Repetitive behaviors in monkeys are linked to specific striatal activation patterns. *J Neurosci* 24:7557–7565.

Tippett LJ, Waldvogel HJ, Thomas SJ, Hogg VM, van Roon-Mom W, Synek BJ, Graybiel AM, Faull RLM (2007) Striosomes and mood dysfunction in Huntington's disease. *Brain* 130:206–221.

White NM, Hiroi N (1998) Preferential localization of self-stimulation sites in striosomes/patches in the rat striatum. *Proc Natl Acad Sci USA* 95:6486–6491.

13

Subthalamo-Pallidal Circuit

Charles J. Wilson

The subthalamo-pallidal system constitutes the second layer of circuitry in the basal ganglia, lying downstream of the striatum. It consists of four nuclei. Two of them, the external segment of the globus pallidus (GPe) and subthalamic nucleus (STN), make their connections primarily within the basal ganglia. The other two, the internal segment of the globus pallidus (GPi) and the substantia nigra pars reticulata (SNr), are the output nuclei of the basal ganglia. Collectively their axons distribute collaterals to all the targets of the basal ganglia, including several thalamic nuclei, the superior colliculus, pedunculopontine nucleus, Forel field H, and lateral habenular nucleus. All the nuclei consist primarily of principal cells. Rare interneurons have been reported in each of them from studies of Golgi-stained preparations but have not so far been confirmed using more modern methods. The circuit as described here is based primarily on studies of the axonal arborizations of neurons stained individually by intracellular or juxtacellular labeling.

Basic Microcircuit

Figure 13.1 shows the organization of the subthalamo-pallidal complex. The dominant feature of the network is the inhibitory feedforward connection from the GPe to the GPi and SNr. Principal neurons in all three of these structures are GABAergic (Wilson, 2003).

The Globus Pallidus

The cells of GPe make feedforward inhibitory connections to the output neurons of the GPi and SNr. They also make inhibitory connections with each

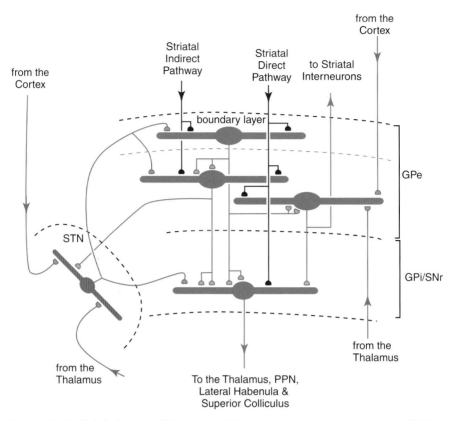

Figure 13–1. Subthalamo-pallidal circuit. The most numerous inputs are the GABAergic input from the direct and indirect striatal pathways, which are shown in dark blue. Pallidal and nigral GABAergic neurons and their synapses are shown in light blue. The GPi and SNr have similar synaptic arrangements, and they are shown combined, but they have different efferent projections. GPe, external segment of the globus pallidus; GPi, the internal segment of the globus pallidus; PPN, pedunculo-pontine nucleus; SNr, substantia nigra pars reticulate; STN, subthalamic nucleus.

other, and with the neurons of the STN. GPe neurons receive a very dense innervation that invests their dendrites in a mosaic of presynaptic boutons, most of which are inhibitory. Most presynaptic boutons in the GPe arise from the striato-pallidal pathway, are GABAergic, and are distributed on the somata and dendrites of the GPe cells (Shink and Smith, 1995). Intrinsic connections among GPe cells constitute a very small proportion of the total innervation, but they are especially focused on somata and proximal dendrites. Together, inhibitory intrinsic and striatal synapses account for more than 80% of all synapses in the GPe. The remaining synapses in the GPe are excitatory inputs from the STN, cerebral cortex, and the intralaminar nuclei of the thalamus (Kita, 2007). The relative proportion of inputs from these three pathways has not been measured, but the STN input is regarded to be the largest excitatory input. The striato-pallidal and striato-nigral projections arise from the two categories of striatal principal cells, which are called the direct and indirect pathway neurons. Both of these are the axonal projections of striatal spiny principal cells. Striatal cells of the indirect pathway project to the GPe only, whereas the direct pathway cells project to both the GPe and the GPi/SNr. The dendritic trees of neurons of the GPe are roughly planar and oriented normal to the trajectory of descending striatal axons. The nucleus is divided into two parts on the basis of striatal axonal arborizations (Kawaguchi et al., 1990). One part is a thin layer subjacent to the boundary with the striatum, and the other is the remainder of GPe. Striatal axons entering the GPe each make an initial collateral arborization confined to this boundary layer, and then make one or more other arborizations deeper in the nucleus. Both sets of striatal arborizations form topographical representations. Thus, in the boundary layer there is a highly compressed projection of the striatal cells of origin, collapsed along the direction normal to the striato-pallidal boundary, whereas in the remainder of the GPe the striato-pallidal projection makes connections that approximate a three-dimensional map of the striatum. Direct pathway neurons make about 50% fewer synapses in the GPe than indirect pathway neurons, so that the striatal innervation of the GPe is approximately two-thirds from the indirect and one-third from the direct pathway.

Projections from the STN to the GPe are also topographically organized but are much more divergent because of the dramatic difference in the number of cells contributing to these projections. For example, in the rat, there are about 3 million striatal spiny neurons on each side, each of which makes between 70 and 250 (average about 150) synapses in the GPe, resulting in 450 million striato-pallidal synapses formed on about 46,000 GPe cells (about 9800 per GPe neuron). In contrast, there are only 13,600 STN neurons on average in each side of the rat brain (Oorschot, 1996). The total proportion of GPe synapses that arises from the STN is not known. Ten percent to 20% of GPe synapses are morphologically similar to those from STN. Some of these arise from the cortico-pallidal and cortico-thalamic pathways. Assuming as few as 5% of all GPe synapses were derived from the STN, making them only one-sixteenth as numerous as striato-pallidal synapses, these would still arise from only about 1/220 the number of afferent cells, so they would be about 14 times more divergent. Taking excitatory synapses in GPe to be 10% of the total, each GPe neuron receives about 1000 excitatory inputs. Even if only half of these are from the STN, there will be about 23 million STN synapses in the GPe arising formed by 13,600 STN cells. If this is correct, each STN cells must make about 1700 synapses in the GPe. This is a large number compared with other basal ganglia axonal arborizations, and it is larger than expected from examination of available drawings of subthalamopallidal axons (Sato et al., 2000a), but no thorough quantitative study of the subthalamopallidal pathway is currently available.

Intrinsic connections follow a spatial pattern similar to that of the striatal axons. Most GPe neurons have local collateral axon arborizations approximating their dendritic trees, and additional discrete axonal arborizations located deeper in the nucleus. GPe neurons on average make and receive about 400 intrinsic contacts, for a total of about 18 million synapses. Unlike the striatal projection, intrinsic axon collaterals often make series of boutons, arranged on the axon like beads on a string, and all making synapses on the soma and proximal dendrites of a single neuron. Thus, the divergence of the intrinsic GPe connections is very low, with each GPe having a strong influence on a small number of other neurons in the nucleus (Sadek et al., 2007).

GPe neurons can be categorized on the basis of their targets outside the GPe. Many cells have collateralized axons that project to the STN and the GPi and/or SNr (Sato et al., 2000b). A subset of cells has ascending axonal branches that preferentially innervate interneurons in the striatum (Bevan et al., 1998).

The Subthalamic Nucleus

The STN is a small nucleus densely packed with glutamatergic, sparsely spined, projection neurons whose axons branch and arborize in the GPe, GPi, and SNr. They receive a dense excitatory innervation from the frontal cortex, including motor, premotor, and prefrontal areas, from the intralaminar thalamic nuclei, and from the pedunculopontine nuclei. In addition, they receive a large inhibitory innervation from the much more numerous cells of the globus pallidus. Presynaptic boutons of thalamic and cortical origin are similar in appearance and are located on the dendrites and dendritic spines (Chang et al., 1983; Bevan et al., 1995). Inhibitory pallidal inputs are about as equally numerous as excitatory ones, but they are present on somata as well as dendrites. STN cell axons branch to innervate combinations of its target structures (Levy et al., 2002).

The Globus Pallidus and the Substantia Nigra Pars Reticulata

GPi neurons are morphologically similar to those in the GPe, and like them they have dendrites and somata densely covered with synaptic boutons of extrinsic origin. Unlike the GPe cells, those of GPi usually do not have local axon collateral arborizations. Most cells in GPi are principal cells, but two kinds of principle cells can be distinguished based on synaptic target. The axons of one group of cells project through the ansa lenticularis and/or lenticular fasciculus and innervate some combination of the ventral tier thalamic nuclei, the intralaminar nuclei of the thalamus, Forel field H, the retrorubral area, and the pedunculopontine nucleus (Parent et al., 2001). The axons of the other cell type project through the stria medullaris to innervate the lateral habenula. Inputs to the GPi arise primarily from the striato-pallidal (direct) pathway, the GPe, pedunculopontine nucleus, and the STN, and like the GPe, inhibitory inputs far outnumber excitatory ones. SNr principal neurons are similar in many ways to those of GPi, but with different axonal targets. SNr axons collateralize to ventral tier thalamic nuclei (but to ones not innervated by the GPi), intralaminar thalamic nuclei (also mostly different areas from GPi), the substantia nigra pars compacta, the pedunculopontine nuclei, and to deep layers of the superior colliculus (Cebrian et al, 2005). In both the GPi and the SNr, striatal inputs are many and divergent, whereas individual axons from the GPe make multiple synaptic contacts with somata and proximal dendrites of a few neurons. The STN and pedunculopontine nuclei are the primary sources of glutamatergic excitatory input in both structures.

Firing Patterns and Synaptic Integration

Despite the predominantly inhibitory nature of synaptic transmission in the circuit, all of these structures are characterized by very high sustained rates of firing. In vivo, both GPe and GPi/SNr cells often maintain firing at mean rates in excess of 40/sec, and the background firing rate of subthalamic neurons averages about 20/sec. Synaptic excitation probably contributes to this high firing rate, but it is not essential (Tachibana et al., 2008). This is because the cells in all of these structures are autonomous oscillators that are rhythmically active in the absence of synaptic input, and they fire at rates comparable to those seen in vivo, even after isolation from their synaptic input (Surmeier et al., 2005). Autonomous activity in all of these nuclei arises primarily because of persistent sodium current that endows the cells with a negative conductance region in their I-V characteristic in the membrane potential range between −60 mV and the action potential threshold. At an early point after subsidence of the afterhyperpolarization currents following each action potential, the persistent sodium current exceeds the combination of outward currents, and this net inward current gradually depolarizes the cell back to the firing threshold. Synaptic interconnections among cells of the subthalamo-pallidal system act in the context of this ongoing activity. Unlike neurons in many other forebrain regions, action potential generation in the GPe or STN cannot be traced to a net excitatory synaptic current immediately preceding action potential generation. Synaptic interactions between neurons in this network do not necessarily cause action potentials but instead shift their times of occurrence one way or the other. In such networks, synaptic inputs may influence action potentials throughout the cycle of firing and need not occur in any particular temporal window relative to action potential generation. The timing of an action potential in the GPe or STN does not signal an immediately preceding synaptic event, but rather an interaction between the time of the previous action potential and the net effect of all synaptic input arriving since the last action potential.

Effects of Convergent Signals on the Output Neurons

A popular model for the basal ganglia depicts a balance between the opposing influences of the direct and indirect pathways in the control of firing rate in the output nuclei. In this view, increased activity in direct pathway striatal neurons (caused by cortical or thalamic inputs) will decrease activity in the output nuclei, whereas increased activation of indirect pathway cells will enhance basal ganglia output. The output of the basal ganglia is itself inhibitory in its targets, and in this model enhanced output (via the indirect pathway) suppresses activity in the motor thalamus and can lead to Parkinsonian symptoms like bradykinesia. Decreased output caused by direct pathway activity would release the motor thalamus and promote movement.

Although this model is successful in predicting the symptoms of Parkinson disease (Wichmann and Delong, 1996), there are multiple routes by which cortical, thalamic, or brainstem inputs might alter the rate or timing of action potentials in the GPi or SNr neurons. For signals arising in the frontal cortex, the most direct routes are via the cortico-subthalamic pathway or the cortical inputs to GPe. These two would differ in sign, as well as timing, and do not rely on the striatum at all. Two other potential routes are through the cortico-striatal direct pathway or the cortico-striato-pallidal indirect pathway, again with a difference in sign and timing. There are still other routes through, and even the possibility of reverberant activity via, the reciprocal subthalamo-pallidal connections. Thus, the net effect of an input to the basal ganglia on firing rate at the output nuclei is not easily predicted from the anatomical relationships between cells.

Studies of the convergent pathways responsible for the effects of cortical stimulation on GPi neurons in awake monkeys show a sequence of excitation, inhibition, and subsequent excitation over a 100 ms period following motor cortex stimulation (Tachibana et al., 2008). Firing of STN under the same stimulation also shows a brief initial excitation, a pause, and a late excitation; and inactivation of the STN with muscimol abolishes both the early and late excitation in GPi cells, leaving only the inhibitory response. GPe cells show a similar pattern of excitation, inhibition, and excitation, so they cannot be directly responsible for the inhibitory component of the response. Inactivation of the GPe by injection of muscimol increased the background firing rate of GPi cells and blocked the late component of excitation, suggesting that the late excitation may be primarily due to GPe–STN interactions. The inhibitory component corresponds to activation of striatal direct and indirect pathways, and at the output nuclei it is dominated by the direct pathway.

Synchrony and Parkinson's Disease

In networks of coupled oscillatory neurons, it is common for synchronous rhythms to emerge. However, studies of simultaneously recorded GPe, GPi, or subthalamic neurons have shown that these cells fire independently of each other, despite the presence of common inputs and direct and indirect synaptic interconnections (Raz et al., 2000). The independence of activity is lost in experimental models of Parkinson disease, however, as cells in both structures fire in synchrony with low-frequency oscillations in the 5–15 Hz range. More than any change in firing rate, the increase in oscillatory firing within cells and increased correlation across cells is associated with the development of Parkinsonism in these preparations, and even in human cases studied during surgery (Levy et al, 2002). These results suggest that the striato-pallidal circuit may function in part to actively maintain the statistical independence of activity across neurons, and that this mechanism is impaired in Parkinson's disease (Terman et al., 2002).

REFERENCES

Bevan MD, Francis CM, Bolam JP (1995) The glutamate-enriched cortical and thalamic input to neurons in the subthalamic nucleus of the rat: convergence with GABA-Positive terminals. *J Comp Neurol* 361:491–511.

Bevan MD, Booth PAC, Eaton SA, Bolam JP (1998) Selective innervation of neostriatal interneurons by a subclass of neuron in the globus pallidus of the rat. *J Neurosci* 18:9438–9452.

Cebrián C, Parent A, Prensa L (2005) Patterns of axonal branching of neurons of the substantia nigra pars reticulata and pars lateralis in the rat. *J Comp Neurol* 492: 349–369.

Chang HT, Kita H, Kitai ST (1983) The fine structure of the rat subthalamic nucleus: an electron microscopic study. *J Comp Neurol* 221:113–123.

Kawaguchi Y, Wilson CJ, Emson PC (1990) Projection subtypes of rat neostriatal matrix cells revealed by intracellular injection of biocytin. *J Neurosci* 10:3421–3438.

Kita H (2007) Globus pallidus external segment. *Prog Brain Res* 160:111–133.

Levy R, Hutchison WD, Lozano AM, Dostrovsky JO (2002) Synchronized neuronal discharge in the basal ganglia of parkinsonian patients is limited to oscillatory activity. *J Neurosci* 22:2855–2861.

Oorschot DE (1996) Total number of neurons in the neostriatal, pallidal, subthalamic, and substantia nigral nuclei of the rat basal ganglia: a stereological study using the Cavalieri and optical dissector methods. *J Comp Neurol* 366:580–599.

Parent M, Lévesque M, Parent A (2001) Two types of projection neurons in the internal pallidum of primates: single axon tracing and three-dimensional reconstruction. *J Comp Neurol* 439:162–175.

Raz A, Vaadia E, Bergman H. (2000) Firing patterns and correlations of spontaneous discharge of pallidal neurons in the normal and the tremuluous 1-methyl-4-phenyl-1,2,3,6-tetrahydropyridine Vervet model of Parkinsonism. *J Neurosci* 20:8559–8571.

Sadek AR, Magill PJ, Bolam JP (2007) A single-cell analysis of intrinsic connectivity in the rat globus pallidus. *J Neurosci* 27:6352–6362.

Sato F, Parent M, Levesque M, Parent A (2000a) Axonal branching pattern of neurons of the subthalamic nucleus in primates. *J Comp Neurol* 424:142–152.

Sato F, Lavallée P, Lévesque M, Parent P (2000b) Single-axon tracing study of neurons of the external segment of the globus pallidus in primate. *J Comp Neurol* 417: 17–31.

Shink E, Smith Y (1995) Differential synaptic innervation of neurons in the internal and external segments of the globus pallidus by the GABA- and glutamate-containing terminals in the squirrel monkey. *J Comp Neurol* 358:119–141.

Surmeier DJ, Mercer JN, Chan CS (2005) Autonomous pacemakers in the basal ganglia: who needs excitatory synapses anyway. *Curr Op Neurobiol* 15:312–318.

Tachibana Y, Kita H, Chiken S, Takada M, Nambu A (2008) Motor cortical control of internal pallidal activity through glutamatergic and GABAergic inputs in awake monkeys. *Eur J Neurosci* 27:238–253.

Terman D, Rubin JE, Yew AC, Wilson CJ (2002) Activity patterns in a model for the subthalamopallidal network of the basal ganglia. *J Neurosci* 22:2963–2976.

Wichmann T, DeLong MR (1996) Functional and pathophysiological models of the basal ganglia. *Curr Op Neurobiol* 6:751–758.

Wilson CJ (2003) The basal ganglia. In: Shepherd GM, ed. *The Synaptic Organization of the Brain*, 5th ed., pp. 361–413. New York: Oxford University Press.

Section 5

Limbic Systems

14

Microcircuits of the Amygdala

Luke R. Johnson and Joseph E. LeDoux

The amygdala was first recognized as a distinct brain region in the early nineteenth century (Swanson and Petrovich, 1998; LeDoux, 2007). The word *amygdala*, derived from the Greek, was meant to denote an almond-shaped structure identified deep in the medial temporal lobe rostral to the ventral reaches of the hippocampus. Like most brain regions, the amygdala is not a single mass; rather, it is composed of distinct subareas or nuclei, each with anatomical and functional subdivisions (Pitkanen et al., 1997). These subdivisions are extensively connected with each other and other brain areas. The almond-shaped region that gives the amygdala its name was actually only one of these nuclei, the basal nucleus, rather than the whole structure. Today, the amygdala is best known for its role in emotional functions, especially fear, but it also contributes to memory and attention (McGaugh et al., 1996; Phelps and LeDoux, 2005). Unique microcircuits within the subdivisions of the amygdala are beginning to be identified (Samson et al., 2003; Johnson and LeDoux, 2004; Samson and Pare, 2006; Johnson et al., 2008; Johnson et al., 2009). These are the subject of this review. First, though, we will discuss some organizational issues.

WHAT IS THE AMGYDALA?

Traditionally, the amygdala was viewed as consisting of an evolutionarily primitive division associated with the olfactory system (the cortico-medial region) and an evolutionarily newer division associated with the neocortex (the basolateral region) (Swanson and Petrovich, 1998; LeDoux, 2007). The cortico-medial region includes the cortical, medial, and central nuclei, while the basolateral region consists of the lateral, basal, and accessory basal nuclei. However, a recent proposal by Swanson argues that the amygdala is neither a structural nor a functional unit, and instead consists of regions that belong to other regions or systems of the brain. For example, in this scheme, the lateral and basal amygdala are viewed as nuclear extensions of the cortex (rather than amygdala regions related to the cortex), while the central and medial amygdala are said to be ventral extensions of the striatum. This scheme has merit, but the present review focuses on the organization and function of the nuclei and subnuclei that are traditionally said to be part of the amygdala since most of the functions of the amygdala are understood in these terms.

For example, extensive evidence suggests that Pavlovian fear conditioning depends on the amygdala. However, only select regions of the amygdala are involved. Specifically, the amygdala centric circuits underlying classical fear conditioning are well characterized (LeDoux, 2000; LeDoux, 2007). In this associative learning paradigm a conditioned stimulus (CS), usually an auditory tone, comes to elicit behavioral and autonomic signs of fear after occurring in conjunction with an aversive unconditioned stimulus (US), typically mild footshock. The CS and US converge in the lateral amygdala (Romanski et al., 1993; LeDoux, 2003). Fear conditioning is dependent on nociceptive inputs from the spinothalamic tract that terminate in the amygdala. These inputs may enter the amygdala via the thalamus or via the cortex (Shi and Davis, 2001). Upon later exposure to the CS, fear responses are expressed via connections from the lateral (LA) to the central (CE) amygdala. The LA and CE would still be important in fear conditioning even if the overall concept of the amygdala were eliminated.

NEURONS OF THE AMYGDALA

Different amygdala nuclei have unique neuron types. The medial areas of the amygdala, including the CE and medial amygdala (M), are comprised of GABAergic projection neurons. These neurons are generally like the medium-sized spiny neurons (MSSNs) of the striatum in the basal ganglia. In contrast, the laterally located nuclei of the basolateral complex (lateral [LA], basal [B], accessory basal [AB]) and the cortical amygdala nuclei (COA) are comprised predominantly of glutamatergic projection neurons (Fig. 14.1). These projection neurons are similar to cortical projection neurons in morphology and electrophysiology. Unlike the cortex, the amygdala is not laminated, with the exception of the cortical nuclei. The other glutamatergic neuron nuclei (LA, B, AB) do not show an immediately obvious structural organization. Given that the CE receives inputs from the LA and B, one immediate feature of amygdala microcircuits is connectivity between cortical-like and basal ganglia–like neurons.

The amygdala is more complex than the simple dichotomy of glutamatergic and GABAergic projection neurons. The LA and B nuclei have been the

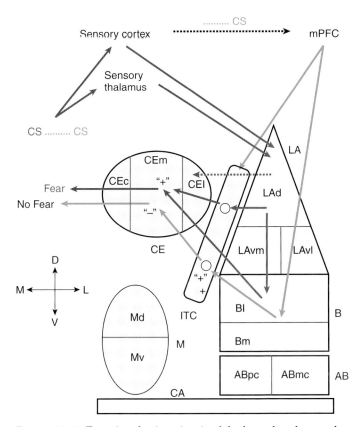

FIGURE 14–1. Functional microcircuit of the lateral and central amygdala, which regulates fear and its extinction. Sensory and nociceptive circuits reach the lateral amygdala (LA) where synaptic plasticity occurs, storing key aspects of classically conditioned fear memory. For this memory to be behaviorally expressed, the amygdala output nucleus (medial division of central amygdala) CEm is activated indirectly via the GABAergic intercalated neurons (ITC). Aspects of the stored memory are transmitted from the LA to the basal (B) nucleus. In the B nucleus, neurons regulate the switch between the behavioral expression of fear or of its extinction. Behavioral expression is again routed through the CEm such that the original fear memory can remain intact but is overridden via the B to CEm circuit. GABAergic ITC play a key role in the gating amygdala microcircuits which express fear and fear extinction.

most extensively studied (McDonald and Culberson, 1981; Rainnie et al., 1993; Pape et al., 2001; Sosulina et al., 2006; Mascagni et al., 2009). Earlier work reported up to seven separate populations of LA neurons (Faulkner and Brown, 1999). These potentially different populations of LA neurons were based on firing characteristics and morphology. More commonly reported are three unique populations: glutamatergic pyramidal and stellate projection neurons; and GABAergic local interneurons (Rainnie et al., 1991a; Rainnie et al., 1991b). Some data have suggested that only one population of glutamatergic principal neurons can be identified in the LA. Pyramidal neurons may look stellate-like when viewed from their ventral surface (Faber et al., 2001). With the exception of the cortical-amygdala nuclei, the amygdala has an apparent lack of neuron orientation and layering. However, important organization may be present in selective axonal connectivity.

Within the amygdala as a whole there are at least three populations of GABAergic neurons. These are *(1)* GABAergic local interneurons throughout the amygdala. These interneurons are located within both the medial GABAergic nuclei and the lateral glutamatergic nuclei (Fig. 14.1). *(2)* The intercalated neurons (ITC), a population of small soma neurons positioned both medially and laterally (in the rat) to the LA that also use GABA as a neurotransmitter and topographically are part of the LA (the role of these neuron in an amygdala functional microcircuit is discussed in the section "The Intercalated Gate." *(3)* The GABAergic projection neurons of the CE and M. These neurons are believed to be GABAergic medium sized spiny projection neurons (MSN) of same kind that comprise sections of the basal ganglia especially the caudate and putamen (or striatum) (McDonald, 1982; Swanson and Petrovich, 1998). These neurons are different from the GABAergic local circuit neurons. Glutamatergic principal and GABA interneurons of the LA and B differ in their spontaneous and maximal firing rates (Rainnie et al., 1991a; Rainnie et al., 1991b; Pare and Gaudreau, 1996; Royer and Pare, 2003; Sosulina et al., 2006). Principal neurons tend to be quiescent in vivo and in vitro unless activated by thalamic or cortical input or current injection. Following synaptic activation, LA principal neurons can fire action potentials up to 20 Hz. In contrast, presumed GABAergic interneurons are often spontaneously firing both in vivo and in vitro and can be induced to fire up to 100 Hz. ITC neurons can fire up to 30 Hz (Rainnie et al., 1991a; Rainnie et al., 1991b; Pare and Gaudreau, 1996; Sosulina et al., 2006; Woodruff and Sah, 2007).

Like the cortex and hippocampus, the LA and B nuclei contain unique populations of local circuit neurons. The LA and B are known to contain excitatory feedforward and feedback circuits (Samson et al., 2003; Johnson and LeDoux, 2004; Johnson et al., 2008). Increasing evidence indicates likely GABA circuits modulated by the known neuromodulators in a cell-specific manner (Rainnie, 1999; Pape, 2005; Rainnie et al., 2006; Mascagni and McDonald, 2007; Muller et al., 2007a; Muller et al., 2007b; Pinard et al., 2008). Moreover, these interneurons, especially the parvalbumin-positive GABA neurons, form interconnected networks of inhibition and disinhibition (Muller et al., 2007a; Muller et al., 2007b; Woodruff and Sah, 2007; Sosulina et al., 2008).

The Intercalated Gate

Amygdala ITC neurons gate the flow of excitatory projections from the LA and BA to the CE. Discrete in vitro electrophysiological studies have identified that LA and BA excitatory projection neurons make synaptic contact with

ITC GABA neurons, which in turn synapse onto GABAergic CE neurons (Pare and Smith, 1993; Pare and Smith, 1994; Pare et al., 2004). Importantly this ITC GABA input on CE neurons is able to shunt excitatory input from other external sources (Delaney and Sah, 2001; Sah and Westbrook, 2008). This means that even if CE neurons receive direct excitatory projections from other relevant nuclei, for example, the hippocampus or the medial prefrontal cortex (mPFC), or direct from the LA and BA themselves, all the excitatory input will be unable to drive the CE neurons to action potential if they are simultaneously receiving synaptic input from the ITC. This places the ITC as a powerful gate to regulate the behavioral output of condition fear stimuli.

The role of the ITC in gating between the LA and CE is best understood in terms of fear extinction, a process of new learning whereby the CS comes to no longer elicit the previously learned response (CR) (Myers and Davis, 2007). Several lines of evidence implicate the amygdala and mPFC (Sotres-Bayon et al., 2006; Myers and Davis, 2007; Sotres-Bayon et al., 2007). Interestingly, the ITC that lie between lateral areas (LA and BA) and the CE appear to be particularly densely innovated by the mPFC and neuromodulators (McDonald et al., 1996; Pape, 2005; Likhtik et al., 2008). This anatomical arrangement is important because it gives clues to the functional anatomy that may allow for the mPFC repressing conditioned responding to fear-inducing stimuli following extinction. Given the apparent role of the ITC as the gate-controlling amygdala output, and also given the extensive afferent input from the mPFC to the ITC (McDonald et al., 1996; Pare et al., 2004), the potential mechanism for fear extinction becomes apparent. A logical and elegant fear extinction circuit was proposed by Pare, Quirk, and LeDoux (2004). Sensory synaptic inputs to the LA, previously strengthened, increase LA neuron firing rates. Likewise infralimbic (IL) neurons of the mPFC also increase their firing rate in response to the CS. Whereas prior to extinction the increased LA neuron firing would result in an output signal from the LA via the central nucleus of the amygdala resulting in fear behavior, after extinction the increased firing of the mPFC neurons drives the ITC GABAergic gate to inhibit CE output. As a result, the CS now drives an extinguished CR that behaviorally manifests as a lost or forgotten fear memory. However, the memory is not forgotten; it is repressed. Important questions remain about the cellular and network mechanisms within the amygdala that drive the switching on and off of fear. Recent data show that, while the LA and IL contain neurons that appear to store fear and extinction memories, respectively, the BA controls the switch between the two states via the ITC (Herry et al., 2008).

INTRALATERAL AMYGDALA MICROCIRCUITS

In contrast to the cortex and the hippocampus, much less is known about the intraamygdala microanatomy and circuitry. Of all the amygdala nuclei, the microcircuits of the LA have been the most extensively studied. The LA is divided into three subnuclei—dorsal (LAd), ventromedial (LAvm), and ventrolateral (LAvl)—based on histologic appearance (Pitkanen et al., 1997). The LAd is further subdivided into superior and inferior based on network behavior and functional properties (Repa et al., 2001; Johnson and LeDoux, 2004; Johnson et al., 2008). The most progress has been made in the LA in part because this nucleus has been a focus for its role in conditioned fear memory. The LA most likely contains a network of interconnected neurons (Johnson and LeDoux, 2004). This network is activated by major cortical and subcortical afferents (Repa et al., 2001; Johnson et al., 2008). For example, thalamic afferents enter the LAd from a dorsomedial direction and then apparently course in a ventral direction through the LA (LeDoux, 1990; LeDoux et al., 1991; Doron and Ledoux, 1999; Doron and Ledoux, 2000). This is the direction of neural activity and information flow considered in most models of the LA (LeDoux, 2003; Maren and Quirk, 2004; Pare et al., 2004; Samson and Pare, 2006; Herry et al., 2008; Sah et al., 2008). However, recent studies have provided more details of the intralateral network (Samson, 2003; Johnson and LeDoux, 2004; Johnson et al., 2008; Johnson et al., 2009).

According to Samson, et al. (2003) and Samson and Pare (2006), intralateral excitatory connectivity predominates in the transverse plane. This is because excitatory activity in the dorsal to ventral plane is transected by a GABA inhibitory network that limits excitatory conduction in this axis. In the dorsal to ventral plane the predominant evoked synaptic responses are inhibitory. This inhibition is mediated by GABA A receptors that are presumed to be postsynaptic to a network of local GABAergic interneurons. In contrast, the excitatory network activity is less inhibited in the transverse plane. Moreover, they find large excitatory activity can be propagated in the transverse plane, suggesting an intrinsic organization of the intralateral network, which is divided in transverse sections along the ventral to dorsal axis (see Fig. 14.2).

Consistent with the findings of Samson and Pare, recent data have also shown that intra LAd excitatory networks are inhibited by GABA A receptors in the dorsoventral plane (Johnson et al., 2008; Johnson et al., 2009). In the brain slice in the coronal plane, thalamic evoked polysynaptic potentials are strongly inhibited, with only the monosynaptic and residual polysynaptic negative potentials present riding on a slow positive potential. In the presence of GABA A antagonism, the slow positive wave is removed and excitatory polysynaptic activity increases (Johnson et al., 2008; Johnson et al., 2009). The same excitatory polysynaptic network activity is also seen in vivo, suggesting the transverse inhibitory modules may be disinhibited in the awake animal, presumably by the neuromodulatory transmitter system (Rainnie, 1999; Muller et al., 2005; Woodruff and Sah, 2007; Johnson et al., 2008; Sosulina et al., 2008; Muller et al., 2009).

Importantly, the intra LAd network also includes an apparent recurrent feedback from the inferior to superior parts of LAd (Fig. 14.2) (Johnson and

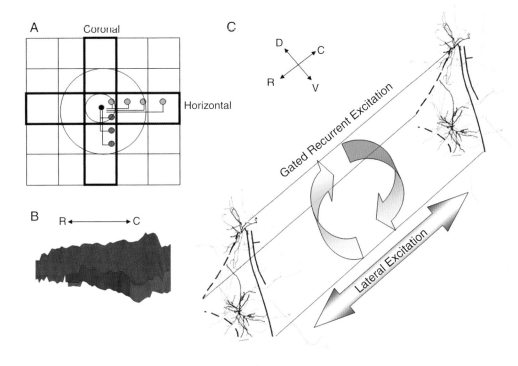

Figure 14–2. Internal microcircuits of the lateral amygdala. Knowledge of microcircuits within the distinct amygdala nuclei is beginning to emerge. (*A*) Three-dimensional reconstruction of the lateral amygdala showing its subdivisions (red, dorsal; green, ventromedial; blue, ventrolateral). (*B*) Horizontal brain slices of the lateral amygdala reveal lateral excitation; in contrast, coronal brain slices reveal extensive lateral inhibition (green and red neurons, respectively). (*C*) Three-dimensional representation of the dorsal lateral amygdala with reconstructed glutamatergic principal neurons (black, soma and dendrites; gray, local axon collaterals, superior; red, local axon collaterals, inferior). Excitatory axon collaterals project multidirectionally, including both dorsally and ventrally. With GABA antagonists, polysynaptic recurrent (inferior to superior) activity is detected. Thus, the dorsal lateral amygdala microcircuit contains excitatory lateral excitation in the rostral-caudal dimension, together with reverberant excitation in the ventral-dorsal dimension, which is gated by GABA circuits. This dorsal lateral amygdala microcircuit may allow the coordination of multisensory signals reaching the lateral amygdala.

LeDoux, 2004; Johnson et al., 2008; Johnson et al., 2009). The LAd network appears to behave in a structured manner with time-locked polysynaptic potentials observed (Johnson et al., 2008). This network can reverberate when triggered by stimulation of thalamic to LA afferents. The same pattern of polysynaptic excitatory events appears in vivo in the awake rat. Upon stimulation of the medial geniculate nucleus of the thalamus, both mono and polysynaptic peaks are observed in the extracellular field potential. Activity can be observed to last for up to 40 ms in vitro, and in the awake rat apparent polysynaptic activity can be observed to continue beyond 240 ms. The function of the reverberant microcircuit in the LAd is not yet fully understood. Based on the intrinsic time signature of the reverberating and recurrent intra LAd network, one potential function proposed is that the network provides a mechanism to facilitate coincident interaction between cortical and subcortical information (Johnson and LeDoux, 2004; Johnson et al., 2008; Johnson et al., 2009). This may allow convergence across time of the thalamic and cortical sensory information (Humeau et al., 2003; Humeau et al., 2005). This converging information will include elements of the same original sound that is processed directly to LA via the thalamus, as well as via the cortex. In addition, it is also possible that other sensory information, for example, tone and light, as well as aspects of tone and shock, could also show temporal convergence via the reverberating LA network (Johnson and LeDoux, 2004; Johnson et al., 2008; Johnson et al., 2009).

SUMMARY

Microcircuits of the amygdala remain somewhat of an enigma. The amygdala itself is comprised of laterally located glutamatergic projection neuron structures, which are cortical-like, and medially located GABA projection neuron structures, which resemble neurons of the striatum. Within these are many nuclei and subnuclei that are distinguished on histologic, hodologic, and functional criteria. Significant progress has been made in understanding the organization of microcircuits in the LA and CE, which play important roles in fear learning and memory, and the ITC, which regulate fear extinction (Pare et al., 2004; Herry et al., 2008; Likhtik et al., 2008). Emerging data indicate a structured excitatory microcircuit within the LA (Samson et al., 2003; Johnson and LeDoux, 2004; Samson and Pare, 2006; Johnson et al., 2008; Johnson et al., 2009). Local axon collaterals of excitatory projection neurons are regulated by transverse modules of local GABA inhibition, which control excitation in the dorsal to ventral and ventral to dorsal planes (Samson et al., 2003; Johnson and LeDoux, 2004; Samson and Pare, 2006; Johnson et al., 2008; Johnson et al., 2009). Bidirectional excitation in this plane may form recurrent networks that contribute to the temporal coordination of sensory signals integrated into the LAd network (Johnson and LeDoux, 2004; Johnson et al., 2008; Johnson et al., 2009 Pare). Future work on amygdala microcircuits will continue to yield important data on how microcircuits regulate learning, memory, and behavior.

REFERENCES

Delaney AJ, Sah P (2001) Pathway-specific targeting of GABA(A) receptor subtypes to somatic and dendritic synapses in the central amygdala. *J Neurophysiol* 86:717–723.

Doron NN, Ledoux JE (1999) Organization of projections to the lateral amygdala from auditory and visual areas of the thalamus in the rat. *J Comp Neurol* 412:383–409.

Doron NN, Ledoux JE (2000) Cells in the posterior thalamus project to both amygdala and temporal cortex: a quantitative retrograde double-labeling study in the rat. *J Comp Neurol* 425:257–274.

Faber ES, Callister RJ, Sah P (2001) Morphological and electrophysiological properties of principal neurons in the rat lateral amygdala in vitro. *J Neurophysiol* 85:714–723.

Faulkner B, Brown TH (1999) Morphology and physiology of neurons in the rat perirhinal-lateral amygdala area. *J Comp Neurol* 411:613–642.

Herry C, Ciocchi S, Senn V, Demmou L, Muller C, Luthi A (2008) Switching on and off fear by distinct neuronal circuits. *Nature* 454:600–606.

Humeau Y, Shaban H, Bissiere S, Luthi A (2003) Presynaptic induction of heterosynaptic associative plasticity in the mammalian brain. *Nature* 426:841–845.

Humeau Y, Herry C, Kemp N, Shaban H, Fourcaudot E, Bissiere S, Luthi A (2005) Dendritic spine heterogeneity determines afferent-specific Hebbian plasticity in the amygdala. *Neuron* 45:119–131.

Johnson LR, LeDoux JE (2004) The anatomy of fear: microcircuits of the lateral amygdala. In: Gorman JM, ed. *In Fear and Anxiety: The Benefits of Translational Research*, pp. 227–250. Washington DC: APPA Press.

Johnson LR, Hou M, Ponce-Alvarez A, Gribelyuk LM, Alphs HH, Albert L, Brown BL, Ledoux JE, Doyere V (2008) A recurrent network in the lateral amygdala: a mechanism for coincidence detection. *Front Neural Circuits* 2:3.

Johnson LR, Ledoux JE, Doyere V (2009) Hebbian reverberations in emotional memory micro circuits. *Front Neurosci* doi:10.3389/neuro.01.027.2009.

LeDoux JE (1990) Information flow from sensation to emotion: plasticity in the neural computation of stimulus value. In: Gabriel M and Moore J, eds. *Learning and Computational Neuroscience: Foundations of Adaptive Networks*, pp. 3–52. Cambridge, MA: MIT Press.

LeDoux JE (2000) Emotion circuits in the brain. *Annu Rev Neurosci* 23:155–184.

LeDoux J (2003) The emotional brain, fear, and the amygdala. *Cell Mol Neurobiol* 23: 727–738.

LeDoux J (2007) The amygdala. *Curr Biol* 17:R868–874.

LeDoux JE, Farb CR, Romanski LM (1991) Overlapping projections to the amygdala and striatum from auditory processing areas of the thalamus and cortex. *Neurosci Lett* 134:139–144.

Likhtik E, Popa D, Apergis-Schoute J, Fidacaro GA, Pare D (2008) Amygdala intercalated neurons are required for expression of fear extinction. *Nature* 454:642–645.

Maren S, Quirk GJ (2004) Neuronal signalling of fear memory. *Nat Rev Neurosci* 5: 844–852.

Mascagni F, McDonald AJ (2007) A novel subpopulation of 5-HT type 3A receptor subunit immunoreactive interneurons in the rat basolateral amygdala. *Neuroscience* 144:1015–1024.

Mascagni F, Muly EC, Rainnie DG, McDonald AJ (2009) Immunohistochemical characterization of parvalbumin-containing interneurons in the monkey basolateral amygdala. *Neuroscience* 158:1541–1550.

McDonald AJ (1982) Cytoarchitecture of the central amygdaloid nucleus of the rat. *J Comp Neurol* 208:401–418.

McDonald AJ, Culberson JL (1981) Neurons of the basolateral amygdala: a Golgi study in the opossum (Didelphis virginiana). *Am J Anat* 162:327–342.

McDonald AJ, Mascagni F, Guo L (1996) Projections of the medial and lateral prefrontal cortices to the amygdala: a Phaseolus vulgaris leucoagglutinin study in the rat. *Neuroscience* 71:55–75.

McGaugh JL, Cahill L, Roozendaal B (1996) Involvement of the amygdala in memory storage: interaction with other brain systems. *Proc Natl Acad Sci USA* 93:13508–13514.

Muller JF, Mascagni F, McDonald AJ (2005) Coupled networks of parvalbumin-immunoreactive interneurons in the rat basolateral amygdala. *J Neurosci* 25:7366–7376.

Muller JF, Mascagni F, McDonald AJ (2007a) Serotonin-immunoreactive axon terminals innervate pyramidal cells and interneurons in the rat basolateral amygdala. *J Comp Neurol* 505:314–335.

Muller JF, Mascagni F, McDonald AJ (2007b) Postsynaptic targets of somatostatin-containing interneurons in the rat basolateral amygdala. *J Comp Neurol* 500:513–529.

Muller JF, Mascagni F, McDonald AJ (2009) Dopaminergic innervation of pyramidal cells in the rat basolateral amygdala. *Brain Struct Funct* 213:275–288.

Myers KM, Davis M (2007) Mechanisms of fear extinction. *Mol Psychiatry* 12:120–150.

Pape HC (2005) GABAergic neurons: gate masters of the amygdala, mastered by dopamine. *Neuron* 48:877–879.

Pape HC, Driesang RB, Heinbockel T, Laxmi TR, Meis S, Seidenbecher T, Szinyei C, Frey U, Stork O (2001) Cellular processes in the amygdala: gates to emotional memory? *Zoology (Jena)* 104:232–240.

Pare D, Smith Y (1993) The intercalated cell masses project to the central and medial nuclei of the amygdala in cats. *Neuroscience* 57:1077–1090.

Pare D, Smith Y (1994) GABAergic projection from the intercalated cell masses of the amygdala to the basal forebrain in cats. *J Comp Neurol* 344:33–49.

Pare D, Gaudreau H (1996) Projection cells and interneurons of the lateral and basolateral amygdala: distinct firing patterns and differential relation to theta and delta rhythms in conscious cats. *J Neurosci* 16:3334–3350.

Pare D, Quirk GJ, Ledoux JE (2004) New vistas on amygdala networks in conditioned fear. *J Neurophysiol* 92:1–9.

Phelps EA, LeDoux JE (2005) Contributions of the amygdala to emotion processing: from animal models to human behavior. *Neuron* 48:175–187.

Pinard CR, Muller JF, Mascagni F, McDonald AJ (2008) Dopaminergic innervation of interneurons in the rat basolateral amygdala. *Neuroscience* 157:850–863.

Pitkanen A, Savander V, LeDoux JE (1997) Organization of intra-amygdaloid circuitries in the rat: an emerging framework for understanding functions of the amygdala. *Trends Neurosci* 20:517–523.

Rainnie DG (1999) Serotonergic modulation of neurotransmission in the rat basolateral amygdala. *J Neurophysiol* 82:69–85.

Rainnie DG, Asprodini EK, Shinnick-Gallagher P (1991a) Inhibitory transmission in the basolateral amygdala. *J Neurophysiol* 66:999–1009.

Rainnie DG, Asprodini EK, Shinnick-Gallagher P (1991b) Excitatory transmission in the basolateral amygdala. *J Neurophysiol* 66:986–998.

Rainnie DG, Asprodini EK, Shinnick-Gallagher P (1993) Intracellular recordings from morphologically identified neurons of the basolateral amygdala. *J Neurophysiol* 69:1350–1362.

Rainnie DG, Mania I, Mascagni F, McDonald AJ (2006) Physiological and morphological characterization of parvalbumin-containing interneurons of the rat basolateral amygdala. *J Comp Neurol* 498:142–161.

Repa JC, Muller J, Apergis J, Desrochers TM, Zhou Y, LeDoux JE (2001) Two different lateral amygdala cell populations contribute to the initiation and storage of memory. *Nat Neurosci* 4:724–731.

Romanski LM, Clugnet MC, Bordi F, LeDoux JE (1993) Somatosensory and auditory convergence in the lateral nucleus of the amygdala. *Behav Neurosci* 107:444–450.

Royer S, Pare D (2003) Conservation of total synaptic weight through balanced synaptic depression and potentiation. *Nature* 422:518–522.

Sah P, Westbrook RF (2008) Behavioural neuroscience: the circuit of fear. *Nature* 454: 589–590.

Sah P, Westbrook RF, Luthi A (2008) Fear conditioning and long-term potentiation in the amygdala: what really is the connection? *Ann N Y Acad Sci* 1129:88–95.

Samson RD, Pare D (2006) A spatially structured network of inhibitory and excitatory connections directs impulse traffic within the lateral amygdala. *Neuroscience* 141: 1599–1609.

Samson RD, Dumont EC, Pare D (2003) Feedback inhibition defines transverse processing modules in the lateral amygdala. *J Neurosci* 23:1966–1973.

Shi C, Davis M (2001) Visual pathways involved in fear conditioning measured with fear-potentiated startle: behavioral and anatomic studies. *J Neurosci* 21:9844–9855.

Sosulina L, Meis S, Seifert G, Steinhauser C, Pape HC (2006) Classification of projection neurons and interneurons in the rat lateral amygdala based upon cluster analysis. *Mol Cell Neurosci* 33:57–67.

Sosulina L, Schwesig G, Seifert G, Pape HC (2008) Neuropeptide Y activates a G-protein-coupled inwardly rectifying potassium current and dampens excitability in the lateral amygdala. *Mol Cell Neurosci* 39:491–498.

Sotres-Bayon F, Cain CK, LeDoux JE (2006) Brain mechanisms of fear extinction: historical perspectives on the contribution of prefrontal cortex. *Biol Psychiatry* 60:329–336.

Sotres-Bayon F, Bush DE, LeDoux JE (2007) Acquisition of fear extinction requires activation of NR2B-containing NMDA receptors in the lateral amygdala. *Neuropsychopharmacology* 32:1929–1940.

Swanson LW, Petrovich GD (1998) What is the amygdale?. *Trends Neurosci* 21:323–331.

Woodruff AR, Sah P (2007) Networks of parvalbumin-positive interneurons in the basolateral amygdala. *J Neurosci* 27:553–563.

15

Hippocampus: Intrinsic Organization

Peter Somogyi

The hippocampus (CA1, CA2, and CA3 areas and the dentate gyrus), together with the subiculum, represents an associational area of the cerebral cortex intimately involved in mnemonic processes. Through its connections with other areas of the temporal lobe, the hippocampus contributes to the encoding, association, consolidation, and recall of representations of the external and internal world in the combined firing rates and spike timing of glutamatergic pyramidal and granule cells. The hippocampus is thought to associate specific life events (items, episodes), on several time scales, in temporally determined firing sequences of neuronal assemblies (see Chapter 16). A single pyramidal cell can be part of several cell assemblies with different partners and contribute to different representations. Pyramidal cell assemblies are thought to be kept together and segregated from other assemblies by the dynamic strengthening and weakening of glutamatergic synaptic weights as well as by GABAergic interneurons. Interneurons generate cell domain and brain state–dependent rhythmic changes in excitability, which are key for the formation, consolidation, and recall of representations. Unsurprisingly, interneurons show intricate spatiotemporal diversity; for example, the CA1 area is served by at least 21 types of resident GABAergic cell. I will attempt to allocate explicit roles for some of them, based on their previously published firing patterns in vivo as observed in identified neurons recorded in anesthetized rats and on their putative equivalents in nonanesthetized animals (Freund and Buzsaki, 1996; Somogyi and Klausberger, 2005; Klausberger and Somogyi, 2008).

Interneurons provide multiple modulatory operations, such as changing threshold, synchronization, gain control, input scaling, and so on, and assist the network in the selection of pyramidal cells for cell assemblies. The spatiotemporal architecture of the network is beginning to be deciphered, but the computational roles of most of the specific synaptic links are not known beyond some general concepts. I concentrate on the spatiotemporal architecture of the *rat* hippocampus, which is by far the most extensively studied species anatomically and physiologically, although genetic engineering methods have provided key system-level insights in the mouse. Events in the hippocampus need to be explained in the context of its interactions with input and output structures (Fig. 15.1). I often emphasise the limits of our knowledge in the hope of generating further interest in exploration and specific tests.

The Hippocampus in the Cortex

A parallel scheme of the hippocampus recognizes a main reciprocal loop formed by projection from mainly layer III pyramidal cells of the entorhinal cortex to the subiculum and CA1 areas, which project back to entorhinal cortex layer V (Fig. 15.1). This primary loop is supplemented by the unidirectional loop of entorhinal mainly layer II projection to the dentate gyrus and the CA2/3 areas, the latter heavily innervating the CA1 area bilaterally (van Strien et al., 2009). The ventral hippocampus and the subiculum also innervate the prefrontal cortex, the perirhinal cortex and the amygdala, in addition to widespread subcortical projections (Amaral and Lavenex, 2007; Cenquizca and Swanson, 2007). In order to analyze the factors that influence the integration of inputs by principal cells in multiple network states (for details, see Chapter 16), it is necessary to have clarity about the cell types (Soltesz, 2006).

Neuron Types

Two individual neurons belong to the same cell type if they deliver the same neuroactive substances to the same range of postsynaptic targets in the same temporal patterns in a brain state–specific manner. Implicit in this definition is the similarity of synaptic input, which allows the same input to output transformation. In an ideal case, the use of this definition requires a knowledge of the inputs, outputs, released neuroactive substances, and temporal behavior of a cell in major brain states before it can be recognized. In most cases, all of this knowledge is not available for individual cells; only population data exist (Soltesz, 2006; Bota and Swanson, 2007). The population of cells, however, is often a mixture of distinct cell types, which means that no circuit-level explanation can be obtained from the population data. Two examples illustrate this point. 1. The population of CA1 pyramidal cells project to at least 10 target areas outside CA1 (Cenquizca and Swanson, 2007), some of them are place cells, increase their firing rate during sharp wave/ripple events, and some but not all of them express calbindin. It is not known whether place cells include both

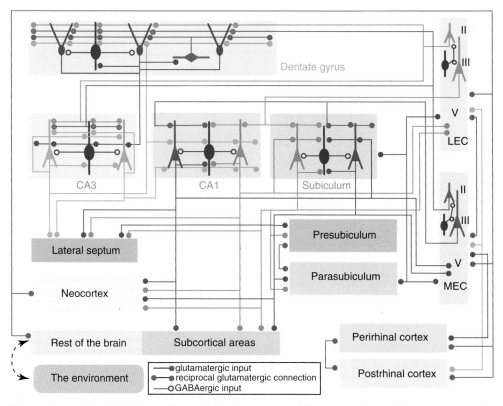

FIGURE **15–1.** Cortical relationships of the hippocampal formation. Simplified schematic diagram based on data reviewed by van Strien et al. (2009). The main features are as follows: 1. Glutamatergic inputs of different origins are segregated on the dendrites. 2. All external and internal glutamatergic pathways (shades of red) innervate both principal cells (pyramidal and granule) and GABAergic interneurons (blue). 3. The dentate gyrus and the CA3 area receive radially segregated layer II inputs from both the medial (MEC) and lateral (LEC) entorhinal cortex. 4. The combined and dentate/CA3 processed MEC and LEC information is transmitted to CA1 pyramidal cells. 5. The MEC and LEC have reciprocal connections with different segments of the CA1 area and the subiculum, which however receive processed information from both MEC and LEC via the CA3 input. 6. Segments of the CA1 area innervate appropriate segments of the subiculum and either the lateral or medial entorhinal cortex. 7. Interareal GABAergic projections from the hippocampus to temporal lobe and the septum, from the perirhinal cortex to LEC, and from the presubiculum to MEC are not shown in order to reduce density. The key input from the septum is not indicated here, but it is described in the text. Only one schematic interneuron is indicated and only some of the pyramidal cells and a limited number of connections are shown. Recurrent connections between pyramidal cells are only shown in the CA3 area. A mossy cell in the dentate gyrus is shown (red diamond). Numbers II, III, and V in MEC and LEC denote layers.

calbindin-negative and calbindin-positive pyramidal cell, whether they project to one, several, or all of the 10 target areas, and if they all can participate in ripple oscillations (see the section "Organizing Principles of GABAergic Interneurons"). 2. The population of parvalbumin (PV)-expressing GABAergic neurons innervate all postsynaptic domains of CA1 pyramidal cells from the axon initial segment to the distal tips of the apical dendritic tuft, but individual cells very specifically terminate on a restricted part of the pyramidal cell surface. Target domain selectivity is accompanied by a similar selectivity for distinct GABA release times during network oscillations. It is not possible to state whether an observed result of perturbing parvalbumin cell activity is a consequence of altering a particular cell type and the resulting changed GABAergic action on a particular subcellular domain of pyramidal cells. These examples illustrate the need for caution in circuit-level interpretation of network events with our rudimentary current knowledge.

Principal Neurons and Glutamatergic Inputs

Glutamatergic inputs of different origin segregate to different parts of the dendritic tree of principal cells (van Strien et al., 2009). All glutamatergic extrinsic and intrinsic hippocampal axons innervate principal cells on dendritic spines and numerically fewer GABAergic interneurons in parallel. The CA1–3 areas have common organizational features and cellular properties with the supragranular neocortical layers, whereas the subiculum shares properties with the infragranular layers. The *CA3 pyramidal cells* (2.5×10^5 per hemisphere in rat) lie at the heart of hippocampal spatiotemporal organization due to their very extensive axonal collateral system (Fig. 15.1), each cell innervating a large and unknown number (estimated in the tens of thousands) of CA3, CA2, and CA1 pyramidal cells and some of them also the dentate gyrus, in a topographically organized manner, bilaterally (van Strien et al., 2009). These synapses on the spines of principal cells show NMDA receptor–dependent synaptic plasticity, and in the CA3 area have been proposed to play the role of pattern completion. This extensive axonal system is thought to form and store associations (memories) in synaptic weights. Pyramidal cells in CA3 and CA2 receive layered, radially segregated lateral and medial perforant path inputs from *layer II glutamatergic cells of the entorhinal cortex* ($\sim 1.2 \times 10^5$ per hemisphere in rat), which also innervate dentate granule cells. In turn, *granule cells* ($\sim 1.2 \times 10^6$ in each hemisphere in rat) provide the mossy fiber input to the proximal apical dendrite of CA3, but not CA2 and CA1 pyramidal cells (Fig. 15.1). Granule cells are exceptional among principal cells in that they do not form recurrent connections with each other and also store GABA in their terminals in addition to glutamate. The CA3 pyramidal cells closest to the hilus may not receive significant entorhinal input, and they may represent a separate cell type under my definition. The dentate hilus contains the *mossy cells* that receive glutamatergic granule cell input, input from each other, and input from hilar-projecting CA3 pyramidal cells (Fig. 15.1). Mossy cells innervate granule cells in the inner dentate molecular layer bilaterally.

Numerically the largest output of CA3 pyramidal cells is the bilateral Schaffer collateral/commissural pathway to *CA1 pyramidal cells* ($\sim 3.8 \times 10^5$ in each hemisphere in rat) terminating in stratum oriens and radiatum in a topographical and sublaminar order (Amaral and Lavenex, 2007; van Strien et al., 2009). In stratum lacunosum moleculare (LM), the distal apical dendritic tufts of pyramidal cells receive medial (in a septotemporal band on the CA3 side) or lateral (in a band toward the subiculum) direct input from *layer III pyramidal cells of the entorhinal cortex* ($\sim 2.5 \times 10^5$ per hemisphere in rat) (Fig. 15.1) and the reuniens nucleus of the thalamus. Thus, CA1 pyramidal cells in different mediolateral positions associate different entorhinal cortical information with parallel processed, and already combined, medial and lateral entorhinal cortical layer II input via CA3 pyramidal cells. How these two major glutamatergic inputs cooperate in sculpting representations and discharging CA1 pyramidal cells is still not resolved, although the issue has been addressed by numerous stimulating hypotheses and models. The rhythmic firing of entorhinal and CA3 pyramidal cells in cooperative action with interneurons provides a rhythmic change in excitability of CA1 pyramidal cells during theta network oscillations. As a result, in each theta cycle (100–150 ms), the highest firing cell assembly at the trough of the pyramidal layer local field potential (LFP) represents the immediate future (next item) in a prospective manner, for example, the next position of the animal (see Chapter 16).

The main outputs of the hippocampal formation are to the entorhinal, perirhinal, and prefrontal cortices; the amygdala; the ventral striatum; the hypothalamus; and the lateral septum (Amaral and Lavenex, 2007; van Strien et al., 2009). Although cognitive roles of the hippocampus are studied very extensively, a major output to the hypothalamus and the measurable functions henceforth may deserve more experimental scrutiny.

SUBCORTICAL INPUTS

The main thalamic input derives from the nucleus reuniens innervating the molecular layer of the subiculum and stratum LM of the CA1 area. Its role in hippocampal activity states has not been well defined. The medial septum sends major cholinergic and GABAergic input to the hippocampus and is often considered as a pacemaker for one of the major network states, the theta rhythm. The cholinergic input synaptically innervates both interneurons and pyramidal cells and also exerts its action via nonsynaptic muscarinic and nicotinic receptors. The rhythmically firing septal GABAergic neurons only innervate GABAergic interneurons of diverse types and receive GABAergic input from hippocampo-septal neurons (Freund and Buzsaki, 1996).

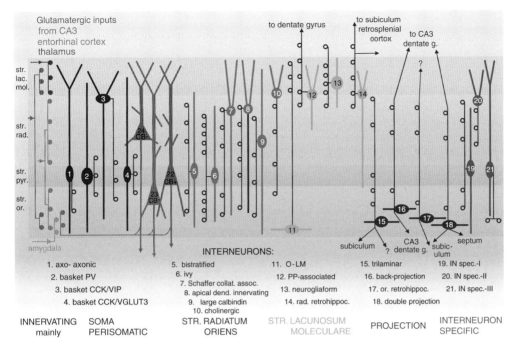

FIGURE 15–2. Three types of pyramidal cell are served by at least 21 types of GABAergic interneuron in the CA1 area. The laminar termination of five glutamatergic inputs is indicated on the left (boutons, filled circles). Pyramidal cells are red; those located closer to str. radiatum express calbinin (CB+); the larger more loosely arranged cells toward str. oriens are mainly calbindin negative as are those in str. radiatum that project to the accessory olfactory bulb. The somata and dendrites of interneurons are blue, cells 1–18 innervate pyramidal cells and interneurons to some degrees; cells 19–21 innervate mainly or exclusively other interneurons. Axons and the main synaptic terminations (boutons, open circles) are dark blue. Note the association of the output synapses of different sets of interneuron with the perisomatic region of pyramidal cells (*left*), and either the Schaffer collateral/commissural or the entorhinal pathway termination zones (*right*), respectively. Projection GABAergic neurons send long-range axons to related areas of the temporal lobe or the septum. Not all interneurons fit these categories (not shown). g., gyrus; IN spec., interneuron specific; lac. mol., lacunosum moleculare; O-LM, oriens lacunosum-moleculare; or., oriens; PP, perforant path; pyr., pyramidale; retrohippoc., retrohippocampal projecting; str., stratum.

Individual septal GABAergic neurons show various firing phases relative to the hippocampal theta rhythm and are thought to be responsible for the rhythmic inhibition of interneurons and the resulting rhythmic disinhibition of pyramidal cells (Borhegyi et al., 2004). A major outstanding question is whether single septal GABAergic neurons innervate only those GABAergic neurons that share similar firing phases during theta. If this were the case, separate septal neuron populations with firing preference for one of at least four theta phases (trough, ascending, descending, and peak) would be expected. Alternatively, or supplementing the above mechanism, a single septal GABAergic axon may innervate interneurons with diverse theta firing phases, and further intra-hippocampal interactions may produce the shifts of the firing phases of different cell types. Another unresolved issue is the degree of divergence of single septal GABAergic axons to different areas of the hippocampus and the temporal lobe.

Like many other cortical areas, the hippocampal formation receives noradrenergic, dopaminergic, serotonergic, histaminergic, and some orexyn-containing subcortical inputs, which cannot be reviewed here. Within the serotonergic input, in addition to activating G protein–coupled 5-HT receptors on principal cells and some interneurons, a subset of axons from the median raphe nucleus strongly innervate CCK/calbindin-containing GABAergic neurons through excitatory 5-HT3 ionotropic receptors (Freund and Buzsaki, 1996). It is not yet clear how 5-HT3 receptor activation contributes to interneuronal network roles.

Organizing Principles of GABAergic Interneurons

The cooperative action of interneurons and glutamatergic inputs results in rhythmic change of excitability during theta, gamma, and ripple frequency network oscillations, which provides windows for coordinated pyramidal cell discharge enabling the formation of cell assemblies and representations. The term *interneuron* is used both for cells with exclusively local axons and for those that in addition project outside the area of their cell body location. Most interneurons receive inputs from the same extrinsic afferents that innervate their target principal cells, as well as, in a recurrent manner, from the principal cells. The weight of these two inputs may vary significantly from cell type to cell type. It is customary to describe extrinsic inputs as feedforward and recurrent inputs as feedback, but the presence of such connections does not mean that the action of one or the other alone leads to firing of the interneuron at any one time in the intact system. Interneurons having exclusively feedforward (e.g., neurogliaform cell) or mainly feedback (e.g., O-LM cell) inputs are exceptional. During oscillations in a rhythmic, cyclical system with both excitatory and inhibitory inputs patterning discharge, it is not possible to delineate pure feedback or feedforward influences.

During a single theta cycle, a pyramidal cell assembly fires at the highest rate at the trough of pyramidal layer LFP, when perisomatic inhibition is minimal but GABA release to dendrites is maximal. However, other pyramidal cell assemblies representing past and future items in temporal sequences also fire at a lower rate at earlier or later theta phases when the balance of perisomatic and dendritic GABA action is different from that at the trough of the LFP. The sequential firing of assemblies representing past, present, and future items is then replayed in a time-compressed manner during high-frequency ripple oscillations (see Chapter 16). Interneurons make multiple

TABLE 15–1. Distribution of Interneuron Types across Hippocampal Areas and Some Molecules That in Combination Are Useful for Their Grouping and/or Recognition

GABAergic Interneuron

Group	No.	Suggested Name (1)	Localized Proteins Useful for Recognition (2, 3)	CA1	CA3	Dentate Gyrus	To Mossy Cells
Soma, perisomatic innervating cells	1	Axo-axonic	Parvalbumin, sd, a; GABA-A-R-a1 low, sdm	P	P	P	P
	2	Basket	Parvalbumin, sd, a; GABA-A-R-a1 high, sdm	P	P	P	not
	3	Basket	CCK, s, a; VIP, s, a; CB1, a	P	P	P	P
	4	Basket	CCK, s, a; VGLUT3, a; CB1, a	P	P	P	P
Stratum radiatum and oriens innervating cells	5	Bistratified	Parvalbumin, sd SM, s, a; NPY s, a	P	NP	?	?
	6	Ivy	nNOS, sd, a; NPY, s, a GABA-A-R-a1, sdm	P	P	?	?
	7	Schaffer collateral-associated	CCK, s, a; calbindin, s; CB1, a	P	?	NA	NA
	8	Apical dendrite innervating	CCK, s, a; VGLUT3, a; CB1, a	P	?	NA	NA
	9	Large calbindin	Calbindin, sd, a	P	?	?	?
	10	Cholinergic	ChAT, sd, a; vAChT, a	P	P	?	?
Stratum lacunosum moleculare innervating cells	11	O-LM	Parvalbumin, sd, a, SM, s, a; mGluR1a strong, sdm mGluR7a, IT	P	P	P, HIPP (4)	?
	12	Perforant path associated	CCK, s, a; calbindin, s; CB1, a	P	NP	NP, MOPP (5)	NA
	13	Neurogliaform	NPY, s; nNOS, s; alpha-actinin-2, s	P	P	NP	?
Projection cells	14	Radiatum retrohippocampal	M2 receptor, sdm; mGluR1a weak, sdm	P	?	?	?
	15	Trilaminar	M2 receptor strong, sdm; mGluR8a, IT	P	P	?	?

(*continued*)

Table 15–1. Continued

GABAergic Interneuron

Group	No.	Suggested Name (1)	Localized Proteins Useful for Recognition (2, 3)	CA1	CA3	Dentate Gyrus	To Mossy Cells
	16	Backprojection cell	?	P	NP	?	?
	17	Oriens retrohippocampal	Calbindin, sd; M2 rec., sdm	P	?	?	?
	18	Double projection	Calbindin, sd; SM, s; NPY, s; mGluR1a, sdm; mGluR7a, it	P	P	?	?
Interneuron-specific cells	19	Interneuron specific I.	Calretinin, sd; mGluR1a weak, sdm	P	NP	?	?
	20	Interneuron specific II.	VIP, s, a; mGluR1a weak, sdm	P	NP	?	?
	21	Interneuron specific III.	VIP, s, a; calretinin, sd, a; mGluR1a weak, sdm	P	NP	?	?
	22	Enkephalin expressing	ENK, s; VIP, s; mGluR1a weak, sdm	P	NP	?	?
Regional projection cells	23	Densely spiny hippocampal-septal	Calretinin, sd; NPY, s; SM, s	NA	P	P	?
	24	Large nNOS positive	nNOS strong, sd, a; NPY, s, a	P	?	?	?
	25	OML, outer molecular layer (6)	?	NA	NA	P	NA
CA3/dentate specialized cells	26	CA3 hilar projection (7)	SM, s; NPY, s; mGluR1a, sdm	NP	NP	P	?
	27	Mossy fiber associated	CCK, s, a; CB1, a	NA	P	?	NP
	28	HICAP, hilar commissural associational path associated	?	NA	NA	P	?

P, strong evidence as a cell type; NP, suggestive evidence, not proven; ?, not known; NA, not applicable; sdm, somato-dendritic membrane; sd, soma and dendrite; a, axon; it, input terminals on soma and dendrite.

Notes. Molecular combinations alone without information on synaptic input–output relationships are weak predictors of a cell type. The presence of some molecules in a single cell is mutually exclusive, but this is not indicated here. The same names and numbering are used as in Figure 15.1; additional cell types are added here. The CA2 region is not listed separately, as the axons of many interneurons in the CA3 and to a lesser extent those in the CA1 area also innervate CA2. Interneurons mostly innervating only CA2 pyramidal cells also exist. Due to restrictions on references, individual papers cannot be cited describing each result. Numbers in parentheses represent the following: (1) Cell types with very partial characterization may be absorbed into other cell types with further analysis; (2) The subcellular locations of the highest concentration of molecular markers are indicated, but in some cases they may be present in other compartments as well; (3) The listed molecules may not be detectable in every individual member of a cell type; (4) I suggest that the HIPP cell (hilar perforant path associated) is homologous to the O-LM cells in the CA1 and CA3 regions in its hippocampal spatiotemporal position and role, although, unlike the O-LM cells, HIPP cells also project to the contralateral dentate gyrus; (5) The MOPP cell (molecular layer perforant path associated) may correspond to perforant path–associated cells of the CA1 area; (6) This neuron projects from the dentate outer molecular layer to the subiculum across the fissure; (7) This neuron projects from the CA3 area and the hilus to the septum with no other known long-range projection.

and essential contributions to temporal order, and some predictions can be made about which cell types may be responsible for particular actions.

The main features of organization are as follows:

1. Interneurons are highly selective in their postsynaptic target domains, reflected in characteristic axonal shapes and laminar patterns (Fig. 15.2). In addition to GABA, they release cell type–specific neuroactive substances, such as neuropeptides, nitric oxide, and endocannabinoids. Each layer of the hippocampus contains the cell bodies of multiple groups of interneurons with different domain selectivity and molecular composition.

2. Interneurons having similar axons fire in a stereotyped brain and network state–dependent manner that differs from the firing of interneurons with different axons in at least one network state; this explains the need for independent cell types.

3. The axon initial segment of all pyramidal, granule, and mossy cells receives GABAergic innervation from parvalbumin-positive axo-axonic cells that generally do not innervate other domains of principal cells. Only axo-axonic cells provide significant innervation to the axon initial segment of principal cells. This design is unique to the cerebral cortex, and it points to a specialized control of action potential generation and backpropagation. Axo-axonic cells are inhibited during sharp waves, but their inhibitory inputs are unknown.

4. The entire somato-dendritic domain is covered by two distinct sets of GABAergic synapses from neurons expressing either PV or cholecystokinin (CCK), but there are additional cell types that innervate only restricted domains and express neither of these molecules (Fig. 15.2). The PV- or CCK-expressing cells release GABA at different preferred phases during theta oscillations.

5. Interneurons expressing CCK have high levels of CB1 cannabinoid receptors on their axons and terminals, which suppress GABA release upon the postsynaptic release of endocannabinoids evoked by depolarization and calcium entry. Therefore, a firing postsynaptic neuron suppresses its GABAergic input from CCK-expressing cells, which continue to release GABA to other innervated cells that do not fire. Such a selective reduction in inhibition of active cells increases contrast between active and inactive cell assemblies.

6. The soma and proximal dendrites of principal cells receive GABAergic innervation from three types of basket cells, which express PV, CCK/VIP, or CCK/VGLUT3 (Fig. 15.2). The consequences of VIP or VGLUT3 expression in the terminals of CCK-positive GABAergic cells and possible differences in activity between these cells are not known.

7. Interneurons innervating the dendritic domain show the greatest diversity, pointing to a sophisticated GABAergic pre- and postsynaptic regulation of dendritic inputs and excitability (Fig. 15.2). Different types of interneuron innervating the same dendrites may cooperate by synchronized action, or they may provide time-differentiated inputs in a given network state.

8. In the CA1 area, dendrite-innervating interneurons associate their synapses mainly with one of the major glutamatergic input zones, the Schaffer collateral/commissural pathway in stratum oriens and radiatum (bistratified cell, apical dendrite innervating cell, ivy cell, some projection cells), or the entorhinal/thalamic input zone in stratum LM (O-LM cell, perforant path associated cell, neurogliaform cell) (Fig. 15.2). This pairing of glutamatergic and GABAergic inputs provides a basis for pathway-specific regulation of glutamate release by presynaptic GABA and/or neuropeptide receptors.

9. The glutamatergic input zone segregation of GABAergic axons seems to hold for the dentate gyrus, as HIPP, MOPP, and OML cells innervate the medial and lateral entorhinal input layers (Freund and Buzsaki, 1996; Ceranik et al., 1997). In turn, the HICAP cell and some CCK-expressing cells innervate the associational input zone innervated by mossy and CA3 pyramidal cells.

10. Less data are available on dendritic GABAergic innervation in the CA3 area. Interneurons have been shown having axonal output associated with stratum oriens and radiatum (possible bistratified cell), which are the CA3 pyramidal axonal input layers, or the mossy fiber input zone in stratum lucidum (mossy fiber associated cell), or the entorhinal input zone (O-LM cell).

11. Some GABAergic neurons with cell bodies in CA1 also innervate the dentate gyrus, their axons freely crossing the hippocampal fissure, pointing to shared modulation of pyramidal and granule cells in as yet undefined ways (Fig. 15.2).

12. Long-range projection GABAergic cells either cut across the boundaries of hippocampal areas (backprojection cells) or project outside the hippocampus proper, particularly to the subiculum (oriens-retrohippocampal cells, double projection cells, trilaminar cells, enkephalin-expressing cells). Double projection cells innervating septal GABAergic cells, as well as retrohippocampal areas, also innervate pyramidal cells in stratum oriens and radiatum showing similar spike timing to bistratified cells both during theta and ripple oscillations (Fig. 15.2).

13. All GABAergic cells innervating pyramidal cells, except the axo-axonic cell, also make synapses with other interneurons, which may form a few percent (basket cells, bistratified cells) or up to half (trilaminar cell, enkephalin-expressing cell) of their postsynaptic targets.

14. Three types of interneuron-specific GABAergic cell were reported to innervate only other interneurons and express calretinin and/or VIP in the CA1 area (Freund and Buzsaki, 2006) (Fig. 15.2). Their network state–dependent firing patterns are not known.

15. Interneurons are electrically coupled through somato-dendritic gap junctions. The selectivity of gap junctional coupling between different interneuron types remains to be tested; examples of coupling among PV-expressing, CCK-expressing, or neurogliaform cells have been documented.

INTERNEURON TYPE-SPECIFIC CONTRIBUTION TO RHYTHMIC
CHANGE IN EXCITABILITY

Assuming that information is stored in the synapses of principal cell spines, recalled, and carried in their firing, it would be useful to explain how inputs are integrated to achieve the firing of hippocampal principal cells. Is the integration of several excitatory input pathways needed, or can one pathway produce suprathreshold responses in a given network state or phase? Which interneurons contribute to the temporal structure of activity at a given time? A large amount of lesion, electrical stimulation, multisite recording, and modeling work have addressed these questions and is available in numerous reviews. Here, I consider possible mechanistic conditions necessary for pyramidal cell assembly activation in two well-recognized network states (Fig. 15.3).

The theta oscillatory state (4–10 Hz) is associated with movement of the animal and REM sleep; it is thought to enable encoding and recall of information and also modulates the amplitude of simultaneously occurring gamma oscillations (30–90 Hz) (O'Keefe and Nadel, 1978). Principal cells fire at very low average frequencies, and they are silent for long periods, indicating tonic inhibitory suppression. This has been demonstrated by in vivo intracellular recording and by the presence of theta-on interneurons (Freund and Buzsaki, 1996). When a pyramidal cell fires, for example, because the animal entered the cell's place field, it can fire with high-frequency bursts of action potentials modulated by the theta rhythm and phase precessing on subsequent theta cycles. The firing of entorhinal cortical layer II and III pyramidal cells is also theta modulated, some of them show phase precession, and their outputs to CA3/dentate and CA1, respectively, are phase shifted. Furthermore, theta modulated CA3 pyramidal cell firing is also phase shifted relative to average CA1 pyramidal cell firing. Therefore, there is no simple explanation of the relative contribution of the different excitatory pathways to the firing of a single principal neuron (see Chapter 16 for hypothesis). It is clear, however, that principal cells are inhibited periodically with their lowest firing probability at the peak of the pyramidal layer theta LFP. Major contributors to this inhibition are the axo-axonic cell with maximum firing probability at the peak of theta, as observed in the anesthetized rat. Theta modulation of basket cells has a broad tuning; the more numerous PV basket cells have maximum firing probability at the descending phase, whereas CCK basket cells fire at the ascending phase of theta. The combined output of basket cells weighted by their relative numbers also has a maximum at the peak of theta. Thus, a cooperative action of the four perisomatically terminating GABAergic cells reduces pyramidal cell firing at the peak of theta, the axo-axonic cells probably playing the major role (Fig. 15.3) (Klausberger and Somogyi, 2008).

In contrast to the perisomatic innervating cells, dendrite innervating bistratified, O-LM, ivy and projection GABAergic cells fire maximally at the

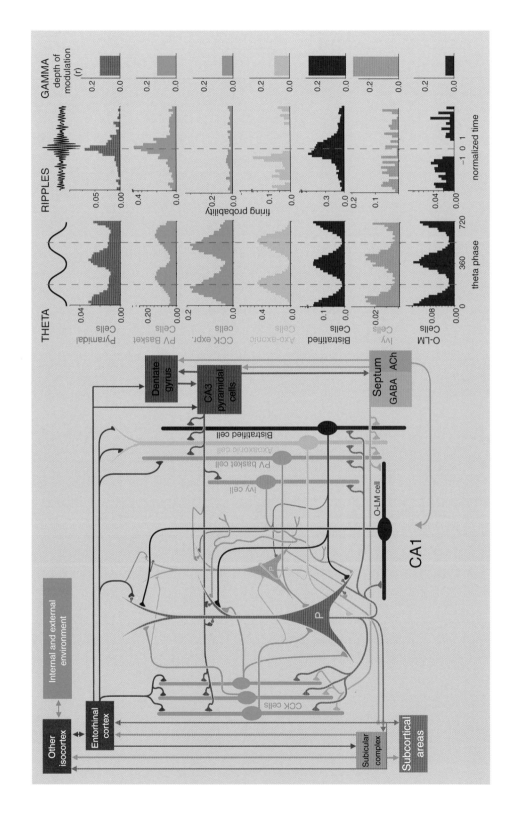

FIGURE **15–3.** Spatiotemporal interaction between pyramidal cells and eight types of interneuron during network oscillations. Schematic summary of the main synaptic connections of pyramidal cells (P), three types of CCK-expressing cells, ivy cells and PV-expressing basket, axo-axonic, bistratified and O-LM interneurons. The firing probability histograms are averages from several cells of the same type; note different scales for the Y axis. Interneurons innervating different domains of pyramidal cells fire with distinct temporal patterns during theta and ripple oscillations. Their spike timing is coupled to field gamma oscillations to varying degrees (averages of several cells of each type) The same somatic and dendritic domains receive differentially timed input from several types of GABAergic interneuron, for example, CCK- and PV-expressing cells. Note that pyramidal cell firing probability is lowest at the peak of the pyramidal layer *theta* local field potential (LFP), when axo-axonic cells, which have the highest mean peak firing probability, fire maximally and the sum of CCK basket and PV basket cell firing probabilities are maximal. Therefore, the cooperative action of these three cell types causes a rhythmic lowering of pyramidal cell firing at the peak of theta and an increase in LFP *gamma oscillation*. Note also the similar theta phase coupling of dendrite innervating cells, roughly counter-phased with the perisomatic innervating cells, and the high gamma coupling of bistratified and ivy cells innervating basal and small oblique pyramidal cell dendrites. During *ripple oscillations*, axo-axonic cell GABA release to the axon initial segments is withdrawn, allowing maximal pyramidal cell discharge synchronized by PV basket and bistratified cells. Synaptic and electrical connections between interneurons are not shown for clarity, but most interneurons innervate other interneurons in addition to pyramidal cells (Somogyi and Klausberger, 1995; Freund and Buzsaki, 1996; Klausberger and Somogyi, 2008).

trough of theta when pyramidal cells also have the highest firing probability (Fig. 15.3). The dendritic GABAergic inputs increase the threshold, provide gamma frequency phasing of excitatory inputs, synchronize dendritic spikes, and scale glutamatergic inputs via post- and presynaptic receptors. An example of the latter is NPY expression and likely release by bistratified and ivy cells acting on inhibitory presynaptic Y2 receptors on CA3 pyramidal cell terminals. The theta modulated GABAergic input to the most distal dendrites in stratum LM by O-LM cells contributes to the phase reversal of dendritic oscillations in the apical dendrites relative to the soma. The maximal discharge of double projection cells at the trough of theta innervating septal GABAergic neurons may help to set up the reciprocal oscillatory loop with the septum. Long-range theta modulated GABAergic output to the subiculum, the retrosplenial cortex, and other related cortical areas by oriens- and radiatum-retorhippocampal cells contributes to the coherence of oscillations across different areas.

The second well-defined network state is the synchronized discharge of pyramidal cells producing the *sharp wave* field potential (see Chapter 16) most easily seen in stratum radiatum of the CA1 area and an associated ripple oscillation (140–200 Hz) in the pyramidal cell layer for about 50–100 ms (O'Keefe and Nadel, 1978). The sharp wave/ripple event is driven by the synchronous discharge of CA3 pyramidal cells. The frequency of the event is modulated by the cortical up and down states, being more frequent in the first half of the up state. Sharp wave/ripple events also occur during awake consummatory behavior and in short gaps in theta activity. What initiates sharp wave ripple events is not known, but Buzsaki (1989) suggested that a sudden drop in inhibition allows initiator pyramidal cells to trigger the replay of cell assembly sequences via glutamatergic synaptic links potentiated during encoding. Indeed, withdrawal of inhibition by the silencing of axo-axonic and O-LM GABAergic neurons has been demonstrated during ripple oscillations in CA1 (Klausberger and Somogyi, 2008). Some CCK-expressing and other GABAergic cells are also silent during some of the ripples (Fig. 15.3). Particularly, the withdrawal of GABAergic inhibition from the axon initial segment may be crucial and causally linked to the development of synchronized pyramidal cell discharge in CA1. It remains to be tested whether such a disinhibitory mechanism by the silencing of specific interneurons might also operate in other cortical areas involved in the sharp wave/ripple event. During the event, pyramidal cell spikes are phase locked to the trough of the ripple, which results in synchronized glutamate release to downstream targets. Such synchronization is probably achieved by the ripple frequency phase–locked firing of PV-expressing basket and bistratified, as well as by trilaminar, double projection, and some CCK-expressing cells (Klausberger and Somogyi, 2008). The highest firing probability of PV basket and bistratified cells is just after the pyramidal cells on the rising phase of the ripple cycle. Thus, pyramidal cell maximal firing at the trough is linked to a ripple periodic reduction in GABA release to the soma, the small oblique dendrites in stratum radiatum and the basal dendrites. The rhythmic phase-locked firing of the PV basket and bistratified interneuron is consistent with their ability to temporally structure pyramidal cell discharge with millisecond accuracy.

The participation of the dendrite-innervating bistratified cells is of particular significance in view of the ability of small oblique and basal dendrites to generate sodium spikes (Freund and Buzsaki, 1996; Losonczy et al., 2008). An oblique dendrite may receive about 600 glutamatergic and 25 GABAergic synapses; the latter number is about a quarter of GABAergic synapses on the entire soma (Megias et al., 2001) and comes from several cell types. The bistratified cells, which start to increase their firing during ripples earlier than the pyramidal cells that they innervate, are in prime position to synchronize dendritic electrogenesis within the same cell and between cells during ripples. The main glutamatergic inputs of bistratified cells are from CA3 and CA1 pyramidal cells (Ali et al., 1998). Bistratified cells probably release somatostatin and NPY with inhibitory effects; the latter, suppressing glutamate release from CA3 terminals via presynaptic Y2 receptors, which may contribute to the termination of ripples. Interestingly, among interneurons in the CA1 area, bistratified cells are most strongly coupled to gamma-frequency LFP oscillations (Klausberger and Somogyi, 2008), the latter initiated by CA3 pyramidal cell discharge. Thus, bistratified cells mediate CA3-driven fast rhythmic GABAergic phasing of basal and small oblique dendrite excitability at both gamma and ripple frequencies. During ripples, bistratified cells are joined by trilaminar and double projection cells, which also discharge at ripple frequencies and innervate basal and oblique dendrites. This is a further example of cooperative action by several types of interneuron during a particular network state.

The above examples illustrate the need for the existence of numerous specific types of interneuron supporting the network to deliver temporally modulated pyramidal cell activity patterns. Complex though it may seem, our understanding of the system is still sketchy and most components have been described only qualitatively. For example, the inputs of most interneurons are poorly known and some of them are completely unknown; consequently, many of the proposals mentioned earlier are speculative and require imaginative experimental testing, which provides fertile ground for discovery.

REFERENCES

Ali AB, Deuchars J, Pawelzik H, Thomson AM (1998) CA1 pyramidal to basket and bistratified cell EPSPs: dual intracellular recordings in rat hippocampal slices. *J Physiol (Lond)* 507:201–217.

Amaral D, Lavenex P (2007) Hippocampal neuronatomy. In: Andersen P, Morris R, Amaral D, Bliss T, O'Keefe J, eds. *The Hippocampus Book*, pp. 37–114. New York: Oxford University Press.

Borhegyi Z, Varga V, Szilagyi N, Fabo D, Freund TF (2004) Phase segregation of medial septal GABAergic neurons during hippocampal theta activity. *J Neurosci* 24:8470–8479.

Bota M, Swanson LW (2007) The neuron classification problem. *Brain Res Rev* 56:79–88.

Buzsaki G (1989) Two-stage model of memory trace formation: a role for 'noisy' brain states. *Neuroscience* 31:551–570.

Cenquizca LA, Swanson LW (2007) Spatial organization of direct hippocampal field CA1 axonal projections to the rest of the cerebral cortex. *Brain Res Rev* 56:1–26.

Ceranik K, Bender R, Geiger JRP, Monyer H, Jonas P, Frotscher M, Lubke J (1997) A novel type of GABAergic interneuron connecting the input and the output regions of the hippocampus. *J Neurosci* 17:5380–5394.

Freund TF, Buzsaki G (1996) Interneurons of the hippocampus. *Hippocampus* 6:347–470.

Klausberger T, Somogyi P (2008) Neuronal diversity and temporal dynamics: the unity of hippocampal circuit operations. *Science* 321:53–57.

Losonczy A, Makara JK, Magee JC (2008) Compartmentalized dendritic plasticity and input feature storage in neurons. *Nature* 452:436–441.

Megias M, Emri Z, Freund TF, Gulyas AI (2001) Total number and distribution of inhibitory and excitatory synapses on hippocampal CA1 pyramidal cells. *Neuroscience* 102:527–540.

O'Keefe J, Nadel L (1978) *The Hippocampus as a Cognitive Map*. Oxford, England: Oxford University Press.

Soltesz I (2006) *Diversity in the Neuronal Machine. Order and Variability in Interneuronal Circuits*. Oxford, England: Oxford University Press.

Somogyi P, Klausberger T (2005) Defined types of cortical interneurone structure space and spike timing in the hippocampus. *J Physiol* 562:9–26.

van Strien NM, Cappaert NL, Witter MP (2009) The anatomy of memory: an interactive overview of the parahippocampal-hippocampal network. *Nat Rev Neurosci* 10: 272–282.

16

Hippocampus: Network Physiology

György Buzsáki

Information is propelled mainly forward in the multisynaptic feedforward loops of the entorhinal-hippocampal system, with each stage adding unique features (Fig. 16.1). A general principle of this complex circuit is that strongly recurrent excitatory networks (i.e., layers 2 and 5 of entorhinal cortex, CA3) are sandwiched between layers with largely parallel organization (i.e., layer 3 of entorhinal cortex, dentate gyrus, CA1). The advantage of such organization is that in successive layers the neuronal representations can be iteratively *segregated* (at parallel stages) and *integrated* (at recursive stages). These operations require time for local computation before the results of neuronal processing are transmitted forward to the next computational stage. The speed of local processing and layer-to-layer transfer is largely determined by the "traffic-controlling" inhibitory interneurons, or more precisely, by the temporal dynamics set by the interactions between principal cells and interneurons (Klausberger and Somogyi, 2008; see also Chapter 18 of this volume). Such dynamics, often in the form of network oscillations, allow for intrahippocampal processing and enable the hippocampal system to communicate effectively with various domains of the neocortex in a temporally discrete manner. Oscillations, in general, impose a spatiotemporal structure on neural ensembles within and across networks. In each cycle, recruitment of principal neurons is temporally protracted and terminated by the build-up of inhibition. During the short periods of fast oscillations, only a small group of local neurons can be recruited, whereas slow oscillations allow for the involvement of large neuronal pools across structures. Because slower oscillations can phase-bias the power of faster oscillations, slow oscillations can coordinate local circuit computations. The periods of the various network patterns, therefore, constrain the time devoted for local processing and the speed of information transmission through multiple layers.

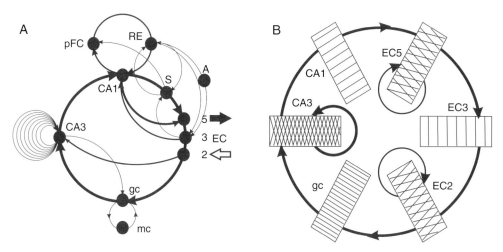

FIGURE 16–1. (*A*) Multiple loops of the hippocampal-entorhinal (EC) circuits. (*B*) A simplified illustration of the wiring and expected computation in successive layers of the EC-hippocampus (mainly) feedforward loop. Parallel-organized local circuits (gc, granule cells; CA1 and layer 3 [EC3] pyramidal neurons) alternate with recurrent circuits characterized by varying densities of recurrent connections (CA3 and principal cells of layers 5 and 2 of EC). Such organization allows for repeated segregation and integration of information in successively coupled parallel and recurrent circuits, respectively.

TEMPORAL PACKAGING OF NEURONAL INFORMATION BY THETA, GAMMA, AND "RIPPLE" OSCILLATIONS

Three major network patterns dominate the hippocampal system: theta (4–10 Hz), sharp waves and associated ripples (140–200 Hz), and gamma (30–100 Hz) oscillations. These patterns also define states of the hippocampus, with the theta state associated with exploratory ("preparatory") movement and REM sleep, while intermittent sharp waves mark immobility, consummatory behaviors, and slow-wave sleep. These two competing states also largely determine the main direction of information flow, with neocortical-hippocampal transfer taking place mainly during theta oscillations and hippocampo-neocortical transfer during sharp waves. These two states also affect the regularity of gamma oscillations.

Gamma frequency oscillations are present in all cortical and other structures where fast inhibition is provided by soma-targeting interneurons. In the simplest case, an interconnected network of basket interneurons can generate sustained gamma oscillations, provided that their depolarization and spiking are secured by some means (such as subcortical neurotransmitters). In the intact brain, gamma oscillations are mainly generated by the interaction between principal cells and interneurons. In both scenarios, the frequency of oscillations is largely determined by the time course of GABAa receptor–mediated inhibition. Neurons that discharge within the time period of the gamma cycle (10–30 msec) define a cell assembly (Harris et al., 2003).

Given that the membrane time constant of pyramidal neurons in vivo is also within this temporal range, recruiting neurons into this assembly time window is the most effective mechanism for discharging the downstream postsynaptic neuron(s) on which the assembly members converge. Although gamma oscillations can emerge in each hippocampal region, they can be coordinated across regions by either excitatory connections or by long-range interneurons. The CA1–CA3 regions appear to form a large coherent gamma oscillator, due to the interaction between the recurrently excited CA3 pyramidal cells and their interneuron targets in both CA3 and CA1 regions. This "CA3 gamma generator" is normally under the suppressive control of the entorhinal-dentate gamma generator, and its power is enhanced severalfold when the entorhinal/dentate input is attenuated (Bragin et al., 1995). Entorhinal circuits generate their own gamma oscillations by largely similar rules and these (generally faster) rhythms can be transferred and detected in the hippocampus.

The extracellularly recorded *theta oscillation* is the result of coherent membrane potential oscillations across neurons in all hippocampal subregions (Buzsaki, 1989; O'Keefe and Recce, 1993; Montgomery et al., 2009). Theta currents derive from multiple sources, including synaptic currents, intrinsic currents of neurons, dendritic Ca^{2+} spikes, and other voltage-dependent membrane oscillations. Theta frequency modulation of perisomatic interneurons provides an outward current in somatic layers and phase-biases the power of gamma oscillations, the results of which is a theta-nested gamma burst. Excitatory afferents form active sinks (inward current) at confined dendritic domains of the cytoarchitecturally organized layers of all regions. Since each layer-specific input is complemented by one or more families of interneurons with similar axonal projections (Freund and Buzsáki, 1996; also see Chapter 15 of this volume), such layer-specific inhibitory dipoles can compete with the excitatory inputs. The resulting rich consortium of theta generators in hippocampal and parahippocampal regions is coordinated by the medial septum and a network of long-range interneurons. Although theta oscillations are generally coherent throughout the hippocampal system, the power, coherence, and phase of theta oscillators can fluctuate significantly in a layer-specific manner as a function of overt behavior and/or the memory "load" to support task performance (Montgomery et al., 2009).

When the subcortical modulatory inputs decrease in tone, theta oscillations are replaced by large-amplitude field potentials, or *"sharp waves" (SPWs)*. SPWs are initiated by the self-organized population bursts of the CA3 pyramidal cells. The CA3-induced depolarization of CA1 pyramidal cell dendrites is reflected by an extracellular negative wave, that is, the SPW, most prominently in stratum radiatum. These CA1 SPWs are associated with fast-field oscillations (140–200 Hz), or *"ripples"* confined to the CA1 pyramidal cell layer (O'Keefe and Nadel, 1978; Buzsaki et al., 1992). At least two factors contribute to the field ripples. First, the rhythmic positive "wave" components reflect synchronously occurring oscillating inhibitory postsynaptic potentials (IPSPs) in the pyramidal cells because the CA3–CA1 pyramidal cells strongly drive perisomatic interneurons during the SPW. Second, the synchronous discharge of pyramidal neurons generates repetitive "mini population spikes" that are responsible for the spike-like appearance of the troughs of ripples in the pyramidal cell layer. In the time window of SPWs, 50,000–100,000 neurons discharge synchronously in the CA3–CA1–subicular complex–entorhinal axis. The population burst is characterized by a three- to five-fold gain of network excitability in the CA1 region, preparing the circuit for synaptic plasticity. SPWs have been hypothesized to play a critical role in transferring transient memories from the hippocampus to the neocortex for permanent storage (Buzsaki, 1989), and this hypothesis is supported by numerous experiments demonstrating that the neuronal content of the SPW-ripple is largely determined by recent waking experiences (Wilson and McNaughton, 1994). Furthermore, selective elimination of SPW ripples during postlearning sleep impairs memory consolidation (Girardeau et al., 2009).

PROPAGATION OF ACTIVITY THROUGH MULTIPLE STAGES OF THE HIPPOCAMPUS IS STATE DEPENDENT

Propagation of neuronal signals across multiple anatomical regions is frequently explained by "box-and-arrow" illustrations, where large populations of neurons in each layer or region are replaced by a single "mean neuron," representing a homogeneously behaving population (Fig. 16.1). It is tempting to designate circumscribed and specific computations for each layer or region. However, such a simplified picture cannot adequately describe information processing and propagation because a good deal of computation takes place at the interface between layers and because global hippocampal states exert an exquisite control on local circuit computations. Furthermore, representation of an initiating event is not merely transferred from one layer to the next but changes progressively, largely determined by top-down influences and global states. Depending on the previous history between the brain and the event, each layer may add unique information to the representation.

A critical factor that limits the addition of novel information to the representation is time. For example, a strongly synchronous input, such as an artificial electrical pulse or an epileptic interictal spike, may propagate through multiple layers at a high speed, limited primarily by axon conduction and synaptic delays. However, propagation of physiological information rarely occurs at this high speed. The fastest physiological speed of spike transmission in hippocampal networks occurs during SPW-ripples. During SPWs, the CA3-initiated population burst propagates through the CA1, subiculum, entorhinal layer 5, and layers 2/3 in just 15 to 20 msec. While the pattern is propelled through these feedforward layers, the large SPW-related gain of excitation in the hippocampus is balanced by the progressive build-up of inhibition in successive layers. In layer 5, inhibition balances the SPW-induced excitation and inhibition in layers 2/3 overcomes the excitation. Because of the progressively increasing inhibitory gain in successive layers, SPW activity rarely reverberates in the hippocampal-entorhinal cortex loop, although multiple reverberations can occur in the epilepsy.

The situation is dramatically different from SPWs in the theta oscillation state. The delay between the population peaks in the entorhinal input layers (layers 2 and 3) and their respective target populations in dentate/CA3 and CA1 is severalfold longer during the theta state than during SPW. Typically, the delays correspond to approximately one half theta cycle (50–70 msec). Importantly, the current sinks in dentate/CA3 and CA1 pyramidal cell dendrites occur within 10–15 msec after the peak of the population activity in entorhinal layers 2 and 3, as expected by the conduction velocities of the entorhinal afferents. However, the build-up of population activity in these hippocampal regions takes another 50 msec or so (Fig. 16.2).

Addressing the potential causes of the delayed spiking activity during theta oscillations requires a thorough understanding of the temporal evolution of spike patterns of principal cells. As described earlier, the hippocampal theta oscillation is not a single entity but a consortium of multiple oscillators. Hippocampal principal cells can be activated by either environmental landmarks ("place cells"; O'Keefe and Nadel, 1978) or internal memory cues (Pastalkova et al., 2008). During its active state, a principal cell oscillates faster than the local field potential (LFP) theta, and the frequency difference between the neuron and the LFP gives rise to phase interference (known as "phase precession"; O'Keefe and Recce, 1993). As an example, entering the place field of a typical CA1 place cell by the rat is marked by a single spike on the peak of the locally derived theta LFP. As the animal moves into the field, the spikes occur at progressively earlier phases. The "lifetime," that is, the duration of activity, of pyramidal neurons in the septal part of the hippocampus corresponds to 7 to 12 theta cycles, during which a full-wave phase advancement (360°) takes place. In addition to spike phase advancement, the number of spikes emitted by the neuron also increases and decreases, with the maximum probability of spiking at the trough of theta in the center of the place field. In short, spikes can occur at all phases of the theta cycle but with the highest probability at the trough. Neither the phase advancement of spikes nor the increased probability of spiking in the firing field of the CA1 pyramidal cell can be explained by simple integration of the direct entorhinal layer 3 inputs because spikes of layer 3 pyramidal cells are phase-locked to the positive peak of the CA1 pyramidal cell layer theta (as reflected by the sink in the stratum lacunosum moleculare; Fig. 16.2), that is, at the theta phase with the least probability of spiking for CA1 neurons. Therefore, the entorhinal input cannot be the sole cause for each spike, especially not for those occurring in the earlier phases of the theta cycle. The situation is similar in the entorhinal

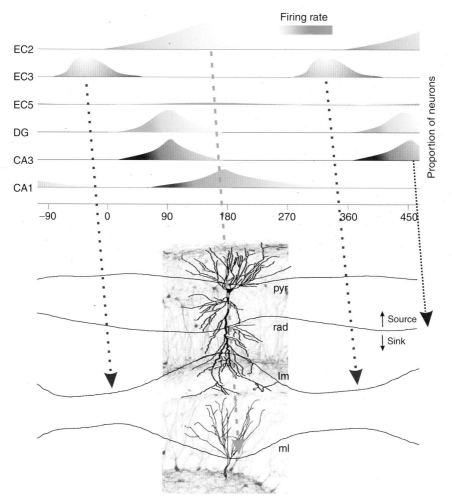

Figure 16–2. Temporal relationship between layer/region-specific population firing patterns and theta current sinks in the hippocampus. In each region and layer, most neurons are silent or fire at low rates, with only a minority of neurons discharging at high frequency. The preferred theta phase of low and high firing neurons is different (advancing to earlier phase in EC2, DG, CA3, and CA1 neurons). Firing rate is illustrated by color intensity. The height of the histograms reflects the proportion of neurons with different discharge rates. m, median of the entire population (based on Fig. 16.3). Note that the preferred phases of the low firing rate (<4 Hz) and high firing rate (>40 Hz) subpopulations are different in most layers. (*Below*) Current-source density (CSD) theta traces are superimposed on a histological section in the CA1-dentate gyrus axis, with highlighted pyramidal cell and granule cell. Note phase-reversed sinks in CA1 str. lacunosum-moleculare (lm) and dentate molecular layers (ml) and phase-shifted sink (relative to lm sink) in str. radiatum (rad). pyr, pyramidal layer. Tilted arrows indicate the temporal (phase) offsets between the peak of population firing in an upstream layer and the theta sinks in the target layers with the expected delays (based on axonal conduction velocity; 30° or ~10 msec). Note that while the population patterns correctly predict the timing of the dendritic sinks in their respective target layers, the propagation of activity between upstream and downstream neuronal populations cannot be deduced from feed-forward excitation between populations. (*A* and *B* are adapted from Mizuseki et al., 2009 and Montgomery et al., 2009, respectively.)

layer 2–dentate granule cell/CA3 cell network because peak firing of these neuronal populations is also delayed by approximately one half theta phase (Mizuseki et al., 2009). It is also important to emphasize that the evolution of spike discharge activity and the associated theta phase precession of spikes are not necessarily controlled by environmental inputs because identical patterns can also occur during memory recall, route planning, and even REM sleep (Pastalkova et al., 2008).

Although the exact cause of the additional spikes that occur at earlier phases of the theta cycle is not understood, they may derive from local circuit mechanisms, according to the following hypothesis. Hippocampal neurons, initially discharged by the entorhinal input, begin to interact with each other in the form of transient assemblies, which oscillate faster than the ongoing theta LFP. The oscillation frequency of the active cell assembly, relative to the frequency of the theta LFP, determines the magnitude of spike phase advancement and the "lifetime" of the assembly (i.e., the size of the firing field in spatial metric). From this perspective, the role of the entorhinal input is to add new members to the perpetually shifting and oscillating cell assemblies rather than to "drive" each spike in the hippocampus. The selected assembly members then begin to interact with each other for a limited time period. Such local interactions require time, which is secured by the theta cycle. Within each theta cycle, multiple (7 to 9) assemblies interact with each other, and the results of this interaction ("local computation") are transmitted to downstream targets.

The above considerations should make it clear that synaptic interactions within and across the circuit (i.e., the "dynamics") are slowed down during theta oscillations. Active neurons in the hippocampus, such as place cells, are speed-controlled oscillators because both the firing rate and the theta oscillation frequency of the principal cells increase with increased running speed of the rat (Geisler et al., 2007). Remarkably, every place cell in the hippocampus oscillates faster than the ongoing LFP, posing the important and general question of how a large group of oscillating neurons can produce a population output slower than its constituent cells. The answer lies in the strict temporal delays between active neurons.

Although a large number of studies have examined the firing patterns of single neurons, much less is known about their interactions and the influence of the networks in which they are embedded. For hippocampal cell pairs with overlapping place fields, the temporal structure of spike trains within a theta cycle reflects the distances between the place-field centers and their sequential activation during the run (Skaggs et al., 1996; Dragoi and Buzsaki, 2006). Within the theta cycle, the relative timing of neuronal spikes reflects the upcoming sequence of locations in the path of the rat, with larger time lags representing larger distances. The time delays between neurons are independent of the running speed of the rat and are not affected by other environmental manipulations (Diba and Buzsaki, 2008). The temporal lags between neurons are specific to theta dynamics because the same sequences can be observed during SPWs but with shorter time delays between neurons.

The theta dynamics-driven "fixed" temporal delays have important consequences for mechanisms of hippocampal coding. The first is a sigmoid relationship between within-theta time lags and distance representations, because the natural upper limit of distance coding by theta-scale time lags is set by the duration of the theta cycle (120–150 msec). As a result, upcoming locations that are more proximal are given better representation, with poorer resolution of locations in the distant future; distances larger than 50 cm are poorly resolved by neurons in the dorsal hippocampus because their expected temporal lags would be longer than the theta cycle (i.e., they fall on the plateau part of the sigmoid). The behavioral consequence of this sigmoid relationship is that objects and locations far away are initially less distinguishable, but as the animal approaches, they are progressively better resolved by the theta-scale code. Another consequence of the relatively fixed within-theta temporal lags is that distance representations should scale with the size of the environment; the temporal lags that represent very fine spatial resolution in small enclosures correspond to much coarser distance representations in large environments. Assuming that locations can be regarded as analogous to individual items in a memory buffer (Lisman and Idiart, 1995; Dragoi and Buzsaki, 2006), this temporal compression mechanism limits the "register capacity" of the "memory buffer." By the same analogy, the sigmoid relationship suggests that the spatiotemporal resolution of an episodic recall is high for the context that surround a recalled event, whereas the relationships among items representing the far past or far future, relative to the recalled event, are progressively less resolved.

How can the mechanism responsible for maintaining theta-scale time delays be protected from firing rate changes, environmental modifications, and other factors, which constantly affect hippocampal neurons? A working model is illustrated in Figure 16.3. The simple hypothesis is that interneuron-mediated inhibition provides a window of opportunity during which a postsynaptic neuron may spike. The timing of this window may be established by the combined effect of presynaptic excitatory activity and inhibition. Through recurrent and feedforward connections, changes in the drive from the leading assembly (represented by neuron 1 in Fig. 16.3) may modify the timing of interneurons inhibiting the trailing assembly (represented by neuron 2), which in turn establish the time lag. In short, the stability of time lags between neurons arises from the theta network dynamics (Fig. 16.3).

With this hypothetical mechanism at hand, we can now return to the "paradox" of how fast oscillating neurons can generate a slower frequency population output: time delays. Consider 100 identical, partially overlapping, evolving place cell assemblies while the rat navigates. With zero time delays between the neurons, the population frequency would be identical to the frequency of place cells. However, inserting temporal lags between cell pairs, in proportion to their distance representations of the environment, can slow

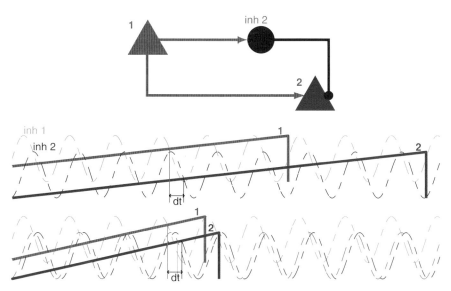

FIGURE 16–3. Model for temporal lag stability on long and short maze tracks. Pyramidal cell (1) excites a second pyramidal cell (2) and an interneuron (inh 2; other sources of depolarization are not shown). In this model, cells fire when excitation exceeds inhibition. The middle panel depicts the excitatory drives for the two interdependent place cells 1 (green) and 2 (blue) on the long track, with inhibition for each superimposed with a dashed lined. The inhibition of cell 2 is delayed relative to cell 1, resulting in net time lag *dt*. When the track length is shortened (*bottom*), the rise in excitatory drives occurs over a shorter duration (i.e., fewer theta cycles), and the place fields are shifted closer relative to each other. Inhibition in the place cells preserves timing with an appropriate shift, relative to that of the long track (superimposed in cyan), thus maintaining the time lag. (From Diba and Buzsaki, 2008)

the momentary population firing frequency. The mean population frequency, also reflected by the LFP, is equal to the mean of the oscillation frequencies of the individual neurons plus the mean time lag (Geisler et al., 2010). In summary, the period of theta oscillations is largely determined by the time lags between active neuron pairs. Conversely, the ensuing theta dynamics constrains the propagation of activity across neurons.

Our discussion on the temporal dynamics was largely confined to hippocampal networks, which reside in the dorsal (septal) part of the structure. Although recent findings point to quantitative differences in place representations of more ventral hippocampal neurons (Maurer et al., 2005; Kjelstrup et al., 2008) the discussed mechanisms may apply to the entire hippocampus (Geisler et al., 2010). Since the hippocampal theta oscillations are coherent along the entire septotemporal axis of the hippocampus, they may serve as a temporal integration mechanism for combining local computations taking place at all segments and representing both spatial and nonspatial information. Furthermore, the computational principles discussed for the operations of the hippocampal circuits likely apply to other systems with similar forms of oscillatory dynamics.

REFERENCES

Bragin A, Jandó G, Nádasdy Z, Hetke J, Wise K, Buzsáki G (1995) Gamma (40–100 Hz) oscillation in the hippocampus of the behaving rat. *J Neurosci* 15:47–60.

Buzsáki G (1989) Two-stage model of memory trace formation: a role for "noisy" brain states. *Neurosci* 31:551–570.

Buzsáki G, Horváth Z, Urioste R, Hetke J, Wise K (1992) High-frequency network oscillation in the hippocampus. *Science* 256:1025–1027.

Diba K, Buzsáki G (2008) Hippocampal network dynamics constrain the time lag between pyramidal cells across modified environments. *J Neurosci* 28:13448–13456.

Dragoi G, Buzsáki G (2006) Temporal encoding of place sequences by hippocampal cell assemblies. *Neuron* 50:145–157.

Freund TF, Buzsáki G (1996) Interneurons of the hippocampus. *Hippocampus* 6:347–470.

Geisler C, Diba K, Pastalkova P, Mizuseki K, Royer S, Buzsáki G (2010) Temporal delays among place cells determine the frequency of population theta oscillations in the hippocampus. *Proc Natl Acad Sci*.

Geisler C, Robbe D, Zugaro M, Sirota A, Buzsáki G (2007) Hippocampal place cell assemblies are speed-controlled oscillators. *Proc Natl Acad Sci USA* 104:8149–8154.

Girardeau G, Benchenane K, Wiener SI, Buzsáki G, Zugaro MB (2009) Selective suppression of hippocampal ripples perturbs learning in a spatial reference memory task. *Nature Neuroscience* 12:1222–1223.

Harris KD, Csicsvari J, Hirase H, Dragoi G, Buzsáki G. (2003) Organization of cell assemblies in the hippocampus. *Nature* 424:552–556.

Kjelstrup KB, Solstad T, Brun VH, Hafting T, Leutgeb S, Witter MP, Moser EI, Moser MB (2008) Finite scale of spatial representation in the hippocampus. *Science* 321(5885): 140–143.

Klausberger T, Somogyi (2008) Neuronal diversity and temporal dynamics: the unity of hippocampal circuit operations. *Science* 321:53–57.

Lisman JE, Idiart MA (1995) Storage of 7 +/− 2 short-term memories in oscillatory subcycles. *Science* 267:1512–1515.

Maurer AP, Vanrhoads SR, Sutherland GR, Lipa P, McNaughton BL (2005) Self-motion and the origin of differential spatial scaling along the septo-temporal axis of the hippocampus. *Hippocampus* 15:841–852.

Mizuseki K, Sirota A, Pastalkova E, Buzsáki G (2009) Theta oscillations provide temporal windows for local circuit computation in the entorhinal-hippocampal loop. *Neuron* 64:267–280.

Montgomery SM, Betancur MI, Buzsáki G (2009) Behavior-dependent coordination of multiple theta dipoles in the hippocampus. *J Neurosci* 29:1381–1394.

O'Keefe J, Nadel L (1978) *The Hippocampus as a Cognitive Map*. New York: Oxford University Press.

O'Keefe J, Recce ML (1993) Phase relationship between hippocampal place units and the EEG theta rhythm. *Hippocampus* 3:317–330.

Pastalkova E, Itskov V, Amarasingham A, Buzsáki G (2008) Internally generated cell assembly sequences in the rat hippocampus. *Science* 321:1322–1327.

Skaggs WE, McNaughton BL, Wilson MA, Barnes CA (1996) Theta phase precession in hippocampal neuronal populations and the compression of temporal sequences. *Hippocampus* 6:149–172.

Wilson MA, McNaughton BL (1994) Reactivation of hippocampal ensemble memories during sleep. *Science* 265:676–679.

17

Entorhinal Cortex

Edvard I. Moser, Menno P. Witter, and May-Britt Moser

In his seminal studies on the anatomy of the nervous system, Ramón y Cajal drew attention to "the sphenoidal cortex" or "the angular ganglion" (Ramón y Cajal, 1902), now commonly known as entorhinal cortex (EC). Struck by the massive bundle of entorhinal fibers perforating the subiculum on its way to the CA fields and dentate gyrus (DG) of the hippocampus, Cajal suggested that the functional significance of the EC had to be related to that of the hippocampus. The first detailed description of the architecture of EC, based on Golgi-impregnated material, was published in 1933 by one of Cajal's students, Lorente de Nó (1933). Over the years, connectional details and new functional cell types have been added, resulting in the contemporary view of EC as a multimodal association cortex, with a unique contribution to high-order cognitive functions such as spatial navigation (Moser et al., 2008). In the present chapter we shall review the intrinsic wiring of the EC as well as the possible computational functions of cell types in this brain region.

Entorhinal Microcircuit

Although species differences are apparent, cytoarchitectonic and connectional data support a subdivision of EC into two functionally different areas generally referred to as medial and lateral entorhinal cortex (MEC and LEC, respectively; Witter and Amaral, 2004). Input from the presubiculum, parahippocampal-postrhinal, and retrosplenial cortices are defining features of MEC. Olfactory, perirhinal, and amygdala inputs are characteristic for LEC (Kerr et al., 2007; Insausti & Amaral, 2008). The two portions of EC do not differ with respect to their laminar organization, and although there are some electrophysiological differences between cell types in layers II and III, the similarities are more striking (Canto et al., 2008; see also Buckmaster et al., 2004; Fig. 17.1).

Layer I contains mainly two types of morphologically defined GABAergic interneurons, horizontal and multipolar neurons, embedded in a dense plexus of axons originating from several afferent areas (Fig. 17.1). Both cell types are almost spineless, and multipolar cells are often positive for calretinin, whereas the remaining cells express a variety of other markers. Horizontal cells give rise to a noncollateralizing axon that enters the white matter, whereas axons of multipolar cells innervate principal cells in layers II and III (Wouterlood, 2002; Canto et al., 2008).

Layer II of MEC contains mainly stellate cells (Klink and Alonso, 1997), which in LEC are replaced by the comparable fan cells (Tahvildari and Alonso, 2005). The deeper part of layer II contains pyramidal-like cells. Stellate, fan, and pyramidal cells of layer II differ slightly in electrophysiological properties, but all do show a characteristic I_h current, which is absent in layer III neurons (Klink and Alonso, 1997; van der Linden and Lopes da Silva, 1998; Tahvildari and Alonso, 2005).

Principal cells in LEC receive excitatory olfactory input on their layer I dendrites (Wouterlood and Nederlof, 1983) and in MEC they are innervated by excitatory input from the presubiculum (van Haeften et al., 1997). For other inputs that distribute to layers I and II (Fig. 17.1; Canto et al., 2008), the synaptic organization and postsynaptic targets have not yet been established. Principal cells in LEC further receive input from the non-calretinin-positive layer I multipolar interneurons, innervated by olfactory inputs (Wouterlood et al., 1985), whereas unidentified GABAergic interneurons in layer I of MEC receive inhibitory inputs from the presubiculum (van Haeften et al., 1997).

All layer II principal cells issue an axon coursing straight toward the angular bundle, where the axon continues to its main targets in the DG and CA3/CA2 (Fig. 17.1). Axons give off thin collaterals in layers I and II. The extent of the local axonal arbor spans about 400 µm (Tamamaki and Nojyo, 1993; Klink and Alonso, 1997). Early in vitro electrophysiological data suggested that the likelihood of direct connections between stellate cells is rather low (Dhillon and Jones, 2000), but more recent studies diverge on this matter (Kumar and Buckmaster, 2006; Couey and Witter, 2009). More work is needed to quantify the extent of recurrent excitatory interconnectivity in layer II, but it is clear that local axon collaterals of stellate cells target interneurons (Jones, 1994; see below). Only sparse axon collaterals have been reported in deep layers III–VI (Canto et al., 2008; but see Garden et al., 2008).

Layer III of MEC and LEC is mainly populated by pyramidal neurons with comparable morphological and electrophysiological characteristics in the two regions. In MEC, axons from the presubiculum provide strong excitatory and inhibitory inputs (van Haeften et al., 1997; Tolner et al., 2005), and throughout EC principal cells are targeted by excitatory inputs from CA1 and subiculum (van Haeften et al., 1995; Kloosterman et al., 2004). Likely, inputs from the postrhinal cortex synapse onto principal cells in MEC. In LEC, inputs from the perirhinal cortex provide a mixture of inhibitory and excitatory inputs (Pinto et al., 2006). Inputs from olfactory domains form excitatory synapses onto apical layer III dendrites, whereas the postsynaptic targets for inputs from the amygdala and insular and medial prefrontal/orbitofrontal cortices remain to be identified.

The main axon of pyramidal cells in layer III projects via the angular bundle to the subiculum and CA1 (Fig. 17.1; Canto et al., 2008). Some layer III cells also project to olfactory and prefrontal areas (Witter and Amaral, 2004). Local collaterals spread within layers III and II but also in the lamina dissecans and layer V (Gloveli et al., 1997; van der Linden and Lopes da Silva, 1998). Among layer III principal cells, collateral innervation may be more common than among layer II stellate cells (Dhillon and Jones, 2000).

There is a large variety of GABAergic interneurons in layers II and III. Most of these are local neurons, although at least two types, multipolar and horizontal cells, have been reported to project to the hippocampus. Interneurons stain positive not only for GABA but also for a variety of other markers such as the calcium-binding proteins parvalbumine, calbindin, and calretinin, as well as VIP, substance-P, CCK, SOM, ENK, and NPY (Wouterlood, 2002). Only two types have been characterized in more detail with respect to their axonal distribution: the fast-spiking neurons and the chandelier cells. Fast-spiking, parvalbumine-positive basket cells form a dense plexus of interconnected neurons that provide a strong inhibitory control onto the soma and dendrites of the principal cells (Wouterlood et al., 1995). Similar to what has been reported for the hippocampus (Klausberger and Somogyi, 2008), a second population of CCK-positive basket cells likely exists (Wouterlood, 2002). Chandelier or axo-axonic cells are parvalbumine positive and provide strong inhibitory control onto the axon initial segment. The axonal spread of both types is rather elaborate—up to 200–300 µm wide and 300–450 µm in the radial direction—and they receive excitatory inputs from principal cells (Soriano et al., 1993). Parvalbumine-positive cells in LEC are innervated by perirhinal axons, whereas in MEC they receive inputs from the presubiculum (van Haeften et el., 1997; Wouterlood, 2002).

Layer IV, also referred to as lamina dissecans, only contains a few neurons that in general can be characterized as either layer III or V neurons as well as a number of interneurons. Details about the connectivity of these neurons are currently lacking, but likely they are innervated by the wide variety of axons present. These include axons that originate in the medial septum/diagonal band complex, pre- and parasubiculum, and monoaminergic brainstem inputs, which all diffusely distribute across all layers of EC (Witter and Amaral, 2004).

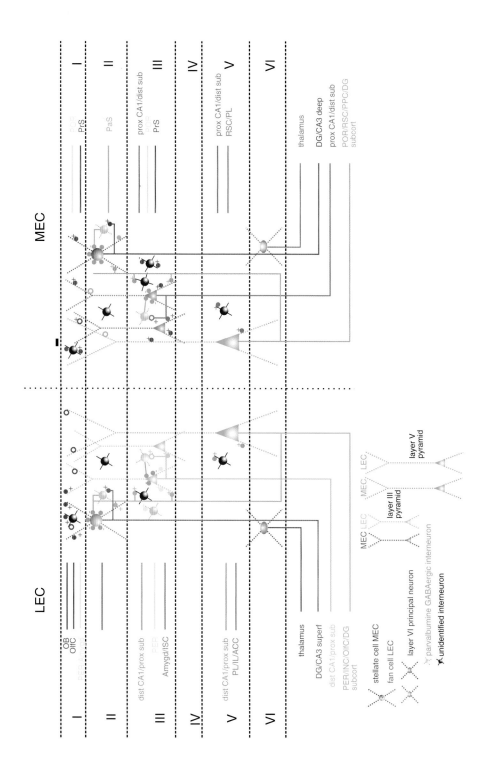

MEC

I

PrS

II

PaS

III
prox CA1/dist sub
PrS

IV

V
prox CA1/dist sub
RSC/PL

VI
thalamus
DG/CA3 deep
prox CA1/dist sub
POR/RSC/PPC/DG
subcort

LEC

I
OB
OlfC
PER/subcort

II
PaS

III
dist CA1/prox sub
PER
Amygd/ISC

IV

V
dist CA1/prox sub
PL/IL/ACC

VI
thalamus
DG/CA3 superf
dist CA1/prox sub
PER/INC/OlfC/DG
subcort

V
dist CA1/prox sub
PL/IL/ACC

VI

layer V
pyramid

layer III
pyramid

MEC LEC

MEC LEC

MEC LEC

stellate cell MEC

fan cell LEC

layer VI principal neuron

parvalbumine GABAergic interneuron

unidentified interneuron

Figure 17–1. Schematic representations of main neuron types and connections of lateral entorhinal cortex (LEC) and medial entorhinal cortex (MEC). Connections are represented as if concentrated into a single columnar module, disregarding available information on topography and divergence of the various extrinsic and intrinsic connections. Inputs and outputs are color coded and presented with respect to their main layers of termination and origin, respectively. Main interlaminar connections are from deep layer V to layers II and III and are known to show extensive spread along the dorsoventral extent of the entorhinal cortex, likely connecting corresponding portions of MEC and LEC, in register with the longitudinal bands which are reciprocally and topographically connected to different parts along the dorsoventral axis of the hippocampus (Dolorfo and Amaral, 1998; Chrobak and Amaral, 2007). Intralaminar connections between principal neurons are most extensive in layers III and V, whereas in layer II the preferential connectivity may be between principal cells and interneurons. Synaptic contacts established anatomically or electrophysiologically are indicated with filled circles and if known the inhibitory or excitatory nature is indicated with an a – or a + sign, respectively. Inferred but not yet established synaptic contacts are indicated with open circles. All cell types and their main dendritic and axonal connections are uniquely color coded. Connectional details of interneurons are limited to parvalbumine-positive GABAergic neurons in layers II and III, which are indicated in dark yellow. Chandelier cells likely have comparable afferent connectivity (see text for further details). Connections with main modulatory systems such as septal complex, monoaminergic systems, and thalamus are not represented. ACC, anterior cingulate cortex; Amygd, amygdaloid complex; CA1-CA3, subfields of hippocampus proper; DG, dentate gyrus; dist, distal; IL, infralimbic cortex; INC, insular cortex; OB, olfactory bulb; OlfC, olfactory cortex; PaS, parasubiculum; PER, perirhinal cortex; PL, prelimbic cortex; POR, postrhinal cortex; PPC, posterior parietal cortex; PrS, presubiculum; prox, proximal; RSC, retrosplenial cortex; sub, subiculum; subcort, subcortical structures such as basal forebrain, amygdala; superf, superficial.

Layer V principal neurons in LEC and MEC are similar both with respect to morphological and electrophysiological features (Hamam et al., 2000; Hamam et al., 2002; Canto et al., 2008). Large pyramidal cells are located immediately below the lamina dissecans, while the deeper part of layer V contains smaller cells. The apical dendrites of the pyramidal cells traverse superficial layers, having a tuft that may reach the pial surface. The basal dendritic tree spreads mainly within deep layers. The smaller cells in the deep part are generally horizontal pyramidal, multipolar, or fusiform cells. They tend to have dendritic trees confined to layers V and VI, although apical dendrites may cross the lamina dissecans into layer III. On their basal dendrites, layer V cells receive inputs from the hippocampal field CA1 and the subiculum (van Haeften et al., 1995; Kloosterman et al., 2004). Other likely inputs are from the retrosplenial cortex (MEC), the medial prefrontal cortex (MEC and LEC), and anterior cingulate cortex (LEC; Jones and Witter, 2007). On their apical dendrites, the pyramidal cells in MEC receive synaptic inputs from the presubiculum (Wouterlood et al., 2004; Tolner et al., 2005). Other inputs to layers I–III—from olfactory structures, perirhinal cortex, and amygdala in the case of LEC, and from postrhinal cortex and parasubiculum in the case of MEC—may likewise target the apical dendrites of layer V pyramidal cells. All layer V principal cells send a main axon into the white matter (Hamam et al., 2000; Gloveli et al., 2001; Hamam et al., 2002), projecting to widespread cortical and subcortical targets (Witter and Amaral, 2004). Additional projections innervate the inner molecular layer of the DG and the hilus (Deller et al., 1996; Gloveli et al., 2001). Local axon collaterals distribute within all layers of EC (Kohler, 1986; Kohler, 1988; Dolorfo and Amaral, 1998; Hamam et al., 2000; Hamam et al., 2002; Kloosterman et al., 2003; Chrobak and Amaral, 2007). The majority of the axons from deep to superficial layers are likely excitatory and target both interneurons as well as principal neurons in almost equal percentages (van Haeften et al., 2003). The overall axonal tree is rather restricted in its transverse extent but is more extensive longitudinally (Kohler, 1986; Kohler, 1988; Van Haeften et al., 2003), providing intrinsic connectivity within the longitudinal bands of EC that project topographically to restricted dorsoventral levels of the hippocampal formation (Dolorfo and Amaral, 1998; Canto et al., 2008).

Layer VI mainly contains multipolar neurons. Their spiny dendrites extend within layer VI, parallel to the layer, occasionally reaching the angular bundle and layer V (Canto et al., 2008). Specific inputs to layer VI are unknown; likely the cells receive inputs similar to those of layer V. The axons join the underlying white matter, projecting to thalamic midline nuclei (Witter and Amaral, 2004). Occasional collaterals reach the subiculum or the superficial layers of EC (Canto et al., 2008). Layers V and VI contain a wide variety of interneurons, including the largest population of NPY-positive ones, whereas the density of parvalbumine-positive interneurons is very low. No connectional details of any of these interneurons are known.

FUNCTION OF THE ENTORHINAL CORTEX

Until very recently the functions of the entorhinal cortex have remained in the dark. It has been known for a while that the profile of impairment after extensive lesions of the entorhinal cortex mirrors that of lesions in the hippocampus, with striking deficits in spatial learning (Olton et al., 1978; Schenk and Morris, 1985) and contextual conditioning (Maren and Fanselow, 1997). However, the exact contribution of the entorhinal cortex has not been well characterized. One way to improve our understanding of neural computation in the entorhinal cortex would be to compare neural firing patterns in entorhinal cortex with firing patterns in the hippocampus, one synapse away. Most principal cells in the hippocampus are place cells, that is, they fire selectively when the animal moves through a certain location of the environment (O'Keefe and Dostrovsky, 1971). Based on the strong spatial firing correlates of hippocampal pyramidal cells and the severe impairments in spatial memory induced by hippocampal lesions, it was suggested early on that the hippocampus is the basis of a Tolmanian "cognitive map" of the environment, consisting of a nontopographical representation of the local space and the animal's experiences in that space (O'Keefe and Nadel, 1978). This suggestion led researchers to ask whether the spatial map expressed by place cells originates in the hippocampus or in regions that project to the hippocampus. An obvious candidate region was the entorhinal cortex, which provides most of the hippocampal cortical input. The first recordings from entorhinal cortex in freely moving rats revealed only weakly place-modulated signals (Barnes et al., 1990; Quirk et al., 1992; Frank et al., 2000), suggesting to many researchers that sharp place signals originated within the hippocampal circuit itself, after the spatial information had passed the entorhinal cortex. This view has been radically challenged during the past 5 years. Recordings from the dorsal parts of MEC, which project to the regions of hippocampus where prototypical place cells are recorded, have identified several cell types with distinct functions in representation of the animal's position, including grid cells, head direction cells, border cells, and cells with conjunctive grid × head direction and border × head direction properties (Moser et al., 2008). Neurons in LEC, in contrast, have limited spatial firing properties (Hargreaves et al., 2005), and the exact nature of the information signaled by those neurons has not been determined. It is known though that outputs from MEC and LEC converge on the same principal cells in DG and CA3, allowing MEC-derived spatial and LEC-derived nonspatial information to be encoded by the same hippocampal cells (Leutgeb et al., 2005; Leutgeb et al., 2008).

GRID CELLS

Grid cells are place-modulated entorhinal neurons whose multiple firing fields define a periodic array across the entire environment available to a walking animal, like the cross-points of widely spaced chicken wire rolled out over the surface of the test arena (Fyhn et al., 2004; Hafting et al., 2005). Approximately 50% of the principal cells in layer II are grid cells (Sargolini et al., 2006). The repeating unit of the grid is an equilateral triangle (Fig. 17.2). The spacing of the grid decreases topographically from dorsal to ventral MEC, with the most dorsal grid cells repeating at a frequency of one field per 30 cm and the most ventral ones spaced at distances of several meters in the rat. Grids of different cells are offset relative to each other such that each place in the environment can be identified from the activity of a limited number of adjacent cells with some variation in spacing. Convergence of inputs from grid cells with overlapping fields but different spacing may generate individual place fields in the hippocampus (Fuhs and Touretzky, 2006; McNaughton et al., 2006; Solstad et al., 2006).

The fact that the position of a moving animal can be reconstructed from activity in assemblies of grid cells (Fyhn et al., 2004) points to grid cells as elements of a dynamic, constantly updated map of the animal's place in the environment. A key property of the map is its apparently universal nature (Fyhn et al., 2007). Unlike place cells in the hippocampus, the entorhinal map is activated in a stereotypic manner across environments, irrespective of particular landmarks in the environment, suggesting that the same map is applied everywhere, although the particular phase and orientation of each grid cell are strongly determined by the geometrical features of the environment (Hafting et al., 2005; Barry et al., 2007; Solstad et al., 2008). Because the grid pattern is immune to changes in the speed and direction of the animal, and because the structure persists after replacement of the external landmarks (Hafting et al., 2005), the firing locations must be derived primarily from information about the animal's self-movement, in a manner that compensates for constantly varying changes in speed and direction. These observations point to MEC as part of a possible substrate for path integration-based spatial navigation (Hafting et al., 2005; McNaughton et al., 2006).

While the basic structure and properties of the entorhinal spatial map are known, the mechanisms for grid formation remain to be determined. One class of models suggests that grid formation is a result of local network activity. In this view, position is represented by an attractor, a stable firing state sustained by recurrent connections with robust performance in the presence of noise (Hopfield, 1982; Amit, 1989). Networks for spatial representation are thought to store several attractor states associated with different locations and retrieve any of them in response to sensory or path-integration cues. When a large number of very close positions are represented, a continuous attractor emerges, which permits a smooth variation of the place representation in accordance with the trajectory of the rat (Tsodyks and Sejnowski, 1995; Samsonovich and McNaughton, 1997; Fuhs and Touretzky, 2006; McNaughton et al., 2006). A second class of models suggests that path integration occurs in individual cells as a result of interference between intracellular oscillators at

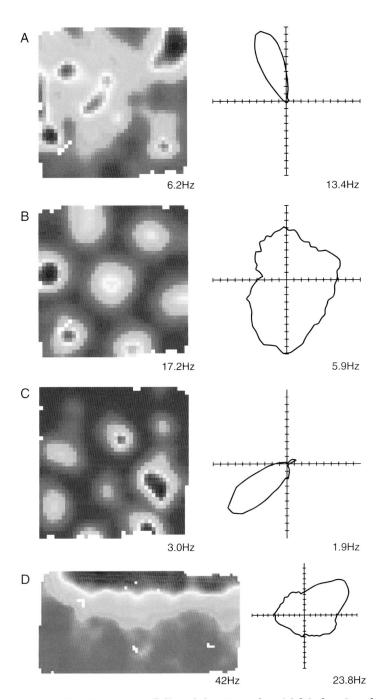

FIGURE 17–2. Rate maps (*left*) and direction plots (*right*) showing characteristic firing properties of a head direction cell (*A*), a grid cell (*B*), a conjunctive grid x head direction cell (*C*), and a border cell (*D*). Rate maps are color-coded from blue to red (low and high rate, respectively). Direction maps show firing rate as a function of direction (360 degrees). Peak firing rates are indicated beneath each map. All cells were recorded in the dorsocaudal part of medial entorhinal cortex.

slightly different theta frequencies (O'Keefe and Recce, 1993; Burgess et al., 2007). If the intrinsic theta frequency increases linearly with speed, the envelope of the two-dimensional interference wave should integrate speed and thus reflect position in the environment (Burgess et al., 2007). Whether grid fields emerge by any of these mechanisms and which cells and cellular interactions are responsible for the elementary computations remain to be determined. An intracellular mechanism for grid formation would not exclude a role for recurrent collaterals and attractors in determining the ensemble properties of grid cells.

Head Direction Cells

Navigation requires representation of the animal's orientation relative to fixed landmarks in the external environment. The entorhinal cortex contains cells that could contribute to this process, referred to as head direction cells (Sargolini et al., 2006). These cells fire whenever the rat faces a certain direction, independently of its location in the environment or its ongoing behaviour. Head direction cells were first recorded in the dorsal presubiculum (Ranck, 1985; Taube et al., 1990), but later research has found them in a number of brain regions, including lateral mammillary nucleus, anterodorsal thalamus, retrosplenial cortex, and parasubiculum (Taube et al., 2007). In MEC, head direction cells are abundant in deep and intermediate layers (layers III–VI; Sargolini et al., 2006). Many of these cells respond conjunctively to head direction and place, that is, they fire when the animal crosses a grid field in a certain direction only. Head direction cells in the presubiculum are thought to provide directional input to nondirectional grid cells in layer II, with which they are likely to form synapses (Tolner et al., 2005), and they may be important for setting the orientation of the grid map (McNaughton et al., 2006; Sargolini et al., 2006).

Border Cells

The strong influence of geometric borders on firing properties of place cells (O'Keefe and Burgess, 1996) and grid cells (Barry et al., 2007) has raised the possibility that the entorhinal cortices or other hippocampal or parahippocampal regions contain cells that specifically respond to boundaries of the local environment (Hartley et al., 2000; Barry et al., 2006). Recently, neurons with border-associated firing fields have been identified in MEC (Savelli et al., 2008; Solstad et al., 2008). Most border cells fire only at one side of a border. Their edge-apposing activity is maintained when the environment is stretched or when the animal is tested in enclosures of different size and shape in different rooms (Solstad et al., 2008). Border cells are more sparse than grid cells and head direction-modulated cells, and they probably comprise less than 10% of the local principal cell population (Solstad et al., 2008). However, this does not rule out a significant role in spatial representation because border cells are widely distributed and can be found in all medial entorhinal cell layers, intermingled with head-direction cells and grid cells. By defining the perimeter of the environment, border cells may determine the firing locations of grid cells as well as place cells; however, this possibility has not been addressed with experiments.

Conclusions

Based on recent studies of the entorhinal microcircuitry and the spatial firing correlates of some of its principal cell populations, it seems reasonable to propose that MEC is part of a neural system for representing the animal's location as it moves around in its proximal environment. The circuit is composed of a number of morphologically distinct cell types, organized in layers with unique intrinsic and extrinsic connections, and cells in different layers respond differently to changes in the rat's location and orientation. We do not know whether grid cells, head direction cells, or border cells have distinct morphologies and whether they correspond to any of the anatomical cell types described in this chapter. The relative predominance of grid cells in layer II (Sargolini et al., 2006) implies that at least some grid cells are stellate cells, but grid cells exist also in deeper layers, which have other cell types, and layer II also contains border cells. If we are to understand how space is represented in the medial entorhinal microcircuit, a vital step will be to map the various cell types onto the variety of morphological and physiological cell types, such that knowledge about their intrinsic properties and their connections with other cells can be used to understand the spatial and temporal interactions of the functional cell types. This will in turn allow us to make firm predictions about the functions of less well understood sister neurons and sister networks in LEC (Fig. 17.1). Understanding these interactions may be rewarding because the existence of cells with reliable firing correlates created within the circuit itself may put us on the track toward identification of some of the most fundamental computational operations of cortical circuits.

References

Amit DJ (1989) *Modelling Brain Function: The World of Attractor Networks*. New York: Cambridge University Press.

Barnes CA, McNaughton BL, Mizumori SJ, Leonard BW, Lin LH (1990) Comparison of spatial and temporal characteristics of neuronal activity in sequential stages of hippocampal processing. *Prog Brain Res* 83:287–300.

Barry C, Lever C, Hayman R, Hartley T, Burton S, O'Keefe J, Jeffery K, Burgess N (2006) The boundary vector cell model of place cell firing and spatial memory. *Rev Neurosci* 17:71–97.

Barry C, Hayman R, Burgess N, Jeffery KJ (2007) Experience-dependent rescaling of entorhinal grids. *Nat Neurosci* 10:682–684.

Buckmaster PS, Alonso A, Canfield DR, Amaral DG (2004) Dendritic morphology, local circuitry, and intrinsic electrophysiology of principal neurons in the entorhinal cortex of macaque monkeys. *J Comp Neurol* 470:317–329.

Burgess N, Barry C, O'Keefe J (2007) An oscillatory interference model of grid cell firing. *Hippocampus* 17:801–812.

Canto CB, Wouterlood FG, Witter MP (2008) What does the anatomical organization of the entorhinal cortex tell us? *Neural Plast* 2008:381243.

Chrobak JJ, Amaral DG (2007) Entorhinal cortex of the monkey: VII. Intrinsic connections. *J Comp Neurol* 500:612–633.

Couey JJ, Moser EI, Witter MP (2009) Development of the entorhinal cortex layer II stellate cell network. *Soc. Neurosci Abstr.*,101 1.

Deller T, Martinez A, Nitsch R, Frotscher M (1996) A novel entorhinal projection to the rat dentate gyrus: direct innervation of proximal dendrites and cell bodies of granule cells and GABAergic neurons. *J Neurosci* 16:3322–3333.

Dhillon A, Jones RS (2000) Laminar differences in recurrent excitatory transmission in the rat entorhinal cortex in vitro. *Neuroscience* 99:413–422.

Dolorfo CL, Amaral DG (1998) Entorhinal cortex of the rat: organization of intrinsic connections. *J Comp Neurol* 398:49–82.

Frank LM, Brown EN, Wilson M (2000) Trajectory encoding in the hippocampus and entorhinal cortex. *Neuron* 27:169–178.

Fuhs MC, Touretzky DS (2006) A spin glass model of path integration in rat medial entorhinal cortex. *J Neurosci* 26:4266–4276.

Fyhn M, Molden S, Witter MP, Moser EI, Moser M-B (2004) Spatial representation in the entorhinal cortex. *Science* 305:1258–1264.

Fyhn M, Hafting T, Treves A, Moser M-B, Moser EI (2007) Hippocampal remapping and grid realignment in entorhinal cortex. *Nature* 446:190–194.

Garden DL, Dodson PD, O'Donnell C, White MD, Nolan MF (2008) Tuning of synaptic integration in the medial entorhinal cortex to the organization of grid cell firing fields. *Neuron* 60:875–889.

Gloveli T, Schmitz D, Empson RM, Dugladze T, Heinemann U (1997) Morphological and electrophysiological characterization of layer III cells of the medial entorhinal cortex of the rat. *Neuroscience* 77:629–648.

Gloveli T, Duglardze T, Schmitz D, Heinemann U (2001) Properties of entorhinal cortex deep layer neurons projecting to the rat dentate gyrus. *Eur J Neurosci* 13:413–420.

Hafting T, Fyhn M, Molden S, Moser M-B, Moser EI (2005) Microstructure of a spatial map in the entorhinal cortex. *Nature* 436:801–806.

Hargreaves EL, Rao G, Lee I, Knierim JJ (2005) Major dissociation between medial and lateral entorhinal input to dorsal hippocampus. *Science* 308:1792–1794.

Hamam BN, Kennedy TE, Alonso A, Amaral DG (2000) Morphological and electrophysiological characteristics of layer V neurons of the rat medial entorhinal cortex. *J Comp Neurol* 418:457–472.

Hamam BN, Amaral DG, Alonso AA (2002) Morphological and electrophysiological characteristics of layer V neurons of the rat lateral entorhinal cortex. *J Comp Neurol* 451:45–61.

Hartley T, Burgess N, Lever C, Cacucci F, O'Keefe J (2000) Modeling place fields in terms of the cortical inputs to the hippocampus. *Hippocampus* 10:369–379.

Hopfield JJ (1982) Neural networks and physical systems with emergent collective computational abilities. *Proc Natl Acad Sci USA* 79:2554–2558.

Insausti R, Amaral DG (2008) Entorhinal cortex of the monkey: IV. Topographical and laminar organization of cortical afferents. *J Comp Neurol* 509:608–641.

Jones RS (1994) Synaptic and intrinsic properties of neurons of origin of the perforant path in layer II of the rat entorhinal cortex in vitro. *Hippocampus* 4:335–353.

Jones BF, Witter MP (2007) Cingulate cortex projections to the parahippocampal region and hippocampal formation in the rat. *Hippocampus* 17:957–976.

Kerr KM, Agster KL, Furtak SC, Burwell RD (2007) Functional neuroanatomy of the parahippocampal region: the lateral and medial entorhinal areas. *Hippocampus* 17:697–708.

Klausberger T, Somogyi P (2008) Neuronal diversity and temporal dynamics: the unity of hippocampal circuit operations. *Science* 321:53–57.

Klink R, Alonso A (1997) Morphological characteristics of layer II projection neurons in the rat medial entorhinal cortex. *Hippocampus* 7:571–583.

Kloosterman F, van Haeften T, Lopes da Silva FH (2004) Two reentrant pathways in the hippocampal-entorhinal system. *Hippocampus* 14:1026–1039.

Kloosterman F, van Haeften T, Witter MP, Lopes da Silva FH (2003) Electrophysiological characterization of interlaminar entorhinal connections: an essential link for re-entrance in the hippocampal-entorhinal system. *Eur J Neurosci* 18:3037–3052.

Kohler C (1986) Intrinsic connections of the retrohippocampal region in the rat brain. II. The medial entorhinal area. *J Comp Neurol* 246:149–169.

Kohler C (1988) Intrinsic connections of the retrohippocampal region in the rat brain: III. The lateral entorhinal area. *J Comp Neurol* 271:208–228.

Kumar SS, Buckmaster PS (2006) Hyperexcitability, interneurons, and loss of GABAergic synapses in entorhinal cortex in a model of temporal lobe epilepsy. *J Neurosci* 26:4613–4623.

Leutgeb S, Leutgeb JK, Barnes CA, Moser EI, McNaughton BL, Moser M-B (2005) Independent codes for spatial and episodic memory in the hippocampus. *Science* 309:619–623.

Leutgeb JK, Henriksen EJ, Leutgeb S, Witter MP, Moser M-B, Moser EI (2008) Hippocampal rate coding depends on input from the lateral entorhinal cortex. *Soc Neurosci Abstr* 34:94.6.

Lorente de Nó R (1933) Studies on the structure of the cerebral cortex. *Journal für Psychologie und Neurologie* 45:381–438.

Maren S, Fanselow MS (1997) Electrolytic lesions of the fimbria/fornix, dorsal hippocampus, or entorhinal cortex produce anterograde deficits in contextual fear conditioning in rats. *Neurobiol Learn Mem* 67:142–149.

McNaughton BL, Battaglia FP, Jensen O, Moser EI, Moser, M-B (2006) Path-integration and the neural basis of the 'cognitive map'. *Nature Reviews Neuroscience*, 7:663–678.

Moser EI, Kropff E, Moser, M-B (2008) Place cells, grid cells and the brain's spatial representation system. *Annu Rev Neurosci* 31:69–89.

O'Keefe J, Dostrovsky J (1971) The hippocampus as a spatial map. Preliminary evidence from unit activity in the freely-moving rat. *Brain Res* 34:171–175.

O'Keefe J, Nadel L (1978) *The Hippocampus as a Cognitive Map*. Oxford, England: Oxford Univ. Press.

O'Keefe J, Recce ML (1993) Phase relationship between hippocampal place units and the EEG theta rhythm. *Hippocampus* 3:317–330.

Olton DS, Walker JA, Gage FH (1978) Hippocampal connections and spatial discrimination. *Brain Res* 139:295–308.

Pinto A, Fuentes C, Paré D (2006) Feedforward inhibition regulates perirhinal transmission of neocortical inputs to the entorhinal cortex: ultrastructural study in guinea pigs. *J Comp Neurol* 495:722–734.

Quirk GJ, Muller RU, Kubie JL, Ranck JB, Jr. (1992) The positional firing properties of medial entorhinal neurons: description and comparison with hippocampal place cells. *J Neurosci* 12:1945–1963.

Ramón y Cajal S (1902) Sobre un ganglio especial de la corteza esfeno-occipital. *Trab del Lab de Invest Biol Univ Madrid* 1:189–206.

Ranck JB, Jr. (1985) Head direction cells in the deep cell layer of dorsal presubiculum in freely moving rats. In: Buzsáki G, Vanderwolf CH, eds. *Electrical Activity of the Archicortex*, pp. 217–220. Budapest, Hungary: Akademiai Kiado.

Samsonovich A, McNaughton BL (1997) Path integration and cognitive mapping in a continuous attractor neural network model. *J Neurosci* 17:272–275.

Sargolini F, Fyhn M, Hafting T, McNaughton BL, Witter MP, Moser M-B, Moser EI (2006) Conjunctive representation of position, direction and velocity in entorhinal cortex. *Science* 312:754–758.

Savelli F, Yoganarasimha D, Knierim JJ (2008) Influence of boundary removal on the spatial representations of the medial entorhinal cortex. *Hippocampus* 18:1270–1282.

Schenk F, Morris RGM (1985) Dissociation between components of spatial memory in rats after recovery from the effects of retrohippocampal lesions. *Exp Brain Res* 58:11–28.

Solstad T, Moser EI, Einevoll GT (2006) From grid cells to place cells: a mathematical model. *Hippocampus* 16:1026–1031.

Solstad T, Boccara CN, Kropff E, Moser M-B, Moser EI (2008) Representation of geometric borders in the entorhinal cortex. *Science* 322:1865–1868.

Soriano E, Martinez A, Farinas I, Frotscher M (1993) Chandelier cells in the hippocampal formation of the rat: the entorhinal area and subicular complex. *J Comp Neurol* 337:151–167.

Tahvildari B, Alonso A (2005) Morphological and electrophysiological properties of lateral entorhinal cortex layers II and III principal neurons. *J Comp Neurol* 491:123–140.

Tamamaki N, Nojyo Y (1993) Projection of the entorhinal layer II neurons in the rat as revealed by intracellular pressure-injection of neurobiotin. *Hippocampus* 3:471–480.

Taube JS (2007) The head direction signal: origins and sensory-motor integration. *Annu Rev Neurosci* 30:181–207.

Taube JS, Muller RU, Ranck JB Jr (1990) Head-direction cells recorded from the postsubiculum in freely moving rats. I. Description and quantitative analysis. *J Neurosci.* 10:420–435.

Tolner EA, Kloosterman F, van Vliet EA, Witter MP, Silva FH, Gorter JA (2005) Presubiculum stimulation in vivo evokes distinct oscillations in superficial and deep entorhinal cortex layers in chronic epileptic rats. *J Neurosci* 25:8755–8765.

Tsodyks M, Sejnowski T (1995) Associative memory and hippocampal place cells. *Int J Neural Syst* 6(Suppl.):81–86.

van der Linden S, Lopes da Silva FH (1998) Comparison of the electrophysiology and morphology of layers III and II neurons of the rat medial entorhinal cortex in vitro. *Eur J Neurosci* 10:1479–1489.

van Haeften T, Jorritsma-Byham B, Witter MP (1995) Quantitative morphological analysis of subicular terminals in the rat entorhinal cortex. *Hippocampus* 5:452–459.

van Haeften T, Wouterlood FG, Jorritsma-Byham B, Witter MP (1997) GABAergic presubicular projections to the medial entorhinal cortex of the rat. *J Neurosci* 17:862–874.

van Haeften T, Baks-Te-Bulte L, Goede PH, Wouterlood FG, Witter MP (2003) Morphological and numerical analysis of synaptic interactions between neurons in deep and superficial layers of the entorhinal cortex of the rat. *Hippocampus* 13:943–952.

Witter MP, Amaral DG (2004) Hippocampal formation. In: Paxinos G, ed. *The Rat Nervous System*, 3rd ed., pp. 635–704. San Diego, CA: Elsevier Academic Press.

Wouterlood FG (2002) Spotlight on the neurones (I): cell types, local connectivity, micro-circuits and distribution of markers. In: Witter MP, Wouterlood FG, eds. *The Para-hippocampal Region. Organization and Role in Cognitive Function*, pp. 61–88. Oxford, England: Oxford University Press.

Wouterlood FG, Nederlof J (1983) Terminations of olfactory afferents on layer II and III neurons in the entorhinal area: degeneration-Golgi-electron microscopic study in the rat. *Neurosci Lett* 36:105–110.

Wouterlood FG, Mugnaini E, Nederlof J (1985) Projection of olfactory bulb efferents to layer I GABAergic neurons in the entorhinal area. Combination of anterograde degen-eration and immunoelectron microscopy in rat. *Br Res* 343:283–296.

Wouterlood FG, Hartig W, Bruckner G, Witter MP (1995) Parvalbumin-immunoreactive neurons in the entorhinal cortex of the rat: localization, morphology, connectivity and ultrastructure. *J Neurocytol* 24:135–153.

Wouterlood FG, van Haeften T, Eijkhoudt M, Baks-Te-Bulte L, Goede PH, Witter MP (2004) Input from the presubiculum to dendrites of layer-V neurons of the medial entorhinal cortex of the rat. *Brain Res* 1013:1–12.

Section 6

Visual System

18

Retina: Microcircuits for Daylight, Twilight, and Starlight

Jonathan B. Demb

Over the course of the day, light intensity shifts by *10 billion-fold*, but a ganglion cell's spike rate can vary only by *100-fold*. To cover the huge intensity range, two fundamentally different circuits are required: a *cone bipolar circuit* for graded photoreceptor signals, and a *rod bipolar circuit* for binary signals. By using gap junctions, the two circuits can share key components (Fig. 18.1).

MICROCIRCUIT FOR DAYLIGHT

Daylight, of course, activates cones and the cone bipolar circuit (Fig. 18.1, left). Cones release glutamate onto ionotropic glutamate receptors (iGluRs) of OFF cone bipolar cells and metabotropic receptors (mGluRs) of ON cone bipolar cells (Miller, 2008). Thus, by a postsynaptic mechanism, a cone's glutamate release generates opposite responses in ON and OFF cone bipolar cells; this explains why the ON and OFF pathways are excited by either light increments or decrements, respectively (see Chapter 19). Both cones and bipolar cells signal by graded changes in membrane potential. Several key features of the daylight circuit serve to efficiently transfer these graded signals: cone terminals are coupled to reduce noise by signal averaging (DeVries et al., 2002); cone and bipolar cell terminals receive inhibitory feedback from lateral positions across the retina (via horizontal and amacrine cells) to reduce response redundancy (Srinivasan et al., 1982; see Chapter 19); and photoreceptors and bipolar cells use ribbon synapses to enable high vesicle release rates for encoding finely graded signals (Sterling and Demb, 2004).

The ON and OFF cone bipolar cells release glutamate onto iGluRs and thereby excite either ON or OFF ganglion cells (Miller, 2008). Furthermore,

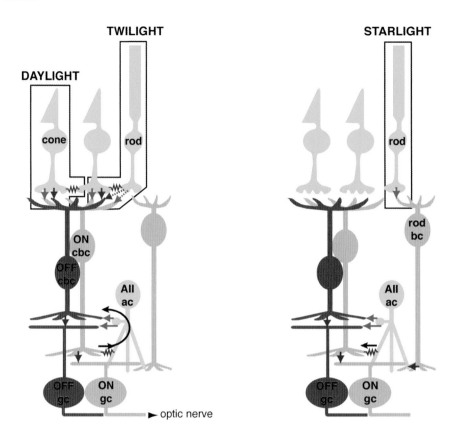

FIGURE 18–1. Microcircuits for daylight, twilight, and starlight. To convey the full range of light intensities efficiently (i.e., minimizing neural noise and retinal thickness) requires three different circuits that partially overlap, plus gap junctions (red resistor symbol) to switch between them. For each circuit, active synapses are indicated. Daylight and twilight circuits use the same pathways in the inner retina (bipolar and amacrine cell synapses). Photoreceptors, OFF cone bipolar cells (cbc), and OFF ganglion cells depolarize to light decrements; other cells depolarize to light increments. The general color scheme for synaptic actions used in this handbook must be adapted for the retina because a given cell type under different conditions may have either an excitatory or inhibitory action. Simplest are bipolar cells (bc) and ganglion cells (gc), which are clearly excitatory (they release glutamate and depolarize the postsynaptic neuron) and are therefore colored red (different shades of red for ON and OFF cells). Photoreceptors and AII amacrine cells (ac) are not strictly inhibitory or excitatory because they each depolarize and hyperpolarize postsynaptic neurons, depending on the synapse; they have been arbitrarily colored green. For example, the photoreceptor glutamate released onto the mGluR6 receptor of an ON bipolar cell closes a cation channel, causing hyperpolarization; this cannot be described as inhibitory because it is different from opening a Cl or K channel. The convention in this case is therefore that a red/blue arrow signifies a depolarizing/hyperpolarizing action on the postsynaptic neuron. Gap junctions are also red, because they are excitatory.

the ON cone bipolar cell couples electrically to the AII amacrine cell, which makes inhibitory glycinergic synapses with OFF cone bipolar terminals and OFF ganglion cell dendrites (Kolb, 1979; see Bloomfield and Dacheux, 2001). Thus, in the daylight circuit the AII cell plays a role in "crossover inhibition," whereby one pathway (ON, in this case) inhibits the other (OFF; Fig. 18.1, left; black arrow) (Manookin et al., 2008); the AII cell plays a different role in the starlight circuit.

Microcircuit for Twilight

At twilight, when ambient light intensity falls below cone threshold (<100 photoisomerizations [Rh*] cone^{-1} sec^{-1}) (DeVries and Baylor, 1995; Yin et al., 2006), the daylight circuit fails. However, rods are now increasing in sensitivity. Since there are ~20 rods per cone, and since a rod, at this light level, transduces ~30 Rh* within its (~300 msec) integration time, signals are available from ~600 photons transduced by rods (Sterling and Demb, 2004; Borghuis et al., 2009). The 30 rod terminals immediately surrounding each cone terminal couple to it via gap junctions and thus inject this graded signal to be carried forward by the cone bipolar circuit (Schneeweis and Schnapf, 1999; Tsukamoto et al., 2001; Hornstein et al., 2005) (Fig. 18.1, left). Rod–cone coupling is controlled by a circadian rhythm, strengthening at night when dopamine levels decline and fail to stimulate D2-like receptors (Ribelayga et al., 2008). Rods also make chemical synapses with certain types of cone bipolar cell to drive responses at twilight (Fig. 18.1., left; dashed arrow) (Soucy et al., 1998; Hack et al., 1999; Tsukamoto et al., 2001; Li et al., 2004; Tsukamoto et al., 2007). Downstream of the photoreceptor synapses, the twilight and daylight circuits use the same pathways in the inner retina.

Microcircuit for Starlight

When ambient intensity falls to one Rh* rod^{-1} integration time^{-1}, photons are spread too thinly for the rods to provide a graded signal to the cones. Thus, the cone bipolar cell pathways are useless and vision now depends on the starlight circuit. The starlight circuit's first task is to transfer a binary signal: 0 or 1 Rh*. "0" is represented by tonic vesicle release from the rod's single ribbon synapse, and "1" is represented by a pause in release. But assuming release is temporally random, some extra-long intervals between quanta will occur by chance that the bipolar cell might "mistake" for a pause. The release rate should be high enough to prevent this source of spurious single-photon signals. A model of the circuit suggests that 50–100 vesicles/s might be sufficient (van Rossum and Smith, 1998; Sterling and Demb, 2004). This fits measured release rates for ribbon synapses and suggests why the rod bipolar circuit can transmit such an elementary signal with only a single synapse at the first stage (Sheng et al., 2007).

However, at the next stage, 20 to 120 rods converge on a rod bipolar cell, and this exposes another problem for processing starlight signals (Sterling and Demb, 2004). Only one of these rods is likely to carry an Rh* response, but the others will carry noise, which if transferred to the rod bipolar cell, would increase its noise (as the square root of the convergence) and swamp the single Rh* response in the bipolar cell (van Rossum and Smith, 1998). To prevent this, the rod synapse acts nonlinearly, amplifying large signals more than small ones. This amplification removes the small, noisy events by thresholding (Field and Rieke, 2002). The thresholding arises from saturation in the transduction cascade that couples glutamate receptors to ion channels in the rod bipolar dendrites (Sampath and Rieke, 2004). The Rh* events in a rod vary in amplitude, many hardly rising above the noise. Thus, thresholding removes the noise but also these small photon signals. Discarding photon events in dim light might seem like a bad strategy. But because small events are much more likely to be noise than photons, it actually proves to be an excellent computational bet; by one estimate, thresholding improves signal/noise ratios by more than 350-fold (Field and Rieke, 2002). At higher intensities, where every rod captures at least one photon, linear amplification is a better bet, and indeed background light attenuates the threshold nonlinearity at the rod-to-rod bipolar synapse (Sampath and Rieke, 2004). Furthermore, a direct pathway from rods to OFF cone bipolar cells acts linearly (twilight circuit; Fig. 18.1, left) (Field and Rieke, 2002).

At all but the lowest intensities, the rod bipolar cell's large convergence gives it, not a binary signal, but a coarsely graded one. Consistent with this, instead of using one ribbon (as at the rod output) the rod bipolar axon uses 30 ribbon synapses at its output (Sterling and Demb, 2004). Rapid depolarization of the rod bipolar cell generates a transient burst of vesicle release that depresses with repeated stimulation; the synaptic depression prevents response saturation in the postsynaptic AII amacrine cell (Singer and Diamond, 2003; Dunn and Rieke, 2008; Snellman et al., 2009). The AII cell collecting from about 30 rod bipolar cells needs to transfer a more finely graded signal, and for this it signals cone bipolar terminals that contribute 150–2000 synapses to a ganglion cell (Sterling and Demb, 2004). Thus, the overall pattern of the rod bipolar circuit is a stepwise expansion in number of ribbon synapses to match the stepwise increase in signal pooling.

At the final stage, the rod bipolar circuit connects into the cone bipolar circuit using appropriate synapses for signaling the ON and OFF pathways (Kolb, 1979; Strettoi et al., 1992). The AII cell receives excitatory input from the rod bipolar cell (an ON-type cell) and so inherits an ON response (Xin and Bloomfield, 1999; Pang et al., 2007). The AII cell signals the downstream ON pathway via its gap junctions with the ON cone bipolar cell (which then signals ON ganglion cells as described earlier) and signals the downstream

OFF pathway via glycinergic synapses through two routes: an indirect route, with OFF cone bipolar terminals (which then signal OFF ganglion cells, as described earlier); and a direct route, with OFF ganglion cell dendrites (Fig. 18.1, right). Thus, the same AII cell that serves crossover inhibition in the daylight circuit forms dual forward synapses onto parallel cone bipolar pathways in the starlight circuit. The rod bipolar circuit's "parasitic" use of the cone bipolar circuit at the final input to ganglion cells saves space, which, in a tissue constrained to be thin (≤ 250 µm), is at a premium.

Tuning the Circuits

Neighboring AII cells are coupled to each other by gap junctions (not shown in Fig. 18.1). The apparent purpose of the coupling is to reduce noise by signal averaging (Bloomfield and Dacheux, 2001; Veruki and Hartveit, 2002). Coupling is modulated by dopaminergic synapses on the AII, which uncouple AII–AII junctions (Hampson et al., 1992). Since retinal dopamine declines in darkness, AII coupling should strengthen; however, the coupling seems to peak at dim backgrounds above absolute darkness (Bloomfield and Dacheux, 2001). Furthermore, AII–ON cone bipolar cell junctions uncouple when cGMP rises within the bipolar cell in response to nitric oxide production (Mills and Massey, 1995). This AII–ON cone bipolar cell uncoupling might occur under some bright lighting conditions where the spread of ON cone bipolar signals through the AII network would degrade spatial acuity. Of course, once neuromodulators of coupling are identified, such as NO and dopamine, the following question arises: What signals and effectors modulate the modulators? This is a question for the future.

References

Bloomfield SA, Dacheux RF (2001) Rod vision: pathways and processing in the mammalian retina. *Prog Retin Eye Res* 20:351–384.

Borghuis BG, Sterling P, Smith RG (2009) Loss of sensitivity in an analog neural circuit. *J Neurosci* 29:3045–3058.

DeVries SH, Baylor DA (1995) An alternative pathway for signal flow from rod photoreceptors to ganglion cells in mammalian retina. *Proc Natl Acad Sci USA* 92:10658–10662.

DeVries SH, Qi X, Smith R, Makous W, Sterling P (2002) Electrical coupling between mammalian cones. *Curr Biol* 12(22):1900–1907.

Dunn FA, Rieke F (2008) Single-photon absorptions evoke synaptic depression in the retina to extend the operational range of rod vision. *Neuron* 57:894–904.

Field GD, Rieke F (2002) Nonlinear signal transfer from mouse rods to bipolar cells and implications for visual sensitivity. *Neuron* 34:773–785.

Hack I, Peichl L, Brandstatter JH (1999) An alternative pathway for rod signals in the rodent retina: rod photoreceptors, cone bipolar cells, and the localization of glutamate receptors. *Proc Natl Acad Sci USA* 96:14130–14135.

Hampson EC, Vaney DI, Weiler R (1992) Dopaminergic modulation of gap junction permeability between amacrine cells in mammalian retina. *J Neurosci* 12:4911–4922.

Hornstein EP, Verweij J, Li PH, Schnapf JL (2005) Gap-junctional coupling and absolute sensitivity of photoreceptors in macaque retina. *J Neurosci* 25:11201–11209.

Kolb H (1979) The inner plexiform layer in the retina of the cat: electron microscopic observations. *J Neurocytol* 8:295–329.

Li W, Keung JW, Massey SC (2004) Direct synaptic connections between rods and OFF cone bipolar cells in the rabbit retina. *J Comp Neurol* 474:1–12.

Manookin MB, Beaudoin DL, Ernst ZR, Flagel LJ, Demb JB (2008) Disinhibition combines with excitation to extend the operating range of the OFF visual pathway in daylight. *J Neurosci* 28:4136–4150.

Miller RF (2008) Cell communication mechanisms in the vertebrate retina. The proctor lecture. *Invest Ophthalmol Vis Sci* 49:5184–5198.

Mills SL, Massey SC (1995) Differential properties of two gap junctional pathways made by AII amacrine cells. *Nature* 377:734–737.

Pang JJ, Abd-El-Barr MM, Gao F, Bramblett DE, Paul DL, Wu SM (2007) Relative contributions of rod and cone bipolar cell inputs to AII amacrine cell light responses in the mouse retina. *J Physiol* 580:397–410.

Ribelayga C, Cau Y, Mangel SC (2008) The circadian clock in the retina controls rod-cone coupling. *Neuron* 59:790–801.

Sampath AP, Rieke F (2004) Selective transmission of single photon responses by saturation at the rod-to-rod bipolar synapse. *Neuron* 41:431–443.

Schneeweis DM, Schnapf JL (1999) The photovoltage of macaque cone photoreceptors: adaptation, noise, and kinetics. *J Neurosci* 19:1203–1216.

Sheng Z, Choi SY, Dharia A, Li J, Sterling P, Kramer RH (2007) Synaptic Ca^{2+} in darkness is lower in rods than cones, causing slower tonic release of vesicles. *J Neurosci* 27:5033–5042.

Singer JH, Diamond JS (2003) Sustained Ca^{2+} entry elicits transient postsynaptic currents at a retinal ribbon synapse. *J Neurosci* 23:10923–10933.

Snellman J, Zenisek D, Nawy S (2009) Switching between transient and sustained signaling at the rod bipolar-AII amacrine synapse of the mouse retina. *J Physiol* 587:2443–2455.

Soucy E, Wang Y, Nirenberg S, Nathans J, Meister M (1998) A novel signaling pathway from rod photoreceptors to ganglion cells in mammalian retina. *Neuron* 21:481–493.

Srinivasan MV, Laughlin SB, Dubs A (1982) Predictive coding: a fresh view of inhibition in the retina. *Proc R Soc Lond B Biol Sci* 216:427–459.

Sterling P, Demb JB (2004) Retina. In: Shepherd GM, ed. *The Synaptic Organization of the Brain*, 5th ed., pp. 217–269. New York: Oxford University Press.

Strettoi E, Raviola E, Dacheux RF (1992) Synaptic connections of the narrow-field, bistratified rod amacrine cell (AII) in the rabbit retina. *J Comp Neurol* 325:152–168.

Tsukamoto Y, Morigiwa K, Ueda M, Sterling P (2001) Microcircuits for night vision in mouse retina. *J Neurosci* 21:8616–8623.

Tsukamoto Y, Morigiwa K, Ishii M, Takao M, Iwatsuki K, Nakanishi S, Fukuda Y (2007) A novel connection between rods and ON cone bipolar cells revealed by ectopic metabotropic glutamate receptor 7 (mGluR7) in mGluR6-deficient mouse retinas. *J Neurosci* 27:6261–6267.

van Rossum MC, Smith RG (1998) Noise removal at the rod synapse of mammalian retina. *Vis Neurosci* 15:809–821.

Veruki ML, Hartveit E (2002) AII (Rod) amacrine cells form a network of electrically coupled interneurons in the mammalian retina. *Neuron* 33:935–946.

Xin D, Bloomfield SA (1999) Comparison of the responses of AII amacrine cells in the dark- and light-adapted rabbit retina. *Vis Neurosci* 16:653–665.

Yin L, Smith RG, Sterling P, Brainard DH (2006) Chromatic properties of horizontal and ganglion cell responses follow a dual gradient in cone opsin expression. *J Neurosci* 26:12351–12361.

19

Evolution of Retinal Circuitry: From Then to Now

Frank S. Werblin and John Dowling

Looking back now at our original paper in 1969, it is remarkable how little we knew then about retinal circuitry (Werblin & Dowling 1969). Figure 19.1 summarizes our understanding of retinal connectivity then. We knew that there were ON and OFF pathways, and our work showed that ON and OFF activity was initiated at the bipolar cell level. We did not know what the excitatory transmitter was, or how the visual signal got inverted at the ON bipolar cell dendrites. With reference to John's electron microscopy, we inferred that the antagonistic surround measured in bipolar cells was mediated by horizontal cells, and that the antagonistic signal was fed back to photoreceptors, but we did not understand the synapse that mediated this feedback. Remarkably the feedback synapse remains mysterious today (Fahrenfort et al 2004; Jouhou et al 2007; Kamermans et al 2001; Verweij et al 1996; Xin & Bloomfield 2000). We missed the distinction between rods and cones. Our photoreceptor recordings were taken from rods, as indicated by their slow return to baseline following a light flash. Because we identified rods, but not cones, and because feedback to rods is difficult to measure electrically as shown by Thoreson et al. (2008), we missed the feedback signal to rods, measured the following year by Baylor, Fuortes, and O'Brien in turtle cones (Baylor et al 1971). A sketch of the primitive circuitry we uncovered in 1969 is shown in Figure 19.1.

We probably did record from cones but misinterpreted these recordings as originating from OFF bipolar cells. Both responses are sustained and hyperpolarizing with fast rise and fall times, and with a small overshoot at light OFF. We now know that the OFF bipolar cells express a more robust antagonistic surround response. By the same token, we missed the sustained amacrine cell responses because they too resembled closely the ON and OFF bipolar responses. We missed these cell types because our cell identification technique was terribly crude. We had learned from Aki Kaneko in Tomita's

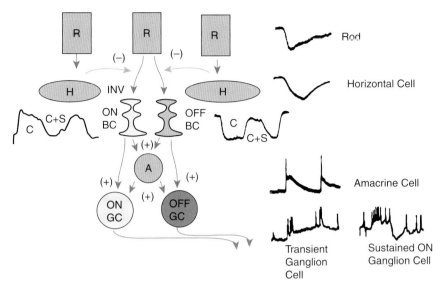

FIGURE 19–1. Abstract of the circuitry derived from Werblin and Dowling (1969). Photoreceptor responses were from rods; we did not identify the cone response. Horizontal cell responses were sluggish and hyperpolarizing. We found ON and OFF bipolar cells with antagonistic surrounds, but we did not identify the pathway for surround antagonism but suggested that the horizontal cells mediated the antagonistic surround response shown with the green arrows. We found ON and OFF ganglion cells that we attributed to inputs from ON and OFF bipolar cells. We also found a robust ON OFF spiking transient amacrine cell that we believed drove the ON OFF ganglion cells. We had not yet appreciated the intricate inhibitory systems that were later identified, mediated by a variety of amacrine cell types. All red arrows represent excitatory pathways except the arrow from rods to ON bipolar cells designated by INV. We still do not know what to call this pathway: we now know that the transmitter is glutamate, the reversal potential is near zero, but the synapse is sign inverting.

lab (Kaneko 1970; 1971) that it was possible to fill a recorded cell with Niagra Sky blue, but it was a major challenge to get the dye into the pipette. At that time we had not learned that one could run a fiber down the pipette and simply backfill it. So at the time we had to boil the solution with the pipettes submerged in it, hoping to drive out the bubbles that accumulated near the tip. Even then, it was a challenge to eject the Niagara Sky blue from the pipette. We increased the voltage on the electrode in delicate increments, but that did not work. Finally, we settled upon a technique in which we passed close to 100 volts (not millivolts) to the electrode and literally blew the tip off. This deposited some of the dye at the electrode tip, but more frequently destroyed the cell we were trying to stain. (We later learned that Aki used an equally and absurdly high voltage!)

The basic structure of parallel pathways and lateral inhibition was in place, but we were not yet aware of the complex inhibitory networks that lie at the core of retinal processing.

CIRCUITRY NOW

We did not learn about ON OFF stratification in the inner plexiform layer until the beautiful work of Famiglietti, Nelson, and Kolb (Famiglietti et al 1977; Famiglietti & Kolb 1976; Nelson et al 1978), who used a much more refined staining technique with Lucifer Yellow to show the stratification of ON and OFF cells in the inner plexiform layer (IPL). We also learned that the excitatory transmitter in the retina was glutamate, and that horizontal cells could feedforward to bipolar cells as well as back to cones.

Years later we learned that the transient ON OFF amacrine cells we had identified were wide field neurons similar to the polyaxonal cells described by Famiglietti and studies physiologically by Stewart Bloomfield's lab (Volgyi et al 2001). Work from Dick Masland's lab (MacNeil et al 1999; MacNeil & Masland 1998) has shown us that there exists a bewildering array of amacrine cells—close to 30 different morphological types. Some of these are endowed with general properties, others with quite specialized function. Among the general classes are the many vertically oriented diffuse amacrine cells. Most of these are glycinergic and serve to carry signals from the ON to the OFF sublamina, an activity that is called "crossover inhibition." When ON inhibition is combined with OFF excitation, the resulting voltage response is more linear than either of the inputs. This linearization of signals occurs at every retinal level and is found in the majority of bipolar amacrine and ganglion cells. Many of Masland's vertically oriented amacrine cells are involved in this activity. These amacrine cells have been identified as glycinergic by Wassle's lab (Menger et al 1998). While the glycinergic amacrine cells run vertically through the retina, the GABAergic amacrine cells run, for the most part, horizontally. Many of these GABAergic amacrine cells respond transiently at ON and OFF and are monostratified like the polyaxonal cells mentioned earlier. Others are narrower in lateral dimension and generate a more sustained response. All inhibition within the amacrine cell community is glycinergic. GABAergic amacrine cells feed back to bipolar cells and forward to ganglion cells, but they do not generally inhibit other amacrine cells.

There are many other specific amacrine cell circuits, but the general pattern of inhibitory activity follows a few fundamental rules as shown in Figure 19.2. Glycinergic inhibition travels vertically and mainly from ON to OFF or OFF to ON layers, serving to linearize signals (Hsueh et al 2008; Molnar & Werblin 2007) (or superimpose rod activity on the ON and OFF cone pathways). GABAergic activity travels laterally, often within a single sublamina of the IPL. In some cases these lateral pathways form an additional layer of lateral inhibition, superimposed on the lateral antagonistic pathway formed at the outer retina through horizontal cell activity. But a narrow, local form of GABAergic inhibition also exists that may be involved in modulating the temporal properties of signals arriving at the ganglion cells. Starburst to starburst GABAergic inhibition follows the lateral GABA rule, although it

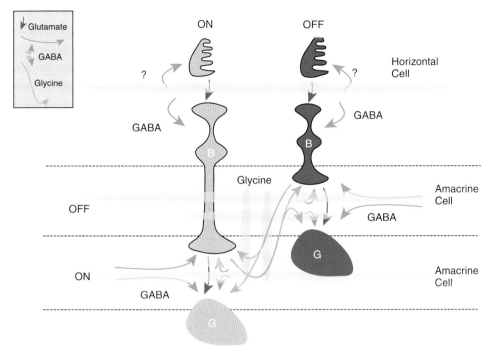

FIGURE 19–2. A general view of the circuitry now identified in the mammalian retina. Horizontal cells contain and release GABA. They have been shown to feed forward to both ON and OFF bipolar cells. Horizontal cells also feed back to cones by a mechanism that remains controversial. Horizontal cell feedback to rods has recently been identified as well. At the inner retina there appear to be three main types of amacrine cells. Wide-field GABAergic amacrine cells are horizontally oriented, often lying completely within a single stratum of the inner plexiform layer. Their responses are transient and often occur at ON and OFF. Narrow-field GABAergic amacrine cells also lie in horizontal orientation, and their response tends to be either ON or OFF and more sustained. Glycinergic amacrine cells have a vertical orientation and carry signals across sublamina, often from ON to OFF or OFF to ON. They are the essential elements in "crossover inhibition," a circuitry that compensates for nonlinear synaptic transmission. Red arrows designate glutamate pathways, green arrows designate GABA pathways, and blue arrows designate glycine pathways. The feedback pathway to photoreceptors designated by the question mark (?) in the Fig. 19.2, remains controversial.

violates the rule that amacrine cell to amacrine cell communication is strictly glycinergic (see Fig. 19.3). A sketch of the full complement of inhibitory pathways is shown in Figure 19.2.

The mechanism of feedback to cones remains controversial as described earlier. Feedforward to bipolar cells appears to be mediated by GABA. The ON bipolar cells appear to have a more depolarized chloride reversal potential than the OFF bipolars (Miller & Dacheux 1983; Vardi et al 2000). A feedforward electrical antagonistic signal from horizontal cells to ON bipolar has also been described (Zhang & Wu 2009). We now have good evidence that the

IPL is stratified into about 10 different functional layers, and the wiring diagram shown in Figure 19.2 is repeated at each stratum. Each stratum is subserved by a distinct population of bipolar cells (Brown & Masland 1999; MacNeil et al 2004). The strata are also defined by the ramification of populations of ganglion cell dendrites (Rockhill et al 2002). Roughly speaking, each stratum is driven by a separate population of bipolar cells and "read out" by a distinct population of ganglion cells. Inhibition at the IPL is carried by a bewildering array of amacrine cells. Amacrine cells carry activity between strata via glycine, and within each stratum via GABA. As a consequence of this stratum by stratum wiring, each ganglion cell type expresses a unique set of excitatory (via bipolar) and inhibitory (via amacrine) patterns (Roska et al 2006; Roska & Werblin 2001; Werblin & Roska 2007; Werblin et al 2001). The challenge remains to understand how the elaborate circuitry at the IPL leads to the functional properties of each of the 12 retinal output channels.

CIRCUITRY MEDIATING DIRECTIONAL SELECTIVITY

There are a number of significant specialized amacrine cell networks in the retina. One of the most thoroughly studied is the AII amacrine cell that piggybacks the rod pathway onto the ON and OFF cone ganglion cells. Among the specialized amacrine cells are the starburst. Starburst cells are the key elements in generating directional selectivity in ganglion cells (Euler et al 2002; Fried et al 2002; 2005; Lee & Zhou 2006; Oesch et al 2005). The starburst cells have some unique properties: *(1)* they release GABA, but only at the distal tips at their circumference; *(2)* they are radially directional such that they release GABA in response to centrifugal but not centripetal movement; *(3)* they are mutually inhibitory, enhancing their directional polarization; and *(4)* their connections to ganglion cells is asymmetric such that they inhibit from the null, but not the preferred side of the ganglion cell's receptive field. A circuit encompassing these properties is shown in Figure 19.3.

   When John and I first completed our work with the 1969 paper we assumed that the retina was solved, and it was time to explore other areas of neuroscience. Nothing could have been further from the truth. We had answered a few initial questions about retinal structure and function, but had not yet formulated the myriad of questions that would eventually arise as a result of our work and that of others. In the 40 years since we originally published our work on retinal organization, our understanding of retinal processing and function has increased by orders of magnitude. Although some key questions remain unresolved, it is likely that most of the questions have not even been addressed yet. One can imagine that in 40 years from now, today's understanding of the retina today will look as primitive as our original work of 40 years ago.

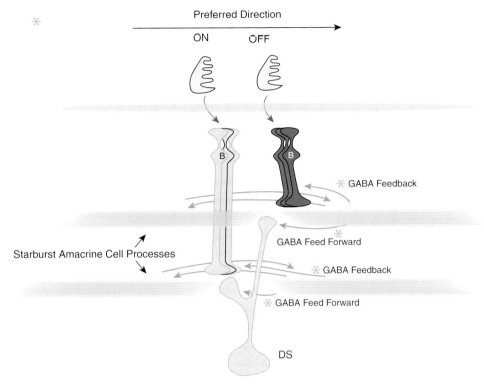

FIGURE 19–3. Circuitry for directional selectivity. This specialized circuitry is embedded in the more general circuitry shown in Figure 19.2. The DS cell responds to movement in the preferred direction (left to right) but is silent for movement in the opposite null direction. It is a bistratified cell with processes in both the ON and OFF sublamina. This asymmetry is mediated by a complex set of interactions. We list here some of the main components: (1) Starburst amacrine cells are the key players. There can be as many as 70 starburst amacrine cells lying within the dendritic reach of a single DS cell. The starburst cells are themselves directional: they respond more strongly to movement away from the cell body, lying at the center of the dendritic field, than to movement toward the cell body. Furthermore, starburst cells are mutually inhibitory via GABA synapses shown by the longer green arrows in the figure. As a consequence of this directionality and mutual inhibition, the starburst cells on the left will be inhibited by movement from right to left, and the starburst cells on the right will be inhibited by movement from left to right. There are two more essential asymmetries. Bipolar and ganglion cells are inhibited by starburst cells on the right, but not inhibited by starburst cells on the left. These four pathways are designated by asterisks. As a consequence of these asymmetries, the DS cell receives directionally selective excitation and directionally selective inhibition. Then a third layer of DS interaction occurs at the DS cell itself, since the DS cell is excited from the preferred direction but inhibited from the null direction. This circuitry shows how a series of antagonistic interactions, operating in the directional modality, enhance the directional selectivity of the overall circuit.

REFERENCES

Baylor DA, Fuortes MG, O'Bryan PM. 1971. Receptive fields of cones in the retina of the turtle. *J Physiol* 214:265–94.

Brown SP, Masland RH. 1999. Costratification of a population of bipolar cells with the direction-selective circuitry of the rabbit retina. *J Comp Neurol* 408:97–106.

Euler T, Detwiler PB, Denk W. 2002. Directionally selective calcium signals in dendrites of starburst amacrine cells. *Nature* 418:845–52.

Fahrenfort I, Sjoerdsma T, Ripps H, Kamermans M. 2004. Cobalt ions inhibit negative feedback in the outer retina by blocking hemichannels on horizontal cells. *Vis Neurosci* 21:501–11.

Famiglietti EV, Jr., Kaneko A, Tachibana M. 1977. Neuronal architecture of on and off pathways to ganglion cells in carp retina. *Science* 198:1267–69.

Famiglietti EV, Jr., Kolb H. 1976. Structural basis for ON-and OFF-center responses in retinal ganglion cells. *Science* 194:193–95.

Fried SI, Munch TA, Werblin FS. 2002. Mechanisms and circuitry underlying directional selectivity in the retina. *Nature* 420:411–14.

Fried SI, Munch TA, Werblin FS. 2005. Directional selectivity is formed at multiple levels by laterally offset inhibition in the rabbit retina. *Neuron* 46:117–27.

Hsueh HA, Molnar A, Werblin FS. 2008. Amacrine-to-Amacrine Cell Inhibition in the Rabbit Retina. *J Neurophysiol* 100:2077–88.

Jouhou H, Yamamoto K, Homma A, Hara M, Kaneko A, Yamada M. 2007. Depolarization of isolated horizontal cells of fish acidifies their immediate surrounding by activating V-ATPase. *J Physiol* 585:401–12.

Kamermans M, Fahrenfort I, Schultz K, Janssen-Bienhold U, Sjoerdsma T, Weiler R. 2001. Hemichannel-mediated inhibition in the outer retina. *Science* 292:1178–80.

Kaneko A. 1970. Physiological and morphological identification of horizontal, bipolar and amacrine cells in goldfish retina. *J Physiol* 207:623–33.

Kaneko A. 1971. Physiological studies of single retinal cells and their morphological identification. *Vision Res* Suppl 3:17–26.

Lee S, Zhou ZJ. 2006. The synaptic mechanism of direction selectivity in distal processes of starburst amacrine cells. *Neuron* 51:787–99.

MacNeil MA, Heussy JK, Dacheux RF, Raviola E, Masland RH. 1999. The shapes and numbers of amacrine cells: matching of photofilled with Golgi-stained cells in the rabbit retina and comparison with other mammalian species. *J Comp Neurol* 413:305–26.

MacNeil MA, Heussy JK, Dacheux RF, Raviola E, Masland RH. 2004. The population of bipolar cells in the rabbit retina. *J Comp Neurol* 472:73–86.

MacNeil MA, Masland RH. 1998. Extreme diversity among amacrine cells: implications for function. *Neuron* 20:971–82.

Menger N, Pow DV, Wassle H. 1998. Glycinergic amacrine cells of the rat retina. *J Comp Neurol* 401:34–46.

Miller RF, Dacheux RF. 1983. Intracellular chloride in retinal neurons: measurement and meaning. *Vision Res* 23:399–411.

Molnar A, Werblin FS. 2007. Inhibitory Feedback Shapes Bipolar Cell Responses in the Rabbit Retina. *J Neurophysiol* 98:3423–35.

Nelson R, Famiglietti EV, Jr., Kolb H. 1978. Intracellular staining reveals different levels of stratification for on- and off-center ganglion cells in cat retina. *J Neurophysiol* 41:472–83.

Oesch N, Euler T, Taylor WR. 2005. Direction-selective dendritic action potentials in rabbit retina. *Neuron* 47:739–50.

Rockhill RL, Daly FJ, MacNeil MA, Brown SP, Masland RH. 2002. The diversity of ganglion cells in a mammalian retina. *J Neurosci* 22:3831–43.

Roska B, Molnar A, Werblin FS. 2006. Parallel processing in retinal ganglion cells: how integration of space-time patterns of excitation and inhibition form the spiking output. *J Neurophysiol* 95:3810–22.

Roska B, Werblin F. 2001. Vertical interactions across ten parallel, stacked representations in the mammalian retina. *Nature* 410:583–87.

Thoreson WB, Babai N, Bartoletti TM. 2008. Feedback from horizontal cells to rod photoreceptors in vertebrate retina. *J Neurosci* 28:5691–95.

Vardi N, Zhang LL, Payne JA, Sterling P. 2000. Evidence that different cation chloride cotransporters in retinal neurons allow opposite responses to GABA. *J Neurosci* 20:7657–63.

Verweij J, Kamermans M, Spekreijse H. 1996. Horizontal cells feed back to cones by shifting the cone calcium-current activation range. *Vision Res* 36:3943–53.

Volgyi B, Xin D, Amarillo Y, Bloomfield SA. 2001. Morphology and physiology of the polyaxonal amacrine cells in the rabbit retina. *J Comp Neurol* 440:109–25.

Werblin F, Roska B. 2007. The movies in our eyes. *Sci Am* 296:72–9.

Werblin F, Roska B, Balya D. 2001. Parallel processing in the mammalian retina: lateral and vertical interactions across stacked representations. *Prog Brain Res* 131:229–38.

Werblin FS, Dowling JE. 1969. Organization of the retina of the mudpuppy, Necturus maculosus. II. Intracellular recording. *J Neurophysiol* 32:339–55.

Xin D, Bloomfield SA. 2000. Effects of nitric oxide on horizontal cells in the rabbit retina. *Vis Neurosci* 17:799–811.

Zhang AJ, Wu SM. 2009. Receptive fields of retinal bipolar cells are mediated by heterogeneous synaptic circuitry. *J Neurosci* 29:789–97.

Section 7

Auditory System

20

Cochlea

Peter Dallos

The most significant microcircuits of the mammalian cochlea are very different from all others in the nervous system in that they are based on electro–mechanical and mechano–mechanical interactions. These are feedback systems, whose function is to facilitate high-frequency, high-sensitivity, and highly frequency-selective operation of the hydro-mechanical components of the cochlea. As the sequential diagrams of Figure 20.1 reveal, pressure waves delivered into the fluid-filled cochlea by the middle-ear bones, set the cochlear partition (basilar membrane [BM], organ of Corti [OC], and tectorial membrane [TM]) into vibration. Due to the mechanical properties of these structures, their vibratory pattern manifests in a traveling wave, which performs a crude spectral analysis. High frequencies produce vibratory maxima at the beginning and low frequencies toward the far end of the partition. Resulting local deformations of the partition produce radial shear between the bottom of the TM and the top of the OC. The sensory receptors of the cochlea are hair cells, polarized epithelial cells, whose mechano-sensitive organelles are highly organized actin-filled villi, misnamed stereocilia. In the mammal there are two distinct groups of hair cells, inner hair cells (IHCs) and outer hair cells (OHCs). Inner hair cells are functionally and morphologically similar to hair cells throughout the animal kingdom—receiving some 95% of afferent innervation, they are the true sensory receptors of the mammalian ear. Outer hair cells are a mammalian innovation. They are innervated by only ~5% of nonmyelinated afferents, along with prominent efferents. Their sensory role is questionable, while they appear to have evolved motor functions. The sensory transducer apparatus of OHCs is "classical"; just as IHCs, they possess mechanotransducer (MET) channels near the tips of their ciliary bundles.

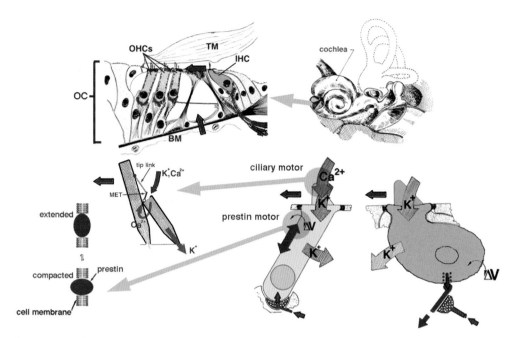

FIGURE 20–1. Increasing physiological detail of the peripheral auditory apparatus, read counterclockwise from top right. See text for detailed description. Abbreviations: OC: organ of Corti; OHC: outer hair cell; IHC: inner hair cell; TM: tectorial membrane; BM: basilar membrane.

These channels are mechanically opened by bundle deflection, upon which cationic current flow (K^+ and Ca^{2+}) enters the cell. Potassium current is largely responsible for depolarization of the OHC, while calcium current facilitates local regulatory functions (for reviews see Dallos, 1992; Dallos et al., 2006).

Two distinct motor processes are identified with OHCs. The first is apparently present in *all* hair cells. This is a two-component operation of Ca^{2+} on the MET channel itself and on myosin–actin interactions in the stereocilia. The process produces adaptation of the transducer current. Coincidentally, there is a resultant mechanical force on the bundle, which, in some circumstances, is such as to enhance bundle displacement that caused it in the first place. Thus, there can be positive feedback on the bundle and consequent amplification of its displacement (for reviews see Fettiplace, 2006; Hudspeth, 2008). The second motor process is unique to OHCs. This is a voltage-dependent elongation-contraction of the cylindrical cell body itself, termed *electromotility*. As the cell changes its length, it feeds back mechanical force onto its environment. A consequence of this feedback is a very significant enhancement of vibratory amplitude of the BM occurring in a narrow spatial extent (frequency range). Thus, the mammalian cochlea's sensitivity and frequency selectivity is entirely dependent on OHC electromotility. This motile response is produced by the concerted action of millions of membrane-spanning molecules,

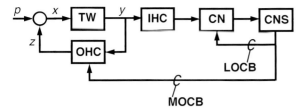

Figure 20–2. Block diagram of the various feedback paths of the peripheral auditory system.

the novel motor protein prestin (SLC26A5). Prestin knockout mice, just as animals with destroyed OHCs, have approximately 50 dB hearing-threshold elevation and they lack frequency selectivity. It is likely that in mammals both OHC feedback mechanisms combine their operation in some (as yet unknown) fashion, while in nonmammalian vertebrates only the ciliary mechanism is effective (for reviews see Fettiplace, 2006; Ashmore, 2008; Dallos, 2008).

The two mechanical feedback processes from OHCs to BM-OC-TM are combined in a single feedback loop in the block diagram of Figure 20.2. The input variable, p, is the pressure gradient across the cochlear partition. The signal x is a modified pressure variable that reflects the contribution of the feedback from OHCs. The block TW represents the passive traveling wave filter, while its output, y, is the traveling wave measured in the live, active cochlea. If OHC feedback is deactivated, such as in the prestin knock-out animal, then y is simply a filtered, passive, and linear version of the input x. Mechanical disturbances evoked by the TW stimulate IHCs and, via their rich afferent connections (CN: cochlear branch of the eighth cranial nerve), they convey sound-related information to a series of way stations, from cochlear nucleus to auditory cortex.

Aside from ascending information, a rich efferent system characterizes the auditory pathway. Our interest here is in the final descending tract from the olivary complex to the cochlea. The olivo-cochlear bundle's two principal branches originate in different cellular complexes and have different destinations. The lateral branch (LOCB) terminates with axo-dendritic synapses on IHC afferents. The medial branch (MOCB) delivers prominent cholinergic synapses on the bottoms of OHCs. Activation of both branches produces predominantly inhibitory effects. The MOCB feedback to OHCs operates by modulating the local OHC-to-cochlear partition amplificatory feedback. This occurs on two timescales, using two mechanisms. The fast mechanism results in hyperpolarization of the cell due to K^+ efflux through Ca^{2+}-activated K-channels. The controlling Ca^{2+} enters via nicotinic ACh receptors. The incoming calcium has a number of downstream effects, ultimately affecting cytoskeletal proteins and prestin itself via protein phosphorylation. These slow effects modify prestin's ability to produce mechanical feedback.

References

Ashmore JF (2008) Cochlear outer hair cell motility. *Physiol Rev* 88:173–210.

Dallos P (1992) The active cochlea. *J Neurosci* 12:4575–4585.

Dallos P (2008) Cochlear amplification, outer hair cells and prestin. *Cur Op Neurobiol* 18:370–376.

Dallos P, Zheng J, Cheatham MA (2006). Prestin and the cochlear amplifier. *J Physiol (London)* 576:37–42.

Fettiplace R (2006) Active hair bundle movements in auditory hair cells. *J Physiol (London)* 576:29–36.

Hudspeth JA (2008) Making an effort to listen: mechanical amplification in the ear. *Neuron* 59:530–545.

21

Cochlear Nucleus

Eric D. Young and Donata Oertel

Neuronal circuits in the brainstem convert the output of the ear, which carries the acoustic properties of ongoing sound, to a representation of the acoustic environment that can be used by the thalamocortical system. Most important, brainstem circuits reflect the way the brain makes use of acoustic cues to determine where sounds arise and what they mean. The circuits merge the separate representations of sound in the two ears and stabilize them in the face of disturbances such as loudness fluctuations or background noise. Embedded in these systems are some specialized analyses that are driven by the need to resolve tiny differences in the time and intensity of sounds at the two ears and to resolve rapid temporal fluctuations in sounds like the sequence of notes in music or the sequence of syllables in speech.

At the first stage of the auditory pathway, in the cochlear nucleus, the auditory nerve (AN) feeds multiple parallel pathways through the brainstem that perform separate computations and ultimately converge in the inferior colliculus. These pathways are formed through as few as one to as many as three synaptic stages (see Winer and Schreiner, 2005, for a summary). The brainstem pathways originate in separate types of principal cells in the cochlear nucleus that have distinct patterns of inputs and axonal projections. The principal cells differ in their biophysical properties and are tied together by three sets of interneuronal connections, two inhibitory and one excitatory.

The wiring diagram in Figure 21.1 shows one isofrequency sheet of the mammalian cochlear nucleus. The nucleus consists of a stack of such circuits, each one innervated by AN fibers sensitive to a particular frequency; across the tonotopically organized stack, the entire frequency range of the animal is represented (see Chapter 20). The dendrites of neurons that are sharply tuned to frequency are confined within one isofrequency sheet, whereas the dendrites of those that are broadly tuned spread across sheets.

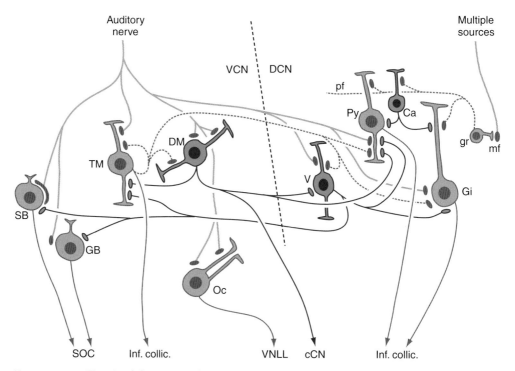

FIGURE 21–1. Circuit of the mammalian cochlear nucleus. Auditory nerve inputs are shown in gray originating at top left. Inputs from other sources, including somatosensory nuclei, vestibular neurons, and others originate at top right and form mossy fiber terminals (mf) in the dorsal cochlear nucleus (DCN; Ryugo et al. 2003). Auditory system targets of principal-cell projections are shown at the bottom. Three inhibitory interneurons (blue) are shown; others are excitatory (red). Excitatory interneurons include T-multipolars (TM), whose axon branches throughout the nucleus, and granule cells (gr), whose parallel-fiber (pf) axons terminate in the DCN (both dashed red). Neurons in both the ventral cochlear nucleus (VCN) and DCN are distinct structurally and functionally. The DCN is layered. The outermost layer is defined by the pf, the second by the cell bodies of the pyramidal cells (Py), and the third containing the vertical (V) and giant (Gi) cells. Three cell types in DCN—Golgi, stellate, and unipolar brush cells—are not shown to simplify the drawing. Excitatory neurons are glutamatergic; inhibitory neurons use glycine, although cartwheel cells also contain GABA. Ca, cartwheel cell; cCN, contralateral cochlear nucleus; DM, D-multipolar; GB, globular bushy cell; Gi, giant cell; gr, granule cell; Inf. colic., central nucleus of the inferior colliculus; mf, mossy fiber; Oc, octopus cell; pf, parallel fiber; Py, pyramidal cell; SB, spherical bushy cell; SOC, superior olivary complex; TM, T-multipolar; V, vertical cell; VNTB, ventral nucleus of the trapezoid body. (Redrawn from Young and Oertel, 2004)

All parts of the cochlear nucleus receive inputs from AN fibers (see top left of Fig. 21.1). AN terminals contact cells through large, dense clusters of somatic endings at some cells and through smaller isolated dendritic endings at others, but all act through postsynaptic AMPA receptors with rapid kinetics and high unitary conductances that produce sharply timed excitatory postsynaptic currents (EPSCs; Raman et al., 1994; Gardner et al., 1999).

Bushy Cells and Temporal Representations

The pathways originating in the bushy cells (SB and GB in Fig. 21.1) use small differences between the two ears in the time of arrival of sounds (the interaural time difference [ITD]) and in the intensity of the sounds (the interaural level difference [ILD]) to determine the direction of a source of sound. Spatial hearing is important not only to locate interesting or dangerous features but also to separate individual sounds from a complex mixture (Bronkhorst and Plomp, 1988). The ITDs and ILDs are computed in the medial and lateral superior olivary nuclei of the superior olivary complex (SOC; see Chapter 22), the inputs to which are provided by the spherical bushy cells (SBs) and globular bushy cells (GBs). There are about 36,600 SBs and 6300 GBs in the ventral cochlear nucleus (VCN) of cat (Osen, 1970).

The computation of ITD requires temporal precision. The ability of humans to localize sound sources to within 1° in azimuth depends on the ability to resolve ITDs as small as 10–30 μs. Detection of such tiny time differences requires accurate encoding of the stimulus waveform in AN fibers, which must be carried by action potentials in SBs to the principal cells of the medial superior olive. The cellular properties of bushy cells reflect their ability to convey precisely timed signals. Spherical bushy cells receive large presynaptic terminals from a small number of AN fibers (<5; Sento and Ryugo, 1989). These synapses can deliver up to 10 nA of current postsynaptically, producing large EPSPs that are sharpened by the fast membrane time constant of the bushy cell (1–2 ms; Fig. 21.2A). The fast time constant is produced by large low-voltage-activated potassium (g_{KL}) and hyperpolarization-activated mixed cation conductances (g_h; Cao et al., 2007). In addition, g_{KL} repolarizes EPSPs rapidly, giving them sharp early peaks and making them brief.

Globular bushy cells share the short membrane time constants of SBs, although they receive many more AN synaptic inputs through smaller synapses (tens of fibers per cell; Ostapoff and Morest, 1991). They project to a nucleus in the SOC that provides inhibitory inputs to the superior olivary nuclei for both ILD (Joris and Yin, 2007) and ITD (Brand et al., 2002) processing.

T-Multipolar Cells and Sound-Level Representation

The array of T-multipolar neurons (TM; a.k.a. T-stellate) forms a direct pathway from the cochlear nucleus to the inferior colliculus (Adams, 1983). Through collateral branches of axons, they excite other neurons within their own isofrequency laminae both in the VCN and dorsal cochlear nucleus (DCN; red dashed line in Fig. 21.1; Smith and Rhode, 1989). Indeed, although AN fibers terminate in the DCN (Ryugo and May, 1993), the denser projection of TMs to the DCN may provide the major source of the DCN's

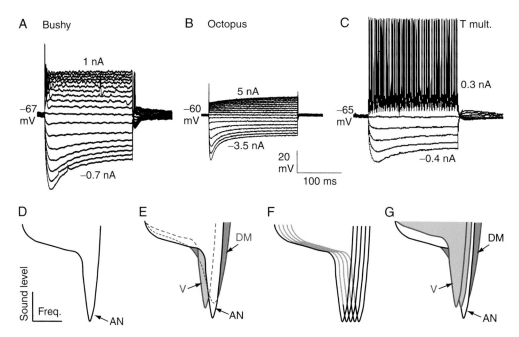

FIGURE 21–2. Some distinctions among cochlear nucleus neurons. (*A–C*) Differences in biophysical properties of principal cells of the ventral cochlear nucleus (VCN) are revealed in responses to injected pulses of current in patch-clamp recordings. The largest positive or negative current is labeled on the voltage responses. Bushy and octopus cells fire transiently at the onset of pulses. In bushy and octopus cells, characteristic clustering of depolarizing traces reflects activation of low-voltage-activated potassium channels (g_{KL}). The sag toward rest in responses to hyperpolarizing pulses reflects the activation of hyperpolarization-activated channels (g_h). Note that current pulses are larger in octopus than bushy cells. T-multipolar cells have little or no g_{KL} and fire steadily in response to depolarization. (*D–G*) Tuning curves for neurons in the cochlear nucleus, shown as sound level just loud enough to elicit a response (ordinate) plotted as a function of frequency (abscissa) for presentations of brief tones. Tuning curves of the cells' inputs are shown superimposed. (*D*) Bushy cells show tuning like AN fibers (AN); inhibitory inputs are weak and often missed. (*E*) TM neurons show excitatory AN inputs with inhibitory inputs from V (green) and DM (blue) cells. Inhibition is weaker than excitation and is thus occluded by it; the occluded parts of the inhibitory inputs are dashed. The detailed arrangement of the inhibitory areas is not known, and it is drawn here the same as the arrangement in the dorsal cochlear nucleus (DCN) (*G*). (*F*) DM and Oc cells receive convergent input across frequency from multiple ANs and are broadly tuned. (*G*) Py and Gi cells in DCN receive the same inhibitory connections as TM cells, but the input from V cells is very strong, giving a large inhibitory area that includes the best frequency of the neuron. Inputs from DM cells are weak and are occluded.

excitatory input (Zhang and Oertel, 1994). There are about 9400 multipolar cells in cochlear nucleus, most of which (~14/15) are TMs.

TM neurons fire tonically when they are depolarized (Fig. 21.2C). Their membrane time constants (~5–10 ms) allow EPSPs to sum and produce steady depolarizations and thus to fire steady trains of action potentials in which the fine timing structure of sounds is poorly preserved.

The array of sharply tuned, tonotopically organized TM neurons carries an ongoing representation of the frequency content, or spectrum, of the stimulus. Each TM neuron receives input from a few AN fibers (~6; Ferragamo et al., 1998), with excitation enhanced through other TM neurons in the same isofrequency band. Inhibition from DM neurons, which are broadly tuned, and V neurons, which are sharply tuned, interacts with the excitation. The summation of sharply tuned excitation and broadly tuned inhibition produces overlapping excitatory and inhibitory response areas (Caspary et al., 1994; Paolini et al., 2005). Although the excitatory and inhibitory areas overlap (Fig. 21.2E), responses to simple sounds such as tones show a central, AN-like, excitatory area and inhibitory "sidebands."

In responses to more complex natural sounds, one would expect TM neurons whose frequency sensitivity corresponds to spectral peaks in the stimulus to have strong excitation from the AN and TM inputs, and the excitation should overcome inhibition; by contrast, in TM neurons that respond to low-energy parts of the spectrum between the peaks, the inhibition should dominate. This combination of excitation and inhibition means that the response rates in these neurons correspond well to the frequency content of the stimulus (May et al., 1998). The representation of speech sounds in AN fibers degenerates at high sound levels because of the limited dynamic ranges of the fibers (Sachs and Young, 1979); this effect is not seen in responses of TM neurons, where the response is stable in form as sound-level changes (Blackburn and Sachs, 1990). The TM neurons complement the role of bushy cells by representing the frequency content of sounds, rather than their temporal characteristics.

Octopus Cells Detect Broadband Transients

Octopus cells (see Fig. 21.1) have dendrites that cross the incoming array of AN fibers and sum inputs from a wide range of frequencies. Octopus cells respond to coincident activation of AN fibers across this array and respond sensitively to broadband transients (such as clicks), having broad tuning and high thresholds (Fig. 21.2F; Rhode and Smith, 1986). There are only about 1500 octopus cells in cats.

Octopus cells have large g_{KL} and g_h conductances and thus have what are among the shortest time constants described in any mammalian neuron (<200 ms; Fig. 21.2B). Octopus cells receive inputs from a large number of AN fibers on their somata and dendrites (>60 per cell; Golding et al., 1995). Because of the low input resistances of octopus cells, synaptic currents from AN fibers produce only small voltage changes. The very large g_{KL} makes octopus cells act as cellular voltage-differentiators; Octopus cell neurons fire only when AN inputs sum to exceed a threshold *rate* of depolarization (Ferragamo and Oertel, 2002). In vivo responses to onset transients are precise to within tens of microseconds (Rhode and Smith, 1986). Octopus cells are themselves excitatory but their targets in the superior paraolivary nucleus and ventral nucleus of the lateral lemniscus inhibit the inferior colliculus (Zhang and Kelly, 2006). They seem to provide a marker of the times of important acoustic events, like the onsets of sounds.

COMPLEMENTARY INHIBITORY SYSTEMS: VERTICAL AND D-MULTIPOLAR CELLS

Two inhibitory systems with contrasting properties affect principal cells in both the VCN and DCN as well as one another (Fig. 21.1; Nelken and Young, 1994). D-multipolar (DM) neurons provide poorly tuned (broadband) inhibition with a strong phasic component (Palmer et al., 1996); vertical (V; a.k.a. tuberculoventral) neurons provide sharply tuned inhibition that is tonic (Spirou et al., 1999). There are 630 DM neurons, located in the VCN; the number of V neurons is not known, but they are all located in the deep layer of the DCN.

The contrast in tuning is evident in V and DM cell morphology. DM neurons have dendrites that cross the incoming array of AN fibers in the VCN (Fig. 21.2F), whereas V neurons have dendrites that are confined to an isofrequency lamina in the deep layer of the VCN (Oertel et al., 1990; Zhang and Oertel, 1993). Individual DM neurons terminate over wide regions of the VCN, DCN, and contralateral cochlear nucleus, whereas individual V neurons terminate within their own isofrequency laminae in the DCN and VCN.

Both groups of interneurons respond tonically to depolarizing current pulses (Oertel et al., 1990; Zhang and Oertel, 1993).

DORSAL COCHLEAR NUCLEUS INTEGRATES COMPLEX ACOUSTIC AND NONACOUSTIC CUES

Filtering of sounds by the ears, head, and shoulders alters the spectrum of sounds at the eardrum. Because the filtering is sensitive to the angle of incidence and ears are asymmetrical, spectral cues are used to learn whether sounds arise from below or above the nose and to distinguish sounds coming from behind from those from the front. Important in these spectral cues are deep spectral notches, narrow frequency regions where there is little sound energy (Musicant et al., 1990). All that is known about the DCN is consistent with the idea that the DCN helps to identify spectral characteristics that are highly correlated with some body movement and therefore are likely to be self-generated and uninformative (Oertel and Young, 2004).

Pyramidal (Py) and giant (Gi) neurons in the DCN integrate two systems of inputs. Their basal dendrites in the deep layer of the DCN receive acoustic input from AN fibers and TM neurons. Their apical dendrites in the cerebellum-like molecular layer receive input from parallel fibers (pf), the axons of granule cells (gr). These neurons convey information from many modalities, including auditory, somatosensory, and vestibular that can inform the animal about its own position and movements (Ryugo et al., 2003). Each of the systems of inputs is shaped by inhibition. The acoustic input in the deep layer is shaped by DM and V neurons (Fig. 21.2G); the multimodal input in the molecular layer is shaped by cartwheel (Ca) and stellate cells (not shown in Fig. 21.1). Synaptic plasticity is associated with only one of these systems of inputs, that through the pf (Fujino and Oertel, 2003; Tzounopoulos et al., 2004).

Because acoustic responses are generally recorded when the head is fixed, those responses reflect mainly the inputs from the deep layer on basal dendrites. The synaptic organization leads to the tuning shown in Figure 21.2G, where the two excitatory inputs are collapsed into the single "AN" excitatory input and the V and DM inhibitory inputs are shown separately. The V cell inputs are narrowly tuned and strong enough to counteract excitation so they dominate the responses of the principal cells when activated (Young and Davis, 2002). The DM inputs can be demonstrated using broadband acoustic stimuli, but their effects are weak and subtle. The combination of these inputs makes Py and Gi neurons sensitive to spectral notches and causes them to produce a characteristic inhibitory/excitatory response to spectral notches in DCN principal neurons (Reiss and Young, 2005).

Responses to the cerebellar-like system of inputs through gr cells are only incompletely understood. Activation of these inputs results in both inhibition and excitation in Py and Gi cells (Shore, 2005), which is predominantly associated with the muscles that move the pinna in cats (Kanold and Young, 2001).

REFERENCES

Adams JC (1983) Multipolar cells in the ventral cochlear nucleus project to the dorsal cochlear nucleus and the inferior colliculus. *Neurosci Lett* 37:205–208.

Blackburn CC, Sachs MB (1990) The representation of the steady-state vowel sound/ε/ in the discharge patterns of cat anteroventral cochlear nucleus neurons. *J Neurophysiol* 63:1191–1212.

Brand A, Behrend O, Marquardt T, McAlpine D, Grothe B (2002) Precise inhibition is essential for microsecond interaural time difference coding. *Nature* 417:543–547.

Bronkhorst A, Plomp R (1988) The effect of head-induced interaural time and level differences on speech intelligibility in noise. *J Acoust Soc Am* 83:1508–1516.

Cao XJ, Shatadal S, Oertel D (2007) Voltage-sensitive conductances of bushy cells of the mammalian ventral cochlear nucleus. *J Neurophysiol* 97:3961–3975.

Caspary DM, Backoff PM, Finlayson PG, Palombi PS (1994) Inhibitory inputs modulate discharge rate within frequency receptive fields of anteroventral cochlear nucleus neurons. *J Neurophysiol* 72:2124–2133.

Ferragamo MJ, Oertel D (2002) Octopus cells of the mammalian ventral cochlear nucleus sense the rate of depolarization. *J Neurophysiol* 87:2262–2270.

Ferragamo MJ, Golding NL, Oertel D (1998) Synaptic inputs to stellate cells in the ventral cochlear nucleus. *J Neurophysiol* 79:51–63.

Fujino K, Oertel D (2003) Bidirectional synaptic plasticity in the cerebellum-like mammalian dorsal cochlear nucleus. *PNAS* 100:265–270.

Gardner SM, Trussell LO, Oertel D (1999) Time course and permeation of synaptic AMPA receptors in cochlear nuclear neurons correlate with input. *J Neurosci* 19:8721–8729.

Golding NL, Robertson D, Oertel D (1995) Recordings from slices indicate that octopus cells of the cochlear nucleus detect coincident firing of auditory nerve fibers with temporal precision. *J Neurosci* 15:3138–3153.

Joris PX, Yin TCT (2007) A matter of time: internal delays in binaural processing. *Trends Neurosci* 30:70–78.

Kanold PO, Young ED (2001) Proprioceptive information from the pinna provides somatosensory input to cat dorsal cochlear nucleus. *J Neurosci* 21:7848–7858.

May BJ, LePrell GS, Sachs MB (1998) Vowel representations in the ventral cochlear nucleus of the cat: Effects of level, background noise, and behavioral state. *J Neurophysiol* 79:1755–1767.

Musicant AD, Chan JCK, Hind JE (1990) Direction-dependent spectral properties of cat external ear: New data and cross-species comparisons. *J Acoust Soc Am* 87:757–781.

Nelken I, Young ED (1994) Two separate inhibitory mechanisms shape the responses of dorsal cochlear nucleus type IV units to narrowband and wideband stimuli. *J Neurophysiol* 71:2446–2462.

Oertel D, Young ED (2004) What's a cerebellar circuit doing in the auditory system? *Trends Neurosci* 27:104–110.

Oertel D, Wu SH, Garb MW, Dizack C (1990) Morphology and physiology of cells in slice preparations of the posteroventral cochlear nucleus of mice. *J Comp Neurol* 295: 136–154.

Osen KK (1970) Afferent and efferent connections of three well-defined cell types of the cat cochlear nucleus. In: Anderson P, Jansen JKS, eds. *Excitatory Synaptic Mechanisms*, pp. 295–300. Oslo, Norway: Universitetsforlaget.

Ostapoff EM, Morest DK (1991) Synaptic organization of globular bushy cells in the ventral cochlear nucleus of the cat: A quantitative study. *J Comp Neurol* 314:598–613.

Palmer AR, Jiang D, Marshall DH (1996) Responses of ventral cochlear nucleus onset and chopper units as a function of signal bandwidth. *J Neurophysiol* 75:780–794.

Paolini AG, Clarey JC, Needham K, Clark GM (2005) Balanced inhibition and excitation underlies spike firing regularity in ventral cochlear nucleus chopper neurons. *Eur J Neurosci* 21:1236–1248.

Raman IM, Zhang S, Trussell LO (1994) Pathway-specific variants of AMPA receptors and their contribution to neuronal signaling. *J Neurosci* 14:4998–5010.

Reiss LAJ, Young ED (2005) Spectral edge sensitivity in neural circuits of the dorsal cochlear nucleus. *J Neurosci* 25:3680–3691.

Rhode WS, Smith PH (1986) Encoding timing and intensity in the ventral cochlear nucleus of the cat. *J Neurophysiol* 56:261–286.

Ryugo DK, May SK (1993) The projections of intracellularly labeled auditory nerve fibers to the dorsal cochlear nucleus of cats. *J Comp Neurol* 329:20–35.

Ryugo DK, Haenggeli C-A, Doucet JR (2003) Multimodal inputs to the granule cell domain of the cochlear nucleus. *Exp Brain Res* 153:477–485.

Sachs MB, Young ED (1979) Encoding of steady-state vowels in the auditory nerve: representation in terms of discharge rate. *J Acoust Soc Am* 66:470–479.

Sento S, Ryugo DK (1989) Endbulbs of held and spherical bushy cells in cats: morphological correlates with physiological properties. *J Comp Neurol* 280:553–562.

Shore SE (2005) Multisensory integration in the dorsal cochlear nucleus: unit responses to acoustic and trigeminal ganglion stimulation. *Eur J Neurosci* 21:3334–3348.

Smith PH, Rhode WS (1989) Structural and functional properties distinguish two types of multipolar cells in the ventral cochlear nucleus. *J Comp Neurol* 282:595–616.

Spirou GA, Davis KA, Nelken I, Young ED (1999) Spectral integration by type II interneurons in dorsal cochlear nucleus. *J Neurophysiol* 82:648–663.

Tzounopoulos T, Kim Y, Oertel D, Trussell LO (2004) Cell-specific, spike timing-dependent plasticities in the dorsal cochlear nucleus. *Nat Neurosci* 7:719–725.

Winer JA, Schreiner CE (2005) *The Inferior Colliculus*. New York: Springer.

Young ED, Davis KA (2002) Circuitry and function of the dorsal cochlear nucleus. In: Oertel D et al., eds. *Integrative Functions in the Mammalian Auditory Pathway*, pp. 160–206. New York: Springer Verlag.

Young ED, Oertel D (2004) The cochlear nucleus. In: Shepherd GM, ed. *The Synaptic Organization of the Brain*, pp. 125–163. New York: Oxford Press.

Zhang S, Oertel D (1993) Tuberculoventral cells of the dorsal cochlear nucleus of mice: intracellular recordings in slices. *J Neurophysiol* 69:1409–1421.

Zhang S, Oertel D (1994) Neuronal circuits associated with the output of the dorsal cochlear nucleus through fusiform cells. *J Neurophysiol* 71:914–930.

Zhang H, Kelly JB (2006) Responses of neurons in the rat's ventral nucleus of the lateral lemniscus to monaural and binaural tone bursts. *J Neurophysiol* 95:2501–2512.

22

Nucleus Laminaris

Yuan Wang, Jason Tait Sanchez, and Edwin W. Rubel

Our ability to detect subtle acoustic cues in noisy environments, like a conversation in a crowded restaurant, is one benefit of binaural hearing. Binaural hearing is also essential for sound localization. The ability to localize the source of a sound is dependent on time disparities in the arrival of low-frequency signals between the two ears, referred to as interaural time difference (ITD). In all vertebrates, ITDs are encoded by distinct neural circuits in the central auditory nervous system specialized for the temporal processing of sound at the network, synaptic, and cellular levels. Coding of ITDs is first performed by the medial superior olive (MSO) of mammals and by the nucleus laminaris (NL) in birds and some reptiles.

This chapter will focus on the microcircuitry of the chicken NL, which is an excellent example of neural architecture exquisitely tailored for its specialized function in sound localization. Neurons in NL are coincidence detectors, encoding temporal information of sound arriving at the two ears by responding maximally when resulting action potentials (APs) arrive simultaneously, a unique feature responsible for the coding of ITDs in the microsecond range. We will discuss the important structural and functional specializations of NL that optimize this specialized ability, fundamental for binaural hearing in most birds and mammals.

SPECIALIZED FEATURES OF NUCLEUS LAMINARIS

Nucleus laminaris neurons are bipolar; dendrites extend dorsally and ventrally from the soma to form two segregated dendritic domains (Fig. 22.1A). Cell bodies of NL neurons align into a single sheet, resulting in separate

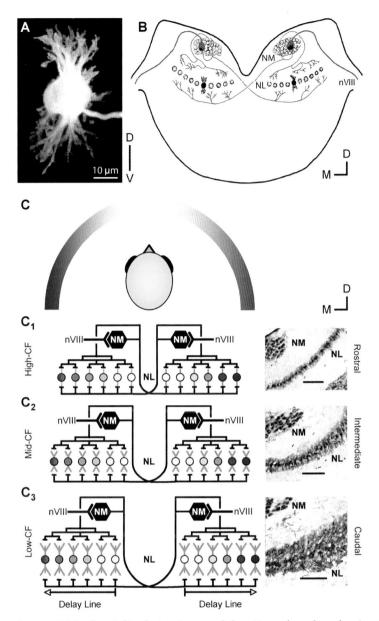

FIGURE 22–1. Specialized structure and function of nucleus laminaris (NL). (*A*) Bipolar NL neuron filled with Alexa 488 clearly shows dorsal (D) and ventral (V) segregated dendrites. (*B*) Schematic drawing of the organization of excitatory inputs (red lines) to NL (coronal view). The dorsal and ventral dendritic domains of NL neurons receive glutamatergic inputs from the ipsilateral and contralateral NM, respectively. The axon lengths to the dorsal NL are equal along an iso-frequency dimension, while the contralateral axon to the ventral NL produces a delay line that compensates for interaural time delays (M, medial; D, dorsal; nVIII, eighth cranial nerve). (*C*) Schematic representation from (*B*) (modified with permission from Dr. Armin Seidl). Blue-white-red arc represents different sound source positions along the horizon. NL neurons respond best to the sound source of the same color and maintain this topographic representation of interaural time delays across iso-frequency lamina (C₁₋₃). Sounds arriving from straight ahead are encoded by NL neurons located medially (white), while sounds originating from the left (blue) and right (red) sides are encoded by neurons located most laterally from the contralateral NL. Note that the three schematics include NL neurons with differing dendritic lengths. Representative MAP2 immunoreactivity sections of NL from the rostral (high-CF), intermediate (mid-CF), and caudal (low-CF) regions are shown in the far right panel to show the dendritic gradient. Scale bars, 100 μm. (From Wang and Rubel, 2008)

dorsal and ventral dendritic neuropil laminas. Structurally, the bipolar configuration and single-cell body layer set the ground rules for the organization of afferent inputs, which is critical for coincidence detection in binaural hearing. Functionally, several physiological mechanisms account for coincidence optimization and are due in part to both intrinsic and synaptic properties of NL neurons. Important anatomical and physiological mechanisms are highlighted in this chapter.

GLUTAMATERGIC EXCITATORY INPUTS

Nucleus laminaris receives glutamatergic excitatory inputs solely from the nucleus magnocellularis (NM), a primary cochlear target, on both sides of the brain. This projection is arranged in a strict topographic manner, resulting in a precise map of sound frequencies (tonotopic) in NL (Rubel and Parks, 1975). That is, NL neurons responsible for encoding high-frequency information are located in the rostro-medial pole, and neurons that are optimally activated by lower frequency acoustic signals are positioned progressively in the caudo-lateral region of the nucleus. Ipsilateral and contralateral terminals from the same parent neuron in NM target identical tonotopic positions in NL but segregate onto dorsal and ventral dendritic domains, respectively (Fig. 22.1B). These terminal arbors are highly anisotropic; oriented orthogonal to the tonotopic axis and confined within a specific iso-frequency band (Young and Rubel, 1983). With this arrangement, individual NL neurons receive information about acoustic signals at the same specific sound frequencies from both ears.

Binaural hearing is further enhanced by substantial convergence and divergence of NM innervation on NL neurons within the same iso-frequency band. In general, NM terminals converge onto about 42% of the somatic surface and 63% of the dendritic surface of NL neurons, as well as the axon hillock and initial segment (Parks et al., 1983). The axon of individual NM neurons appears to terminate onto 20–35 NL neurons, and the composite character of excitatory postsynaptic potentials (EPSPs) by intracellular recordings suggests that multiple NM axons converge onto each NL neuron (Hackett et al., 1982). These integrations occur within iso-frequency bands and thus maintain tonotopic specificity.

To perform coincidence detection between the two ears, ipsilateral and contralateral terminal arbors of NM neurons form radically different and highly stereotyped morphologies. The ipsilateral axon bifurcates to provide equivalent axon length to dorsal dendrites of NL neurons along an iso-frequency dimension (Young and Rubel, 1983). The contralateral axon, however, from NM neurons in the same tonotopic position, extends across the midline and bifurcates several times to create an orderly, serial set of axonal branches to ventral dendrites along a matching iso-frequency lamina of NL on the opposite side of the brain. This arrangement of innervation, predominately along the medial to lateral dimension, results in medial NL neurons receiving contralateral inputs from the shortest input axons and lateral NL neurons from the longest axons. This systematic increase in axonal length across the ventral dendritic field effectively establishes a series of delay lines that compensate for time delays between the two ears and is an excellent example of the modified Jeffress model for processing ITDs (Fig. 22.1C) (Jeffress, 1948).

In concert with these specialized structural features, physiologic properties of individual NL neurons also play a key role in coding time delays between the two ears, improving coincidence detection and ITD processing. One such mechanism involves both pre- (NM) and post- (NL) factors that mediate the depression of synaptic transmission. These factors include the following: the depletion of releasable neurotransmitter vesicles; inactivation of presynaptic Ca^{2+} channels; and the desensitization of postsynaptic AMPA-receptors (AMPA-Rs). In NL, where intrinsic and synaptic responses are extremely brief in order to accommodate rapid transmission, excitatory postsynaptic currents (EPSCs) help control the time course of EPSPs, thus minimizing distortion (Trussell, 1997, 1999).

For example, evoked AMPA-R-mediated EPSCs in NL are extremely fast, having an average decay time constant of less than 1 ms (Fig. 22.2A), due in part to the expression of the GluR3 and GluR4 subunits. Furthermore, there is a considerable reduction in EPSC amplitudes following a train of stimuli (Fig. 22.2B). This synaptic depression (i.e., the reduction of glutamatergic neurotransmitter released from NM) plays a key role in improving coincidence detection in NL (Funabiki et al., 1998). As first reported by Funabiki et al. (1998), a positive correlation exists between the optimal response window for coincidence detection and the EPSP decay time constant. However, the observed importance of EPSP amplitude, namely, the response window obtained by subthreshold stimuli was narrower than those obtained using intense stimuli. This observation was later confirmed to be critical for coincidence detection and ITD processing (Kuba et al., 2002). Using a stimulus train at a fixed intensity and frequency, Kuba et al. (2002) reported significantly reduced EPSPs that corresponded to a decrease in the width of the response window for coincidence detection. Consequently, the accuracy of coincidence detection is improved not only by the faster decay time course of evoked responses but also by the smaller amplitudes, reducing the confounding effects of stimulus-intensity-related information (Cook et al., 2003).

In addition to the aforementioned synaptic mechanisms, the extremely fast EPSP of NL neurons are also due to specialized intrinsic properties. Namely, the expression of strong voltage-activated K^+ conductances (K^+_{VA}), which consists of high-threshold and low-threshold K^+ channel currents (K^+_{HVA} and K^+_{LVA}, respectively). The expression of the Kv3.1 channel is responsible for the K^+_{HVA} currents and is activated at depolarizing potentials (~ -20 mV), while the Kv1.1 channel is responsible for K^+_{LVA} currents, which are

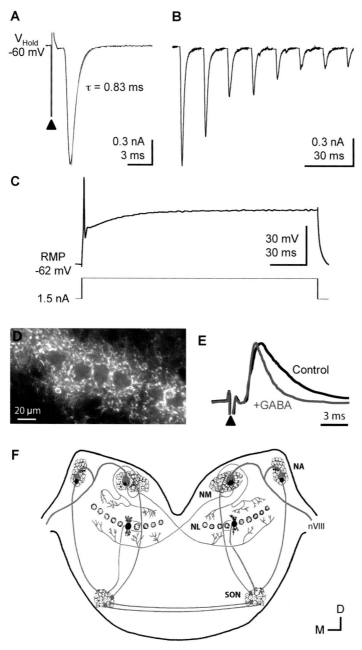

FIGURE 22–2. Specialized synaptic, intrinsic, and GABAergic features of nucleus laminaris (NL). (A) Voltage-clamp recordings showing the decay time constant of an isolated AMPA-R evoked excitatory postsynaptic current (τ, tau). τ was fit with an single exponential (superimposed red-trace). (B) Synaptic depression in NL following an 80 Hz stimulus train. V_{Hold} = –60 mV for traces in (A) and (B). (C) Current-clamp recording (top trace) of a NL neuron showing a single AP following a prolonged current injection into the soma (1.5 nA, 100 ms, bottom trace). RMP, resting membrane potential. (D) GABA immunoreactivity on both the soma and dendrites of NL neurons. (E) Local GABA application (10 μM) shortens the excitatory postsynaptic potential

Figure 22–2. continued
(EPSP) decay time constant (blue trace) compared to control EPSP (black trace). Arrowheads in
(*A*) and (*E*) indicate stimulus artifact. Stimulus artifacts in (*B*) removed for clarity. Normalized
traces in (*E*) are modified with permission from Funabiki et al. (1998). The rapid AMPA-R
kinetics (*A*), synaptic depression (*B*), single AP firing (*C*), and GABAergic depolarization
(*E*) are just a few examples of how NL neurons optimize coincidence detection. (*F*) Schematic
drawing of the inhibitory feedback loop (coronal view). Red and blue lines indicate excitatory
and inhibitory projections, respectively. The reciprocal connections between superior olivary nuclei (SONs) are believed to preserve interaural time difference (ITD) processing in NL
by equalizing input strength from nucleus magnocellularis (NM). For the purpose of clarity,
projections from SON to nucleus angularis (NA) and ascending projections to more rostral
regions of the brain, are not included.

activated at or near the resting membrane potential. The strong expressions
of these K$^+$ channels have several important influences on the function of NL
neurons acting as coincidence detectors. First, K$^+_{HVA}$ conductances allow neurons the ability to repolarize quickly following the generation of an AP.
Second, strong outward K$^+_{LVA}$ conductances reduce input resistance while
increasing the membrane time constant, thus shortening the time window
inputs can summate. Finally, this K$^+_{LVA}$ conductance increases the neurons
threshold for AP generation and as a result, NL neurons elicit only a brief,
single AP at the onset of a prolonged current injection (Fig. 22.2C) (Reyes
et al., 1996). Thus, only strong excitatory inputs excite NL neurons. Taken
together, fast AMPA-R kinetics, significant synaptic depression, and the
strong expression of K$^+$ channels allow NL neurons to be highly sensitive to
simultaneous bilateral inputs from NM, a required prerequisite for coincidence detector neurons and for optimizing ITDs.

Furthermore, the distribution of the site of AP generation in NL neurons is
specialized for the accuracy of ITD detection. In NL, the site of AP initiation
in the axon is arranged at a distance from the soma and is dependent on the
frequency tuning of the neuron. That is, in high characteristic frequency (CF)
neurons, Na$^+$ channels are located 20–50 μM from the soma, while low-CF
neurons have Na$^+$ channels clustered in longer segments of the axon closer to
the soma. This distribution of Na$^+$ channels in NL has functional benefits:
first, APs are initiated at more remote sites as the CF of NL neurons increase,
optimizing ITD processing at each CF; second, the reduction of Na$^+$ channel
inactivation is crucial in detecting ITDs with accuracy during high-frequency
inputs (Kuba et al., 2006).

GABAergic Inhibitory Inputs

In addition to the glutamatergic excitatory inputs from NM, NL neurons are
innervated by GABAergic inhibitory terminals, which cover about 31% of the

somatic surface and 10% of distal dendritic surface of NL neurons (Parks et al., 1983). The major origin of this inhibitory input is from GABAergic neurons located in the superior olivary nucleus (SON). In contrast to NM inputs, the projection from SON to NL apparently lacks precise terminal arbor specificity related to the tonotopic organization of NL, suggesting that the activity of this projection is not highly tuned to sound frequency.

The role of SON input on coincidence detection of NL neurons underlies the general pattern of SON innervation with a number of auditory nuclei involved. First, SON receives excitatory inputs from the ipsilateral NL and in turn projects back to NL and NM on the same side, forming an inhibitory feedback loop (Yang et al., 1999). Second, this loop is accompanied by two additional inputs to SON, one from the ipsilateral nucleus angularis (NA), and the other from the contralateral SON, which provides a negative coupling between feedback loops on each side of the brain (Fig. 22.2F). The ipsilateral circuitry is thought to maintain physiological function near a neuron's threshold, while the contralateral circuitry is believed to provide a cellular substrate that preserves ITD processing in NL by equalizing input strength from NM (Burger et al., 2005).

Indeed, inhibitory inputs from SON to NL plays a critical role in preserving the stability of ITD processing across a broad dynamic range of sound intensity by providing one of the most unique and uncommon properties in all of the central nervous system. $GABA_A$-receptor ($GABA_A$-R) activation results in a depolarizing response in NL neurons due to unusually high internal Cl^- concentration that persists into maturity. GABAergic depolarization activates K^+_{LVA} conductances, which, as mention previously, lowers the membrane input resistance and shortens the EPSP decay time constant (Fig. 22.2E). This GABAergic effect is attributed to the shunting conductances of postsynaptic $GABA_A$-Rs (Funabiki et al., 1998). This shunting mechanism inactivates voltage-gated Na^+ channels, increasing the threshold for AP generation and provides yet another unique mechanism for optimizing coincidence detection. Taken together, inhibitory inputs enhance intrinsic membrane properties and provide binaural control of system gain for NL neurons.

DENDRITIC GRADIENT OF NUCLEUS LAMINARIS NEURONS

The contribution of dendrite geometry for the functioning of single neuron subtypes has remained elusive despite the fact that it is widely accepted that dendritic form plays an important role in neuronal computation. The unique dendritic structure of NL neurons, as well as their known functional role as coincidence detectors, provides an ideal circuit for understanding the specific role of such dendrites in neuronal computation.

In addition to their bipolar configuration, another specialization of NL structure is its highly stereotyped dendritic gradient (Fig. 22.1C) (Smith and

Rubel, 1979). This gradient is evident not only in the total dendritic branch length of individual neurons but also in the distance from the most distal dendritic branches to the soma, that is, the width of the dendritic band. The dendritic gradient conforms precisely to the tonotopic axis with an 11-fold increase in total dendritic branch length and approximately 5-fold increase in the width of the dendritic band from high- to low-CF neurons. The fact that dendritic length increases with decreasing frequency of the optimal sound stimulus suggests a computational role for the dendrites. In contrast, the soma of NL neurons does not show any such gradient.

This dendritic length gradient in NL appears to be an adaptation for ITD processing for particular sound frequencies. Low- and mid-CF neurons, which contain the longer dendrites, may exhibit dendritic filtering resulting from their large surface area (Kuba et al., 2005). A possible advantage of this filtering property is that it may enhance the electrical isolation of dorsal and ventral dendrites and, thus, the inputs from each ear. Indeed, using basic biophysical modeling of known electrophysiological and structural features of NL neurons, recent studies have demonstrated that dendrites improve coincidence detection by allowing a nonlinear summation between the segregated inputs from the ipsi- and contralateral NM (Agmon-Snir et al., 1998). Furthermore, dendrites act as current sinks for each other and modulate the nonlinear integration of inputs. Their results confirm that one aspect of the unique morphology of NL neurons, the spatial segregation of the inputs to different dendrites, enhances the computational power of these neurons to act as coincidence detectors. However, the precise role of this 11-fold gradient in dendritic branch length remains to be convincingly shown.

DEVELOPMENTAL SPECIALIZATIONS

Chickens can hear well before they hatch (about 21 days of incubation) and the major developmental events in the auditory system occur in ovo. In fact, most if not all, intrinsic and synaptic properties appear mature at the time of hatching with minimal refinement thereafter. The development of these highly specialized physiological, organizational, and morphological features of the NM-to-NL projection is temporally correlated with the establishment of synaptically driven neuronal activity (Jackson et al., 1982) and may be dynamically sculpted by synaptic input. However, NM is not a passive receiver of extrinsic influences; the topography and organization of NM projection onto NL are determined by cues intrinsic to the nucleus. Early unilateral destruction of the otocyst (embryonic precursor of the inner ear) induces formation of a functional aberrant axonal projection to the ipsilateral NM from the contralateral NM, which maintains the tonotopic map in NM and NL on both sides of the brain. In addition, following cochlea removal, the deinnervated NL dendritic domains are innervated by the afferents from the opposite NM, which are normally restricted to the opposite domain of NL dendrites (Rubel et al., 1981).

Similar to the development of NM terminals in NL, dendritic growth and dendritic gradient in NL are affected by both intrinsic and extrinsic cues. The formation and sharpening of the dendritic gradient is temporally correlated with the onset and maturation of auditory function. Although dendritic length in NL is dramatically affected by synaptic inputs from NM, the dendritic gradient is preserved and appears to be determined by intrinsic properties as well (Parks, 1981). Recent studies have shown that the segregation of inputs to NL can be partially altered by disrupting expression of single genes such as Eph receptors (Cramer et al., 2006), but the basic pattern seems to remain intact, suggesting regulation by multiple pathways.

Summary and Conclusion

The brain develops specialized neurons and circuits to perform particular functions. In birds, NL neurons are part of a circuit responsible for sound localization. The timing challenges of this task, resolving microsecond differences in the arrival time of sound to the two ears, has imposed important constraints on the morphophysiology of the circuitry. Neuronal architecture, dendritic morphology, and afferent organization determine how information is received, while intrinsic and synaptic properties define how the neurons respond to binaural sound information. The chicken NL is an excellent example of a highly specialized brain region where the reasons for these specialization can be understood in terms of the functions the system subserves.

Acknowledgments

The authors would like to thank former and current Rubel Lab members and Dr. Armin Seidl and Dr. Harunori Ohmori for permission to modify previous work. This work is supported by National Institutes on Deafness and Other Communication Disorders grants DC-03829, DC-04661, and DC-00018.

References

Agmon-Snir H, Carr CE, Rinzel J (1998) The role of dendrites in auditory coincidence detection. *Nature* 393:268–272.

Burger RM, Cramer KS, Pfeiffer JD, Rubel EW (2005) Avian superior olivary nucleus provides divergent inhibitory input to parallel auditory pathways. *J Comp Neurol* 481:6–18.

Cook DL, Schwindt PC, Grande LA, Spain WJ (2003) Synaptic depression in the localization of sound. *Nature* 421:66–70.

Cramer KS, Cerretti DP, Siddiqui SA (2006) EphB2 regulates axonal growth at the midline in the developing auditory brainstem. *Dev Biol* 295:76–89.

Funabiki K, Koyano K, Ohmori H (1998) The role of GABAergic inputs for coincidence detection in the neurones of nucleus laminaris of the chick. *J Physiol* 508 (Pt 3): 851–869.

Hackett JT, Jackson H, Rubel EW (1982) Synaptic excitation of the second and third order auditory neurons in the avian brain stem. *Neuroscience* 7:1455–1469.

Jackson H, Hackett JT, Rubel EW (1982) Organization and development of brain stem auditory nuclei in the chick: ontogeny of postsynaptic responses. *J Comp Neurol* 210:80–86.

Jeffress LA (1948) A place theory of sound localization. *J Comp Physiol Psychol* 41:35–39.

Kuba H, Koyano K, Ohmori H (2002) Synaptic depression improves coincidence detection in the nucleus laminaris in brainstem slices of the chick embryo. *Eur J Neurosci* 15:984–990.

Kuba H, Ishii TM, Ohmori H (2006) Axonal site of spike initiation enhances auditory coincidence detection. *Nature* 444:1069–1072.

Kuba H, Yamada R, Fukui I, Ohmori H (2005) Tonotopic specialization of auditory coincidence detection in nucleus laminaris of the chick. *J Neurosci* 25:1924–1934.

Parks TN (1981) Changes in the length and organization of nucleus laminaris dendrites after unilateral otocyst ablation in chick embryos. *J Comp Neurol* 202:47–57.

Parks TN, Collins P, Conlee JW (1983) Morphology and origin of axonal endings in nucleus laminaris of the chicken. *J Comp Neurol* 214:32–42.

Reyes AD, Rubel EW, Spain WJ (1996) In vitro analysis of optimal stimuli for phase-locking and time-delayed modulation of firing in avian nucleus laminaris neurons. *J Neurosci* 16:993–1007.

Rubel EW, Parks TN (1975) Organization and development of brain stem auditory nuclei of the chicken: tonotopic organization of n. magnocellularis and n. laminaris. *J Comp Neurol* 164:411–433.

Rubel EW, Smith ZD, Steward O (1981) Sprouting in the avian brainstem auditory pathway: dependence on dendritic integrity. *J Comp Neurol* 202:397–414.

Smith DJ, Rubel EW (1979) Organization and development of brain stem auditory nuclei of the chicken: dendritic gradients in nucleus laminaris. *J Comp Neurol* 186:213–239.

Trussell LO (1997) Cellular mechanisms for preservation of timing in central auditory pathways. *Curr Opin Neurobiol* 7:487–492.

Trussell LO (1999) Synaptic mechanisms for coding timing in auditory neurons. *Annu Rev Physiol* 61:477–496.

Yang L, Monsivais P, Rubel EW (1999) The superior olivary nucleus and its influence on nucleus laminaris: a source of inhibitory feedback for coincidence detection in the avian auditory brainstem. *J Neurosci* 19:2313–2325.

Young SR, Rubel EW (1983) Frequency-specific projections of individual neurons in chick brainstem auditory nuclei. *J Neurosci* 3:1373–1378.

Section 8

Touch System

23

Spinal Cord: Dorsal Horn

Tiphaine Dolique, Marc Landry, and Frédéric Nagy

The dorsal horn (DH) of the spinal cord is the first central site where incoming somatosensory signals are processed, and it is an essential stage of synaptic integration in the pain pathway. Neuronal circuitry in the DH comprises four principal elements (Todd and Koerber, 2006): *(1)* central axonal terminals of primary afferent sensory neurons; *(2)* intrinsic spinal cord interneurons (INs); *(3)* projection neurons (PNs) ascending to the brain; and *(4)* terminals of axons descending from the brain. Actually, dorsal horn neurons (DHNs) comprise a very heterogeneous population.

CYTOARCHITECTONIC ORGANIZATION AND PRIMARY AFFERENT PROJECTIONS

The global structure and cyto-architectonic organization of the DH, and the nature and projections of its primary afferent input, have been comprehensively reviewed elsewhere (Millan, 1999; Grant and Koerber, 2004; Ribeiro-da-Silva, 2004; Willis and Coggeshall, 2004; Todd, 2006) and will be only briefly recalled here. The dorsal horn network extends from thin superficial lamina I of Rexed (Rexed, 1952) to deep lamina VI (Fig. 23.1). Lamina II, or substantia gelatinosa (SG), is subdivided into outer lamina II (IIo) and a thicker inner lamina II (IIi). Laminas I–IIi and III–VI comprise the superficial (SDH) and deep (DDH) dorsal horn, respectively. Laminas I, IIo, V, and VI are predominantly implicated in the processing of nociceptive information, whereas laminas III–IV are principally concerned with the treatment of nonnociceptive somatosensory inputs.

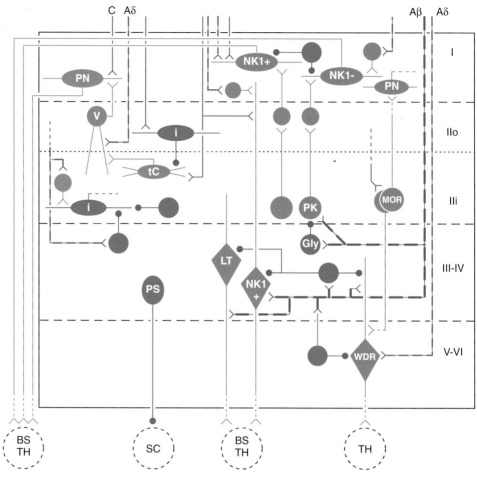

Figure 23–1. The dorsal horn microcircuit: major cell types and synaptic connections. Excitatory neurons are in red, inhibitory neurons are in blue, excitatory synapses are shown as V-shapes, and inhibitory synapses are shown as filled circles. Lines in green represent primary afferent fibers (Aβ, Aδ, C). Horizontal striped lines specify Rexed's laminae (I–VI). Superficial dorsal horn encompasses laminae I–IIi, whereas deep laminae comprise laminae III–VI. Dashed circles illustrate regions of the nervous system targeted by projection neurons (PN): spinal cord (SC), brainstem (BS), and thalamus (TH). Labeling in white indicates neurons with characterized electrophysiological, morphological, and/or neurochemical phenotypes. Subgroups of projection neurons in laminae I and III express the neurorokinin 1 receptor (NK1+), while others do not (NK1−). Low-threshold neurons (LT) integrate innocuous Aβ input fibers, whereas wide dynamic range neurons (WDR) integrate both innocuous and noxious (Aδ, C) inputs. A large majority of inhibitory interneurons are GABAergic and/or glycinergic (Gly). Propriospinal (PS) inhibitory interneurons project to other segments of the spinal cord. Excitatory interneurons release glutamate. In lamina II, those expressing either the protein kinase Cγ (PK) or μ-opioid receptors (MOR) comprise distinct subgroups. An example is given of a connectivity module made of a specific combination of identified neurons: islet (i), transient central (tC), vertical (V) cells, and a projection neuron. Presynaptic inhibition and inputs descending from the brain are not represented.

All primary input neurons have cell bodies in the dorsal root ganglia (DRG). Large myelinated Aß fibers responding to innocuous stimuli project from ventral margin of lamina IIi to lamina V. Thin myelinated A∂ nociceptors primarily project to laminas I–IIo, ventral part of IV, and lamina V. Thin unmyelinated, substance P–containing, nociceptive C fibers project mainly in SDH (I–IIo), with a small set in DDH, while nonpeptidergic IB4-positive C fibers preferentially project to the dorsal half of lamina IIi. In each DH lamina, projections of primary input fibers are somatotopically organized in the mediolateral axis.

MORPHOLOGICAL CHARACTERISTICS OF DORSAL HORN NEURONS

Projection Neurons

The highest density of PNs is found in lamina I (5% of the neurons) (Spike et al., 2003). The rest are scattered in laminas III–VI. Most of them have an axon ascending through the contralateral ventro- or dorsolateral tracts of the white matter, and they project to various regions of the brainstem and thalamus (Dostrovsky and Craig, 2006; Al-Khater et al., 2008). About 80% of PNs in lamina I, and a small population in laminas III–IV, express the NK1 receptors (Fig. 23.1, NK1R+) and transmit nociceptive information (Spike et al., 2003). Many PNs in laminas III–IV convey innocuous somatosensory information to the dorsal column nuclei, whereas in laminas V–VI most of them relay nociceptive input through the spinothalamic tract. Lamina I PNs may belong to three morphological classes: fusiform, pyramidal, or multipolar (Lima and Coimbra, 1986; Spike et al., 2003; Yu et al., 2005), whereas in DDH, they show large variations in size and morphology (Willis and Coggeshall, 2004).

Interneurons

Spinal cord INs may project into other segments (propriospinal INs), but they are in majority intrasegmental (local INs). In lamina I, INs are fusiform, pyramidal, or multipolar (Prescott and De Koninck, 2005). Lamina II neurons were also categorized into four major morphologically distinct cell groups: islet, central, radial, and vertical cells (Grudt and Perl, 2002; Eckert et al., 2003; Yasaka et al., 2007). Dendrites of islet and central cells (also referred to as "small islet"; Beal et al., 1989) are elongated preferentially in the rostrocaudal direction. Radial cells extend dendrites in all directions and are comparable to the human "stellate" cells (Schoenen, 1982). Vertical cells, which include the previously described "stalked" neurons (Gobel, 1978), have dendrites with a significant dorsoventral spread, predominantly ventral to the soma. A fifth class of "medial-lateral" neurons has a dendritic arbor preferentially arranged in a transverse plane (Grudt and Perl, 2002). In DDH, the neuronal dendritic arbors are preferentially sagittal in laminas III–IV and transversal in lamina V. Lamina III–IV neurons may project dendrites in SDH (Willis and Coggeshall, 2004).

The ratio of inhibitory/excitatory INs in the SDH is ~ 30%/70% (Polgar et al., 2003). Islet cells are GABAergic and/or glycinergic inhibitory INs (Maxwell et al., 2007). Inhibitory INs also include some vertical and radial cells (Heinke et al., 2004; Maxwell et al., 2007), but a good proportion of vertical cells are excitatory glutamatergic INs (Lu and Perl, 2005; Maxwell et al., 2007). Another important local source of inhibition comes from cholinergic INs, whose somata were mapped predominantly in laminas III–V (Stewart and Maxwell, 2003; Zhang et al., 2009).

CELLULAR PROPERTIES OF DORSAL HORN NEURONS

Based on their firing patterns in response to sustained current injection, DHNs were categorized into tonic firing, initial bursting, delayed firing, and single-spike discharge. These patterns characterize all types of DHN in all DH laminas without a strict correlation with neuronal morphology, location, or afferent inputs (Yoshimura and Jessell, 1989a; Lopez-Garcia and King, 1994; Jo et al., 1998; Schneider, 2005). Yet, in lamina I, tonic cells were reported to be typically fusiform; phasic cells, pyramidal; and delayed-onset and single-spike cells, preferentially multipolar (Prescott and De Koninck, 2002). Some identified PNs in lamina I also display gap firing (long first inter-spike interval) and bursting (Ruscheweyh et al., 2004). In lamina II, the tonic firing neurons are abundant (Grudt and Perl, 2002; Melnick et al., 2004b), although all types are encountered in the different cell classes (but see Hantman et al., 2004; Heinke et al., 2004). Interestingly, in lamina II, firing diversity appears much reduced when the neuronal discharge is elicited by a "facsimile" current profile mimicking natural response to a peripheral pinch (Graham et al., 2007b). In addition to the four basic patterns, deep DHNs and some lamina I GABAergic INs may produce intrinsic regenerative plateau potentials and bistability (Morisset and Nagy, 1996; Dougherty and Hochman, 2008; Reali et al., 2008), which profoundly alter their output properties, yielding intense firing, afterdischarge, and nonlinear input–output relationships (Morisset and Nagy, 1998; Derjean et al., 2003; Reali and Russo, 2005).

A complex repertoire of membrane conductance was identified in diverse DHNs, which specify neuronal firing and integrative properties (Thomson et al., 1989; Yoshimura and Jessell, 1989a; Yoshimura and Jessell, 1989b; Morisset and Nagy, 1999; Grudt and Perl, 2002; Ruscheweyh and Sandkuhler, 2002; Melnick et al., 2004a; Melnick et al., 2004b; Prescott and De Koninck, 2005).

NEURAL CIRCUITRY

Primary afferent terminals make synaptic contacts on DHNs' dendrites and soma. In turn, all of them, except most of the peptidergic C-fibers, receive axo-axonic contacts mediating presynaptic inhibition. In SDH, many of these contacts are comprised in glomeruli, that is, complex synaptic arrangements in which a primary afferent axon terminal interacts with several dendrites from INs or PNs, and axonal boutons from intrinsic INs (Ribeiro-da-Silva and Coimbra, 1982; Todd, 2006).

In SDH, the transfer of sensory information from A∂ and C fibers to PNs is achieved by parallel monosynaptic and polysynaptic excitatory pathways, in which glutamate is the major excitatory neurotransmitter (Yoshimura and Jessell, 1989a, 1990; Yoshimura and Nishi, 1993). In DDH, PNs are monosynaptically activated by myelinated afferent fibers A∂ (nociceptive specific neurons, NS), Aß (low threshold, LT), or both (wide dynamic range, WDR), whereas C fibers activate NS and WDR mostly via excitatory INs (Millan, 1999; Willis and Coggeshall, 2004; Todd, 2006). Regarding SDH INs, GABA- and/or glycine-releasing inhibitory INs are activated by all three types of primary input (Yoshimura and Nishi, 1995; Daniele and MacDermott, 2009) and in turn exert a direct postsynaptic inhibition on PNs, in addition to presynaptic inhibition of primary afferent terminals (Todd, 2006). These data conducted to assume a prototypal circuit in which an inhibitory IN activated by innocuous afferent fibers controls nociceptive activation of a WDR projection neuron, supporting the gate theory of pain (Melzack and Wall, 1965). A comprehensive report of primary afferent evoked IPSCs in identified SDH interneurons indicated a more complex arrangement (Yasaka et al., 2007).

As stressed in a recent review (Graham et al., 2007a), new technical approaches begin to unravel the specific organization of SDH circuits. Simultaneous paired recordings of DHNs indicated that SG is not a broadly interconnected region (~10% of the recorded pairs) (Lu and Perl, 2003; Lu and Perl, 2005; Santos et al., 2007; Labrakakis et al., 2009), and that 85% of connected INs are excitatory and glutamatergic (Santos et al., 2007), although inhibitory INs represent 30% of the DHNs (Polgar et al., 2003). A similar study in DDH (Schneider, 2008) yielded a higher connectivity (~ 30% of the tested pairs), a larger proportion of inhibitory connections (~ 69%), and a relatively low reliability of synaptic transmission. A common feature of the preceding studies is the predominance of unidirectional connections, precluding an important contribution of feedback loops to information transfer in the DH. Paired recordings also evidenced that particular types of SDH neurons are specifically linked. A first pattern recurrently identified consisted in a monosynaptic GABAergic inhibition linking inhibitory islet cells to excitatory central neurons in lamina II (Lu and Perl, 2003). Specific glutamatergic excitatory connections were also recurrently observed (Lu and Perl, 2005). Transient central neurons of lamina IIi excited vertical neurons of lamina IIo, which in turn excited projection and nonprojection lamina I neurons. Because both the inhibitory and excitatory circuits involve transient central lamina II neurons, it is conceivable that, as a whole, they comprise a four-neuron circuit linking specific primary inputs to lamina I Pns (Fig. 23.1, i, tC, V, PN). The recurrent identification of such arrangements leads up to the important concept of the SDH being organized into connectivity modules comprised of specific combinations of neurons engaged into a common processing function (Lu and Perl, 2005).

Another important step was made by examining spatial organization of the global SDH connectivity (Kato et al., 2007; Kato et al., 2009). Systematic mapping of all local sites that are monosynaptically connected to a single postsynaptic SDH neuron revealed that excitatory interneuronal connections exhibit a predominantly sagittal orientation, fitting the orientation of both the primary afferent terminal fields and the DHN's dendritic arbors. This pattern of connectivity appears suitable for preserving and sharpening the primary afferent somatotopy. Moreover, there is a predominantly upward (dorsal) flow of excitatory transmission within the superficial laminas, which could support a polysynaptic transmission of deep mechanoreceptive input to nociceptive neurons.

Besides primary afferent fibers and DHNs, another cardinal component of DH circuitry consists of the synaptic contacts by abundant inputs descending from the brain, (Millan, 2002; Gebhart and Proudfit, 2005), of which the monoaminergic systems, originating predominantly at the brainstem, exert a major inhibitory control (see review of Yoshimura and Furue, 2006), with differential effects on neuronal subtypes (Lu and Perl, 2007). Interestingly, there is also an inhibitory pathway from the brainstem (RVM), mediated by GABAergic and glycinergic monosynaptic connections on more than 50% of the recorded SG neurons (Kato et al., 2006).

Plasticity of the Dorsal Horn Microcircuits

The DH network turns out to be extremely plastic, especially in conditions of central sensitization to pain (Woolf and Salter, 2006). The mechanisms are diverse (see reviews of Woolf and Salter, 2000; Scholz and Woolf, 2007; Sandkuhler, 2009) and include functional modifications of the DH circuitry. Amplification of neuronal responsiveness may result from alterations of the DHN's intrinsic properties (Hains et al., 2003; Dougherty and Hochman, 2008), or from the reinforcement of excitatory pathways, through the activation of silent synapses (Li and Zhuo, 1998), induction of LTP (Sandkuhler, 2007), or increased excitatory synaptic drive to excitatory INs (Lu et al., 2009). But a major substrate for DH hyperexcitability is a reduction of synaptic inhibition (Sivilotti and Woolf, 1994; Baba et al., 2003). Modifications of DH inhibitory synapses play a pivotal role. In conditions of chronic inflammatory pain, disinhibition may result from blockade of inhibitory glycine receptors in the SDH (Ahmadı et al., 2002; Harvey et al., 2004). It may also result from modification of the transmembrane gradient for chloride ions in nociceptive lamina I neurons, due to a reduction in the expression of the potassium–chloride exporter KCC2 following nerve injury (Coull et al., 2003; Coull et al., 2005). The change in chloride gradient may cause paradoxical excitation of lamina I PNs, which acquire the ability to relay innocuous inputs (Keller et al., 2007; Lu et al., 2008).

This indicates that in the spinal DH, disinhibition may activate "sleeping circuits", that is, pathways that are normally maintained silent by inhibition. In vitro blockade of local GABAergic and glycinergic inhibition revealed significant Aß fiber–mediated innocuous polysynaptic input to lamina I NK1R-expressing neurons that otherwise only respond to high-threshold A∂/C fiber input (Torsney and MacDermott, 2006). In vivo a similar phenomenon was reported for NK1R-negative lamina I NS neurons (Miraucourt et al., 2007). The local circuit comprise lamina IIi PKCγ-expressing cells (Fig. 23.1, PK), which excite other lamina II excitatory INs, activating in turn lamina I NS neurons (Miraucourt et al., 2009). The majority of PKCγ cells are excitatory INs (Polgar et al., 1999) activated by medium- and large-diameter myelinated afferents that transmit nonnoxious information (Neumann et al., 2008). Tactile myelinated Aß fibers activate in parallel the PKCγ neurons and local glycinergic INs, which normally inhibit the PKCγ neurons. Blockade of this inhibition gates the innocuous input up to the nociceptive lamina I PN, and to pain pathways ascending to the brain. Once unmasked by glycinergic disinhibition, this circuit induces a dynamic mechanical allodynia, a condition wherein pain is produced by innocuous light pressure stimuli (Miraucourt et al., 2009).

Concluding Remarks

A rapidly growing amount of data about identification of cells, connectivity, and in vivo operation now sheds light on neuronal modules that substantiate the functional hypothesis about DH microcircuits operation. Of particular interest is the fact that the role of neuronal subsystems might only become apparent under specific pathophysiological conditions (Hunt and Mantyh, 2001).

Acknowledgments

The authors wish to thank R. Dallel (*Inserm U929, Clermont-Ferrand*) for helpful discussion. This work was supported by grants from the Agence Nationale de la Recherche (ANR-07-NEURO-015-01), the Conseil Régional d'Aquitaine (2008/30/023), and by a PhD grant from the Ministère de l'Education Nationale, de la Recherche et de la Technologie to TD.

References

Ahmadi S, Lippross S, Neuhuber WL, Zeilhofer HU (2002) PGE(2) selectively blocks inhibitory glycinergic neurotransmission onto rat superficial dorsal horn neurons. *Nat Neurosci* 5:34–40.

Al-Khater KM, Kerr R, Todd AJ (2008) A quantitative study of spinothalamic neurons in laminae I, III, and IV in lumbar and cervical segments of the rat spinal cord. *J Comp Neurol* 511:1–18.

Baba H, Ji RR, Kohno T, Moore KA, Ataka T, Wakai A, Okamoto M, Woolf CJ (2003) Removal of GABAergic inhibition facilitates polysynaptic A fiber-mediated excitatory transmission to the superficial spinal dorsal horn. *Mol Cell Neurosci* 24:818–830.

Beal JA, Nandia KN, Knight DS (1989) Characterization of long ascending tract projection neurons and non-tract neurons in the superficial dorsal horn (SDH). In: Cervero F, Bennett GJ and Headley PM, eds. *Processing of Sensory Information in the Superficial Dorsal Horn of the Spinal Cord*, pp.181–197. New York: Plenum.

Coull JA, Boudreau D, Bachand K, Prescott SA, Nault F, Sik A, De Koninck P, De Koninck Y (2003) Trans-synaptic shift in anion gradient in spinal lamina I neurons as a mechanism of neuropathic pain. *Nature* 424:938–942.

Coull JA, Beggs S, Boudreau D, Boivin D, Tsuda M, Inoue K, Gravel C, Salter MW, De Koninck Y (2005) BDNF from microglia causes the shift in neuronal anion gradient underlying neuropathic pain. *Nature* 438:1017–1021.

Daniele CA, MacDermott AB (2009) Low-threshold primary afferent drive onto GABAergic interneurons in the superficial dorsal horn of the mouse. *J Neurosci* 29:686–695.

Derjean D, Bertrand S, Le Masson G, Landry M, Morisset V, Nagy F (2003) Dynamic balance of metabotropic inputs causes dorsal horn neurons to switch functional states. *Nature Neurosci* 6:274–281.

Dostrovsky JO, Craig AD (2006) Ascending projection systems. In: McMahon SB and Koltzenburg M, eds. *Wall and Melzack's Textbook of Pain*, 5th ed., pp.187–203. Philadelphia: Elsevier Limited.

Dougherty KJ, Hochman S (2008) Spinal cord injury causes plasticity in a subpopulation of lamina I GABAergic interneurons. *J Neurophysiol* 100:212–223.

Eckert WA, 3rd, McNaughton KK, Light AR (2003) Morphology and axonal arborization of rat spinal inner lamina II neurons hyperpolarized by mu-opioid-selective agonists. *J Comp Neurol* 458:240–256.

Gebhart GF, Proudfit H (2005) Descending control of pain processing. In: Hunt SP and Koltzenburg M, eds. *The Neurobiology of Pain*, pp. 289–309. Oxford, England: Oxford University Press.

Gobel S (1978) Golgi studies of the neurons in layer II of the dorsal horn of the medulla (trigeminal nucleus caudalis). *J Comp Neurol* 180:395–413.

Graham BA, Brichta AM, Callister RJ (2007a) Moving from an averaged to specific view of spinal cord pain processing circuits. *J Neurophysiol* 98:1057–1063.

Graham BA, Brichta AM, Callister RJ (2007b) Pinch-current injection defines two discharge profiles in mouse superficial dorsal horn neurones, in vitro. *J Physiol* 578:787–798.

Grant G, Koerber HR (2004) Spinal cord cytoarchitecture. In: Paxinos G, ed. *The Rat Nervous System*, 3rd ed., pp. 121–128. San Diego, CA: Elsevier Acad Press.

Grudt TJ, Perl ER (2002) Correlations between neuronal morphology and electrophysiological features in the rodent superficial dorsal horn. *J Physiol* 540:189–207.

Hains BC, Klein JP, Saab CY, Craner MJ, Black JA, Waxman SG (2003) Upregulation of sodium channel Nav1.3 and functional involvement in neuronal hyperexcitability associated with central neuropathic pain after spinal cord injury. *J Neurosci* 23:8881–8892.

Hantman AW, van den Pol AN, Perl ER (2004) Morphological and physiological features of a set of spinal substantia gelatinosa neurons defined by green fluorescent protein expression. *J Neurosci* 24:836–842.

Harvey RJ, Depner UB, Wassle H, Ahmadi S, Heindl C, Reinold H, Smart TG, Harvey K, Schutz B, Abo-Salem OM, Zimmer A, Poisbeau P, Welzl H, Wolfer DP, Betz H, Zeilhofer HU, Muller U (2004) GlyR alpha3: an essential target for spinal PGE2-mediated inflammatory pain sensitization. *Science* 304:884–887.

Heinke B, Ruscheweyh R, Forsthuber L, Wunderbaldinger G, Sandkuhler J (2004) Physiological, neurochemical and morphological properties of a subgroup of GABAergic spinal lamina II neurones identified by expression of green fluorescent protein in mice. *J Physiol* 560:249–266.

Hunt SP, Mantyh PW (2001) The molecular dynamics of pain control. *Nat Rev Neurosci* 2:83–91.

Jo YH, Stoeckel ME, Schlichter R (1998) Electrophysiological properties of cultured neonatal rat dorsal horn neurons containing GABA and met-enkephalin-like immunoreactivity. *J Neurophysiol* 79:1583–1586.

Kato G, Yasaka T, Katafuchi T, Furue H, Mizuno M, Iwamoto Y, Yoshimura M (2006) Direct GABAergic and glycinergic inhibition of the substantia gelatinosa from the rostral ventromedial medulla revealed by in vivo patch-clamp analysis in rats. *J Neurosci* 26:1787–1794.

Kato G, Kawasaki Y, Ji RR, Strassman AM (2007) Differential wiring of local excitatory and inhibitory synaptic inputs to islet cells in rat spinal lamina II demonstrated by laser scanning photostimulation. *J Physiol* 580:815–833.

Kato G, Kawasaki Y, Koga K, Uta D, Kosugi M, Yasaka T, Yoshimura M, Ji RR, Strassman AM (2009) Organization of intralaminar and translaminar neuronal connectivity in the superficial spinal dorsal horn. *J Neurosci* 29:5088–5099.

Keller AF, Beggs S, Salter MW, De Koninck Y (2007) Transformation of the output of spinal lamina I neurons after nerve injury and microglia stimulation underlying neuropathic pain. *Mol Pain* 3:27.

Labrakakis C, Lorenzo LE, Bories C, Ribeiro-da-Silva A, De Koninck Y (2009) Inhibitory coupling between inhibitory interneurons in the spinal cord dorsal horn. *Mol Pain* 5:24.

Li P, Zhuo M (1998) Silent glutamatergic synapses and nociception in mammalian spinal cord. *Nature* 393:695–698.

Lima D, Coimbra A (1986) A Golgi study of the neuronal population of the marginal zone (lamina I) of the rat spinal cord. *J Comp Neurol* 244:53–71.

Lopez-Garcia JA, King AE (1994) Membrane properties of physiologically classified rat dorsal horn neurons *in vitro*: correlation with cutaneous sensory afferent input. *Eur J Neurosci* 6:998–1007.

Lu Y, Perl ER (2003) A specific inhibitory pathway between substantia gelatinosa neurons receiving direct C-fiber input. *J Neurosci* 23:8752–8758.

Lu Y, Perl ER (2005) Modular organization of excitatory circuits between neurons of the spinal superficial dorsal horn (laminae I and II). *J Neurosci* 25:3900–3907.

Lu Y, Perl ER (2007) Selective action of noradrenaline and serotonin on neurons of the spinal superficial dorsal horn in the rat. *J Physiol* 582:127–136.

Lu Y, Zheng J, Xiong L, Zimmermann M, Yang J (2008) Spinal cord injury-induced attenuation of GABAergic inhibition in spinal dorsal horn circuits is associated with downregulation of the chloride transporter KCC2 in rat. *J Physiol* 586:5701–5715.

Lu VB, Biggs JE, Stebbing MJ, Balasubramanyan S, Todd KG, Lai AY, Colmers WF, Dawbarn D, Ballanyi K, Smith PA (2009) Brain-derived neurotrophic factor drives the changes in excitatory synaptic transmission in the rat superficial dorsal horn that follow sciatic nerve injury. *J Physiol* 587:1013–1032.

Maxwell DJ, Belle MD, Cheunsuang O, Stewart A, Morris R (2007) Morphology of inhibitory and excitatory interneurons in superficial laminae of the rat dorsal horn. *J Physiol* 584:521–533.

Melnick IV, Santos SF, Safronov BV (2004a) Mechanism of spike frequency adaptation in substantia gelatinosa neurones of rat. *J Physiol* 559:383–395.

Melnick IV, Santos SF, Szokol K, Szucs P, Safronov BV (2004b) Ionic basis of tonic firing in spinal substantia gelatinosa neurons of rat. *J Neurophysiol* 91:646–655.

Melzack R, Wall PD (1965) Pain mechanisms: a new theory. *Science* 150:971–979.

Millan MJ (1999) The induction of pain: an integrative review. *Prog Neurobiol* 57:1–164.

Millan MJ (2002) Descending control of pain. *Prog Neurobiol* 66:355–474.

Miraucourt LS, Dallel R, Voisin DL (2007) Glycine inhibitory dysfunction turns touch into pain through PKCgamma interneurons. *PLoS ONE* 2:e1116.

Miraucourt LS, Moisset X, Dallel R, Voisin DL (2009) Glycine inhibitory dysfunction induces a selectively dynamic, morphine-resistant, and neurokinin 1 receptor-independent mechanical allodynia. *J Neurosci* 29:2519–2527.

Morisset V, Nagy F (1996) Modulation of regenerative membrane properties by stimulation of metabotropic glutamate receptors in rat deep dorsal horn neurons. *J Neurophysiol* 76:2794–2798.

Morisset V, Nagy F (1998) Nociceptive integration in the rat spinal cord: Role of nonlinear membrane properties of deep dorsal horn neurons. *Eur J Neurosci* 10:3642–3652.

Morisset V, Nagy F (1999) Ionic basis for plateau potentials in deep dorsal horn neurons of the rat spinal cord. *J Neurosci* 19:7309–7316.

Neumann S, Braz JM, Skinner K, Llewellyn-Smith IJ, Basbaum AI (2008) Innocuous, not noxious, input activates PKCgamma interneurons of the spinal dorsal horn via myelinated afferent fibers. *J Neurosci* 28:7936–7944.

Polgar E, Fowler JH, McGill MM, Todd AJ (1999) The types of neuron which contain protein kinase C gamma in rat spinal cord. *Brain Res* 833:71–80.

Polgar E, Hughes DI, Riddell JS, Maxwell DJ, Puskar Z, Todd AJ (2003) Selective loss of spinal GABAergic or glycinergic neurons is not necessary for development of thermal hyperalgesia in the chronic constriction injury model of neuropathic pain. *Pain* 104:229–239.

Prescott SA, De Koninck Y (2002) Four cell types with distinctive membrane properties and morphologies in lamina I of the spinal dorsal horn of the adult rat. *J Physiol* 539:817–836.

Prescott SA, De Koninck Y (2005) Integration time in a subset of spinal lamina I neurons is lengthened by sodium and calcium currents acting synergistically to prolong subthreshold depolarization. *J Neurosci* 25:4743–4754.

Reali C, Russo RE (2005) An integrated spinal cord-hindlimbs preparation for studying the role of intrinsic properties in somatosensory information processing. *J Neurosci Methods* 142:317–326.

Reali C, Fossat P, Landry M, Russo RE, Nagy F (2008) Intrinsic membrane properties of dorsal horn neurons modulate nociceptive information processing in the spinal cord. *Soc Neurosci Abstr* 34:771–774.

Rexed B (1952) The cytoarchitectonic organization of the spinal cord in the cat. *J Comp Neurol* 96:414–495.

Ribeiro-da-Silva A (2004) Substantia gelatinosa of the spinal cord. In: Paxinos G, ed. *The Rat Nervous System*, 3rd ed., pp. 129–148. San Diego, CA: Elsevier Acad Press.

Ribeiro-da-Silva A, Coimbra A (1982) Two types of synaptic glomeruli and their distribution in laminae I-III of the rat spinal cord. *J Comp Neurol* 209:176–186.

Ruscheweyh R, Sandkuhler J (2002) Lamina-specific membrane and discharge properties of rat spinal dorsal horn neurones in vitro. *J Physiol* 541:231–244.

Ruscheweyh R, Ikeda H, Heinke B, Sandkuhler J (2004) Distinctive membrane and discharge properties of rat spinal lamina I projection neurones in vitro. *J Physiol* 555:527–543.

Sandkuhler J (2007) Understanding LTP in pain pathways. *Mol Pain* 3:9.

Sandkuhler J (2009) Models and mechanisms of hyperalgesia and allodynia. *Physiol Rev* 89:707–758.

Santos SF, Rebelo S, Derkach VA, Safronov BV (2007) Excitatory interneurons dominate sensory processing in the spinal substantia gelatinosa of rat. *J Physiol* 581:241–254.

Schneider SP (2005) Mechanosensory afferent input and neuronal firing properties in rodent spinal laminae III-V: re-examination of relationships with analysis of responses to static and time-varying stimuli. *Brain Res* 1034:71–89.

Schneider SP (2008) Local circuit connections between hamster laminae III and IV dorsal horn neurons. *J Neurophysiol* 99:1306–1318.

Schoenen J (1982) The dendritic organization of the human spinal cord: the dorsal horn. *Neuroscience* 7:2057–2087.

Scholz J, Woolf CJ (2007) The neuropathic pain triad: neurons, immune cells and glia. *Nat Neurosci* 10:1361–1368.

Sivilotti L, Woolf CJ (1994) The contribution of GABAA and glycine receptors to central sensitization: disinhibition and touch-evoked allodynia in the spinal cord. *J Neurophysiol* 72:169–179.

Spike RC, Puskar Z, Andrew D, Todd AJ (2003) A quantitative and morphological study of projection neurons in lamina I of the rat lumbar spinal cord. *Eur J Neurosci* 18: 2433–2448.

Stewart W, Maxwell DJ (2003) Distribution of and organisation of dorsal horn neuronal cell bodies that possess the muscarinic m2 acetylcholine receptor. *Neuroscience* 119:121–135.

Thomson AM, West DC, Headley PM (1989) Membrane characteristics and synaptic responsiveness of superficial dorsal horn neurons in a slice preparation of adult rat spinal cord. *Eur J Neurosci* 1:479–488.

Todd AJ (2006) Anatomy and neurochemistry of the dorsal horn. In: Cervero F and Jensen TS, eds. *Handbook of Clinical Neurology*, 3rd ed., pp. 61–76. Amsterdam: Elsevier Science Publisher.

Todd AJ, Koerber RH (2006) Neuroanatomical substrates of spinal nociception. In: McMahon SB and Koltzenburg M, eds. *Wall and Melzack's Textbook of Pain*, 5th ed., pp. 73–90. Philadelphia: Elsevier Limited.

Torsney C, MacDermott AB (2006) Disinhibition opens the gate to pathological pain signaling in superficial neurokinin 1 receptor-expressing neurons in rat spinal cord. *J Neurosci* 26:1833–1843.

Willis WD, Coggeshall RE (2004) *Sensory Mechanisms of the Spinal Cord*, 3rd ed., Vol. 1. *Primary Afferent Neurons and the Spinal Dorsal Horn*. New York: Kluwer Academic/ Plenum.

Woolf CJ, Salter MW (2000) Neuronal plasticity: increasing the gain in pain. *Science* 288:1765–1769.

Woolf CJ, Salter MW (2006) Plasticity and pain: role of the dorsal horn. In: McMahon SB and Koltzenburg M, eds. *Wall and Melzack's Textbook of Pain*, 5th ed., pp. 91–105. Philadelphia: Elsevier Limited.

Yasaka T, Kato G, Furue H, Rashid MH, Sonohata M, Tamae A, Murata Y, Masuko S, Yoshimura M (2007) Cell-type-specific excitatory and inhibitory circuits involving primary afferents in the substantia gelatinosa of the rat spinal dorsal horn in vitro. *J Physiol* 581:603–618.

Yoshimura M, Jessell TM (1989a) Primary afferent-evoked synaptic responses and slow potential generation in rat substantia gelatinosa neurons *in vitro*. *J Neurophysiol* 62: 96–108.

Yoshimura M, Jessell TM (1989b) Membrane properties of rat substantia gelatinosa neurons *in vitro*. *J Neurophysiol* 62:109–118.

Yoshimura M, Jessell TM (1990) Amino acid-mediated EPSPs at primary afferent synapses with substantia gelatinosa neurones in the rat spinal cord. *J Physiol London* 430:315–335.

Yoshimura M, Nishi S (1993) Blind patch-clamp recordings from substantia gelatinosa neurons in adult rat spinal cord slices: pharmacological properties of synaptic currents. *Neuroscience* 53:519–526.

Yoshimura M, Nishi S (1995) Primary afferent-evoked glycine- and GABA-mediated IPSPs in substantia gelatinosa neurones in the rat spinal cord in vitro. *J Physiol London* 482:29–38.

Yoshimura M, Furue H (2006) Mechanisms for the anti-nociceptive actions of the descending noradrenergic and serotonergic systems in the spinal cord. *J Pharmacol Sci* 101:107–117.

Yu XH, Ribeiro-da-Silva A, De Koninck Y (2005) Morphology and neurokinin 1 receptor expression of spinothalamic lamina I neurons in the rat spinal cord. *J Comp Neurol* 491:56–68.

Zhang HM, Chen SR, Cai YQ, Richardson TE, Driver LC, Lopez-Berestein G, Pan HL (2009) Signaling mechanisms mediating muscarinic enhancement of GABAergic synaptic transmission in the spinal cord. *Neuroscience* 158:1577–1588.

Section 9

Olfactory System

24

Olfactory Bulb

*Gordon M. Shepherd, Michele Migliore, and
David C. Willhite*

The olfactory bulb is the site of the first synaptic processing of the olfactory
input from the nose. It is present in all vertebrates (except for cetaceans) and
in the analogous antennal lobe in invertebrates. With its sharply demarcated
cell types and histological layers, and some well-studied synaptic interactions, it was one of the first and is one of the clearest examples of the microcircuit concept in the central nervous system.

The overall organization of the vertebrate olfactory bulb is summarized in
the microcircuit diagram shown in Figure 24.1. We describe the synaptic connections, the functional operations, the multiple parallel pathways involved
in information processing of the sensory input, and its modulation by brain
centers.

Basic Synaptic Connections

Sensory Input

The sensory input is carried by the axons of *olfactory receptor neurons* (ORNs)
in the nasal epithelium. Each neuron expresses one of several hundred to
several thousand (depending on the species) olfactory receptors (Buck and
Axel, 1991). All of the neurons expressing a given receptor constitute a *subset*,
of which most axons converge onto two *glomeruli* in the olfactory bulb, where
they make glutamatergic synapses on the dendritic tufts of mitral, tufted, and
periglomerular cells.

The mammalian olfactory receptors have two phylogenetically distinct
classes, the dorsally represented class I, or "fish-like" receptors, and the

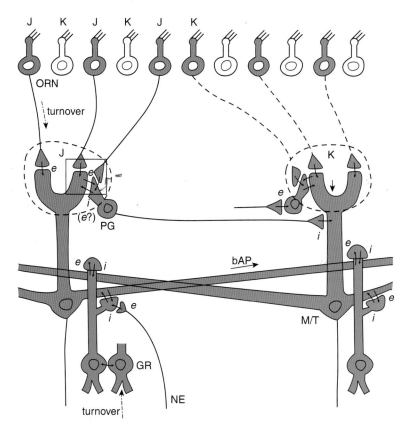

FIGURE 24–1. Diagram of the basic circuit of the olfactory bulb in the rodent. On the left, a glomerular unit consists of a glomerulus receiving inputs from its olfactory receptor neuron (ORN) subset (J) and connecting to a subset of mitral/tufted (M/T) cells and their interneurons: periglomerular (PG) cells at the glomerular layer and granule cells (GR) at the mitral cell body layer. On the right, a backpropagating action potential (bAP) in the mitral cell lateral dendrite provides excitation (e) of granule cells within neighboring or distant glomerular units to mediate lateral inhibition (i) in processing the distributed odor maps laid down in the glomeruli by the activated ORNs. Centrifugal modulatory circuits include norepinephrine (NE). Turnover during adult life occurs in olfactory stem cells in ORNs and the rostral migratory stream in GR and PG cells. (Diagram and legend adapted from Shepherd et al., 2007)

class II, or terrestrial receptors (Zhang and Firestein, 2002). Due to the sequence similarities to the receptors of fish, the class I olfactory receptors are thought to bind more soluble ligands, though this hypothesis has not been rigorously tested. By functionally disrupting the class I receptors, however, it has been shown that this subset may form a distinct behavioral circuit from class II, perhaps dedicated to innate versus learned responses (Kobayakawa et al., 2007).

Output Neurons

There are two types of output cells: relatively large cells called *mitral (M) cells*, and smaller cells called *tufted (T) cells*. Mitral cell bodies lie in a thin layer. In mammals they give rise to a single apical (primary) dendrite that extends for several hundred microns across the external plexiform layer and ends in a terminal tuft within a single glomerulus. The tuft branches are interconnected by gap junctions, and they have synaptic interactions with glomerular interneuron dendrites (see Functional Operations). Each cell body also gives rise to several basal (secondary) dendrites, which branch sparingly and extend through the external plexiform layer for up to 1000 microns. The tufted cells are smaller versions of the mitral cells, though transcriptionally distinct. They may be external, middle, or deep, depending on where their cell bodies lie. The lateral dendrites are smooth and have synaptic interactions with granule cell dendritic spines. Mitral and tufted cells are glutamatergic.

The mitral/tufted cells process the sensory input through interactions with two layers of *intrinsic neurons* in the glomerular layer and the external plexiform layer. With regard to the overall organization of the olfactory bulb microcircut, the glomerular layer can be regarded as involved in *input processing*, the external plexiform layer in *output processing*.

Intrinsic Neurons

The most numerous *glomerular layer interneuron* is generally called a *periglomerular (PG) cell* by the localization of its cell bodies around the glomeruli. It falls in the general category of *short-axon cell* because its axon remains within the olfactory bulb. It has a short spinous dendritic tuft confined to a single glomerulus, as well as an axon of variable length that can extend across several glomeruli, and terminates on cell bodies and dendrites outside the glomeruli. Periglomerular cells may be GABAergic, dopaminergic, or both. The PG cell dendrites engage in *intraglomerular* dendrodendritic synaptic interactions with M/T dendrites and have *interglomerular* synaptic actions through their axon terminals (Kosaka and Kosaka, 2005). These have inhibitory and disinhibitory effects on glomerular output through acting on the primary dendrites of mitral-tufted cells (see Fig. 24.1). Juxtaglomerular cells send axonal processes to 7 to 10 glomeruli, which may indicate that the glomerular layer plays a significant role in lateral processing (Aungst, et al., 2003). External tufted cell subsets have spontaneous pacemaker activity; their actions on PG and short axon cells activate a glomerular network which is believed to drive mitral cell responses to olfactory input.

The second level of synaptic processing in the olfactory bulb microcircuit occurs through *granule cells*, whose cell bodies lie in the deeper granule cell layer. They have long radial dendrites, which ramify in the external plexiform layer among the mitral and tufted cell dendrites, and short central dendrites.

The dendritic branches are covered with spines. An axon is lacking; these cells are therefore the anaxonal type, similar in this respect to amacrine cells in the retina. Inputs are the glutamatergic dendrodendritic synapses from the M/T cell dendrites, and axodendritic from centrifugal modulator fibers and possibly M/T axon collaterals, in addition to a recently identified inhibitory input from deep short-axon cells. All output is through the dendrodendritic GABAergic synapses on the M/T dendrites.

Output Projections

As the output neurons of the olfactory bulb, the mitral and tufted cells have overlapping but distinct projection sites. Mitral cells send their axons to the lateral olfactory tract, from which they distribute terminal collateral branches to the several subregions of olfactory cortex (Chapter 25). The projections of the tufted cells are more concentrated in the olfactory tubercle.

Centrifugal Fibers

All centrifugal input is thought to be targeted to GABAergic or modulatory interneurons, with no evidence of direct connections with glutamatergic cell types. A direct centrifugal modulalatory input comes from axon collaterals of glutamatergic pyramidal neurons in the olfactory cortex that receive the M/T output of the bulb. Central modulatory centrifugal systems include serotonergic fibers from the dorsal raphe and noradrenergic fibers from the locus ceruleus in the brainstem, and cholinergic fibers from the basal forebrain. Internal modulation occurs through intrinsic dopaminergic PG cells.

FUNCTIONAL OPERATIONS

Phylogenetic comparisons suggest that the *glomerulus* is the organizing principle of the olfactory bulb in all species, essential for the processing that leads to smell perception. It is probably the best example of an anatomically defined *cortical processing unit*. It is equivalent in this respect to cortical columns and barrels in visual and somatosensory cortices; however, their presence varies in different species (Horton and Adams, 2005), whereas the olfactory glomerulus is more nearly universal. It is therefore a structure that clearly has a function, a function that is essential to the perception of smell.

Glomerular Units

A *glomerular unit* consists of all ORNs projecting to a given glomerulus, all M/T and PG cells connected through their dendrites to a given glomerulus, plus all the granule cells making connections at or near the cell bodies of those M/T cells. This connectivity defines a relatively narrow column of cells centered on its home glomerulus (Willhite et al., 2006). Columns so identified by viral tracing from a single injection site form an extended mosaic throughout the olfactory bulb. These distributed processing units have been proposed to reflect processing of the distributed odor maps formed by olfactory stimulation (see below).

Odor Maps

Different ORNs respond with overlapping *molecular receptive ranges* (MRRs) to stimulation by different odor molecules, reflecting different affinities within the binding pockets of their receptors (Mori and Yoshihara, 1995). The MRRs are thus analogous to the *receptive fields* of cells in other sensory systems.

The MRRs of the ORN subsets projecting to their unimodal glomeruli give rise to differential responses of the olfactory glomeruli to different odor molecules. A variety of imaging techniques has revealed these responses as *spatial activity patterns*, called odor maps (odor images) (Xu et al., 2000). Each odor molecule or combination of odor molecules is associated with its unique odor map. A consensus hypothesis is that the function of the olfactory bulb microcircuits is to detect and discriminate these spatial patterns as the basis for odor perception. This discrimination involves *temporal processing* of the patterns as well (Mazor and Laurent, 2005).

The hundreds to thousands of glomeruli in different species mean that processing in the olfactory bulb is *massively parallel* (see Parallel Processing).

Glomerular Mechanisms

The high convergence ratio of ORNs is believed to play an essential role in *signal-to-noise enhancement* within the glomerulus, together with active properties of the dendritic tufts on which they terminate, to enhance odor detection.

Within the glomerulus, the ORN terminals and M/T and PG cell dendrites form *synaptic triads*. This provides for maximum flexibility of synaptic interactions between individual synaptic terminals, with individual terminals and synapses between all members. This may be compared with the thalamus, where the large sensory axon terminal makes synapses onto both output and intrinsic cell dendrites (see Chapter 8), with only feedforward inhibition onto the output cell dendrite and the cerebellum, where a single large granule cell terminal makes simultaneous synaptic inputs onto many postsynaptic dendrites, with no feedback or feedforward relationship (see Chapter 28).

The difference in size between mitral and tufted cells recalls differentiation of output cell size in many brain regions. There is evidence that the *size principle* applies, meaning that the smaller tufted cells with higher input resistances would be more sensitive to sensory input, activated by lower stimulus concentrations, and the mitral cells would be activated more at higher concentrations. This might be correlated with different output targets, the tufted cells to the olfactory tubercle, the mitral cells more broadly to multiple olfactory cortical regions.

Self-Inhibition and Lateral Inhibition

The best-understood processing unit within the olfactory bulb is the *dendrodendritic microcircuit*, mediating mitral-to-granule excitation and granule-to-mitral inhibition (Rall and Shepherd, 1968; Yokoi et al., 1995). It is believed that this circuit mediates *self-inhibition* of activated mitral cells and *lateral inhibition* of other mitral cells. Evidence has recently emerged that lateral inhibition by a given mitral cell through granule cells on mitral cells belonging to other glomerular units is effected by propagating action potentials in the lateral dendrites (Xiong and Chen, 2002), which leads to activation of inhibitory granule cells in other glomerular units (Migliore and Shepherd, 2008). This type of processing is referred to as *distance-independent lateral inhibition*; that is, it is not limited to neighboring cells as in spatial receptive fields (Willhite et al., 2006). (For comparison with other types of center-surround lateral inhibition, see Chapter 19). The active properties of the mitral cell lateral dendrite, together with a columnar arrangement of granule cells applying synaptic inhibition at or near the cell bodies of mitral cells within their glomerular unit column, provide a potential mechanism for distance-independent processing of the extended odor map input (see Fig. 24.2).

The extent of the action potential spread along any given lateral dendrite will depend on the strength of the feedback inhibition from the sequence of glomerular units connected to that dendrite. Recent studies are expanding the possible roles of this dendrodendritic inhibition in the olfactory bulb (Urban and Arevian, 2009).

Glomerular Layer Processing

Glomerular layer cells have complex bursting properties that activate and coordinate the distributed activity between glomeruli. They also engage in lateral connections that are believed to mediate *lateral inhibition*, through inhibitory axo-dendritic synapses onto cellular elements associated with other glomeruli, and also *disinhibitory* (excitatory) *actions* due to inhibition of inhibitory elements. These axonal actions are believed to be *distance independent* through the variable extents of the PG axons. The PG cells are also believed to have other actions, such as *resetting* the activation levels of the glomeruli to make the glomerular response independent of stimulus concentration (Aungst et al., 2003; Cleland et al., 2007).

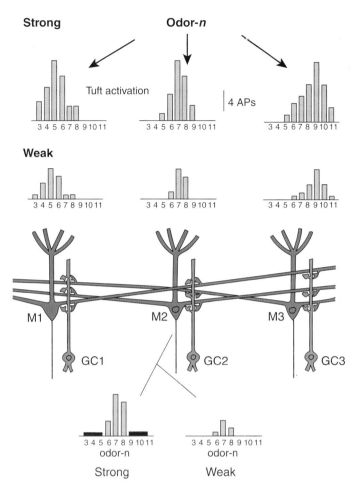

FIGURE 24–2. Computational simulation of lateral inhibition mediated by connectivity of the granule–mitral cell network. M1, M2, and M3 indicate mitral cells; GC1, GC2, and GC3 indicate granule cells. Each histogram represents the number of somatic APs elicited in a given mitral cell by each odor input without (top histograms) or in the presence of granule cells in the network (bottom histograms). The effect of lateral inhibition in suppressing M2 output for flanking odors is schematically represented by the black bars; this is similar to the report of Yokoi et al. (1995). Weak odor stimulation was modeled with a 60% compared with strong odor concentration. Locations of dendrodendritic synaptic contacts are indicated with small open and closed circles. The dendrodendritic synaptic interconnections are similar to those indicated in Figure 24.1. For simplicity, each GC is represented with only the branch including the synapses. (From Migliore and Shepherd, 2008)

Overall, the basic organization of the olfactory bulb has been compared with several other brain regions. The straight through pathways and two levels of lateral interactions are similar to the retina (Chapter 19). The similarities of synaptic triads and inhibitory control have been compared with the thalamus (Chapter 8). The columnar organization has been compared with the primary visual cortex (Chapters 1–3). The relatively simplicity of the unimodal input from the olfactory sensory neurons has suggested that analogous functions of these other regions may be compressed in the olfactory bulb.

Modulation

Modulation is crucial to olfactory processing, with 10 times more centrifugal projections innervating the olfactory bulb than M/T axons projecting from it. These centrifugals influence olfactory bulb processing through the multiple inhibitory systems mentioned earlier. The modulatory neurotransmitters include serotonin, norepinephrine, and acetylcholine; norepineprhinc is represented in Figure 24.1. In addition, the olfactory bulb cells contain receptors for a variety of modulator peptides and hormones. For example, the olfactory bulb has among the highest density of insulin receptors in the brain. The olfactory bulb in fact appears to be one of the most heavily modulated regions in the brain. One function of this modulation is to set the *behavioral state* of the system. For example, mitral cells respond differently depending on whether the animal is hungry or sated.

Neurogenesis

The olfactory bulb is one of only two regions in the brain where extensive neurogenesis takes place in the cells that form its microcircuits. The ORNs turn over during a period of about a month, throughout adult life, replenished by *stem cells* within the olfactory epithelium. In addition, the granule cells and PG cells in the olfactory bulb turn over, replenished by new cells generated in the ventricular zone of the brain and reaching the olfactory bulb through the *rostral migratory stream*. How these new cells are incorporated into the adult microcircuits of the olfactory bulb is under intensive study (Lledo et al., 2008).

PARALLEL PATHWAYS

Parallel processing is an important feature of neural circuits. The olfactory bulb contains several types of parallel pathways for processing olfactory information (summarized in Shepherd et al., 2004)

Main Olfactory Pathway

The most obvious parallel paths consist of the glomerular units. In the vertebrates, these number in the hundreds or thousands. Although traditionally it has been believed that in the vertebrate these ordinary glomeruli and their columns are all similar, anatomically identifiable glomeruli have begun to be recognized. In insects and arthropods, the ordinary glomeruli are distinct and range in number from the tens to a few hundreds.

Modified Glomerular Complex

The first and most clearly identifiable parallel pathway in the main olfactory pathway in the vertebrate is a modified glomerular complex (MGC) that forms a separate "labeled line" within the main olfactory bulb. It is believed to mediate information concerning odor cues related to suckling in young animals. An even clearer analogous set of specialized glomeruli is present in insects; two large glomeruli in *Manduca* have been called the macroglomerular complex, and they have been shown to be involved in processing of pheromone signaling.

Necklace Glomeruli

Related to the MGC in vertebrates are "necklace glomeruli" at the border of the main olfactory bulb facing the accessory olfactory bulb (AOB). In addition to the anatomical specificity of the necklace glomeruli, they are also the recipients of afferent input from a subpopulation of olfactory sensory neurons that express a guanylyl cyclase and a cyclic guanylate monophosphate (GMP)-stimulated phosphodiesterase and do not depend on the cyclic nucleotide gated (CNG) channel to process odor information.

Mitral and Tufted Cells

In addition to these parallel pathways related to the glomeruli, there are also parallel pathways provided by the mitral and tufted cell populations. It is not known whether, at the level of input processing within the glomeruli, the dendritic tufts of the two types receive input from different receptor cell axons, or if they interact with common or different PG cell dendrites. However, there is evidence that the smaller tufted cells may be more excitable than the mitral cells, an expression of the size principle first demonstrated in the spinal cord motoneuron.

At the level of output control in the external plexiform layer (EPL), each type interacts with different subpopulations of granule cells: superficial granule cells (G_S) control superficial and middle tufted cells (T_M), and deep granule cells (G_D) control mitral cells (M_I). Granule cells forming a third subpopulation appear to interact with both tufted and mitral cells.

When differing projection sites of mitral and tufted cells in olfactory cortical areas were first recognized, it was suggested that there might be an analogy in this regard with the different classes of retinal ganglion cells. The differing morphologies of the dendritic trees of these cells further support that analogy. In the retina, the particular sublamina of dendritic ramification of a ganglion cell has been found to be the main morphological feature correlated with the physiological type of its response (see Chapter 19). The fact that both mitral and tufted cells are further divided into subclasses on the basis of dendritic morphology indicates that multiple parallel pathways exist. This may be important in the mediation of different types of information about molecular stimuli.

Accessory Olfactory Bulb

The AOB lies in parallel with the main olfactory pathway in most vertebrate species, though it is largely vestigial in humans. It is receptive to both volatile ligands and ligands in solution. The sensory neurons are found in a tube-like structure, the vomeronasal organ (VNO), located at the base of the nasal septum. There are two main classes: V1R and V2R (the latter have large N-terminal domains resembling metabotropic glutamate receptors). Cells expressing V1Rs are localized in the apical part of the VNO epithelium, whereas those expressing V2Rs are in the basal part.

The V1R cell type projects to the anterior part of the AOB, whereas the V2R type projects to the posterior area. The cytoarchitecture and synaptic organization of the AOB are similar to the main olfactory bulb, although the laminar organization is less distinct. The receptor cell axons terminate in glomerular regions on the dendrites of the primary projection neuron, the mitral cells. Glomeruli receive inputs from more than one subset, and mitral cells also connect to more than one glomerulus. Periglomerular cells are few in number and are also likely to receive direct afferent input.

Intraglomerular circuits appear similar to those described for the main bulb. Modulation of mitral cell output occurs in the EPL, where reciprocal dendrodendritic synapses are formed with granule cells. Despite these similarities, there are also differences. The glomeruli are small and fewer in number. Although the projection neurons are called mitral cells, they are generally smaller and more polymorphic than their counterparts in the main bulb. There are also differences in some neurotransmitters.

In the AOB, the dendrodendritic microcircuit has been implicated as a mechanism for *learning* and *memory*. There is evidence that the dendrodendritic synapses are involved in the learning of the smell of a familiar stud male; after impregnation of a female, the smell of a strange male blocks the pregnancy, known as the Bruce effect. This mechanism involves arousal to activate noradrenaline centrifugal fibers onto the granule spines, which modulate mGluRs involved in mitral-to-granule excitation (Kaba et al., 1994). This has been claimed to be one of the clearest examples of a correlation between a synaptic circuit (the dendrodendritic interactions between mitral and granule cells, modulated by noradrenaline centrifugal fibers) and a specific learned behavior in the nervous system.

The output of the AOB is to the medial anterior, medial posterior, and posterior cortical nuclei of the amygdala and to the bed nucleus of the stria terminalis. From these regions there are multiple pathways to the hypothalamus. Through them, the accessory pathway is believed to be involved in processing signals involved in mating in many mammals, as well as hormonally regulated odor-stimulated behaviors. Increasing studies indicate that many of these functions can be mediated by the main olfactory pathway in higher primates and perhaps humans.

REFERENCES

Aungst JL, Heyward PM, Puche AC, Karnup SV, Hayar A, Szabo G, Shipley MT (2003) Centre-surround inhibition among olfactory bulb glomeruli. *Nature* 426: 623–629.

Buck L, Axel R (1991) A novel multigene family may encode odorant receptors: a molecular basis for odor recognition. *Cell* 65:175–187.

Cleland TA, Johnson BA, Leon M, Linster C (2007) Relational representation in the olfactory system. *Proc Natl Acad Sci USA* 104:1953–1958.

Horton JC and Adams DL (2005) The cortical column: a structure without a function. *Phil Trans R Soc B* 360:837–862.

Kaba H, Hayashi Y, Higuchi T, Nakanishi S (1994) Induction of an olfactory memory by the activation of a metabotropic glutamate receptor. *Science* 265:262–264.

Kobayakawa K, Kobayakawa R, Matsumoto H, Oka Y, Imai T, Ikawa M, Okabe M, Ikeda T, Itohara S, Kikusui T, Mori K, Sakano H (2007) Innate versus learned odour processing in the mouse olfactory bulb. *Nature* 450:503–508.

Kosaka K, Kosaka T (2005) Synaptic organization of the glomerulus in the main olfactory bulb: compartments of the glomerulus and heterogeneity of the periglomerular cells. *Anat Sci Int* 80:80–90.

Lledo PM, Merkle FT, Alvarez-Buylla A (2008) Origin and function of olfactory bulb interneuron diversity. *Trends Neurosci* 31:392–400.

Mazor O, Laurent G (2005) Transient dynamics versus fixed points in odor representations by locust antennal lobe projection neurons. *Neuron* 48:661–673.

Migliore M, Shepherd GM (2008) Dendritic action potentials connect distributed dendrodendritic microcircuits. *J Comput Neurosci* 24:207–221.

Mori K, Yoshihara Y. (1995) Molecular recognition and olfactory processing in the mammalian olfactory system. *Prog Neurobiol* 45:585–619.

Rall W, Shepherd GM (1968) Theoretical reconstruction of field potentials and dendrodendritic synaptic interactions in olfactory bulb. *J Neurophysiol* 31:884–915.

Shepherd GM, Chen WR, Greer CA (2004) Olfactory bulb. In: Shepherd GM, ed. *The Synaptic Organization of the Brain.* 5th ed., pp. 165–216. New York: Oxford University Press.

Shepherd, GM, Chen, WR, Willhite, D, Migliore, M, Greer, CA (2007) The olfactory granule cell: from classical enigma to central role in olfactory processing. *Brain Res Rev* 55(2):373–382.

Urban, NN, Arevian AC (2009) Computing with dendrodendritic synapses in the olfactory bulb. *Ann NY Acad Sci* 1170:264–269.

Willhite DC, Nguyen KT, Masurkar AV, Greer CA, Shepherd GM, Chen WR (2006) Viral tracing identifies distributed columnar organization in the olfactory bulb. *Proc Natl Acad Sci USA* 103:12592–12597.

Xiong W, Chen WR (2002) Dynamic gating of spike propagation in the mitral cell lateral dendrites. *Neuron* 34:115–126.

Xu, FQ, Greer, CA, Shepherd, GM (2000) Odor maps in the olfactory bulb. *J Comp Neurol* 422: 489–495.

Yokoi M, Mori K, Nakanishi S (1995) Refinement of odor molecule tuning by dendrodendritic synaptic inhibition in the olfactory bulb. *Proc Natl Acad Sci USA* 92:3371–3375.

Zhang X, Firestein S (2002) The olfactory receptor gene superfamily of the mouse. *Nat Neurosci* 5:124–133.

25

Olfactory Cortex

Donald A. Wilson and Edi Barkai

The olfactory cortex is defined as those brain areas receiving direct input from mitral and tufted cells of the olfactory bulb (Neville and Haberly, 2004). The olfactory cortex thus includes anterior olfactory nucleus, olfactory tubercle, nucleus of the lateral olfactory tract, cortical nucleus of the amygdala, tenia tecta, piriform cortex, and entorhinal cortex. The largest component of the olfactory cortex is the piriform cortex, and descriptions in this chapter will be limited to the piriform cortex. Piriform cortex is also referred to as pyriform cortex (most common in primates) or prepyrifom cortex (less common usage). Piriform cortex is a phylogenetically old, three-layered paleocortex, as opposed to the six-layered neocortex. The primary afferent to the piriform cortex is mitral cells from the olfactory bulb, which receive direct excitatory input from olfactory sensory neurons within the olfactory epithelium. Thus, piriform cortex is only two synapses from the outside world.

The piriform cortex can be divided into at least three subregions based on intrinsic circuitry and extrinsic connectivity (Ekstrand et al., 2001). These subregions are the dorsal and ventral portions of the anterior piriform cortex and the posterior piriform cortex. The traditional division between anterior and posterior piriform cortex is the end of the myelinated lateral olfactory tract, which runs along the surface of the anterior piriform cortex. Mitral cell axons extend beyond the end of this tract, but the tract as an identifiable structure ends at the caudal end of the anterior subregion. Each of the three subregions has reciprocal connections with the others, though there are differences in density and termination patterns, which will be described later.

The three subregions also differ in their connectivity with other brain areas (Fig. 25.1A). For example, the orbitofrontal cortex projects strongly to the

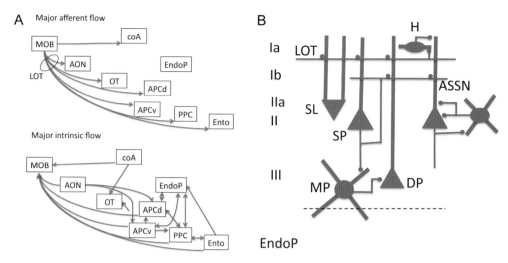

Figure 25–1. (*A*) Major afferent and intrinsic connectivity of olfactory cortical regions. The olfactory tubercle receives input from both anterior and posterior regions of the piriform cortex, though only a single arrow is shown for clarity. AON, anterior olfactory nucleus; APCd, dorsal region of the anterior piriform cortex; APCv, ventral region of the anterior piriform cortex; coA, cortical nucleus of the olfactory tract; EndoP, endopiriform nucleus; Ento, entorhinal cortex; LOT, lateral olfactory tract; MOB, main olfactory bulb; OT, olfactory tubercle; PPC, posterior piriform cortex. (*B*) Major local circuit components of the piriform cortex. (See text for circuit description.) ASSN, association fibers; DP, deep pyramidal cell; H, horizontal cell interneuron; MP, multipolar interneuron; SL, semilunar pyramidal cell; SP, superficial pyramidal cell. Excitatory neurons are shown in red; inhibitory neurons are shown in blue.

ventral anterior piriform and posterior piriform cortices, but only sparsely to the dorsal anterior piriform cortex. A further example shows that the posterior piriform receives a strong input from the basolateral amygdala, while the amygdala input to anterior regions is less dense.

Deep to the piriform cortex lies the endopiriform nucleus, a long thin nucleus that runs nearly the length of the piriform cortex. The endopiriform nucleus has been referred to as piriform cortex Layer IV, though this terminology is controversial. The endopiriform nucleus receives excitatory input from pyramidal cells within piriform cortex, has a strong local excitatory interconnectivity, and projects back throughout the piriform cortex in a widely distributed manner.

The three layers of the piriform cortex are distinct in their neural components (Fig. 25.1B). Layer I includes pyramidal cell apical dendrites along with afferent and intracortical axons and a small number of intrinsic interneurons. The afferent input from the olfactory bulb terminates within superficial Layer I (Ia), while the deeper Layer Ib contains mostly intracortical association fibers. Layer II is a pyramidal cell body layer, and again has a superficial lamina, IIa, which contains the somata of seminlunar pyramidal neurons and the deeper

Layer IIb, which has densely packed pyramidal cell bodies. Finally, Layer III contains basal dendrites and axons of Layer II pyramidal neurons, as well as deep pyramidal neurons that extend apical dendrites into Layer I. Local interneurons are also located within Layer III.

BASIC CONNECTIONS

Cellular Elements

There are three main classes of pyramidal cells within piriform cortex: deep, superficial, and semilunar. Deep pyramidal cells have somata in Layer III, basal dendrites that can extend into the endopiriform nucleus, and apical dendrites that extend into Layer Ia. Superficial pyramidal cells have somata that are tightly packed within Layer IIb, basal dendrites extending into Layer III, and apical dendrites extending into Layer Ia. Finally, semilunar pyramidal cells have somata in Layer IIa and extend multiple apical dendrites into Layer Ia. Semilunar cells have no basal dendrites. All of these different pyramidal cell types have extensive dendritic spines, though the semilunar cells appear to express very large spines selectively on their distal-most dendritic branches, primarily within Layer Ia. Estimates of total pyramidal cell numbers suggests that they outnumber their primary afferent, mitral cells by at least an order of magnitude.

In addition to the pyramidal neurons, there are several classes of interneurons, distinguished both by morphology and laminar location. Most of these interneurons are GABAergic with the exception of large, spiny multipolar cells, which lie in deep layer III and the endopiriform nucleus and may be glutamatergic. The GABAergic interneurons include horizontal cells within Layer Ia and with long dendrites parallel to the cortical surface. Additional multipolar cells (without dendritic spines) lie within Layer II and III, presumably mediating classic inhibitory feedback functions, and a class of small bipolar or bitufted cells has somata in Layer IIa and dendrites extending into both Layers I and III. In addition to GABA, many of the interneurons colocalize neuropeptides, including cholecystokinin, corticotropin-releasing factor, enkephalin, and neurotensin, among others.

Afferent Input

The primary afferent to the piriform cortex is from the mitral and tufted cells of the main olfactory bulb. Mitral cells project throughout the rostral-caudal extent of the piriform cortex. In contrast, tufted cells (middle and internal tufted cells) terminate only in the most anterior regions, with densest projections limited to the anterior olfactory nucleus and olfactory tubercle. The input from individual mitral cells appears broadly dispersed, and termination of mitral cells from different glomeruli (presumably conveying information from olfactory sensory neurons expressing different olfactory receptors) overlaps, allowing convergence.

In contrast to thalamic input to neocortex, which enters orthogonally to the cortical sheet allowing near simultaneous activation of a large cortical region, mitral cell input to the piriform cortex travels as a rostral to caudal wave across the cortical surface. Small axonal collaterals leave the myelinated lateral olfactory tract and terminate in small patches exclusively within Layer Ia. The division between Layers Ia and Ib as evidenced by tracings of individual mitral cell axons is very precise. There is no evidence of any columnar organization to the afferent input termination within piriform cortex.

Modulatory Inputs

The piriform cortex and other olfactory cortical areas are targets of several neuromodulatory inputs from the brainstem and basal forebrain. Cholinergic input to the piriform cortex arises within the horizontal limb of the diagonal band of Broca from a group of neurons distinct from, though interspersed with, those projecting to the main olfactory bulb. Cholinergic fibers are dense throughout the rostral caudal extent of the piriform cortex, with fibers most pronounced within Layers II and III and somewhat more sparse within Layer I. Both muscarinic (m1, m2, m3, m4) and nicotinic receptors are present within the cortex, with muscarinic receptors largely localized in Layers I and II and nicotinic primarily within Layers II and III.

Noradrenergic input from the locus ceruleus extends throughout the rostral-caudal extent of the piriform cortex, though differs in its laminar density from acetylcholine, with most dense fibers located within Layers I and III, and only light innervation of Layer II. Both α and β noradrenergic receptors are expressed within the piriform cortex. Finally, serotonergic and dopaminergic fibers also terminate with the piriform cortex, though these are less well characterized. Receptors for monoamines appear to be expressed by both principle neurons and interneurons.

Intrinsic Connectivity

The piriform cortex can be considered an autoassociative network in that cortical pyramidal cell axons form extensive association fiber connections to other pyramidal cells within the piriform cortex (Johnson et al., 2000). Based on reconstructions of individual pyramidal cell axons, a single pyramidal cell may contact as many as 1000–2000 other pyramidal cells, in addition to sending collaterals to other olfactory cortical regions and back to the olfactory bulb. These association fiber connections terminate on pyramidal cell proximal apical dendrites (Layer Ib) and basal dendrites (Layer III), while as noted earlier the afferent input terminates on the distal apical dendrites within

Layer Ia. Layer Ib and III also contain axons of commissural fibers via the anterior commissure and fibers from other olfactory cortical regions. Both mitral cells and piriform cortex pyramidal cells are glutamatergic, and pyramidal cells express AMPA, NMDA, and metabotropic glutamate receptors. In addition to the laminar separation of afferent and intrinsic fibers, within Layer Ib there is further lamination of intrinsic connections. Feedforward connections from anterior regions to more posterior regions terminate within superficial Layer Ib, while connections from posterior regions to anterior regions terminate within deep Layer Ib.

Interestingly, this spatial lamination of excitatory synapses onto piriform cortical pyramidal cells recapitulates the temporal sequence of activation. Thus, anatomically, inputs are ordered with afferent inputs most distally, then feedforward intrinsic inputs, then feedback intrinsic inputs on the most proximal apical dendrite. Temporally, afferent inputs would be active first, which could then excite pyramidal cells in the anterior piriform cortex, driving action potentials in feedforward intrinsic fibers, and finally posterior pirfirom cortex pyramidal cells would be excited to provide activity in feedback intrinsic fibers.

It should also be noted that connectivity between the subregions of the piriform cortex are not completely reciprocal. The projection from anterior piriform cortex to posterior piriform cortex is primarily (though not exclusively) one way, as is the ventral anterior piriform cortical projection to the dorsal anterior piriform. A similar feedforward asymmetry exists in commissural connections, with the anterior piriform primarily receiving input from the contralateral anterior olfactory nucleus and the posterior piriform cortex primarily receiving commissural input from the contralateral anterior piriform cortex.

Interneurons terminate on pyramidal cells throughout the apical dendritic tree, soma, and initial axon segment, though synaptic targets of different interneuron subpopulations can be segregated. For example, horizontal cells lying in Layer Ia receive direct input from afferent mitral cells and synapse onto pyramidal cell distal apical dendrites. The dendrites of many multipolar cells branch throughout Layer II and III, where they receive excitatory input from pyramidal cells. The axons of these multipolar cells form basket connections around pyramidal cell somata and axon initial segments, thus creating classic feedback and lateral inhibitory circuits.

Together, these circuit components and patterns of input result in an initial wave of excitation in piriform cortical pyramidal cells driven by an odor-evoked mitral cell volley via the lateral olfactory tract and axodendritic synapses with Layer Ia. The excitation of pyramidal cells then drives a second, short-latency wave of excitation via the intrinsic association fibers and their synapses within Layers Ib and III. Excitability of the pyramidal cells is tightly controlled by the strong inhibitory feedback generated by the inhibitory multipolar basket cells.

FUNCTIONAL OPERATION

Afferent and Intrinsic Synaptic Transmission

The simple and well-defined anatomical organization of the piriform cortex enables physiological recordings from a homogenous population of pyramidal cells and specific activation of well-defined synaptic pathways. While both afferent and association fiber synapses are glutamatergic, the difference in location of these synapses on the dendritic tree induces differences in their effects on the pyramidal cell. For example, postsynaptic potentials (PSPs) evoked by stimulation of axons in layer Ib, which terminate close to the cell soma, have significantly shorter rise time compared to PSPs activated by layer Ia axons, which terminate distally (Fig. 25.2A).

These two synaptic pathways also differ in the susceptibility for activity and learning-induced synaptic plasticity. NMDA receptor–dependent long-term potentiation (LTP) can be induced in both pathways, but to a much greater extent in the intracortical association fibers. Olfactory learning–induced long-lasting modulation of synaptic transmission, discussed later, is also much more pronounced in the intrinsic pathway (Barkai, 2005).

Another form of synaptic plasticity expressed by afferent synapses is a presynaptic metabotropic glutamate receptor–dependent depression of glutamate release following intense mitral cell activation. This afferent synaptic depression is both necessary and sufficient for short-term cortical adaptation to odors as well as short-term behavioral odor habituation (Wilson and Linster, 2008).

Network Oscillations

The four oscillatory patterns that are detected in the mammalian cortex (delta, theta, beta, and gamma) are also present in the piriform cortex. Under anesthesia, olfactory bulb and anterior piriform cortex show a peak of neural firing that tends to occur between the end of inspiration and the beginning of expiration phase. Oscillations in the beta and gamma frequency ranges are prominent in field potentials induced by odorants in the olfactory bulb and piriform cortex. Both oscillation frequencies are induced by odorants in a concentration-dependent manner (Neville and Haberly, 2004). The two oscillations patterns are differently affected by surgical interruption of the lateral olfactory tract. While lateral olfactory tract lesions abolish the beta oscillation, gamma oscillation is still induced in the olfactory bulb, confirming that the gamma oscillation is generated within the olfactory bulb and that the beta oscillation requires the participation of piriform cortex (Neville and Haberly, 2004).

Piriform cortex oscillations are strongly dependent on coactivation with related brain regions, such as the olfactory bulb, hippocampus, entorhinal cortex, and orbitofrontal cortex. It has been suggested that beta oscillations seem to be evoked in circumstances where a distinct behavior or reward valence is identified with an odorant, and gamma oscillations with the formation of activation in cell assemblies during exploratory behavior (Kay et al., 2009). However, the biophysical mechanisms underlying the induction of oscillations, the factors which modulate their frequency, and their functional role are yet to be described.

Role of Piriform Cortex in Olfactory Information Processing

The inputs from the olfactory bulb are nontopographically spread across the entire surface of the piriform cortex. Accordingly, presentation of eight different odors to rats results in increased firing rate in 30% of the piriform cortex cells, with each cell responding to at least one of the odors (Schoenbaum and Eichenbaum, 1995). This unique widespread activation results from overlapping afferent inputs and from the hardwired connectivity between pyramidal neurons (Johnson et al., 2000). Individual cells have widespread axonal arbors that extend over nearly the full length of the cerebral hemisphere, with no regularly arranged patchy concentrations like those associated with the columnar organization in other primary sensory areas. Each pyramidal cell makes a small number of synaptic contacts on a large number of other cells in piriform cortex at disparate locations. Based on these findings, Haberly (2001) proposed the intriguing hypothesis that redefines the traditional functional role of the piriform cortex in olfactory information processing. According to this hypothesis, the piriform cortex carries out functions that have traditionally defined association cortex—it detects and learns correlations between olfactory gestalts formed in anterior olfactory cortex and a large repertoire of behavioral, cognitive, and contextual information to which it has access through reciprocal connections with prefrontal, entorhinal, perirhinal, and amygdaloid areas.

Role of Piriform in Perceptual Stability and Discrimination

The distributed input and autoassociative network described earlier leads to odors evoking highly dispersed patterns of neural activity across the piriform cortex, as opposed to the spatially stereotyped patterns of odor-evoked activity within the olfactory bulb. Furthermore, synaptic plasticity within the association fiber network allows previously experienced patterns of olfactory bulb input (combinations of mitral cell activity) to be stored within piriform cortex as a change in synaptic weight between coactive pyramidal cells. This is equivalent to storing a template of past input patterns, against which new patterns can be compared. There are two hypothetical consequences of such pattern recognition circuits. First, if the incoming pattern is noisy or differs only slightly from an existing pattern, the previously strengthened synapses can, in essence, fill in missing components and allow pattern completion. Pattern completion promotes perceptual stability, which may be especially critical in a sensory system dealing with complex mixtures of volatile molecules. As the overlap of an incoming pattern and existing templates diverges, however, the piriform cortex can separate or decorrelate the patterns, enhancing odor discrimination. Recent evidence supports this dual role for ensembles of piriform cortical neurons (Barnes et al., 2008), demonstrating both pattern completion and enhanced pattern separation compared to olfactory bulb mitral cell ensembles dependent on pattern overlap. Furthermore, the cortical ensemble odor-processing capabilities predict behavior odor discrimination.

Role of the Piriform Cortex in Olfactory Learning and Memory

Rats can easily learn to associate odor with reward. When trained with odor-discrimination tasks, they demonstrate a dramatic increase in their capability to acquire memories of new odors, once they have learned the task ("rule learning"). Such rule learning is accompanied by a series of pre- and postsynaptic cellular modifications, which share two major common traits:

1. They are widespread throughout the piriform cortex network. Both physiological and morphological modifications are found in most neurons (Barkai, 2005).
2. The time course in which these modifications appear and disappear is strongly correlated with the time course in which the skill is acquired and decays (Barkai, 2005). However, memories for specific odors outlast these modifications by far.

This evidence suggests that such learning-induced cellular modifications are not the mechanism by which specific memories are stored. Rather, it may be the mechanism by which the cortical network enters into a "learning mode," a state in which it can acquire and store memories for specific odors rapidly and efficiently (Barkai, 2005).

Dynamics of Learning-Induced Cellular Modifications

Two types of learning-related cellular changes occur in the piriform cortex after olfactory learning: modifications in the intrinsic properties of neurons and modifications at the synapses interconnecting these neurons. These modifications differ in the dynamics at which they appear and are maintained (Barkai, 2005). One day after learning, pyramidal neurons show enhanced neuronal excitability. This enhancement results from reduction in a calcium-dependent conductance, the sI_{AHP}, which mediates the late postburst after hyperpolarization and thus controls repetitive spike firing. Such enhanced

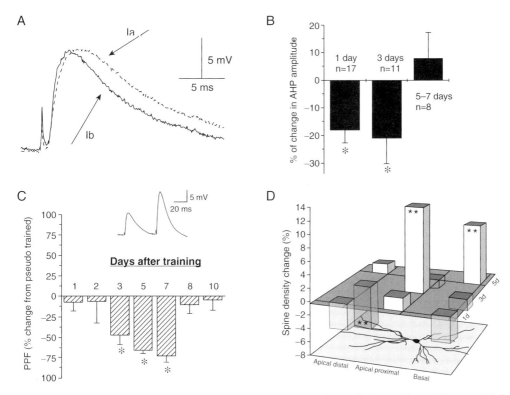

Figure 25–2. (*A*) Example traces of postsynaptic potentials evoked in a layer II pyramidal cell by stimulation of layer Ib (continuous line) and stimulation of layer Ia (dotted line). Note the difference in rise times of postsynaptic potentials evoked at the two different pathways. (*B*) Postburst afterhyperpolarization (AHP) reduction (relative to control) in neuron from trained rats. One and 3 days after the last training session, the AHP amplitudes in neurons from trained rats were significantly smaller than in neurons from pseudo-trained rats. Five days or more after the last training session, AHP amplitudes were not significantly different between groups. (*C*) Paired pulse facilitation is reduced starting 3 days after learning. Dynamics of learning-induced paired-pulse facilitation (PPF) reduction (interstimulus intervals = 50 ms). Decreased PPF appears 3 days after rule learning and lasts for 5 days. (*D*) The effect of odor discrimination learning on spine density. Summary of changes in spine density in trained rats as compared with pseudo-trained rats, at three time points, along apical distal, apical proximal, and basal dendrites. $**p < 0.01$.

excitability lasts for 3 days (Fig. 25.2B) and is followed by a series of synaptic modifications. On the third day after learning, several forms of long-term enhancement in synaptic connections between layer II pyramidal neurons appear. Enhanced synaptic release is indicated by reduced paired-pulse facilitation. Postsynaptic enhancement of synaptic transmission is indicated by reduced rise time of postsynaptic potentials, and formation of new synaptic connections is indicated by increased spine density along dendrites of these neurons. Such modifications last for up to 5 days (Figs. 25.2C and D).

Learning-Induced Modifications in Synaptic Transmission to the Piriform Cortex

Enhanced learning capability is also accompanied by long-term enhancement of synaptic transmission in both the descending synaptic inputs from the orbitofrontal cortex and the ascending inputs from the olfactory bulb (Cohen et al., 2008). Such modifications are likely postsynaptic. Through these descending connections, the orbitofrontal cortex might actively modulate afferent input to the piriform cortex so that cells fire in certain contexts but not others. Moreover, the orbitofrontal cortex could potentially initiate activity in anterior piriform cortex in the absence of any odor stimulation, allowing for recall of odors and odor-related associations. The overall strengthening of the descending pathway suggests that the specificity of the evoked odor memory is achieved not by these inputs, but by specific synaptic connections that were strengthened within the piriform cortex network, during the learning process.

Summary

Although the piriform cortex anatomical structure is simple and well defined compared to neocortical structures, it subserves higher brain function like other associative cortices. While the piriform cortex is engaged in relatively basic functions like identifying familiar odors and discriminating between similar odors, it also has a central role in complex tasks such as integration of information about the identity and the reward value of odors. Such a variegated capability is enabled by its intrinsic hardwiring, which resembles that seen in higher neocortical areas, and by the efficient sets of connections with ascending and descending brain areas with which it generates synchronized activity.

References

Barkai E (2005) Dynamics of learning-induced cellular modifications in the cortex. *Biological Cybernetics* 92:360–6.

Barnes DC, Hofacer RD, Zaman AR, Rennaker RL, Wilson DA (2008) Olfactory perceptual stability and discrimination. *Nat Neurosci* 11:1378–1380.

Cohen Y, Reuvenu I, Barkai E, Maroun M (2008) Olfactory learning-induced long lasting enhancement of descending and ascending synaptic transmission to the piriform cortex. *J Neurosci* 28:6664–6669.

Ekstrand JJ, Domroese ME, Johnson DM, Feig SL, Knodel SM, Behan M, Haberly LB (2001) A new subdivision of anterior piriform cortex and associated deep nucleus with novel features of interest for olfaction and epilepsy. *J Comp Neurol* 434:289–307.

Haberly LB (2001) Parallel-distributed processing in olfactory cortex: new insights from morphological and physiological analysis of neuronal circuitry. *Chem Senses* 26: 551–576.

Johnson DM, Illig KR, Behan M, Haberly LB (2000) New features of connectivity in piriform cortex visualized by intracellular injection of pyramidal cells suggest that "primary" olfactory cortex functions like "association" cortex in other sensory systems. *J Neurosci* 20:6974–6982.

Kay LM, Beshel J, Brea J, Martin C, Rojas-Líbano D, Kopell N (2009) Olfactory oscillations: the what, how and what for. *Trends Neurosci* 32:207–214.

Neville KR, Haberly L (2004) Olfactory cortex. In: Shepherd GM, ed. *The Synaptic Organization of the Brain*, 5th ed., pp. 415–454. New York: Oxford University Press.

Schoenbaum G, Eichenbaum H (1995) Information coding in the rodent prefrontal cortex. I. Single-neuron activity in orbitofrontal cortex compared with that in pyriform cortex. *J Neurophysiol* 74:733–750.

Wilson DA, Linster CL (2008) Neurobiology of a simple memory. *J Neurophysiol* 100:2–7.

Section 10

Taste System

26

Taste Coding and Feedforward/ Feedback Signaling in Taste Buds

Stephen D. Roper and Nirupa Chaudhari

Peripheral sensory organs of gustation—taste buds—are believed to transmit two categories of signals to higher centers in the brain: *(1)* information related to *conscious perceptions* and *palatability* of ingested foodstuff, such as sweetness, bitterness, saltiness, sourness (acidity), and umami (the taste of certain amino acids, notably monosodium glutamate, and related compounds);[1] and *(2)* information pertaining to *energy balance and metabolism*. For instance, taste stimulation leads to a brief, early (cephalic phase) release of insulin, presumably preparing the gastrointestinal tract for the food bolus. Identical mechanisms and circuits within taste buds might encode both of these two categories of signals. However, it would be wise to leave open the possibility that there may be separate but parallel streams of signal processing in taste buds underlying taste perceptions versus triggers for nutrient handling (Roper, 2009). To date, chemosensory researchers have focused greater attention on the first category—transduction and processing of sweet, bitter, salty, sour, and umami tastes. Thus, this overview of peripheral circuits for taste signals will only examine those mechanisms.

GUSTATORY CODING: LABELED LINE OR COMBINATORIAL CODING?

The role of taste buds vis-à-vis conscious perceptions is to parse a complex mixture of food chemicals—everything from protons to proteins—into the five (or more) qualities noted above. A major, and as yet unresolved, controversy in the field is whether this is accomplished by labeled lines or some form of combinatorial coding (Spector and Travers, 2005). A "labeled line," in the strict sense, means first that within the peripheral sensory organ there is a dedicated population of receptor cells for each of the categories (such as a population of sweet-selective receptor cells). Second, each separate population of receptor cells should excite dedicated primary afferent fibers that transmit the same signal to the central nervous system along a "line" devoted only to that taste quality (for example, here "sweetness"). Furthermore, within the central nervous system, relay and projection neurons for these signals would retain the same label ("sweet"). The entire "line" of neuronal connections from periphery to primary gustatory cortex would be "labeled" (sweet, bitter, etc.). Combinatorial coding in the extreme implies that at one or more points in the sequence (or "line") of connections, signals in two or more neurons are combined and compared such that the final output represents coactivation of multiple inputs. Consequently, in combinatorial coding taste qualities (e.g., "sweetness") are constructed from a mosaic of individual signals, none of which on its own necessarily represents a unique basic taste. It is likely that neither of these extremes fully explains gustatory coding.

There is no experimental evidence for any strict labeled "line" in mammalian taste. Although different types of taste receptors are expressed in separate taste bud cells, primary sensory afferent fibers at best show a predominant signal for one quality over others. This is by no means an exclusive "line." The strongest case for "labeled lines" might be made for sweet-sensitive fibers in primates. Some gustatory afferent fibers in the monkey respond remarkably well to low concentrations of sugars and not much to other taste compounds. However, even in these cases, a "sweet-best" afferent fiber (as such a unit is termed) usually will also respond to high concentrations of other taste qualities such a bitter. Thus, on its own, a single fiber is not, in the strict sense, "labeled" for sweet insofar as activity in a "sweet-best" unit taken alone could not distinguish between a low concentration of a sugar versus a high concentration of a bitter compound.

There is stronger experimental evidence for combinatorial coding. Many afferent fibers respond to multiple taste compounds, such as sweet, bitter, or salty. At higher levels in the central nervous system, such as in the hindbrain (nucleus of the solitary tract) or in the gustatory cortex, the multiplicity of responses is even more striking. Any sense of a "labeled line" is lost.

Nonetheless, advocates of "labeled line" coding in taste point to molecular biological studies indicating that any given taste bud sensory cell expresses G protein–coupled taste receptors (taste GPCRs) for only one quality and that this implies a labeled line (Chandrashekar et al., 2006). How, then, is the discrepancy between "labeled receptor cells" versus multiply responsive afferent fibers resolved? The answer may lie in the circuitry underlying signal processing in taste buds, which will be explained later.

THE CELLULAR COMPOSITION OF MAMMALIAN TASTE BUDS

Mammalian taste buds typically consist of a collection of 50–100 cells. Within this population there are four major subcategories: *(1)* Type I cells that are thought to function as glia and help degrade or remove neuro transmitters released by other cell types; *(2)* Type II cells, or receptor cells, that express taste GPCRs and appropriate downstream effectors; Type III cells, or presynaptic cells, that possess synapses and express synaptic proteins; and Type IV cells, which are undifferentiated stem cells for renewing the population of other cell types (Roper, 2007).

Neurotransmitters and Cell-to-Cell Communication in Taste Buds

Recent studies have identified neurotransmitters released by taste buds and, by so doing, have revealed which cells secrete which transmitter(s) and how the transmitters activate the taste bud circuits. Related studies have also revealed how the different types of taste bud cells respond to chemical stimulation. Receptor (Type II) and presynaptic (Type III) cells have been shown to be the principal players in chemosensory stimulation. Isolated receptor cells respond mainly to a single taste quality—sweet, bitter, or umami—and not to combinations. This is consistent with the selective expression of taste GPCRs in these cells (see earlier discussion). In contrast, isolated presynaptic cells respond mainly to KCl depolarization and to sour (acid) taste stimuli (DeFazio et al., 2006; Huang et al., 2007). *In situ*, however, presynaptic cells appear to respond to stimulation from multiple taste categories, including sweet, bitter, salty, umami, and sour (Tomchik et al., 2007). This discrepancy between responses in isolated presynaptic cells and presynaptic cells *in situ* stems from the presence of cell-to-cell communication within taste buds.

Cell-to-cell communication in taste buds was initially proposed based upon ultrastructural studies on fish taste buds and then demonstrated with intracellular recordings in amphibian taste buds. However, the full details have only recently been learned from investigations using biosensor cells and confocal Ca^{2+} imaging in lingual slices from mouse vallate taste buds. These studies have revealed that upon taste stimulation with bitter, sweet, or umami substances, receptor cells secrete adenosine triphosphate (ATP). Taste-evoked ATP acts on three postsynaptic targets: *(1)* P2X2 and P2X3 receptors expressed on gustatory primary afferent fibers (Finger et al., 2005); *(2)* P2Y receptors (probably P2Y4) on presynaptic cells in close vicinity to the activated receptor cells (Roper and Huang et al., 2009); and *(3)* P2Y1 autocrine receptors on receptor cells (Huang et al., 2009). All these actions are excitatory. A robust ecto-ATPase (NTPDase 2) is expressed on the glial-like Type I taste bud cells (Bartel et al., 2006). NTPDase 2 serves to terminate the postsynaptic actions of ATP by degrading the purine to adenosine diphosphate and, to a lesser extent, adenosine monophosphate.

Taste-evoked excitation of receptor cells thus secondarily activates presynaptic cells via ATP. Presynaptic cells in turn secrete serotonin

(5 hydroxytryptamine, or 5-HT) and norepinephrine (NE) (Huang et al., 2008). Postsynaptic targets for these aminergic transmitters are not well characterized, especially for NE. However, 5-HT exerts a strong negative feedback onto receptor cells. 5-HT inhibits receptor cells and diminishes their ability to secrete ATP (Huang et al., 2009). Blocking 5-HT1A receptors relieves this inhibition and leads to a striking enhancement of taste-evoked ATP secretion. These interactions are summarized in Figure 26.1.

Interestingly, the mechanisms involved in the synaptic release of ATP and 5-HT in taste buds are quite unusual. First, receptor cells secrete ATP via pannexin 1 gap junction hemichannels (Huang et al., 2007; Romanov et al., 2007; Dando and Roper, 2009). These gap junction hemichannels are believed to be opened by the combined action of intracellular Ca^{2+} release (from activation of taste GPCRs) and membrane depolarization (from calcium-activated TRPM5 cation channels) (Fig. 26.2).

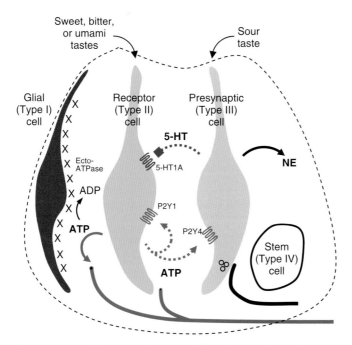

FIGURE 26–1. Schematic diagram of feedforward and feedback signaling in mammalian taste buds. The diagram shows a taste bud (dotted lines) with the four main categories of taste bud cells. Glial (Type I) cells express an ecto-ATPase (NTPDase2) that degrades ATP. Receptor (Type II) taste bud cells express G protein–coupled taste receptors for sweet, bitter, or umami tastes. Taste stimulation evokes ATP secretion from receptor cells. ATP excites gustatory primary afferent fibers (shown at bottom), neighboring presynaptic (Type III) taste bud cells, and via autocrine feedback, presynaptic cells, as shown above in red. Presynaptic cells make synaptic contacts with nerve fibers. Presynaptic cells secrete serotonin (5-HT) and norepinephrine (NE), perhaps at these synapses. 5-HT also exerts negative (paracrine) feedback onto receptor cells, shown above in blue. Stem (type IV) cells are progenitor cells for the taste bud.

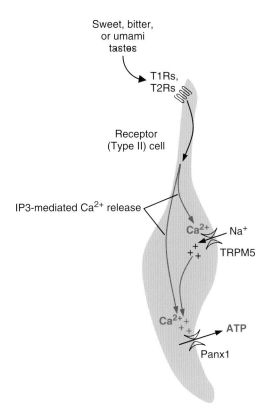

Figure 26–2. Transmitter release mechanism for ATP in taste receptor (Type II) cells. Activating taste G protein–coupled receptors (T1Rs, T2Rs) causes release of intracellular Ca^{2+} via phospholipase C subclass β2 and inositol triphosphate (IP3). Intracellular Ca^{2+} mobilization gates open TRPM5 cation channels and depolarizes the cell membrane. Ca^{2+} also acts on pannexin 1 hemichannels. The combined action of membrane depolarization (via TRPM5 activity) and intracellular Ca^{2+} opens pannexin 1 hemichannels and allows the secretion of ATP.

This novel extra synaptic secretory mechanism explains why conventional ultrastructural features of synapses, including clusters of synaptic vesicles and synaptic membrane thickenings, are absent on receptor cells. It is not known whether there are focal sites for taste-evoked ATP secretion or whether ATP secretion is broadly dispersed across the basolateral membrane of the receptor cell. Release of 5-HT and NE appears to be more conventional insofar as secretion of both these biogenic amines can be triggered by membrane depolarization and Ca^{2+} influx, the usual mechanism for vesicular exocytosis. Further, 5-HT and NE are coreleased in some cases (Huang et al., 2008). However, 5-HT secretion can also be triggered in the absence of Ca^{2+} influx by activating intracellular Ca^{2+} release, indicating that there are at least two different modes for secretion of this amine. Whether this latter mechanism, secretion triggered by Ca^{2+} released from intracellular stores, is vesicular has not been investigated.

Precisely how feedforward excitation of ATP onto presynaptic cells, feedback (autocrine) excitation of ATP onto presynaptic cells, and negative feedback (paracrine) exerted by 5-HT from presynaptic onto receptor cells work together to encode taste qualities and shape signals in the peripheral sensory organs of gustation remains to be worked out in detail. It is clear, however, that cell-to-cell signaling takes place during taste stimulation and that purinergic and aminergic transmitters are involved. The functional organization of taste buds thus can be understood as (a) "labeled" receptor cells (not lines); (b) convergent input onto presynaptic cells (several receptor cells may converge onto a single presynaptic cell; Tomchik et al., 2007); and (c) some extent of feedback (autocrine and paracrine) signal processing. Interestingly, manipulating peripheral circulating levels of 5-HT and NE affects human taste thresholds (Heath et al., 2006), supporting the role of biogenic amines in modulating peripheral gustatory mechanisms. We and others are investigating the details of signal processing in taste buds and determining whether other neurotransmitters and neuromodulators are involved.

NOTES

1. Arguably, there are additional basic taste qualities, such as fatty, metallic, astringent, and so forth. There is considerable debate about whether these qualities are mediated by taste, trigeminal nerves, or both.

REFERENCES

Bartel DL, Sullivan SL, Lavoie EG, Sevigny J, Finger TE (2006) Nucleoside triphosphate diphosphohydrolase-2 is the ecto-ATPase of type I cells in taste buds. *J Comp Neurol* 497:1–12.

Chandrashekar J, Hoon MA, Ryba NJ, Zuker CS (2006) The receptors and cells for mammalian taste. *Nature* 444:288–294.

Dando R, Roper SD (2009) Cell-to-cell communication in intact taste buds through ATP signalling from pannexin 1 gap junction hemichannels. *J Physiol* 587:5899–906.

DeFazio RA, Dvoryanchikov G, Maruyama Y, Kim JW, Pereira E, Roper SD, Chaudhari N (2006) Separate populations of receptor cells and presynaptic cells in mouse taste buds. *J Neurosci* 26:3971–3980.

Finger TE, Danilova V, Barrows J, Bartel DL, Vigers AJ, Stone L, Hellekant G, Kinnamon SC (2005) ATP signaling is crucial for communication from taste buds to gustatory nerves. *Science* 310:1495–1499.

Heath TP, Melichar JK, Nutt DJ, Donaldson LF (2006) Human taste thresholds are modulated by serotonin and noradrenaline. *J Neurosci* 26:12664–12671.

Huang YA, Dando R, Roper SD (2009) Autocrine and paracrine roles for ATP and serotonin in mouse taste buds. *J Neurosci* 29:13909–18.

Huang YJ, Maruyama Y, Dvoryanchikov G, Pereira E, Chaudhari N, Roper SD (2007) The role of pannexin 1 hemichannels in ATP release and cell-cell communication in mouse taste buds. *Proc Natl Acad Sci USA* 104:6436–6441.

Huang YA, Maruyama Y, Roper SD. (2008) Norepinephrine is coreleased with serotonin in mouse taste buds. *J Neurosci* 28:13088–93.

Romanov RA, Rogachevskaja OA, Bystrova MF, Jiang P, Margolskee RF, Kolesnikov SS (2007) Afferent neurotransmission mediated by hemichannels in mammalian taste cells. *EMBO J* 26:657–667.

Roper SD (2007) Signal transduction and information processing in mammalian taste buds. *Pflugers Arch* 454:759–776.

Roper SD (2009) Parallel processing in mammalian taste buds? *Physiol Behav* 97(5): 604–608.

Spector AC, Travers SP (2005) The representation of taste quality in the mammalian nervous system. *Behav Cogn Neurosci Rev* 4:143–191.

Tomchik SM, Berg S, Kim JW, Chaudhari N, Roper SD (2007) Breadth of tuning and taste coding in mammalian taste buds. *J Neurosci* 27:10840–10848.

27

Microcircuitry of the Rostral Nucleus of the Solitary Tract

Joseph B. Travers and Susan P. Travers

Gustatory and somatosensory information from the oral cavity is carried by afferent fibers in the fifth, seventh, and ninth cranial nerves to synapse in the rostral division of the nucleus of the solitary tract (rNST). Two dominant themes characterize afferent organization: topography and convergence (Lundy and Norgren, 2004; Bradley, 2007; Smith and Travers, 2008). Incoming taste afferents from the three cranial nerves follow a rostral-caudal gradient within the nucleus that predisposes second-order neurons to convergence between adjacent or apposing receptors, while maintaining an orotopic representation (Fig. 27.1). Thus, there is a gradual transition from neurons with anterior to posterior mouth receptive fields that extends in an uninterrupted sequence to the representation of the remaining gastrointestinal tract in the caudal NST (cNST). Dendrites of many second-order gustatory neurons extend rostrocaudally and mediolaterally, providing a further substrate for convergence. Although there is little evidence for convergence between seventh and ninth nerve afferents from extracellular recordings, patch recordings indicate considerable interaction between seven and nine, suggesting more subtle modulation by longer-range inputs (Grabauskas and Bradley, 1996). This convergence serves to increase the overall firing rate, receptive field size, and responsiveness to a wider range of taste stimuli of second-order neurons compared to peripheral fibers. An example of afferent convergence that broadens gustatory sensitivity is the emergent responsiveness of second-order neurons to stimulation of the anterior tongue with NaCl, and the hard palate (nasoincisor duct) to sucrose (Travers et al., 1986).

The rNST can be divided into four subdivisions based on cytoarchitectonics: rostral lateral (RL), rostral central (RC), medial (M), and ventral (V) (Whitehead, 1988) (Fig. 27.2). Three morphological cell types can be discerned,

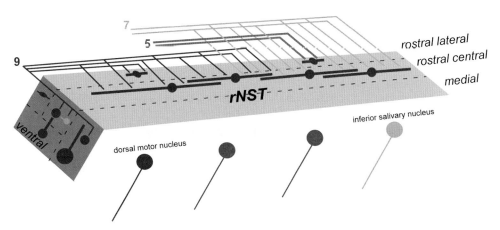

FIGURE 27–1. Topographic organization of the rostral nucleus of the solitary tract. Seventh and ninth nerve primary afferent fibers carrying gustatory afferent information from the anterior tongue and palate versus the posterior tongue synapse in an overlapping rostral-to-caudal and lateral-medial orotopic sequence. Convergent interactions are facilitated by overlap between incoming afferents and dendritic spread in the horizontal plane. Somatosensory fibers from oral branches of the fifth nerve innervating the anterior mouth, and the ninth nerve innervating the posterior mouth, terminate lateral and caudal to the gustatory fibers. There is a topographic organization of preganglionic parasympathetic neurons in the M subnucleus as well. Some neurons in the inferior salivary nucleus that control salivary secretions from the posterior lingual salivary glands (Von Ebners glands) are situated in the most rostral part of the nucleus. Similar neurons with efferents in the superior laryngeal nerve are situated more caudally in the medial subdivision of the most caudal rNST and continuous with the dorsal motor nucleus of the vagus. Note that the map of preganglionic parasympathetics is offset relative to afferent inputs such that afferents represent a more rostral region of the gastrointestinal tract, perhaps preparing the more caudal structures for the passage of ingested substances. Reticular outputs from the ventral subdivision likewise show some preferential termination at a given coronal plane. Ascending projections to parabrachial nucleus likewise exhibit some topography, but increased convergence at this level, both between gustatory afferents and between gustatory and visceral afferents is the more salient feature (not shown). DMN, dorsal motor nucleus of the vagus; ISNve, inferior salivatory nucleus innervating Von Ebner's gland. Subdivisions of the rostral nucleus of the solitary tract: M, medial, RC, rostral central; RL, rostral lateral; V, ventral. (See Fig. 27.2 for details of subdivision circuitry.)

but a major functional distinction is binary: elongate- and stellate-shaped neurons project outside the nucleus but the smaller round/spheroid cells are considered to be GABAergic interneurons (Lasiter and Kachele, 1988). All three cell types occur within each subdivision, but there is a disproportionate representation of elongate neurons in RL, where nongustatory, oro-tactile responsive neurons predominate. Output neurons in RC and V are more evenly divided between stellate and elongate. If preganglionic parasympathetic neurons in M are included, a fourth cell type can be defined.

Most second-order neurons are excited by gustatory stimulation, despite a few instances of frank response suppression. Because primary afferent

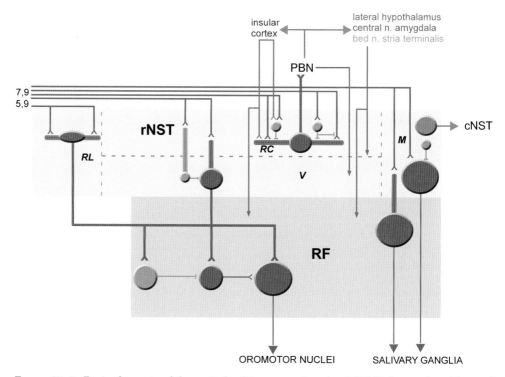

FIGURE 27–2. Basic elements of the rostral solitary complex: the rNST (light gray) and immediately subjacent reticular formation (RF, dark gray). Four rNST subdivisions (RL, rostral lateral; RC, rostral central; M, medial; V, ventral) are depicted with key inputs and outputs. Major differences in cell type distributions within subdivisions are shown; for example, RL has a proclivity toward elongate neurons, and M contains some large preganglionic parasympathetic neurons, along with a population of very small cells. Most gustatory afferent fibers synapse on RC dendrites and there is notable convergence. Afferent fibers also end on GABAergic interneurons (blue) that in turn synapse on cell bodies or more proximal dendrites of the same RC neurons receiving direct afferent input. In addition, GABAergic interneurons exert some of their effects via presynaptic inhibition on primary afferents. Neurons in V likely receive primary input via dendrites extending into RC or interneurons (not shown). Primary afferents may also synapse on preganglionic parasympathetic neurons in M or on dendrites of functionally identical neurons in the subjacent RF. Somatosensory input via the lingual and glossopharyngeal nerves comprise preferential inputs to RL. At the electron microscopic level, synapses from primary afferent fibers in rNST are asymmetric, consistent with evidence for excitatory synapses that utilize glutamate and ionotropic (AMPA/kainate and NMDA) receptors. Outputs of the subdivisions vary: most parabrachial (PBN) contacts originate from RC, whereas RF projections preferentially arise from V and RL. The M subdivision has projections to the caudal solitary nucleus (cNST) and also some preganglionic parasympathetic neurons, suggesting an intimate relationship with autonomic function. Two largely separate RF circuits control (1) salivary secretions or (2) modulate oromotor responses. Like the overlying NST, these circuits rely on glutamate and ionotropic receptors to mediate excitatory responses to oral stimulation; both circuits involve inhibitory interneurons as well. The oromotor circuitry is likely more complex, with an extra layer of processing interposed between the NST and effector neurons. A Hodgkin-Huxley dynamical model of this substrate indicates the importance of inhibition in oromotor consummatory behavior (Nasse et al., 2008). Within the RF,

Figure 27–2. Continued

information flow is from the more lateral parvocellular RF to the more medial intermediate zone. Descending modulatory signals feed back from higher levels of the gustatory and limbic pathway to both the rNST and RF. Little is known of the details of this modulatory circuitry, although inhibitory cortical modulation is known to require GABA^A signaling in NST. Nevertheless, a functional modulation of gustatory responses has been demonstrated for all of the areas depicted; the predominant sign of the influence in NST is denoted by the color of the lettering, red: excitatory, blue: inhibitory, purple: mixed. BNST, bed nucleus of the stria terminalis; CNA, central nucleus of the amygdale; LH, lateral hypothalamus.

signaling is via glutamate, this inhibition likely reflects primary afferent activation of GABAergic interneurons that synapse on elongate/stellate cells. In fact, such synapses are probably more common than apparent in the suppression of action potential discharge, as mixed excitatory/inhibitory currents are ubiquitous to electrical stimulation of primary afferents in vitro (Bradley, 2007). Indeed, GABAergic modulation via GABA$_A$ receptors gives rise to a state of tonic inhibition that suppresses spontaneous and evoked responses and appears to sharpen tuning profiles.

Because outputs to tertiary targets originate from different neuronal populations, it can be argued that one function of the solitary nucleus is to segregate and distribute gustatory signals. A projection to the parabrachial nucleus (PBN) from stellate/elongate neurons in RC is obligate in rodents for relaying information to thalamocortical and other forebrain pathways (Norgren and Leonard, 1973). Brainstem projections to the subjacent reticular formation (RF) (and to a lesser extent hypoglossal nucleus) to mediate oromotor responses are not as anatomically segregated but arise preferentially from stellate/elongate neurons in V and RL (Halsell et al., 1996). The M subdivision contains some preganglionic parasympathetic neurons as well as hosting a caudal intrasolitary projection that may underlie gustatory–visceral interactions. Thus, based on these target destinations, different functions can be assigned to the different subdivisions, but so far, central recording has revealed only subtle differences.

Modulatory influences onto second-order neurons from a wide variety of centrifugal sources are also well established and function to alter gustatory responses, perhaps imparting homeostatic or experiential influences (Lundy and Norgren, 2004; Bradley, 2007; Smith and Travers, 2008). Centrifugal projections to rNST include those from higher order taste structures, the PBN and gustatory cortex, and limbic regions known to receive gustatory input, for example, lateral hypothalamus, central nucleus of the amygdala, and bed nucleus of the stria terminalis. The bed nucleus of the stria terminalis exerts predominantly inhibitory influences; the lateral hypothalamus and amygdala, excitatory influences; and the cortex, mixed influences. Projections from the amygdala are somewhat more dense in M and V (the major motor output subdivisions) compared to the other subdivisions. Inhibitory modulation from cortex requires GABAergic processing in rNST, suggesting mediation by an inhibitory interneuron; the circuitry of the other projections is as yet unknown. A relatively unexplored projection from the caudal NST terminates primarily in RC and M.

Several additional neurotransmitters also constitute potential modulatory influences. Subpopulations of rNST neurons contain (at least) enkephalin, NOS, and dopamine. Likewise the NST has fibers positive for these substances and their receptors, as well as several other ligands, including substance P. Moreover, most of these same neurotransmitters appear to operate in the underlying RF. Substance P and enkephalin are known to exert potent excitatory and inhibitory effects on NST taste cells; substance P has a similar effect on preganglionic salivary neurons in the subjacent RF. Only scant information is available regarding the circuitry of these modulatory systems. However, ganglion cells of primary afferents respond in an excitatory manner to substance P, suggesting that this neurotransmitter may exert its effects in NST presynaptically. In contrast, enkephalin acts postsynaptically via delta receptors on NST neurons that project to PBN (Zhu et al., 2009). In addition, it is presumed that activation of many of these modulatory systems is initiated via the descending influences discussed earlier.

Neurons in the V and M subdivisions of rNST have a strikingly intimate relationship with the subjacent RF. There are some preganglionic parasympathetic salivary neurons in M and this distribution continues in the RF, where such cells are more numerous. In addition, dendrites of many RF salivary neurons extend dorsally to make synaptic contact with primary afferents or interneurons in rNST. Likewise, V contains some prehypoglossal neurons with the distribution continuing and becoming denser in the subjacent RF, where premotor neurons to other oromotor nuclei can also be identified. As with salivary preganglionic parasympathetics in the subjacent RF, dendrites from RF neurons in the oromotor pathway sometimes extend into rNST (Nasse et al., 2008). These anatomical relationships parallel, in large measure, those between the cNST and dorsal motor nucleus of the vagus, which together form the dorsal vagal complex for the control of gut and other visceral reflexes. Indeed, similar to the overlying NST, oromotor and salivary RF circuits contain GABAergic interneurons and are under tonic inhibition. Reinforcing the view of an integrated structure, centrifugal projections to the subjacent RF appear to constitute a continuation of the projection fields evident in the overlying NST. Because of the close functional and anatomical association between the rNST and the subjacent RF in both salivary and somato-(oro)motor control, we depict the circuitry of an expanded "rostral solitary complex" that parallels in large degree, its caudal counterpart, the dorsal vagal complex.

ACKNOWLEDGMENTS

Supported by NIH DC00416 (SPT) and DC00417 (JBT). The authors wish to acknowledge the many contributors to our understanding of the circuitry of the rNST who could not be formally recognized in this short synopsis; more complete citations of primary works can be found in the book edited by Robert Bradley (2007) and the reviews by Lundy and Norgren (2004) and Smith and Travers (2008).

REFERENCES

Bradley BE (2007) *The Role of the Nucleus of the Solitary Tract in Gustatory Processing*. Boca Raton, FL: CRC Press.

Grabauskas G, Bradley RM (1996) Synaptic interactions due to convergent input from gustatory afferent fibers in the rostral nucleus of the solitary tract. *J Neurophysiol* 76(5):2919–2927.

Halsell CB, Travers SP, Travers JB (1996) Ascending and descending projections from the rostral nucleus of the solitary tract originate from separate neuronal populations. *Neuroscience* 72(1):185–197.

Lasiter PS, Kachele DL (1988) Organization of GABA and GABA-transaminase containing neurons in the gustatory zone of the nucleus of the solitary tract. *Brain Res Bull* 21(4):623–636.

Lundy RF, Norgren R (2004) *Gustatory System*, 3rd ed. San Diego, CA: Elsevier Academic Press.

Nasse J, Terman D, Venugopal S, Hermann G, Rogers R, Travers JB (2008) Local circuit input to the medullary reticular formation from the rostral nucleus of the solitary tract. *Am J Physiol Regul Integr Comp Physiol* 295(5):R1391–1408.

Norgren R, Leonard CM (1973) Ascending central gustatory pathways. *J Comp Neurol* 150(2):217–237.

Smith DV, Travers SP (2008) Central neural processing of taste information. In: Basbaum AKAI, Shepherd GM, and Wesheimer G, eds. *The Senses: A Comprehensive References*, Vol. 4, pp. 289–327. San Diego, CA: Academic Press.

Travers SP, Pfaffmann C, Norgren R (1986) Convergence of lingual and palatal gustatory neural activity in the nucleus of the solitary tract. *Brain Res* 365(2):305–320.

Whitehead MC (1988) Neuronal architecture of the nucleus of the solitary tract in the hamster. *J Comp Neurol* 276(4):547–572.

Zhu M, Cho YK, Li CS (2009) Activation of delta-opioid receptors reduces excitatory input to putative gustatory cells within the nucleus of the solitary tract. *J Neurophysiol* 101(1):258–268.

Section 11

Cerebellum

28

Cerebellar Cortex

Masao Ito

The microcircuit in the cerebellum is featured by the relative simplicity, precision, and geometric beauty of its arrangement. Its structure is identical all over the cerebellar cortex except for some regional differences (Fig. 28.1). The cerebellar cortex has three layers (molecular layer, Purkinje cell layer, and granular layer) and can be divided into more than a hundred subareas by horizonal grooves and longitudinal bands. Each subarea can be further subdivided into a number of microzones (there could be 10,000 microzones in the human cerebellum). A microzone, in combination with a small portion of the interior olive and in some regions also with that of parvocellular red nucleus, consists of a microcomplex, a functional unit of the cerebellum (Fig. 28.2). This chapter overviews the current knowledge on neuronal elements and their connections in the cerebellar microcircuit and its functional principles. Most of the relevant references can be found in Ito (2006).

CIRCUIT ELEMENTS

Mossy Fiber

Mossy fiber afferents arise from numerous sources in peripheral nerves, the spinal cord, and the brainstem, and they convey major information to be processed in the cerebellar cortical circuit. Mossy fiber terminals in the granular layer of the cerebellar cortex form a characteristic rosette structure within a glomerulus. Within this structure, a mossy fiber terminal supplies excitatory synapses (mediated by both AMPA and NMDA receptors, but some can be cholinergic) to granule cell dendrites.

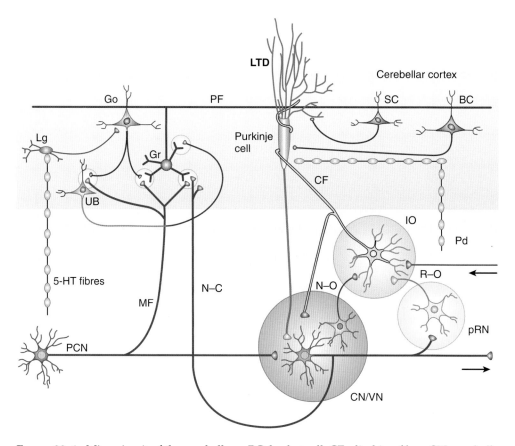

Figure 28–1. Microcircuit of the cerebellum. BC, basket cell; CF, climbing fiber; CN, cerebellar nucleus; GR, granule cell; GL, glomerulus; GO, Golgi cell; IO, inferior olive; LC, Lugaro cell; MF, mossy fiber; N-C, nucleocortical mossy fiber projection; N-O, nucleoolivary inhibitory projection; PCN, precerebellar neuron; PF, parallel fiber; pRN, parvicellular red nucleus; R-O, rubroolivary excitatory projection; SC, stellate cell; 5-HT, serotonergic; UB, unipolar brush cell; and VN, vestibular nucleus. (From Ito, 2008)

Figure 28–2. A microcomplex of the cerebellum. LTD, long-term depression.

Granule Cells

Granule cells are individually the smallest (5–8 μm in diameter), yet the most numerous neurons (10^{10}–10^{11} in humans) in the brain. A large divergence (from one mossy fiber to 400–600 granule cells) and a small convergence (from four to five mossy fibers to a granule cell) characterize the mossy fiber–granule cell pathway. The degree of mossy fiber–granule cell divergence is functionally regulated by Golgi cells, which supply inhibitory synapses to granule cells. Each granule cell issues an ascending axon, which branches in the molecular layer in T shape to form a parallel fiber (PF). Parallel fiber synapses on Purkinje cells are mediated by AMPA and mGluR1 receptors, and those on basket/stellate cells are mediated by AMPA, NMDA, and mGluR1a receptors.

Purkinje Cells

Purkinje cells extending magnificent dendritic trees lie in a single layer of the cerebellar cortex (about 1000 cells per mm^2 in rat). Their dendrites receive excitatory input from numerous PFs (175,000 per Purkinje cell in rat) and inhibitory input from basket and stellate cells. They also receive climbing fibers and beaded fibers. Purkinje cells in turn supply GABA-mediated inhibitory synapses to their target neurons in cerebellar nuclei and certain brainstem neurons. Reciprocal inhibition occurs among Purkinje cells via their recurrent collaterals extending within 300 μm. Axon collaterals of Purkinje cells also inhibit basket cells.

Climbing Fibers

These are unique structures in the cerebellum with no homolog elsewhere in the central nervous system. Each Purkinje cell is innervated by one climbing fiber as a consequence of the postnatal elimination of multiple innervation. Each climbing fiber forms numerous synaptic contacts with the dendrites of a single Purkinje cell (1,300 in proximal dendrites of rat Purkinje cells (Strata, 2002), but a much larger value, 26,000, is derived from the density ratio of climbing and parallel fiber synapses (Nieto-Bona et al., 1997)). This arrangement results in a particularly large excitatory postsynaptic potential (EPSP) superimposed with Ca^{2+} spikes. The major transmitter of climbing fibers is glutamate.

Beaded Fibers

The cerebellar cortex receives not only mossy and climbing fibers but also beaded fibers, which contain various amines (e.g., serotonin, dopamine, acetylcholine, norepinephrine, or histamine) or neuropeptides (angiotensin II, orexin, etc). This third type of afferent extends fine varicose fibers sparsely throughout the granular and molecular layers to form direct contact with Purkinje cells and other cerebellar neurons. On the basis of this morphology, the third type of afferent does not convey specific information to the cerebellar cortex. Rather, its role is modulatory and important in setting the activity level or switching the operational mode of cerebellar microcomplexes to match a behavioral demand.

Unipolar Brush Cells

Unipolar brush (UB) cells are located primarily in the granular layer of the vestibulocerebellum. On their dendritic brush, a UB cell receives a single mossy fiber terminal forming a giant glutamate-mediated synapse. NMDA, kainate, and AMPA receptors are expressed in the synaptic membrane and mGluR1 and mGluR2/3 receptors in the peri- and extrasynaptic parts of the spiny appendages of dendrites. Mossy fiber impulses induce in UB cells an AMPA-mediated fast excitation and a predominantly NMDA-mediated slow excitation. Unipolar brush cell axons branch within the granular layer and give rise to large terminals that synapse with both granule cell and UB cell dendrites within glomeruli. Unipolar brush cells may amplify mossy fiber inputs.

Basket/Stellate Cells

Basket cells are located near the Purkinje cell layer and receive numerous glutamate-mediated excitatory synapses from a bundle of PFs on their dendrites and, in turn, supply GABA-mediated inhibitory synapses to the bottleneck of a Purkinje cell soma, forming a unique complex structure called a "pinceau." Stellate cells are located in the molecular layer and they also supply inhibitory synapses to Purkinje cell dendrites. Basket/stellate cells mediate the feedforward inhibition from PFs to Purkinje cells. A basket cell extends axons perpendicular to PFs and covers an area containing 10 × 7 rows of Purkinje cells, with a probable divergence number of 50. Twenty to 30 basket cell axons may converge onto one Purkinje cell. Basket cells receive also collaterals of climbing fibers and those of Purkinje cell axons.

Golgi Cells

Each Golgi cell receives ~4788 excitatory inputs to its dendrites in the molecular layer from PFs and also ~228 mossy fiber terminals on its descending dendrites. The major excitatory inputs from PFs to Golgi cells are mediated by both AMPA and NMDA receptors and mGluR2. Golgi cells also receive inhibitory synapses from Lugaro cells. A Golgi cell, in turn, extends a broadly branching axon to supply GABA-mediated inhibitory synapses up to ~5700 granule cells (cat).

Lugaro Cells

These cells are inhibitory neurons in the granular layer. Their fusiform cell soma is located in or slightly below the Purkinje cell layer. The cerebellar cortex contains approximately one Lugaro cell and one Golgi cell for every 15 Purkinje cells. Axons of more than 10 Lugaro cells converge onto only one Golgi cell, while the axon of one Lugaro cell diverges onto 150 Golgi cells. Lugaro cells express calretinin but not mGluR2 or somatostatin, whereas Golgi cells express mGluR2 and somatostatin but not calretinin. Lugaro cells in cerebellar slices are silent, but in the presence of serotonin, they discharge regularly at 5–15 Hz and induce inhibition in Golgi cells.

Globular Neurons

These are a newly identified group of large granular layer neurons having a globular soma located at variable depths in the granular layer (Laine and Axelrad, 2002). They extend three to four long radiating dendrites coursing through the three layers of the cortex. Their axons project into the molecular layer and expand a local plexus, with a pattern similar to that of Lugaro cell. The axons of several of these cells give off a collateral that courses for a long distance in the transverse direction, just above the Purkinje cell somata, parallel to PFs. Globular cells may be inhibitory neurons.

CIRCUITS AND MODELS

Synaptic Plasticity

Various synapses in cerebellar microcircuit express activity-dependent changes in synaptic efficacy, that is, synaptic plasticity, typically long-term potentiation (LTP) and long-term depression (LTD). In mossy fiber–granule cell synapses, LTP involving NMDA receptors appears to functionally regulate the degree of mossy fiber–granule cell divergence. In Purkinje cells, whereas repetitive stimulation of PFs alone induces LTP in the PF synapses, either presynaptic or postsynaptic, conjunctive activation of PFs and a CF induces PF-LTD underlain by complex signal transduction. Purkinje cells also exhibit a prolonged potentiation of $GABA_A$ receptor–mediated inhibitory postsynaptic potentials (IPSPs) (i.e., rebound potentiation [RP]) after the activation of climbing fibers, basket/stellate cell–induced IPSPs. Rebound potentiation is input-nonspecific in contrast to the input-specific PF-LTD.

Inhibitory synapses supplied by Purkinje cell axons to cerebellar nuclear neurons exhibit input-nonspecific LTD due to a decreased postsynaptic GABA sensitivity that is caused by an increase in $[Ca^{2+}]$. In basket/stellate cells, PF burst stimulation paired with CF activity induces LTP complementary to PF-LTD in Purkinje cells. Likewise, stimulation of a PF bundle unpaired with CF activity induces LTD complementary to PF-LTP in Purkinje cells. Large stable CF-evoked excitation in Purkinje cells exhibits a weak LTD when repetitively evoked.

Neurocomputing

The forward connections from cells of origin of mossy fibers to granule cells and to Purkinje cells form the three-layered neuronal networks, which have been modeled as simple perceptron, adaptive filter, or liquid-state machine (Yamazaki and Tanaka, 2007). Whereas stimuli are received by the cells of origin of mossy fibers on the first layer, granule cells receive mossy fiber afferent on the second layer, where the small convergence may suggest "sparse" coding; that is, each granule cell represents an integration of a small number of mossy fibers. However, a recent analysis of sensory signals suggests "similar" coding in that each granule cell retains characteristics of each mossy fiber (Bengtsson and Jörntell, 2009). On the third layer, Purkinje cells receive numerous synapses from PFs and so integrate information from numerous granule cells to generate outgoing signals. How a Purkinje cell processes incoming and outgoing signals, by linear algorithm or pause, has been a matter of recent discussions (Steuber et al., 2007; Alviña et al., 2008; Walter and Khodakah, 2009).

Learning

The finding that PF synapses of Purkinje cells undergo LTP or LTD depending on pairing or unpairing with CF provides the basis for learning and memory capability of the cerebellar cortical microcircuit. The more recently found reciprocal pattern of a combination of LTD in Purkinje cells and LTP in basket/stellate cells, or vice versa, may have a synergistic effect to augment the memory storage capacity of the cerebellar cortical microcircuit. The learning in the cerebellar cortex is considered to be primarily "error learning" because CF signals induce LTD in coactivated PF synapses in Purkinje cells like punishment and because climbing fibers convey information regarding errors in a motor performance. The predominance of functionally silent PF synapses in Purkinje cells may imply that 85% or more of PF synapses are long-term depressed during repeated learning, by which the neurocomputing circuits are shaped through experiences from the original preponderantly connecting PF synapses (Dean et al., 2010).

Internal Model

A microcomplex is inserted into various microcircuits in the spinal cord and brainstem and acts as an embedded adaptive component in the motor or autonomic control systems. Furthermore, a microcomplex is connected to the cerebral cortical area and acts as an internal model of the latter. For the motor cortex, a microcomplex may provide a forward model or an inverse model of the lower motor centers and motor apparatus, which enables the motor cortex to perform a precise movement without the external feedback about the movement (Kawato, 2009). A similar idea has been expanded to cognitive functions; a microcomplex would provide an internal model for mental activities going on in the cerebral association cortex (Ito, 2008).

Oscillation

A computer simulation suggests that reciprocal inhibition causes an oscillation at 100–250 Hz in the activity of basket/stellate cells. A computer simulation also predicts that feedback inhibition from Golgi to granule cells induces 10–50 Hz oscillations in spike discharges from the latter. Because of the large divergence from Lugaro cells to Golgi cells, an interesting possibility is that Lugaro cells play a role in synchronizing activity among Golgi cells situated along the parallel fiber beam as observed in anesthetized rats. Lugaro cells may switch the operation of Golgi cells from the individual rhythmic mode to the synchronous mode.

Golgi Cell Clock

The loop connection involving granule cells and Golgi cells (Fig. 28.1) has been interpreted to constitute a phase converter, whose function is to generate a set of multiphase versions of a mossy fiber input. Another unique model proposed by Yamazaki and Tanaka (2005) features a randomly connected granule cell–Golgi cell loop pathway, which operates as a clock to generate granule cell discharge in an ever-changing ensemble of patterns. The patterns do not repeat unless reset by a large mossy fiber input to the pathway. When this Golgi cell clock is incorporated in a neuronal circuit model for eyeblink reflex, it reproduces appropriately timed conditioned responses.

References

Alviña K, Walter JT, Kahn A, Ellis-Davis G, Khodakhah K (2008) Questioning the role of rebound firing in the cerebellum. *Nat Neurosci* 11:1256–1258.
Bengtsson F, Jörntell H (2009) Sensory transmission in cerebellar granule cells relies on similarly coded mossy fiber inputs. *Proc Natl Acad Sci* 106:2389–2394.

Dean. P, Porrill J, Ekerot C-F, Jöntell H (2010) The cerebellar microcircuit as an adaptive filter: experimental and computational evidence. *Nat Rev Neurosci* 11:30–43.

Ito M (2006) Cerebellar circuitry as a neuronal machine. *Prog Neurobiol* 78:272–303.

Ito M (2008) Control of mental activities by internal models in the cerebellum. *Nat Rev Neurosci* 9:304–313.

Kawato M (2009) Cerebellum: models. In: Squire LR, ed. *Encyclopedia of Neuroscience*, pp. 757–767. Oxford, England: Academic Press. Elsevier Limited.

Lainé J, Axelrad H. (2002) Extending the cerebellar Lugaro cell class. *Neuroscience* 115: 363–374.

Nieto-Bona MP, Garcia Sergura LM, Torres-Aleman I (1997) Transsynaptic modulation by insulin-like growth factor I of dendritic spines in Purkinje cells. *Int J Devl Neurosci* 15:749–754.

Steuber V, Mittmann W, Hoebeek FE, Angus Silver R, De Zeeuw CI, Häusser M, De Schutter E (2007) Cerebellar LTD and pattern recognition by Purkinje cells. *Neuron* 54:121–136.

Strata P (2002) Dendritic spines in Purkinje cells. *The Cerebellum* 1:230–232.

Walter JT, Khodakhah K (2009) The advantages of linear information processing for cerebellar computation. *Proc Natl Acad Sci* 17:4471–4476.

Yamazaki T, Tanaka S (2005) Neural modeling of an internal clock. *Neural Computation* 17:1032–1058.

Yamazaki T, Tanaka S (2007) The cerebellum as a liquid state machine. *Neural Networks* 20:290–297.

29

Olivocerebellar System

Rodolfo R. Llinás

The olivocerebellar system is one of the most conserved in vertebrate brain, being present in all forms studied so far. It comprises a set of bilaterally symmetrical inferior olivary nuclei (IO), ventrally located in the bulbar brainstem, and the overlaying cerebellum. These two structures are mutually linked through axonal pathways within the cerebellar peduncles. Thus, axons from the IO neurons traverse the midline at the bulbar region, course up the contralateral cerebellar peduncle, and enter the cerebellar white matter. There they establish excitatory synaptic contacts with the cerebellar nuclear neurons and proceed into the cerebellar cortex reaching the cerebellar cortex. The IO axons establish synaptic contact with Purkinje cells (PCs) (Fig. 29.1A), forming a one-to-one chemical junction known as climbing fibers (CFs) (Fig. 29.1A by Ramon y Cajal, 1904) and are, exclusively, of inferior olive origin (Szentagothai and Rajkovits, 1959). The climbing fiber–Purkinje cell synapse is the largest synaptic junction in the vertebrate central nervous system, establishing hundreds of junctions of the large spines in the main branches of the Purkinje cell dendritic tree. Activation of a climbing fiber elicits an all-or-nothing excitatory response called a complex spike (Fig. 29.1A, right upper trace) in the PC (Eccles et al., 1966), as opposed to the simple spike produced by parallel fiber activation (right lower trace). There are about 10 times more PCs than IO neurons and, on average, each IO neuron generates 10 climbing fibers. The PC axons, the only output of the cerebellar cortex, terminate in the cerebellar and related vestibular nuclei, where they form inhibitory synapses (Ito and Yoshida, 1966) (Fig. 29.2, green terminal).

Cerebellar nuclei neurons are of two types, *excitatory* and *inhibitory* (50% each), and are the only output of the cerebellum. The axons of the *inhibitory* cerebellar nuclei return, in their entirety, to the contralateral olive and contact the IO neurons (Fig. 29.1B) onto structures called glomeruli (Fig. 29.1D).

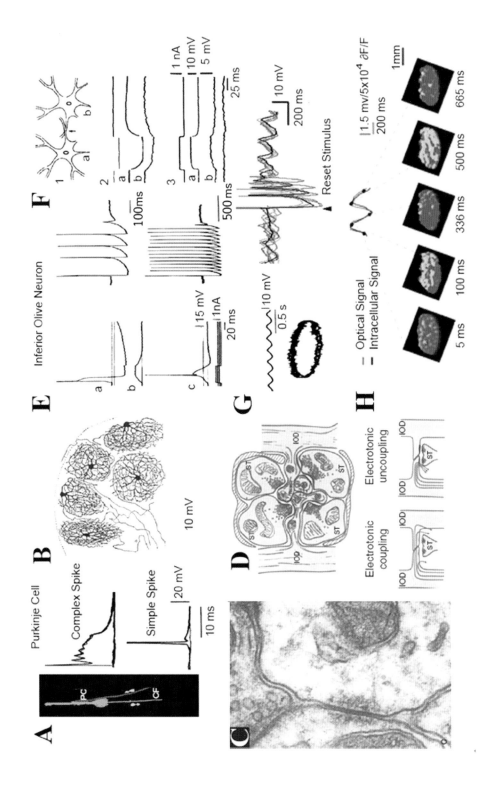

A Purkinje Cell

Complex Spike

Simple Spike

|20 mV

10 ms

B

10 mV

C

E Inferior Olive Neuron

a

b

c

|15 mV

|1nA

20 ms

100ms

500 ms

F

1

a b

2

a
b

| 1 nA

| 10 mV

| 5 mV

3

a
b

25 ms

G

|10 mV

0.5 s

| 10 mV

200 ms

Reset Stimulus

D

IOD

ST

ST

ST

ST

IOP

H 5 ms   100 ms   336 ms   500 ms   665 ms

|1.5 mv/5x10^-4 ∂F/F

200 ms

1mm

— Optical Signal

- - Intracellular Signal

Electrotonic coupling

IOD

IOD

ST

Electrotonic uncoupling

IOD

IOD

ST

FIGURE 29–1. (A) Cerebellar cortex. Green, Purkinje cell; red, climbing fiber (Ramon y Cajal, 1904). (Right) Upper trace, intracellular recording of an all-or-none complex spike; lower trace, simple spike activation. (B) Inferior olivary (IO) neurons. Note their spherical dendritic tree. (C) Electronmicrograph showing gap junction between spines of IO dendrites within IO glomerulus (Llinas et al., 1974). (D) Diagram of IO glomerulus. The center shows spines from IO dendrites (IODs) coupled by gap junctions. (Bottom left) Coupling current path between IO dendrites. (Bottom right) Current flow shunted when GABAergic synapses are active at the gap junction (Llinas, 1974). (E) Left traces: In vitro intracellular recordings from an IO neuron showing high voltage activated by an outward pulse from a depolarized potential with respect to rest (a), the same outward pulse at rest potential (no spike) (b), and the same pulse from a hyperpolarized membrane potential level (c), generating a low-voltage activate spike (Llinas and Yarom, 1981). Right traces: Rebound activated spikes from the same cell following single spike stimulation and associated oscillatory spike train, two different sweep speeds to demonstrate self-sustained oscillatory properties. (F) (1) Simultaneous intracellular recordings from two coupled IO neurons, (2) electrical coupled potential in cell b when cell a is hyperpolarized with a short square current pulse. (3) as in (2) following a depolarizing pulse. Third trace extracellular recording from cell b (Llinas and Yarom, 1981). (G) Left, subthreshold membrane oscillation recorded intracellularly from an IO neuron. Associated Lissajeux image demonstrating oscillatory stability (Llinas and Yarom, 1986). Right, subthreshold oscillation phase reset by neuronal activation following an electrical stimulus (Leznik et al., 2002). (H) Voltage-dependent dye imaging showing spatial organization of coupled cell oscillatory neuronal clusters. Upper trace simultaneously intracellularly recorded neuronal oscillation showing the close spatiotemporol relation between ensemble oscillation (red) to single cell recorded membrane potential (black). Voltage and time calibrations as indicated.

Each olivary glomerulus contains approximately five to eight spines from dendrites of different IO neurons (Fig. 29.1D) and supports IO electrotonic coupling via gap junctions (Fig. 29.1C) (Llinas, 1974; Llinas et al., 1974; Sotello et al., 1974). The glomeruli receive both excitatory and inhibitory synapses of extrinsic origin. The degree of coupling is dynamically modulated by the inhibitory synaptic shunting (Llinas, 1974; Lang et al., 1996) as a feedback from the cerebellar nuclear output (De Zeeuw et al., 1990) (Figs. 29.1c and 29.1d). The other half of the cerebellar nuclear system projects as excitatory afferents to the brainstem nuclei, spinal cord, and thalamus, the last as part of the cerebrocerebellar system, as illustrated in Figure 29.2.

Electrophysiology

Inferior Olive

The function of the olivocerebellar system is closely related to the noncontinuous nature of motor organization (Llinas, 1991) that results in the socalled physiological tremor and to the timing for motor execution. This can be easily observed by measuring, for instance, the velocity of voluntary human finger movements (Vallbo and Wessberg, 1993), which occur at a close to constant of 8–10 Hz steps independently of movement speed. Electrophysiologically, inferior olive neurons are characterized by their ability to generate high-threshold (Fig. 29.1Ea) and low-threshold calcium spike (Fig. 29.1Ec) (Llinas and Yarom, 1981). The latter is generated by the activation of both T-type calcium channels (Cav3.1) and Ih potassium currents (Llinas and Yarom, 1986; Lampl and Yarom, 1997; De Zeeuw et al., 1998), which limit their frequency to 8 to 10 Hz. In addition to such spikes, IO neurons support spontaneous subthreshold membrane potential oscillations near 10 Hz (Fig. 29.1G, left upper trace) (Llinas and Yarom, 1986). These oscillations modulate spike initiation, and so action potentials are normally generated at the crest of such oscillations and fire at 1 to 10 Hz. Thus, this intrinsic rhythm is entrained with speed of movement execution mentioned earlier. Moreover, the phase of subthreshold oscillations may be influenced by activity, as shown in Figure 29.1G (color traces, right). Indeed, extracellular local electrical stimuli or strong excitatory synaptic input will reset the oscillatory phase, but not the amplitude or frequency, of subthreshold oscillations (Leznik et al., 2002).

As electrotonic coupling is ubiquitous across IO neurons, they tend, as a group, toward oscillatory entrainment. Dynamically entrained groups of IO neurons oscillating in phase will synchronously activate a large population of PCs and will, as a result, control patterns of synchronous activity in the cerebellar output during motor coordination. Such entrainment generates inferior olive oscillatory clusters that can be observed using voltage-dependent dye imaging and may be concurrently observed with simultaneous intracellular recordings from IO neurons (Llinas and Muhlethaler, 1988).

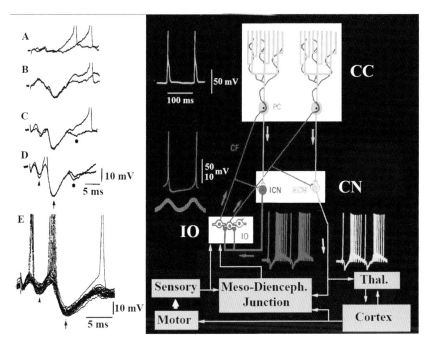

FIGURE 29–2. The olivocerebellar loop circuit. (*Left*) Set of intracellular recordings from cerebellar nuclear neurons indicating the synaptic connectivity with mossy, climbing, and Purkinje cell axons. (*Right*) Diagram represents the main neuronal elements that constitute the connectivity matrix for the olivocerebellar loop system. On the left lower corner in red the inferior olive (IO) nucleus, comprised of electrically neurons (red circles) electrically coupled (small superimposed side bars). These neurons are capable of subthreshold quasi-sinusoidal oscillation (lowest trace on the left, in red) and of action potential generation (upper trace in red); note different voltage gains. Their action potentials are conducted via climbing fibers (CF, in red) to Purkinje cells (PC, in blue) and to excitatory (ECN, in red) and inhibitory (ICN, in blue) cerebellar nucleus neurons. Arrows indicate the direction of action potential conduction. As the system works, subthreshold oscillation in the inferior olive neurons generate action potentials at the crest of the oscillation (upper spikes in red) that activate inhibitory cerebellar nuclear neurons spikes (blue spikes). These spikes return (loop back to the olive) and decouple the electrical connections between IO neurons. The action potentials from the IO also go to the excitatory cerebellar nucleus cells, which send a timing signal to the rest of the brain (red spikes). In the cerebellum, the axons of the IO neurons continue to the cerebellar cortex and activate Purkinje cells (blue spikes), which are inhibitory and so generate powerful inhibitory potentials (down going triangular voltage steps, in both the blue and red traces, that block spike trains for a short time). This inhibitory input from Purkinje cells serve to limit the duration of the IO activation of the cerebellar nuclear cells and so generate a pulsatile output at 10 Hz that controls movement timing. The voltage amplitude (50 mV) and time (100 ms) are indicated in the upper left corner in white and are the same for all traces, except for IO oscillations, which are shown at 5x voltage gain (10 mV) with respect to the spike amplitude. The output of the excitatory cerebellar nuclei is relayed to the thalamus and the cortex and loops back to the mesodiencephalic junction via pontine nuclei to the inferior olive and the cerebellar cortex via mossy fibers and proceed to the spinal cord to activate motoneurons. Sensory system returns to the inferior olive and to the cerebellar nuclei and cerebellar cortex via mossy fibers.

Cerebellar Nuclei

The cerebellar nuclei represent the only output of the cerebellum. The output of the excitatory cerebellar nuclei neurons terminate on the contralateral mesodiencephalic junction (the red nucleus, the nucleus of Darkschewitsch, interstitial of Cajal, nucleus of Bechterew, tegmental field of Forel, zona incerta, subparafascicularis nucleus, and the prerubral reticular formation) (Ruigrok and Voogd, 1995) and the thalamus, while the inhibitory neuron contingent, as described earlier, goes mainly to the inferior olive. These cells receive excitatory input from the collaterals of mossy fibers arising from spinal cord, pontocerebellar and vestibular nuclei, and from collaterals of the olivo-cerebellar input on their way to the cortex. They are also the sole recipient (in conjunction with Deiters nucleus) of the inhibitory terminals originating from the axons of Purkinje cells (Ramon y Cajal, 1904; Ito and Yoshida, 1966).

The functional connectivity into the cerebellar nuclei can be demonstrated by intracellular recordings from cerebellar nuclear neuron in the isolated brainstem–cerebellar preparation. Electrical stimulation (Fig. 29.2A) (Llinas and Muhlethaler, 1988) of the cerebellar white matter between the cerebellar cortex and the cerebellar nuclei, is demonstrated by intracellular recording from the cerebellar nuclear neurons, a set of stereotyped synaptic potentials. As the white matter stimulus is gradually increased (Fig. 29.2A, a to d), there is a gradual appearance of a short latency excitatory postsynaptic potential (EPSP) followed by an inhibitory postsynaptic potential (IPSP). This early sequence is followed by a longer latency response that increases in size as the electrical stimulus is augmented. In Fig. 29.2A(e), 10 superimposed large-amplitude stimuli with little variation in latency or amplitude are illustrated. Of these, the first is a short-latency EPSP (1–2 ms) followed by a short-latency IPSP (2–3 ms) that could be repeatedly activated and corresponded, most likely to direct activation of mossy fiber input to the CN neurons followed by a small short-latency IPSP of probable Purkinje cell origin (Llinas and Muhlethaler, 1988). This early response was followed by a second longer latency EPSP followed by a secondary IPSP of much larger amplitude and duration. This second EPSP–IPSP sequence was the likely product of climbing fiber afferent activation of the CN neuron followed by a climbing fiber activation of Purkinje cells, with a fixed latency and the expected larger amplitude given the synchronous activation of Purkinje cells by the CF afferent system.

IMPLICATION CONCERNING THE ROLE OF THE OLIVOCEREBELLAR SYSTEM IN MOTOR CONTROL

The synchronous, rhythmic firing of the olivocerebellar system has led to the hypothesis that climbing fibers may perform a timing function in motor coordination (Welsh and Llinas, 1997). Thus, IO activity would provide a timing signal for the motor system by generating synchronous and rhythmic activation of the cerebellar nuclei. The synchronized inhibitory barrage originated by climbing fiber activation of PC complex spikes thus controls the timing of the output of the cerebellar nuclei. Evidence to support the motor timing proposal includes the following: synchronous firing of a population of PCs in anesthetized and awake animal; recordings of rhythmic inhibitory potentials in the deep cerebellar nuclei; and temporal correlation between firing of the olivocerebellar system and execution of movements. To this effect, the IO seems to be dynamically organized in clusters of synchronously oscillating neurons that can generate rhythmic and coherent activity of the cerebellar output. Ultimately, the dynamic properties of single neurons and the channel kinetics that generate them seem to hone the "impedance matching" from the molecular to the macroscopic levels to the point that our motor timing is nothing other than the echo of our IO oscillational parameters in time, locality, and number of degrees of freedom. It has been proposed that to meet these requirements, the IO uses its intrinsic oscillatory properties for temporal organization, that is, its ability to organize, support, and change cluster size and location to control spatial organization. In addition, its weakly chaotic properties exhibit the necessary fluidity to control the number of degrees of freedom by rapidly modulating the phase relation between clusters.

The main conclusion from this group of studies is that the IO is a truly integrated system in which the neuronal network and the macroscopic dynamic features (cluster permutation agility) are so well matched that their control properties almost match cellular time constants. This allows the system to effectively control innumerable movements without overloading brain function.

REFERENCES

De Zeeuw C, Holstege JC, Ruigrok TJ, Voogd J (1990) Mesodiencephalic and cerebellar terminals terminate upon the same dendritic spines in the glomeruli of the cat and rat inferior olive. *Neuroscience* 34:645–655.
De Zeeuw Simpson JI, Hoogenraad CC, Galjart N, Koekkoek SK, Ruigrok TJ (1998) Microcircuitry and function of the inferior olive. *TINS* 21:391–400.
Eccles JC, Llinas R, Sasaki K (1966) The excitatory synaptic action of climbing fibres on the Purkinje cells of the cerebellum. *J Physiol* 182:268–296.
Ito M, Yoshida M (1966) The origin of cerebellar-induced inhibition of Deiters neurons. *Exper Brain Res* 2:330–349.
Lampl I, Yarom Y (1997) Subthreshold oscillations and resonant behavior: two manifestations of the same mechanism. *Neuroscience* 78:325–341.
Lang EJ, Sugihara I, Llinas R (1996) GABAergic modulation of complex spike activity by the cerebellar nucleoolivary pathway in rat. *J Neurophys* 76:255–275.
Leznik E, Makarenko V, Llinas R (2002) Electrotonically mediated oscillatory patterns in neuronal ensembles: an in vitro voltage-dependent dye-imaging study in the inferior olive. *J Neuroscience* 22:2804–2815.

Llinas R (1974) Motor aspects of cerebellar control. *Physiologist* 17:19–46.

Llinas R (1991) Noncontinuous nature of movement execution. In: Humphrey DR and Freund H-J, eds. *Motor Control: Concepts and Issues*, pp. 223–242. New York: John Wiley.

Llinas R, Yarom Y (1981) Electrophysiology of mammalian inferior olivary neurones in vitro: different types of voltage dependent ionic conductances. *J Physiol* 315: 549–567.

Llinas R, Yarom Y (1986) Oscillatory properties of guinea pig inferior olivary neurones and their pharmacological modulation: an in vitro study. *J Physiol* 376:163–182.

Llinas R, Muhlethaler M (1988) Electrophysiology of guineapig cerebellar nuclear cells in the in vitro brain stem-cerebellar preparation. *J Physiol* 404:241–258.

Llinas R, Baker R, Sotello C (1974) Electrotonic coupling between neurons in cat inferior olive. *J Neurophysiol* 37:560–571.

Ramon y Cajal S (1904) *La Textura del Sistema Nervioso del Hombre y de los Vertebrados.* Madrid, Spain: Moya.

Ruigrok TJ, Voogd J (1995). Cerebellar influence on olivary excitability in the cat. *Europ J Neuroscience* 7:679–693.

Sotelo C, Llinas R, Baker R (1974) Structural study of inferior olivary nucleus of the cat: morphological correlates of electrotonic coupling. *J Neurophysiology* 37:541–559.

Szentagothai J, Rajkovits K (1959) Ueber den Ursprung der Kletterfasern des kleinhirns. *Zeitschrift für Anatomie und Entwicklungsgeschichte* 121:130–141.

Vallbo A, Wessberg J (1993) Organization of motor output in slow finger movements in man. *J Physiol* 469:673–691.

Welsh JP, Llinas R (1997) Some organizing principles for the control of movement based on olivocerebellar physiology. *Prog Brain Res* 114:449–461.

Section 12

Motor Systems

30

Superior Colliculus

Katsuyuki Kaneda and Tadashi Isa

The superior colliculus (SC) is a midbrain center that integrates various kinds of sensory inputs and transforms the information into a command signal to initiate orienting behaviors such as saccadic eye and head movements. Figure 30.1 depicts a schematic diagram of the SC circuitry involving input–output organization and intrinsic wirings, which have been clarified by recent in vitro studies. In this chapter, we first focus on the intrinsic connectivity in the SC that is crucial to generate and regulate bursts of discharges, which triggers initiation of saccades. Then we describe the role of extrinsic inputs such as GABAergic and cholinergic in modulating the activity of the SC local circuit, based primarily on recent in vitro experimental findings.

INPUT AND OUTPUT ORGANIZATION

The SC is composed of two major functionally segregated layers. The superficial layers of the superior colliculus (sSC), comprising the stratum zonale (SZ), stratum griesum superficiale (SGS), and the stratum opticum (SO), receive visual inputs directly from the retina and indirectly from the primary visual cortex (Fig. 30.1, cells 1–3). The contralateral visual field is represented in a topographically organized map in the sSC. The sSC sends its output signal to the visual thalamus. Deeper layers of the SC (dSC), which are constituted of the stratum griesum intermediale (SGI), the stratum album intermediale (SAI), the stratum griesum profundum (SGP), and stratum album profundum (SAP), receive multimodal sensory, cortical, and basal ganglia inputs and send a command of orienting behavior to brainstem gaze centers (Fig. 30.1, cells 4 and 6). The dSC output neurons exhibit high-frequency burst

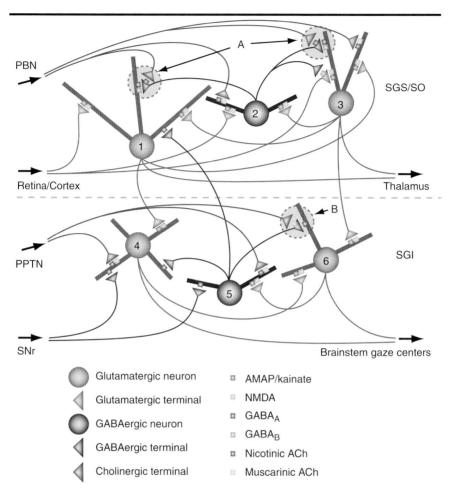

Figure 30–1. Diagram of the fundamental circuit of the superior colliculus. Neurons in the superficial layers (sSC) receive excitatory inputs from the retina and the primary visual cortex (cells 1, 2, and 3) and send the signal to the thalamus and deeper layers (dSC) (cells 1 and 3). Neurons in the dSC receive excitatory inputs from cortex (not shown) and sSC (cells 4 and 6) and send a command for saccade generation to brainstem gaze centers (cells 4 and 6). The dSC also receive inhibitory inputs from the substantia nigra pars reticulata (SNr) (cells 4 and 5). Some GABAergic neurons located in the dSC (cell 5) project back to the sSC. The sSC and dSC receive cholinergic inputs from the parabigeminal nucleus (PBN) and pedunculopontine tegmental nucleus (PPTN), respectively, which might modulate the signal processing therein. Although there are various subclasses of GABAergic neurons, they are drawn as a single cell class in this figure.

activity prior to orienting movements. The dSC contains a vector map of saccades that is in register with the retinotopic map in the overlying sSC.

SIGNAL TRANSMISSION FROM THE SUPERFICIAL TO THE DEEPER LAYERS OF THE SUPERIOR COLLICULUS

Over the past three decades, the existence and function of neural pathways from the sSC to dSC have been an issue of debate (for review, see Isa and Sparks, 2006), but recent electrophysiological and anatomical examinations in in vitro slice preparations obtained from rodents have provided strong evidence that the visual input signals received by the sSC can be directly transmitted to the dSC (Fig. 30.1, cells 1 and 3). Electrical stimulation of the optic nerve evokes monosynaptic excitatory postsynaptic potentials (EPSPs) in sSC neurons. These neurons include excitatory narrow-field vertical (NFV) and wide-field vertical (WFV) cells. The former processes signals from stationary and the latter processes those from moving targets and axons of these project to the dSC (cells 1 and 3). The EPSPs are enhanced by application of a $GABA_A$ receptor antagonist and are partially blocked by an NMDA receptor antagonist. An AMPA/kainate receptor antagonist completely blocks the remaining components. Thus, both types of sSC neurons receive monosynaptic glutamatergic inputs from the optic nerve. On the other hand, optic nerve stimulation evokes di- and/or oligosynaptic EPSPs in dSC neurons (cells 4 and 6). Stimulation of the sSC induces monosynaptic EPSPs in dSC neurons (Lee et al., 1997; Isa et al., 1998), including those projecting to the brainstem gaze centers. Thus, there exists an excitatory pathway from the optic nerve to the dSC mediated by sSC neurons. Such direct connection from the sSC to dSC appears to be a fundamental mechanism underlying the "express saccade," which is triggered with extremely short latency. In vitro studies suggested that the signal transmission from the sSC is considerably enhanced by application of a $GABA_A$ receptor antagonist (Isa et al., 1998) or application of nicotine (Isa and Sparks, 2006), often leading to bursts of discharges. This might parallel the behavioral observation that occurrence of express saccades are context dependent, especially facilitated by insertion of a gap period between fixation offset and target onset in a visually guided saccade task, which might cause reduction in tonic inhibition of presaccadic burst neurons in the dSC. The enhanced signal propagation from the sSC to dSC is largely blocked by local application of an NMDA antagonist into the sSC. Thus, NMDA receptors located in both the sSC and dSC contribute critically to the enhanced signal transmission (Isa et al., 1998; Kaneda et al., 2008a).

In addition to $GABA_A$ receptors, $GABA_B$ receptors also play an important role in regulating burst activity in both sSC and dSC. $GABA_B$ receptors are densely expressed in the SC, especially in the sSC. Activation of postsynaptic $GABA_B$ receptors in NFV cells by synaptically released GABA induces hyperpolarization (cell 3) and that in WFV cells produces shunting inhibition (cell 1) (Kaneda et al., 2008a). In addition, activation of presynaptic $GABA_B$ receptors located in glutamatergic terminals reduces glutamate release (cell 3). Thus, when these $GABA_B$-mediated inhibitions are reduced, duration of burst discharges not only in the sSC but also in the dSC is significantly prolonged. Therefore, $GABA_B$ receptor–mediated inhibition, which exhibits slow-onset and long-lasting metabotropic properties, can shorten the duration of burst activity in the SC so that unnecessary long-lasting bursts should be stopped at appropriate times.

Because the effects of $GABA_A$ and/or $GABA_B$ receptor blockade are observed in isolated SC slice preparations, the source of GABA that activates these receptors is most likely intrinsic GABAergic neurons in the SC. The observation that nearly 50% of neurons are GABAergic in the sSC supports this concept, although GABAergic neurons in the dSC are relatively sparse.

FEEDBACK INHIBITION FROM THE DEEPER TO SUPERFICIAL LAYERS OF THE SUPERIOR COLLICULUS

Combination of advanced technologies for creating transgenic animals and chemical activation of targeted neurons allows us to address the functional significance of anatomically identified GABAergic neurons in the SC. One important example of the combined approaches is feedback inhibition from the dSC to sSC. In slice preparations obtained from GAD67-GFP knock-in mice, in which GABAergic neurons specifically express GFP fluorescence, whole-cell recordings combined with intracellular dye injection has revealed a novel type of GABAergic neurons in the dSC, which have dense axon collaterals projecting back to the sSC (cell 5) (Lee et al., 2007). By using a photostimulation technique to release free glutamate locally from a caged glutamate compound, dSC stimulation evoked inhibitory postsynaptic currents in sSC neurons, including those projecting to the visual thalamus. Thus, GABAergic neurons in the dSC may exert inhibitory influence over the sSC. Such feedback inhibition might be a mechanism for "saccadic suppression," which prevents unwanted saccades from being triggered by visual inputs caused by saccades.

EXTRINSIC INHIBITION

The dSC neurons receive descending inputs from the cortex, multimodal sensory inputs, visual signals from sSC and axon collaterals of local network from other dSC neurons. On the other hand, the main source of inhibitory inputs to the dSC is the substantia nigra pars reticulata (SNr), one of the output structures of the basal ganglia. A recent electrophysiological and anatomical study has revealed that not only excitatory neurons but also local GABAergic interneurons in the dSC are the target of nigral inhibition (Kaneda et al., 2008b). This inhibition is mediated by solely GABA$_A$ receptors. Thus, the inputs from the SNr may not only be related to initiation of burst discharges in the output neurons, but they also might be involved in shaping the spatiotemporal properties of dSC neuron network.

Cholinergic Modulation in the Superficial Layers of the Superior Colliculus

The sSC receives dense cholinergic inputs from the parabigeminal nucleus (PBN). Local application of acetylcholine (ACh), nicotinic receptor agonist, or muscarinic receptor agonist reduces visually evoked responses in sSC neurons of anesthetized rats, although muscarinic activation increases background activity (Binns and Salt, 2000). The reduction in visual response is prevented by prior iontophoretic application of an antagonist for GABA$_B$, but not for GABA$_A$ receptors. These results indicate that local GABAergic neurons in the sSC are excited by ACh through the activation of nicotinic receptors (cell 2), which, in turn, results in increased release of GABA that could activate GABA$_B$ receptors in projection neurons (cells 1 and 3). On the other hand, activation of nicotinic receptors located in presynaptic terminals of GABAergic neurons facilitates GABA release (synapse A in sSC), enhancing GABA$_A$-mediated inhibition of non-GABAergic neurons in the sSC (Endo et al., 2005). In addition, activation of postsynaptic nicotinic receptors expressed in GABAergic neurons gives rise to inward currents (cell 2), thus exciting these GABAergic neurons. Consequently, ACh augments GABAergic inhibition in sSC neurons through both pre- and postsynaptic mechanisms. All these studies suggest that the eventual effect of cholinergic inputs on sSC neuronal activity is suppression. However, activation of nicotinic receptors in projection neurons also induces depolarizing currents (Endo et al., 2005). Thus, such inhibitory and excitatory actions could enhance the contrast of the visual images and the signal-to-noise ratio of the network activity or provide a substrate for a winner-take-all network.

Cholinergic Modulation in the Deeper Layers of the Superior Colliculus

The dSC is innervated by cholinergic afferents arising from the pedunculopontine tegmental nucleus (PPTN). The effects of activation of postsynaptic nicotinic and muscarinic receptors depend on cell types and receptor subtypes. In the majority of dSC neurons including crossed tecto-reticular output neurons in the SGI, application of cholinergic agonists results in nicotinic inward current with muscarinic inward current or nicotinic inward current with muscarinic inward current followed by muscarinic outward current (Sooksawate et al., 2008). Other responses observed in the minority of cells include nicotinic inward only, nicotinic inward with muscarinic outward, or muscarinic outward only. Thus, primary postsynaptic action of cholinergic inputs on dSC neurons is excitatory. On the other hand, activation of presynaptic muscarinic receptors reduces GABAergic transmission in the dSC (synapse B in dSC). Based on these post- and presynaptic mechanisms, therefore, the total effect of cholinergic inputs to the dSC is excitation, which depolarizes the membrane potential of dSC neurons close to the threshold for the NMDA receptor–dependent burst firings (Isa and Sparks, 2006). Indeed, in awake behaving monkeys, local nicotine injection into the dSC leads to reduction in reaction time of saccades toward the movement field of the neurons at the injection site of the SC (Aizawa et al., 1999).

References

Aizawa H, Kobayashi Y, Yamamoto M, Isa T (1999) Injection of nicotine into the superior colliculus facilitates occurrence of express saccades in monkeys. *J Neurophysiol* 82:1642–1646.

Binns KE, Salt TE (2000) The functional influence of nicotinic cholinergic receptors on the visual responses of neurones in the superficial superior colliculus. *Vis Neurosci* 17:283–289.

Endo T, Yanagawa Y, Obata K, Isa T (2005) Nicotinic acetylcholine receptor subtypes involved in facilitation of GABAergic inhibition in mouse superficial superior colliculus. *J Neurophysiol* 94:3893–3902.

Isa T, Sparks D (2006) Microcircuit of the superior colliculus: a neuronal machine that determine timing and endpoint of saccadic eye movements. In: Grillenr S and Graybiel AM, eds. *Microcircuits: The interface Between Neurons and Global Brain Function*, pp. 1–34. Cambridge, MA: MIT Press.

Isa T, Endo T, Saito Y (1998) The visuo-motor pathway in the local circuit of the rat superior colliculus. *J Neurosci* 18:8496–8504.

Kaneda K, Phongphanphanee P, Katoh T, Isa K, Yanagawa Y, Obata K, Isa T (2008a) Regulation of burst activity through presynaptic and postsynaptic GABA$_B$ receptors in mouse superior colliculus. *J Neurosci* 28:816–827.

Kaneda K, Isa K, Yanagawa Y, Isa T (2008b) Nigral inhibition of GABAergic neurons in mouse superior colliculus. *J Neurosci* 28:11071–11078.

Lee PH, Helms MC, Augustine GJ, Hall WC (1997) Role of intrinsic synaptic circuitry in collicular sensorimotor integration. *Proc Natl Acad Sci USA* 94:13299–13304.

Lee PH, Sooksawate T, Yanagawa Y, Isa K, Isa T, Hall WC (2007) Identity of a pathway for saccadic suppression. *Proc Natl Acad Sci USA* 104:6824–6827.

Sooksawate T, Isa K, Isa T (2008) Cholinergic responses in crossed tecto-reticular neurons of rat superior colliculus. *J Neurophysiol* 100:2702–2711.

31

The Mammalian Brainstem Chewing Circuitry

Arlette Kolta and James P. Lund

The act of mastication requires the coordinated activity of jaw, facial, and tongue muscles. The form of movements varies between foods and changes gradually during a single masticatory sequence, in parallel with the physical properties of the food bolus. The basic pattern of mastication is produced by a brainstem central pattern generator (CPG) when tonically activated by inputs from sensory afferents and/or higher brain regions (Dellow and Lund, 1971). Although not required to produce the basic pattern, sensory feedback is largely responsible for adjusting the motor output to changes in food properties, while cortical inputs are thought to play a role in anticipation and determination of the initial patterns of movements.

ORGANIZATION OF THE BRAINSTEM CIRCUITRY FORMING THE MASTICATORY CENTRAL PATTERN GENERATOR

The CPG is formed of interneurons that are tightly interconnected, some of which have an intrinsic burst-generating ability. It receives inputs from specific groups of fifth sensory afferents and from higher centers and projects to all relevant motoneurons (MNs).

Inputs to the Central Pattern Generator

Sensory

Sensory afferents that provide feedback to the CPG during mastication include mechanoreceptors in the skin, hair, and mucosa of the lips and oral cavity (Fig. 31.1A). Most skin and mucosal afferents are not active during mastication unless something touches their receptive field (Fig. 31.1A,B #2),

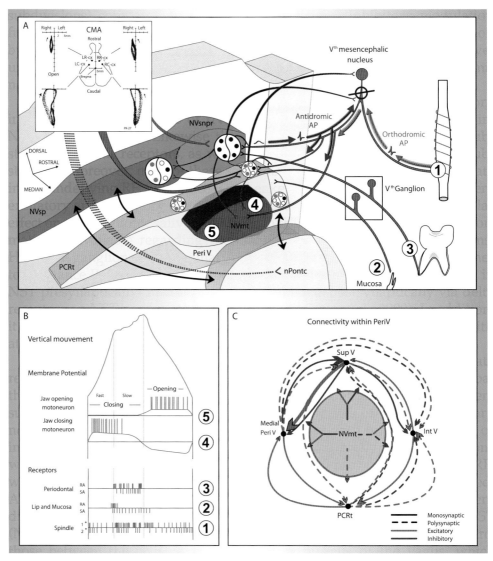

FIGURE 31–1. (*A*) Sagittal view of the brainstem nuclei controlling chewing with their sensory and cortical inputs and their local connections. (*Inset*) Shows that stimulation of distinct locations of the cortical masticatory area (CMA) elicits distinct movement patterns of the jaw. Circles in black, white, and gray indicate cells that are active during all, some, or just one of the cortically evoked movement patterns, respectively. Continuous arrows and projections indicate direct connections. Dashed projections indicate polysynaptic input. Blue connections are GABAergic, while red connections are glutamatergic. (*B*) Firing pattern of sensory afferents and trigeminal motoneurons during a masticatory cycle (top trace). The numbers to the right refer to the numbers in *A*. RA: rapidly adapting; SA: slowly adapting. (*C*) Schematic diagram of the monosynaptic and polysynaptic excitatory and inhibitory connections linking the different divisions of the peritrigeminal area (yellow area in *A*) surrounding the motor nucleus.

while facial hair afferents fire throughout the masticatory cycle at a frequency that is proportional to the velocity of the movement (see Lund, 1991). Mechanoreceptors located in the periodontal ligament are sensitive to force (or its rate of change) on teeth, but they saturate at relatively low loads. They fire only in the later part of the jaw closing (JC) phase of the cycle when pressure is applied to teeth via the food (Fig. 31.1A,B #3). Those that are rapidly adapting (RA) fire short bursts, while slowly adapting (SA) afferents fire until the jaw opening (JO) phase begins. Only the muscles that elevate the mandible (masseter, temporalis, and medial pterygoid) contain significant numbers of muscle spindles. These fire throughout the masticatory cycle. During JO they are stretched, but γ drive causes the strongest firing to occur in the JC phase (Fig. 31.1A,B #1). Most fifth sensory afferents have their cell bodies in the trigeminal ganglion, but the cell bodies of all muscle spindles and about half of the periodontal afferents are found within the brainstem in the trigeminal mesencephalic nucleus (NVmes). The central axons of all afferents give off collaterals to second-order neurons in the V principal sensory nucleus (NVsnpr) and spinal nucleus (NVsp). Those of spindle afferents in NVmes also give off collaterals to the fifth motor nucleus (NVmt) (Fig. 31.1A). Like other large spinal primary afferents, NVmes neurons are pseudounipolar neurons with large I_h currents and subthreshold oscillations that sometimes lead to bursting (Verdier et al., 2004). They differ from other primary afferents by the fact that they are located centrally, that many are electrotonically coupled together and to local multipolar neurons, and that they receive synaptic inputs. Most of these inputs are depolarizing, even those that are GABAergic because their chloride equilibrium potential is above their resting membrane potential. These depolarizing potentials are powerful and often cause firing because they occur directly on the soma and proximal axon (Verdier et al., 2004). Firing generated at the soma invades the stem axon of these afferents and propagates orthodromically along the central branch (Westberg et al., 2000). There is some evidence that it may also propagate antidromically along the peripheral branch and cause either a change of sensitivity of the receptor in periphery and/or release of glutamate. Another important effect of synaptic inputs is an increase in the amplitude and the resetting of the oscillations that persists for several cycles (Verdier et al., 2004). If the same input is received by several NVmes neurons, it will synchronize their oscillations. This, combined with electrical coupling, may cause sustained and synchronous firing in entire subpopulations of NVmes neurons and presumably in the MNs that they control. This could be important during rapid biting and during the mastication of very hard or tough foods.

From Cortex and Other Higher Centers

The jaw, tongue, and facial muscles are represented individually in the primary motor cortex (M1). In humans and other primates, this region is partly overlapped laterally by the cortical masticatory area (CMA) from which coordinated natural-looking mastication can be induced by stimulation (Fig. 31.1A). In lower mammals, the CMA completely overlaps the jaw, face, and tongue representations of M1. Both M1 and CMA project to the trigeminal sensory complex and parvocellular reticular formation directly (Hatanaka et al., 2005). Despite this, long-lasting (hundreds of milliseconds) repetitive stimulation of the CMA is required in order to induce mastication.

Ablation or cooling of these regions does not prevent mastication from occurring, but it does cause difficulty in initiating mastication and increases total masticatory time (Enomoto et al., 1987;Yamamura et al., 2002). This is due in part to a loss of ability to coordinate food manipulation before and during chewing, and in transporting the food bolus toward the pharynx for swallowing.

Subcortical areas known to influence mastication include the amygdala, the hypothalamus, the anterior pretectal nucleus, the red nucleus, the periaqueductal gray, the raphé nuclei, the cerebellum, and various parts of the basal ganglia (nucleus accumbens, subthalamic nucleus, and substantia nigra, and pars reticulata).

Outputs of the Central Pattern Generator

To Jaw, Facial, and Tongue Muscles

Most MNs supplying the jaw muscles are located in NVmt (red structure in Fig. 31.1A) except for those supplying the posterior belly of the digastric (JO muscle). The latter are found with those of the facial muscles in the seventh nucleus, while tongue MNs are located in the twelfth nucleus. Motoneurons innervating the JC muscles occupy the rostral two-thirds of NVmt (Fig. 31.1A #4), whereas those supplying the JO muscles are found in the ventromedial middle third and the caudal portion of NVmt (Fig. 31.1A #5). Isolated populations of small MNs to both groups are found outside its ventral boundaries.

When mastication is elicited by stimulation of the CMA in paralyzed animals, it is termed "fictive" mastication because no movement occurs; therefore, sensory feedback is eliminated. Under such conditions, JO MNs display long depolarizing potentials overridden by bursts of action potentials (APs) alternating with periods (JC phase) in which their membrane potential returns to resting levels (Fig. 31.1B #4). Likewise, XII MNs do not display a phase of hyperpolarization during fictive mastication (Sahara et al., 1988). JC MNs also show slow depolarizing potentials during their active phase, but their firing is much weaker than in JO MNs, and their membrane potentials are hyperpolarized in the opposite phase in (Fig. 31.1B #5) ((Lund, 1991;Yamada et al., 2005). These important differences in membrane potential and in firing rate between the two MN pools reflect the relative importance of sensory feedback in controlling their outputs.

To Interneurons and Sensory Terminals

Central pattern generators control synaptic transmission from the primary afferents to interneurons and MNs and also modulate interneurons directly. One reason for this is to enhance reflexes that assist movement and suppress reflexes that may perturb it, excepting those that have protective functions. For instance, at rest the JO reflex can be triggered by nociceptors and low-threshold mechanoreceptors. However, the latter are stimulated by movement alone (Fig. 31.1B, #2 and #3). Therefore, to prevent disruption, interneurons receiving these inputs are tonically inhibited during mastication, while on the other hand, interneurons receiving inputs from the normally silent high-threshold afferents see their excitability increased during JC when most damage to the soft tissues of the mouth may occur (review in Lund, 1991).

Another method by which the CPG reduces the amplitude of reflexes is GABAergic presynaptic inhibition of transmission from terminals of primary sensory afferents, such as those from periodontal and muscle spindles. GABA causes a depolarization of the terminals, which reduces transmitter release from the synapse, but which is sometimes strong enough to trigger antidromic action potentials (APs) (blue AP in Fig. 31.1A). In the case of JC muscle spindle, these antidromic APs intermingle with orthodromic APs in the JC phase of fictive mastication. Phasic inhibition occurs in the JO phase, producing a rhythmic bursting firing pattern like the one shown in Figure 31.2D (Westberg et al., 1998). Interestingly, induction of a variety of fictive movement patterns by stimulation of separate sites in the CMA (see inset in Fig 31.1A) causes specific patterns of modulation of firing (see Fig. 31.2D), suggesting that muscle spindle central terminals are being differentially modulated by the CPG. Because antidromic APs generated at one terminal can propagate toward other axonal collaterals and invade them orthodromically (blue arrows in Fig. 31.1A), they can contribute to the activation of JC MNs (Westberg et al., 2000). However, propagation of APs into the stem axon, and perhaps into other parts of the axonal tree, is blocked by a shunt caused by second group of GABAergic synapses on the central axon. This divides the central axonal tree, allowing the rostral portion to convey sensory information while the caudal part acts as a local interneuron, carrying APs from one terminal to the other. The axonic GABAergic inputs arise from at least two groups of CPG neurons, one in the supratrigeminal area (SupV) and the other in dorsal NVsnpr (Verdier et al., 2003).

IDENTIFYING CENTRAL PATTERN GENERATOR SUBPOPULATIONS

Transection studies in en bloc brainstem preparations suggest that the essential core of the CPG lies between the rostral poles of the fifth and seventh motor nuclei (Kogo et al., 1996). The lateral portion of this area is occupied by the trigeminal sensory complex (green area in Fig. 31.1A), which is formed of

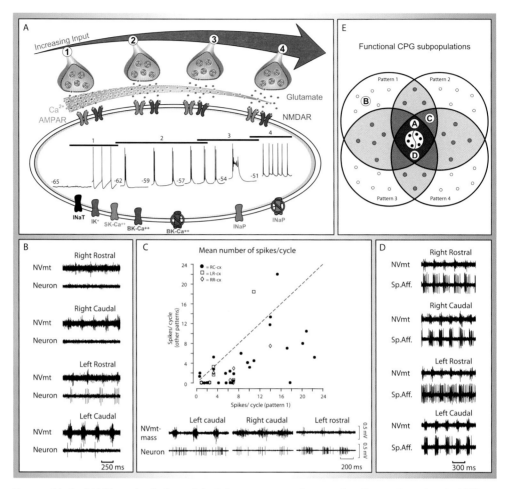

FIGURE 31–2. (A) Hypothetical model of the sequence of events leading to bursting in NVsnpr neurons. The numbers 1 to 4 indicate build up of activity of inputs. The traces inside the cell show the firing pattern in function of the level of activity of the inputs, and the channels underneath are those presumably involved in each of these firing patterns. The green fading band and circles indicate that extracellular Ca^{2+} drops with increasing activity. (B) Shows the example of an interneuron that would be a white cell in Figure 31.1A and 31.2E and that fires rhythmically (second trace of each pair) in only one of four movement patterns (seen in the upper trace of each pair). (C) Lower traces show an example of what would be a gray cell that fires in three movement patterns. The plot above shows that cells that fire in more than one pattern change their firing frequency with each pattern. (D) Rhythmic modulation of firing of the terminal of a spindle afferent during four different patterns. (E) Model of the organization of the central pattern generator (CPG) in which the active population of interneuron may vary constantly depending on the conjunction of sensory and cortical inputs. Only rhythmogenic cells (like in A; black circles) receiving sufficient inputs will fire rhythmically and entrain subpopulations of cells to which they project.

NVsnpr and NVsp. These nuclei contain second-order sensory neurons, many of which project to the thalamus. However, they also project to the fifth, seventh, and twelfth motor nuclei and an adjacent column of last-order interneurons in the parvocellular reticular formation (PCRt), which extends caudally to NXII and to which we refer as the peritrigeminal area (PeriV; yellow area in Fig. 31.1A). Its rostral pole surrounds NVmt and has three subdivisions: SupV, the Intertrigeminal area (IntV), and medial PeriV (dorsal, lateral, and medial to NVmt, respectively). The sensory complex also projects to the medial reticular formation, especially nucleus pontis caudalis (gray area in Fig. 31.1A).

Cortical and sensory inputs reach neurons in the sensory complex and PeriV at short latencies, and those of nPontc at longer latencies. PeriV and some parts of the sensory complex have direct projections to NVmt.

During fictive mastication in vivo, many neurons of PeriV and the sensory complex change their firing rate and a smaller proportion also change their firing pattern to fire rhythmically in phase with either JC or JO MNs. Few bursts occur at phase transitions (Tsuboi et al., 2003). It is significant that rhythmically firing neurons seem to receive sensory inputs only from mucosal, periodontal, or muscle spindle afferents. Firing in many of these neurons (represented by white circles in Figs. 31.1A and 31.2E) is only associated with one pattern of the several movement patterns elicited by changing the site of stimulation in the CMA (example in Fig. 31.2B). Others are linked to more than one (gray circles in Figs. 31.1A and 31.2E, example in Fig. 31.2C bottom), but few are modulated in all movement patterns (black circles in Figs. 31.1A and 31.2E).

Neurons that participate in more than one movement usually change their rhythmical firing pattern (Fig. 31.2C), suggesting that at least two mechanisms are involved in generating the final envelope of the masticatory movements: addition and subtraction of neurons from the active pool, and changes in firing pattern of common elements (like in Fig. 31.2D).

Most neurons of PeriV and NVsp are pattern specific (white or gray cells), while masticatory neurons of dorsal NVsnpr show modulation for all patterns (black cells), suggesting that they may form a crucial element of the CPG. The role of nPontC in mastication is not yet clear. Many neurons in its dorsal half fire in either the JO or JC phases, while most ventral neurons fire tonically. Ventral neurons appear to inhibit lateral CPG subgroups, while dorsal neurons have mixed effects (Scott et al., 1997).

Rhythmic outputs of CPGs can arise from interactions within a network, from intrinsic properties of individual neurons, or a combination of the two. All three divisions shown in Figure 31.1A are extensively interconnected (thick black arrows), and they are also connected to homologous nuclei on the contralateral side. In particular, extensive and robust excitatory and inhibitory interconnections exist between the various subregions of PeriV (Fig. 31.1C; Bourque and Kolta, 2001). These connections may be important in the reorganization of the active pool and in insuring that the rhythm is rapidly propagated throughout the circuitry.

Neurons with rhythmogenic abilities have also been found within the CPG, frequently in the anterodorsal region of NVsnpr and less often in dorsolateral PeriV. Rhythmic bursting in these neurons first appears at the same age as the earliest masticatory movements around P12. It is dependent on a plateau potential mediated by a voltage-dependent persistent sodium current (I_{NaP}) (Brocard et al., 2006). At low and at very high activity levels (periods 1 and 4, respectively, in Fig. 31.2A), the cell fires tonically because its membrane potential is not in the range in which I_{NaP} is active. In between, the cells fire rhythmically, first in short plateaus overridden by two or three APs (Fig. 31.2A period 2) and eventually in larger bursts (Fig. 31.2A, period 3). The firing pattern is also sensitive to the extracellular concentration of Ca^{2+} ($[Ca^{2+}]e$). Reduction of $[Ca^{2+}]e$ increases the duration of the I_{NaP}-mediated plateau and promotes bursting. It is probable that increasing synaptic activity driven by afferent inputs to these neurons gradually reduces $[Ca^{2+}]e$. In many brain areas, sustained neuronal activity is associated to a drop of $[Ca^{2+}]e$. Lowering $[Ca^{2+}]e$ will first inactivate the big Ca^{2+}-dependent K^+ channels (BK-Ca^{2+}), which are very sensitive to $[Ca^{2+}]e$ and which are responsible for part of the afterhyperpolarization (AHP). As a result, the increasing depolarization produced by the inputs will now cause an afterdepolarization (ADP) rather than an AHP (period 2 in Fig. 31.2A). Further increases in synaptic activity will further deplete Ca^{2+}, favoring activation of I_{NaP} and leading to bursting (period 3 in Fig. 31.2A).

This model is supported by the fact that both in vivo fictive mastication and in vitro I_{NaP}-mediated bursting of NVsnpr neurons at physiological $[Ca^{2+}]e$ can only be induced by sustained and repetitive stimulation of sensory and/or cortical afferents, even though these inputs reach all subgroups of the CPG at short latencies.

SUMMARY

The model of the masticatory CPG presented in Figure 31.2E is composed of subpopulations of neurons distributed within the trigeminal sensory complex and adjacent reticular formation. Some of these have rhythmogenic abilities (black circles), but all are tightly interconnected. A differential input from the cortex and other motor centers and from the periphery leads to the activation of movement-specific neuronal subpopulations. Cells endowed with rhythmogenic abilities and receiving sufficient input will generate rhythmic activity and transmit it to other subpopulations, which will relay it to the appropriate MNs and interneurons.

ACKNOWLEDGMENTS

This work was supported by grants from the Canadian Institutes of Health Research and an infrastructure grant from the Fonds de la Recherche en Santé du Québec.

REFERENCES

Bourque MJ, Kolta A (2001) Properties and interconnections of trigeminal interneurons of the lateral pontine reticular formation in the rat. *J Neurophysiol* 86:2583–2596.

Brocard F, Verdier D, Arsenault I, Lund JP, Kolta A (2006) Emergence of intrinsic bursting in trigeminal sensory neurons parallels the acquisition of mastication in weanling rats. *J Neurophysiol* 96:2410–2424.

Dellow PG, Lund JP (1971) Evidence for central timing of rhythmical mastication. *J Physiol (Lond)* 215:1–13.

Enomoto S, Schwartz G, Lund JP (1987) The effects of cortical ablation on mastication in the rabbit. *Neurosci. Lett.* 82:162–166.

Hatanaka N, Tokuno H, Nambu A, Inoue T, Takada M (2005) Input-output organization of jaw movement-related areas in monkey frontal cortex. *J Comp Neurol* 492:401–425.

Kogo M, Funk GD, Chandler SH (1996) Rhythmical oral-motor activity recorded in an *In Vitro* brainstem preparation. *Somatosens Motor Res* 13:39–48.

Lund JP (1991) Mastication and its control by the brain stem. *CRC Crit Rev Oral Biol Med* 2:33–64.

Sahara Y, Hashimoto N, Kato M, Nakamura Y (1988) Synaptic bases of cortically-induced rhythmical hypoglossal motoneuronal activity in the cat. *Neurosci Res* 5:439–452.

Scott G, Westberg KG, Olsson KÅ, Lund JP (1997) Role of medial pontobulbar reticular formation neurons in patterning mastication. *J Dent Res* 76:124.

Tsuboi A, Kolta A, Chen CC, Lund JP (2003) Neurons of the trigeminal main sensory nucleus participate in the generation of rhythmic motor patterns. *Eur J Neurosci* 17: 229–238.

Verdier D, Lund JP, Kolta A (2003) GABAergic control of action potential propagation along axonal branches of mammalian sensory neurons. *J Neurosci* 23:2002–2007.

Verdier D, Lund JP, Kolta A (2004) Synaptic inputs to trigeminal primary afferent neurons cause firing and modulate intrinsic oscillatory activity. *J Neurophysiol* 92:2444–2455.

Westberg K, Clavelou P, Sandstrom G, Lund JP (1998) Evidence that trigeminal brainstem interneurons form subpopulations to produce different forms of mastication in the rabbit. *J Neurosci* 18:6466–6479.

Westberg K-G, Kolta A, Clavelou P, Sandstrom G, Lund JP (2000) Evidence for functional compartmentalization of trigeminal muscle spindle afferents during fictive mastication in the rabbit. *Eur J Neurosci* 12:1145–1154.

Yamada Y, Yamamura K, Inoue M (2005) Coordination of cranial motoneurons during mastication. *Respir Physiol Neurobiol* 147:177–189.

Yamamura K, Narita N, Yao D, Martin RE, Masuda Y, Sessle BJ (2002) Effects of reversible bilateral inactivation of face primary motor cortex on mastication and swallowing. *Brain Res* 944:40–55.

32

The Lamprey Locomotor Central Pattern Generator

Sten Grillner and Peter Wallén

The lamprey network coordinating locomotion has been studied extensively. As in all other vertebrates investigated, the pattern generation is provided at the spinal-cord level, whereas the decision to activate the network is controlled primarily by the locomotor command centers in the brainstem (mesencephalic locomotor region and diencephalic locomotor region), which in turn are controlled from the basal ganglia in the forebrain (Fig. 32.1). We will limit this brief account to the segmental rhythm-generating circuits and the mechanism underlying the intersegmental coordination (see Buchanan, 2001; Grillner, 2003; Dubuc et al., 2008).

THE SEGMENTAL COORDINATION

The core of the segmental locomotor network (unit CPG) is formed by pools of excitatory interneurons that generate burst activity and excite each other. Inhibitory glycinergic or GABAergic neurons are not required in this process (Buchanan and Grillner, 1987; Cangiano and Grillner, 2003, 2005; see Grillner, 2003, 2006). There is thus a core unit CPG on each side of the spinal cord that can operate without inhibitory mechanisms. The latter do, however, play a critical role for the left–right coordination between the two sides.

Figure 32.1A shows a diagram of the lamprey CPG organization. The segmental network is activated by excitatory drive signals via reticulospinal (RS) neurons that excite both excitatory (E) and inhibitory (I) neurons and the output motoneurons (for review, see Grillner, 2003, 2006). The command drive signal turns on the network and controls the level of activity (burst rate between 0.2 and 10 Hz). In each segment there is a pool of around 40–50

326

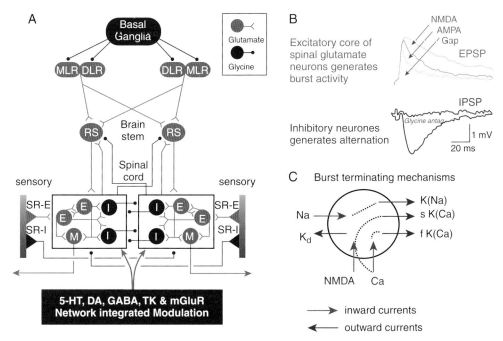

Figure 32–1. Locomotor network of the lamprey. Schematic representation of the segmental neural circuitry that generates rhythmic locomotor activity. (*A*) All neuron symbols denote populations rather than single cells. The reticulospinal (RS), glutamatergic neurons excite all classes of spinal interneurons and motoneurons. The excitatory interneurons (E; red) excite all types of spinal neurons, that is, the inhibitory glycinergic interneurons (I: blue) that cross the midline to inhibit all neuron types on the contralateral side, and motoneurons (M). The stretch receptor neurons are of two types: one excitatory (SR-E), which excites ipsilateral neurons, and one inhibitory (SR-I), which crosses the midline to inhibit contralateral neurons. RS neurons receive excitatory synaptic input from the diencephalic and the mesencephalic locomotor regions (DLR and MLR), which in turn receive input from the basal ganglia as well as visual and olfactory input. In addition, metabotropic receptors are also activated during locomotion and are an integral part of the network (5-HT, GABA, Tachykinins [TKs], and mGluR). (*B*) The descending axons and the E-interneurons of the burst-generating excitatory core provide excitatory postsynaptic potentials (EPSPs), which have three components: AMPA, NMDA, and gap junctions. The glutamatergic interneurons in turn activate glycinergic interneurons (blocked by strychnine) with axons that inhibit neurons on the contralateral side. This results in alternating efferent activity between the left and right side. (*C*) Several factors contribute to burst termination. Calcium entering via voltage-gated calcium channels during the action potential activate calcium-dependent potassium channels (f KCa), which underlie the postspike afterhyperpolarization (AHP). During the burst, AHP summation will lead to frequency adaptation and contribute to burst termination. During NMDA-receptor activation, KCa channels will be activated also by calcium entering through NMDA channels and contribute to NMDA-plateau termination. In addition, sodium entry will activate sodium-dependent potassium channels (KNa), which may also contribute to burst termination.

excitatory interneurons. The connectivity ratio between these excitatory interneurons has been estimated at around 10%, which means that when they become activated, further excitation will be generated within the pool. Both the descending command and the excitatory interneurons utilize glutamate as the transmitter and activate both NMDA and AMPA receptors, and in addition gap junctions provide further excitation at many synaptic sites (Fig. 32.1B). The segmental excitatory interneurons receive additional excitation from the same type of neurons in nearby segments. The excitatory postsynaptic potential (EPSP) amplitudes vary around 1 mV.

The interaction within the pool of excitatory interneurons will thus provide a continuous depolarization to a degree sufficient to generate spike activity in a proportion of the cells, provided that the background excitatory drive is able to make some of the neurons generate action potentials. They will in turn excite motoneurons and also inhibitory commissural interneurons that inhibit the contralateral CPG. When one hemisegment is active, it will also inhibit the contralateral network, and when this inhibition ceases, the other side will take over due to the background excitatory drive from the brainstem.

The excitatory interneurons thus interact to promote a burst, but which process makes the burst terminate in a predictable way? The main reason relates to an accumulation of Ca^{2+} ions during the burst, which in turn activate calcium-dependent potassium channels (K_{Ca}), which will lead to a progressive hyperpolarization, and a closure of voltage-dependent ion channels like NMDA receptors (el Manira et al., 1994; Grillner, 2003). The Ca^{2+} entry (Fig. 32.1C) is via two main sources, NMDA receptors and voltage-dependent Ca^{2+} channels of both the low- and high-voltage activate variety (Ca_v 1.3 and 2.2). The latter is activated during each action potential and triggers the postspike afterhyperpolarization (sAHP) due to K_{Ca}. The sAHP is the main determinant of firing frequency and spike train adaptation. During the burst, the sAHPs are summed, which results in a frequency adaptation, being greater the larger the sAHPs. At higher levels of activity, the accumulation of Na^+ ions will also be able to activate Na^+ activated potassium channels (K_{Na}) that may contribute to burst termination (Wallén et al., 2007).

If the pool of excitatory interneurons in a hemisegment is functionally isolated pharmacologically or mechanically (Cangiano and Grillner, 2003, 2005), it will still be able to generate burst activity, but now at a fairly high burst frequency from 2.5 to over 10 Hz. At lower frequencies, depolarizing plateaus due to the activation of voltage-dependent NMDA receptors contribute to the burst activity. These plateau depolarizations are terminated when K_{Ca} channels have been sufficiently activated, leading to a closure of, for example, voltage-dependent NMDA receptors. A normal network activity (inhibition intact) will generate the entire range of locomotor frequencies from 0.2 to 10 Hz, and a successive blockade of the inhibitory mechanisms not only blocks the alternation but also markedly increases the burst frequency.

The inhibition is thus an important part of the burst frequency control. In addition to this core circuit, there are also a group of segmental inhibitory interneurons with an ipsilateral action that could possibly contribute to burst termination under some conditions. They are, however, not of crucial importance, since their action can be blocked without any effects on burst frequency.

There are also a large number of modulator actions via G protein–activated receptors through neurons releasing glutamate, GABA, 5-HT, dopamine, endocannabinoids, NO, and neuropeptides (Fig. 32.1A; see Grillner, 2006, Kyrikatos and el Manira, 2007). They are all activated as locomotion is turned on, but their action is to fine tune the properties of neurons of the network rather than to take part in the cycle to cycle generation of excitation and inhibition, respectively. This is the task of ionotropic glutamate and glycine receptors.

INTERSEGMENTAL COORDINATION AT THE SPINAL LEVEL

Single segments (Fig. 32.1) can thus generate burst activity, but the intersegmental coordination along the entire spinal cord (100 segments) results in a wave of activity along the entire spinal cord with a phase lag between each segment. In the intact animal this will result in an undulatory wave traveling down the body, thereby pushing the animal forward through the water (Fig. 32.2A). The lag between the different segments is always a certain proportion of the cycle duration and thus represents a constant phase lag. Also the isolated spinal cord will generate the same type of coordination. However, if the lamprey in the wild is caught in a corner, it can instantaneously reverse the direction of the undulatory wave and thereby swim backward for a few cycles. Similarly, in the isolated spinal cord the phase lag can be reversed by inhibiting the rostral spinal segments or by providing extra excitation to the caudalmost segments.

The neural mechanisms underlying the phase lag have been studied both experimentally and through simulations (Matsushima and Grillner, 1992; Kozlov et al, 2009), using detailed neuronal models of each population of neurons in each of the hundred segments with an approximately correct number of cells, appropriate cellular properties, and synaptic connectivity. Altogether 100 cells in each of the 100 segments (n = 10,000) were simulated with a rostrocaudal connectivity as defined by experiments (Fig. 32.2B). The excitatory interneurons are known to have a descending axonal branch extending over around six segments, and a briefer ascending branch. The inhibitory commissural interneurons can extend the caudal axonal branch further than the excitatory axons. These large-scale simulations show that the rostrocaudal pattern of activity with a constant phase lag along the entire spinal cord could be generated as a robust network activity (Fig. 32.2C).

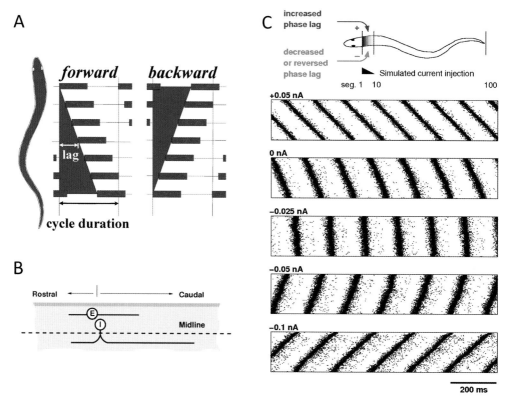

FIGURE 32–2. Intersegmental coordination and control of movement direction. (A) Pattern of neuronal activity during forward and backward swimming in segments along one side of the spinal cord. The intersegmental coordination will result in an undulatory wave travelling along the body, pushing the animal through the water. A positive intersegmental phase lag results in forward swimming, while a reversed (negative) lag corresponds to backward swimming. (B) Rostro-caudal connectivity of excitatory (E) and inhibitory (I) interneurons. The commissural I interneurons extend their axons over many segments caudally on the contralateral side. (C) Large-scale simulations of the intersegmental coordination along 100 segments and with 100 model neurons in each segment. Positive or negative current injection is applied to interneurons in the first 10 segments with a linear decay, producing increased or decreased/reversed intersegmental phase lags (*inset at top*). The magnitude of the phase lag varies continuously with the current injection. Only the left E population of the spinal cord network is shown in five different cases with different stimulation levels from +0.05 nA to −0.1 nA. In each sequence, the most rostral segment is represented in the uppermost position and the neurons in segment 100 in the lowermost position. The positive phase lag is largest with +0.05 nA applied to the 10 most rostral segments and the most negative phase lag is seen with −0.1 nA.

It proved important to incorporate the variation in cellular size and properties in order to have an orderly recruitment and derecruitment in each cell population. Moreover, the phase lag along the entire spinal cord could be modified in a progressive way by adding excitation or inhibition to only a few rostral segments of the spinal cord (inset Fig. 32.2C). Thus, the phase lag along the entire spinal cord can be modified by just tinkering with the excitability of a few rostral segments. This is a very efficient and simple control mechanism that most likely can be applied under a variety of circumstances in different networks.

SENSORY CONTROL OF THE CENTRAL PATTERN GENERATOR

The network activity is also influenced by sensory input from stretch receptor (SR) neurons (Fig. 32.1A) that sense the ongoing movements (see Grillner, 2003). The sensory input can exert a profound effect on the network activity and prolong or shorten an ongoing burst. This sensory action provides a mechanism for adaptation of the network activity to external perturbations. There are two types of SR neurons: ipsilateral excitatory and commissural inhibitory neurons. Their synaptic action on the CPG of the two sides of the spinal cord is indicated in Figure 32.1. In the intact swimming lamprey, the sensory input will also act to stabilize the motor pattern

CONCLUSION

The basic cellular, synaptic, and network mechanisms underlying swimming in lamprey is relatively well understood. This includes both the robust network performance underlying forward swimming and the mechanisms underlying the flexible changes that will occur when the swimming direction is reversed.

REFERENCES

Buchanan JT (2001) Contributions of identifiable neurones and neurone classes to lamprey vertebrate neurobiology. *Prog Neurobiol* 63(4):441–466.
Buchanan JT, Grillner S (1987) Newly identified "glutamate interneurons" and their role in locomotion in the lamprey spinal cord. *Science* 236(4799):312–314.
Cangiano L, Grillner S (2003) Fast and slow locomotor burst generation in the hemispinal cord of the lamprey. *J Neurophysiol* 89(6):2931–2942.
Cangiano L, Grillner S (2005) Mechanisms of rhythm generation in a spinal locomotor network deprived of crossed connections: the lamprey hemicord. *J Neurosci* 25(4): 923–935.
Dubuc R, Brocard F, Antri M, Fénelon K, Gariépy JF, Smetana R, Ménard A, Le Ray D, Viana Di Prisco G, Pearlstein E, Sirota MG, Derjean D, St-Pierre M, Zielinski B, Auclair F, Veilleux D (2008) Initiation of locomotion in lampreys. *Brain Res Rev* 57(1):172–182.
el Manira A, Tegner J, Grillner S (1994) Calcium-dependent potassium channels play a critical role for burst termination in the locomotor network in lamprey. *J Neurophysiol* 72(4):1852–1861.
Grillner S (2003) The motor infrastructure: from ion channels to neuronal networks. *Nat Rev Neurosci* 4(7):573–586.

Grillner S (2006) Biological pattern generation: the cellular and computational logic of networks in motion. *Neuron* 52(5):751–766.

Kozlov A, Huss M. Lansner A, Hellgren Kotaleski J, Grillner S (2009) Simple cellular and network control principles govern complex patterns of motor behavior. *PNAS* 106:20027–20032.

Kyriakatos A, el Manira A (2007) Long-term plasticity of the spinal locomotor circuitry mediated by endocannabinoid and nitric oxide signaling. *J Neurosci* 27(46): 12664–12674.

Matsushima T, Grillner S (1992) Neural mechanisms of intersegmental coordination in lamprey - Local excitability changes modify the phase coupling along the spinal cord. *J Neurophysiol* 67:373–388.

Wallén P, Robertson B, Cangiano L, Löw P, Bhattacharjee A, Kaczmarek LK, Grillner S (2007) Sodium-dependent potassium channels of a Slack-like subtype contribute to the slow afterhyperpolarization in lamprey spinal neurons. *J Physiol* 585:75–90.

33

The Mauthner Cell Microcircuits: Sensory Integration, Decision Making, and Motor Functions

Donald S. Faber and Henri Korn

The teleost and amphibian Mauthner (M) cell is the critical decision-making element in an exquisitely designed brainstem microcircuit that serves the vital function of deciding whether uni- or multimodal stimuli are sufficiently threatening to the animal's survival that they warrant triggering a massive motor response. This behavior, called the startle response or C-start, was initially improperly characterized as a reflex, but, in fact, its governing process involves computations that can occur over a period as short as a few milliseconds or as long as 0.5 s. Because much of what is known about this system was initially derived from studies in goldfish of the C-start evoked by loud abrupt sounds and from the associated basic electrophysiology, this behavior is discussed first. Then we address M cell–initiated escapes triggered by more complex stimuli, as well as the notion that they can be voluntary. A common theme throughout is the central role of inhibitory mechanisms in shaping the behavioral threshold.

The neuronal system that subserves the escape behavior is well understood because there are only two M cells per fish, and they are large reticulospinal neurons that can be identified electrophysiologically *in vivo* and *in situ*. Indeed, the M cell action potential is analogous to a signature, permitting recordings from defined cellular regions, such as the soma and the lateral or ventral dendrites, or chronically from the cell or its axon in freely swimming fish. Thus, it is well established that C-starts are triggered by a single spike in one of the two paired M cells and involve bilateral contraction of cranial muscles (jaw, eyes, opercula) and unilateral contraction of trunk musculature—hence, the "C" shape of the body (see Eaton et al., 1991; Korn and Faber, 2005). In addition, these experimental advantages have made it possible to clarify the critical role of inhibition in determining the firing threshold and to relate membrane properties, dynamic features of synapses, and network connectivity to the C-start and its requirements.

ACOUSTIC STARTLE

The acoustic startle is mediated rapidly, with the latency from stimulus onset to the M spike being about 2 to 4 ms and the first detectable movement occurring ~8 ms later. However, it is not guaranteed to occur at even the highest sound intensities. When the fish is in the center of the aquarium, this response is appropriately aversive, that is, away from the stimulus, in 75% to 80% of all successful trials (Eaton et al., 1991). Thus, it is necessary to account for the speed of the response, its high behavioral threshold, its directionality, and additional observations that although both M cells are excited, only one is activated to threshold and it does not fire repetitively.

Sensory Circuitry: Inhibition Sets Threshold

As shown in Figure 33.1A, the M cell has a characteristic morphology, with a large soma (typically, 50 by 100 μm) and two primary dendrites. This reticulospinal neuron integrates information from a number of sensory systems. Its excitatory inputs are segregated to specific regions, such that the lateral dendrite receives monosynaptic statoacoustic information from the eighth nerve and the lateral line nerves, while the ventral one is activated by polysynaptic visual and somatosensory drives. In addition to the large axon, two features of the acoustic connections serve the requirement for speed: (1) all of the excitatory afferents have mixed electrical and chemical synapses, and the electrical component, mediated by gap junctions, underlies a minimal synaptic delay (while the chemical glutamatergic component prolongs and amplifies the postsynaptic response), and (2) the afferent fibers, especially those from the sacculus that terminate as myelinated club endings, are large and conduct rapidly.

   The organization of afferents to the M cell is highly specified, as individual components of the statoacousitic systems are segregated to defined loci on the lateral dendrite and its daughter branches (Fig. 33.1B; Szabo et al., 2007). One testable possibility is that there is a preferred spatiotemporal order of dendritic activation, with distal signals preceding the more proximal ones. If so, the saccular input, which is located relatively distal, would be active before inputs from the lateral line, and this combination would be most effective in bringing the cell to threshold. In fact, lateral line signals should be slightly delayed, on the basis of both conduction distance and fiber caliber.

   Each sensory drive is paralleled by a feedforward inhibitory input, relayed through a small population (~10 to 20 cells per side) of identified interneurons known as PHP cells. This acronym refers to the finding that these cells

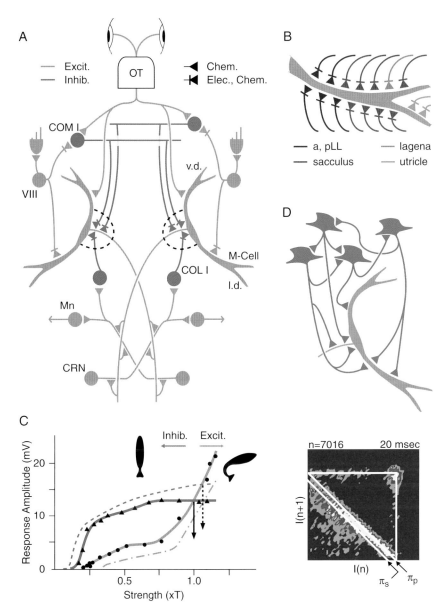

FIGURE 33–1. The Mauthner cell microcircuit and its functional organization. (*A*) Horizontal view of the supraspinal excitatory and inhibitory networks involving the Mauthner cell (M cell) and underlying the generation of the C-start of the teleost escape response. Excitatory cells are in red and inhibitory ones in blue, and their terminals mediate chemical transmission either alone (filled terminals) or in concert with electrical transmission (filled terminals with cross bar). Two sensory pathways are shown, that involving the posterior eighth nerve, which relays auditory information from hair cells of the inner ear to the lateral dendrite (l.d.) of the M cell, and the visual one, which connects the retina to the ventral dendrite (v.d.) via the optic tectum (OT). Not shown are projections of other components of the statoacoustic and somatosensory systems to the lateral and ventral dendrites, respectively. All sensory systems also activate a bilateral feedforward inhibition of the M cell, mediated by convergent input to commissural interneurons (COM Is).

FIGURE 33–1. Continued

Note that inhibitory terminals in the axon cap (dashed lines) mediate both electrical (field effect) and chemical inhibitions. The supraspinal output circuit includes a disynaptic excitation of cranial motoneurons (Mn) relayed via axo-axonic synapses between the M axon and cranial relay neurons (CRNs). The latter also excite recurrent collateral inhibitory interneurons (COL Is), which feedback onto the M cells. The pattern of connections in these output circuits is such that an action potential in one M cell leads to bilateral activation of cranial motorneurons and to collateral inhibition of both M cells (modified from Korn and Faber, 2005). (B) Topography of the terminations of eighth nerve and lateral line nerve afferents on the M cell lateral dendrite, based on electrophysiology and/or morphology. The terminations code input sources as follows: dark red, anterior and posterior lateral line; bright red, sacculus; orange, lagena; green, utricle. (C) Model of interplay between afferent excitation and inhibition and consequences for the escape threshold. Plots of the amplitude of monosynaptic excitation (red) and disynaptic inhibition (blue) versus the strength of stimulation of the posterior eighth nerve, expressed as a fraction of that required to bring the M cell to its firing threshold (1.0 T, solid vertical arrow). Three conditions are illustrated, namely, control (solid lines with data points, ●, O,), extrapolated from intra- and extracellular recordings, and after activity-dependent long-term potentiation of inhibition (- -) or long-term depression of excitation (—•—), both of which shift the behavioral threshold to higher intensities (dotted vertical arrow). Fish silhouettes indicate the intensity domains in which inhibition or excitation dominate, with the latter leading to C-start initiation (modified from Korn and Faber, 2005). (D) Above, schematic representation of connections between interneurons inhibitory to the M cell and underlying the concept that this neuronal population functions as weakly coupled oscillators and generates a nonrandom synaptic noise (from Korn and Faber, 2005). Below, Poincaré return map representing the relationship between the duration, I, of successive intervals (n, n+1) in inhibitory synaptic noise recorded from the M cell. The signal-flag pattern reveals principal (πp) and secondary (πs) frequencies, with intervals equal to 13.3 and 14.4 ms, respectively (from Faure et al., 2000).

can be identified electrophysiologically by the presence of a passive hyperpolarizing potential coincident with, and generated by, the M cell action potential, but with sign inversion. That is, an action potential in one cell hyperpolarizes the other, by a field effect or ephaptic interaction, due to the high resistance of the extracellular space surrounding the M cell's axon hillock and initial segment, which channels action currents across the membrane of neighboring neurons. Since the PHP cells are electronically coupled to the afferent terminations of the eighth nerve, activation of the disynaptic feedforward inhibitory pathway effectively opposes the monosynaptic excitation of the lateral dendrite. In addition, the inhibitory synapses are mixed because the PHP cell terminals are glycinergic.

We previously postulated that there is a fine balance between excitation and inhibition, such that threshold can only be reached after the latter is saturated (see Korn and Faber, 2005). This concept is illustrated in Figure 33.1C, which is based on a comparison of the input–output relation for early electrical inhibition, represented by extracellular fields, with that for the excitatory postsynaptic potentials (EPSPs) recorded in the M cell. More recently, the role of electrical inhibition in regulating M cell threshold has been confirmed by experiments where ineffective acoustic inputs became suprathreshold after this field effect was canceled by an applied opposing extracellular current (Weiss et al., 2008).

Furthermore, the dynamic properties of M cell afferent terminals complement the circuit design. In the case of excitation, the EPSPs exhibit paired pulse facilitation at frequencies typical of effective auditory tones. This feature assists in overcoming inhibition, which, in contrast, rapidly undergoes frequency-dependent depression. Two other features of inhibition maximize its strength. One is the lateral diffusion of transmitter, glycine, to neighboring synapses, which was first described in this system, where it leads to a synergistic action. The second is voltage dependence; inhibition is prolonged as the M cell is depolarized (see Korn et al., 1990).

M Cell Properties

The M axon conducts at ~100 m/s. The cell has a low-input resistance, ~200 KΩ, and an unusually short time constant, of the order of 400 μsec. These three features, combined with a high resting membrane potential which is at least 15 to 20 mV below threshold for spike initiation, guarantee that a powerful input is required to evoke the behavior, but that the processing and conduction time will be brief. For example, the EPSP due to activation of one club ending is ~100 μV at the soma, implying that more than 150 endings are active at threshold. Other important properties include *(1)* an inward rectification in the dendrite that is manifest close to threshold and channels excitation toward the axon hillock, the spike initiating site, and *(2)* a Dendrotoxin-1-sensitive K$^+$ channel that activates rapidly during a spike and effectively blocks repetitive activation of this neuron.

Motor Output Circuitry and Feedback Inhibition

At the supraspinal level, postsynaptic targets are the so-called cranial relay neurons (CRNs). This section of the circuitry guarantees that one M spike reliably results in a bilateral contraction of cranial musculature and that only the first action potential of a train would trigger the behavior. As diagrammed in Figure 33.1A, an M axon has axo-axonic synapses with bilaterally symmetrical sets of CRNs, thereby activating both the motor outputs and feedback inhibition of the two M cells. These junctions are nicotinic and have a high safety factor due to the high quantal content of transmitter released. Their strategic location on excitable axonal membrane gives the M cell priority over any other sensory drive. There is a similarly high safety factor for impulse transmission at the CRN output connections. In addition, these connections exhibit a marked frequency depression, such that a second C bend cannot occur for intervals less than 2 to 5 seconds.

Figure 33.1 illustrates that the CRNs activate a second class of PHP cells, the interneurons that electrically and chemically inhibit the M cell. These collateral interneurons guarantee that only one M cell fires and that it does not do so repetitively. Thus, there are redundant mechanisms that powerfully limit M cell activation.

Not shown in the figure are the output connections of the M axon in the spinal cord where each M axon excites motoneurons and segmental interneurons on just one side of the cord, while activating a crossed inhibition. This unilateral excitation underlies the C-start form of the behavior (see details in Chapter 36).

Directionality of the Acoustic Startle

How can the startle have a high probability of being appropriately aversive, at least with a fish in the open field, since the largest excitatory inputs to the two M cells are due to vibrations of the bilaterally symmetrical swim bladder? Eaton et al. (1991) proposed a major role for the PHP cells in this computation. Recently, Weiss et al. (2009) reported that both the monosynaptic excitation and feedforward inhibition of the M cell are locked to the phase of acoustic tones, independent of intensity. This phase locking could theoretically be the basis of logical operations involving comparisons of the pressure and particle motion components of an acoustic signal at the inhibitory or M cell levels. An underlying mechanism that favors the phase locking of feedforward inhibition, a necessary feature of any model for left–right discrimination, is that a population of PHP cells function as weakly coupled oscillators, biased toward frequencies close to those that most effectively evoke escapes (Fig. 33.1D; Faure et al., 2000). This bias presumably reflects subthreshold oscillations and firing patterns in the gamma range (Marti et al., 2008).

Dynamic Modulation of Escape Threshold and Direction

The probability of evoking a C-start, and its direction, are subject to modulation by multiple factors, including experience. Notably, both the excitatory and feedforward inhibitory inputs activated by the eighth nerve exhibit long-term potentiation (LTP), including the electronic EPSPs, with the underlying mechanisms being comparable to those found in cortical and limbic structures of mammals. The excitation also undergoes long-term depression (LTD). Figure 33.1C illustrates schematically how these activity-dependent changes could be expected to alter the balance of excitation and inhibition and hence the threshold for the C-start, as also documented in vivo (reviewed in Korn and Faber, 2005).

In addition, the synaptic bed of the M cell is subject to modulation by both serotonergic and dopaminergic afferents, which enhance inhibition and excitation, respectively. Another first for M cell studies was the demonstration of silent synapses, both excitatory (club endings of saccular afferents) and inhibitory (terminals of PHP cells), and the demonstration that these connections formed a latent reserve pool of afferents that can be recruited by learning and modify the gain of the sensory processing circuits.

Finally, another dynamic influence derives from the ongoing inhibitory synaptic noise that influences the M cell's input–output relations, as for other central neurons. The strength of this activity, generated by the firing of presynaptic interneurons, alternates between the two M cells (Hatta and Korn, 1999) and could bias the motor behavior, for example, in favor of responses away from a neighboring barrier, and overriding the directional cues that dominate in the open field. Furthermore, analysis of the inhibitory synaptic noise revealed that it is not stochastic but rather has a chaotic structure that favors rapid state transitions initiated by sudden stimuli (Faure et al., 2000).

Variability of Escape Trajectory

Eaton et al. (1991; see also Korn and Faber, 2005) demonstrated that the trajectory of the acoustic C-start is not stereotyped but is variable in its orientation, angular displacement, and distance moved. This uncertainty makes it difficult for a predator to adopt a successful strategy for prey capture. The C-start can be modeled by assuming two major driving forces: *(1)* an initial phase with its direction set by the activation of one M cell, and *(2)* a subsequent direction change set by the "brainstem escape network," which includes M cell analogs and other reticulospinal neurons. The final orientation and angular acceleration of the fish depends upon the location of the threatening stimulus, that is, on its approach angle. Thus, the M cell microcircuits only account for the initiation of the escape.

VISUAL EVOKED C-STARTS

Although it has been experimentally advantageous to dissect the salient features of the M cell system with nerve stimuli and single-cycle auditory tones, attacks by natural predators are clearly more complex. Thus, we have explored the visual pathway with looming stimuli that may simulate approach by a diving bird, with its shadow growing in size on the retina. Looming shadows effectively excite the M cell and trigger C-starts, with latencies of the order of a few 100 msec (Preuss et al., 2006). Comparison of the behavioral and electrophysiological results unraveled the computational transformation of the sensory stimulus into an exponentially scaled function of view angle, with part of the calculations performed in the retina and optic tectum. The scaling function was mediated by the feedforward inhibitory circuit that involves the PHP cells and has a critical role in limiting the duration of excitation in the network. That is, inhibition seems to implement a multiplicative function at the M cell level.

CONCLUSION

The escape behavior is the result of a complex decision-making process and has a more extensive survival function than appreciated previously. For example, the C-start can be viewed as a "fixed action pattern" that is apparently used for voluntary movements with similar motor requirements, such as predation or social aggression. And, the uncertainties introduced by the variability in its execution, and even its initial direction, presumably have significant survival value. This feature may be preserved evolutionarily, even though the startle response is bilateral in mammals and involves activation of a set of large reticulospinal neurons. The switch to such a distributed network may well be under genetic control, as suggested by studies of larval zebrafish with mutations in *the deadly seven/notch 1a (des)* gene which have supernumerary M cells (Liu et al., 2003). These M cells parse up the escape, which is behaviorally normal, revealing a developmental plasticity that might be pivotal in the evolutionary changes in the escape behavior and the networks that mediate it.

REFERENCES

Eaton RC, Di Domenico R, Nissanov J (1991) Role of the Mauthner cell in sensorimotor integration by the brain stem escape network. *Brain Behav Evol* 37:272–285.

Faure P, Kaplan D, Korn H (2000) Synaptic efficacy and the transmission of complex firing patterns between neurons. *J Neurophysiol* 84:3010–3025.

Hatta K, Korn H (1999) Tonic inhibition alternates in paired neurons that set direction of fish escape reaction. *Proc Natl Acad Sci USA* 96:12090–12095.

Korn H, Faber DS (2005) The Mauthner cell half a century later: a neurobiological model for decision-making? *Neuron* 47:13–28.

Korn H, Faber DS, Triller A (1990) Convergence of morphological physiological, and immunocytochemical techniques for the study of single Mauthner cells. In: Björkland A, Hökfelt T, Wouterlood FG, and van Den Pol AN, eds. *Handbook of Chemical Neuroanatomy, Vol. 8: Analysis of Neuronal Microcircuits and Synaptic Interactions*, pp. 403–480. Amsterdam: Elsevier Science Publishers.

Liu KS, Gray M, Otto SJ, Fetcho JR, Beattie CE (2003) Mutations in *deadly seven/notch1a* reveal developmental plasticity in the escape response circuit. *J Neurosci* 23:8159–8166.

Marti F, Korn H, Faure P (2008) Interplay between subthreshold potentials and gamma oscillations in Mauthner cells' presynaptic inhibitory interneurons. *Neuroscience* 151: 983–994.

Preuss T, Osei-Bonsu PE, Weiss SA, Wang C, Faber DS (2006) Neural representation of object approach in a decision-making motor circuit. *J. Neurosci* 26:3454–3464.

Szabo TM, McCormick CA, Faber, DS (2007) Otolith endorgan input to the Mauthner neuron in the goldfish. *J Comp Neurol* 505:511–525.

Weiss SA, Preuss T, Faber DS (2008) A role of electrical inhibition in sensorimotor integration. *Proc Natl Acad Sci USA* 105:18047–18052.

Weiss SA, Preuss T, Faber DS (2009) Phase encoding in the Mauthner system: implications in left-right sound source discrimination. *J Neurosci* 29:3431–3441.

34

Modulation of Lamprey Locomotor Circuit

A. El Manira

The spinal circuitry produces the basic motor pattern underlying locomotion and is also capable of acting as a processing interface to adjust the motor output to descending inputs from the brain and sensory feedback (Rossignol, 1996; Rossignol et al., 2006; Dubuc et al., 2008; Grillner et al., 2008; Jordan et al., 2008). The locomotor pattern is not only the result of a hardwired network of neurons with fast ionotropic synaptic transmission, but it is also the consequence of continuous modulation that regulates the strength of synaptic transmission and the firing of the network's constituent neurons.

In the lamprey, the detailed information available on the organization of the locomotor network has facilitated the assessment of the role of different modulatory systems (Buchanan, 1999; El Manira and Wallen, 2000; Buchanan, 2001; Alford et al., 2003; Grillner, 2003; El Manira et al., 2008; Grillner et al., 2008). Each hemisegment of the spinal cord contains a network of excitatory interneurons that drive the activity of motoneurons and generate rhythmic bursting, while inhibitory commissural interneurons are responsible for the left-right alternation (Buchanan, 1999; Grillner, 2003; Cangiano and Grillner, 2005). These components are responsible for the cycle-to-cycle operation, while a number of modulatory systems regulate the membrane properties of individual network neurons and/or the synaptic properties and thus contribute to generation of the locomotor activity.

This modulation primarily involves activation of G protein–coupled metabotropic receptors. It can be intrinsic, mediated by transmitters released from network neurons during ongoing locomotor activity such as glutamate activating the family of metabotropic glutamate receptors (mGluRs) (El Manira and Wallen, 2000; El Manira et al., 2002; Alford et al., 2003; El Manira et al., 2008). Extrinsic modulation is mediated by transmitters released from neurons that are not part of the network that activate, for instance, 5-HT, dopamine, GABA, or tachykinin receptors (Katz and Frost, 1996). These intrinsic and extrinsic modulatory systems play a fundamental role in the basic network function and in the induction of short-term (seconds or minutes) or long-term (hours or days) plasticity of the locomotor circuit activity. This review summarizes the influence of some of these modulatory systems on the global output of the network and the underlying cellular and synaptic mechanisms.

INTRINSIC GLUTAMATERGIC MODULATION: ROLE OF METABOTROPIC GLUTAMATE RECEPTORS

The glutamate released from excitatory interneurons not only provides the on cycle excitatory drive via activation of ionotropic AMPA and NMDA receptors, but it also activates mGluRs responsible for an ongoing modulatory tone during locomotor activity. All three mGluR groups (I, II, and III) exist in the lamprey spinal cord. The two subtypes of group I (mGluR1 and mGluR5) are located on postsynaptic soma and dendritic membrane, while group II and III mGluRs are located on presynaptic axon terminals where they act as auto-receptors to regulate transmitter release (Krieger et al., 1996).

A significant part of the baseline frequency is dependent on the activation of mGluR1. Blockade of this receptor type decreases the locomotor burst frequency (Krieger et al., 1998; El Manira et al., 2002; El Manira et al., 2008). Conversely, pharmacological activation of mGluR1 induces a short- and long-term increase in the locomotor frequency (Kyriakatos and El Manira, 2007).

The short-term increase in the locomotor frequency is mediated via inhibition of leak channels and potentiation of NMDA receptors that require activation of G proteins and phospholipase C (PLC) (Fig. 34.1) (Kettunen et al., 2003; Nanou et al., 2009). The mGluR1-mediated modulation of leak channels, but not that of NMDA receptors, involves Ca^{2+} release from internal store and protein kinase C (PKC) (Kettunen et al., 2003; Nanou et al., 2009). The fact that the different cellular mechanisms activated by mGluR1 can be separated based on the intracellular pathways allowed us to explore their individual contribution to the overall locomotor network activity. By interfering with the mGluR1-induced modulation of leak channels, we show that this cellular effect indeed plays a role in the mGluR1-induced short-term increase of the locomotor frequency. Thus, mGluR1 uses separate intracellular pathways to modulate leak channels and NMDA receptors that seem to act synergistically to induce the short-term increase in the locomotor frequency.

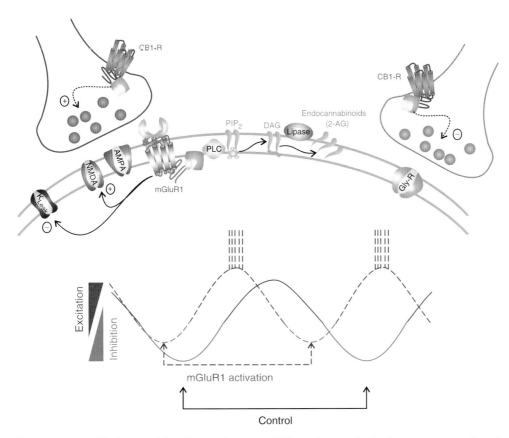

FIGURE **34–1.** Endocannabinoids release within the spinal locomotor network. Endocannabinoids are released during locomotion and set the baseline burst frequency. Their release is also triggered by activation of metabotropic glutamate receptor 1 (mGluR1). This receptor type activates G_q proteins and phospholipase C (PLC) that hydrolyzes phosphatidylinositol 4,5-bisphosphate (PIP2) to diacylglycerol (DAG) and inositol triphosphate (IP3). DAG is in turn hydrolyzed by DAG lipase to the endocannabinoid 2-arachidonoylglycerol (2-AG), which acts as a retrograde messenger and activates presynaptic cannabinoid 1 receptor (CB1-R). This leads to a depression of inhibitory synaptic transmission and potentiation of the excitatory synaptic drive. The shift in the balance between inhibition and excitation mediates the long-term increase in the locomotor frequency induced by mGluR1. The mGluR1-mediated short-term increase in the locomotor frequency is mediated by potentiation of NMDA receptors and inhibition of leak channels.

LONG-TERM PLASTICITY OF THE LOCOMOTOR NETWORK

For the mGluR1 modulation to have any significant physiological role, this receptor type needs to be activated by synaptically released glutamate. During an ongoing locomotor rhythm induced by NMDA, a brief tetanic stimulation of the descending reticulospinal axons results in a long-lasting potentiation of the locomotor frequency. Blockade of mGluR1 at the level of the spinal locomotor circuitry suppresses this synapticallyinduced long-term potentiation, sparing only a short-term increase in the frequency (Kyriakatos and El Manira, 2007). These results demonstrate that excitatory inputs from the brainstem recruit mGluR1 and induce long-term plasticity in the locomotor network activity. This long-term potentiation is mimicked by pharmacological activation of mGluR1 and is associated with a concomitant long-term depression of mid-cycle reciprocal inhibition and long-term potentiation of ipsilateral synaptic excitation arising from locomotor circuit interneurons (Fig. 34.1). The mGluR1-mediated long-term modulation is also dependent on release of endocannabinoids and nitric oxide (Kyriakatos and El Manira, 2007).

Endocannabinoids and Nitric Oxide Are Released within the Locomotor Network

Endocannabinoids and nitric oxide (NO) not only mediate the long-term potentiation of mGluR1, but they also act individually to set the level of activity of the locomotor network. Endocannabinoids are released from somata and dendrites of network neurons and act as retrograde messengers to activate presynaptic cannabinoid 1 receptor (CB1-R); they also modulate inhibitory and excitatory synaptic transmission and, as a result, increase the locomotor frequency (Kettunen et al., 2005; Kyriakatos and El Manira, 2007). Similarly, nitric oxide is also synthesized within the locomotor network and increases the burst frequency. Interfering with NO synthesis by blocking its synthesizing enzyme (NOS) decreases the locomotor burst frequency (Kyriakatos et al., 2008; Sillar et al., 2008). The NO modulation also induces a shift in the balance between excitation and inhibition in the spinal locomotor network (Kyriakatos et al., 2008; Sillar et al., 2008). These two modulatory systems seem to act in a serial manner because the effect NO can be occluded by blocking cannabinoid receptors. Thus, the modulation within the locomotor network does not only involve classical transmitters, but it is also dependent on retrograde signaling involving endocannabinoids and NO.

Endogenous Release of Tachykinins Regulates the Locomotor Frequency

In the lamprey spinal cord, cells with immunoreactivity for tachykinins (TKs), such as substance P, are located in the ventromedial plexus that colocalizes serotonin (5-HT) and dopamine (DA) (Van Dongen et al., 1985a; Van Dongen et al., 1985b; Schotland et al., 1995; Auclair et al., 2004). Tachykinins (TKs) are stored in large dense-core vesicles, which can also contain 5-HT (Pelletier et al., 1981; Van Dongen et al., 1985b). No synaptic specializations have been found in the ventral plexus (Christenson et al., 1990), and therefore TKs and their costored transmitters 5-HT and DA are thought to be released paracrinically from varicosities and act on surrounding dendrites of network neurons. When locomotor activity is induced by NMDA in the isolated spinal cord, blockade of TK receptors results in a decrease in the locomotor frequency, indicating that TKs are released endogenously and set the baseline activity of the locomotor network (Thörn Perez et al., 2007). Extrinsic application of substance P causes a modest long-term increase in the locomotor frequency (Thörn Perez et al., 2007) that was initially overestimated (Parker et al., 1998), since the time to reach a stable frequency in control was underestimated.

Substance P modulation of the locomotor frequency seems to depend on activation of NMDA receptors because no change in the frequency is induced when locomotor activity is induced by kainate in the presence of an NMDA antagonist (Parker et al., 1998). At the cellular level, the increase in the locomotor frequency is associated with a PKC-dependent potentiation of the response to NMDA, but not to AMPA (Parker et al., 1998). Substance P also modulates synaptic transmission from network interneurons in a manner consistent with its effect on the locomotor frequency. Excitatory synaptic transmission is enhanced, whereas reciprocal inhibitory synaptic transmission is depressed (Parker and Grillner, 1998). The modulation of inhibitory synaptic transmission is mediated by release of endocannabinoids that can act as a retrograde messenger to activate presynaptic CB1-R located on axons of inhibitory interneurons and thus decrease synaptic transmission. The activation of CB1-R by endocannabinoid release by substance P is necessary for the increase in the locomotor burst frequency (Thörn Perez et al., 2009). Thus, TKs act directly or via release of endocannabinoids on different cellular and synaptic targets and modulate the locomotor activity.

In many aspects the mechanisms underlying the TK modulation of the locomotor network are similar to those of mGluR1; both receptor types are coupled to G_q proteins and PLC. It is not yet known whether their effects are additive or whether their actions are converging on the same cellular and synaptic targets.

Monaminergic Modulation

In the lamprey spinal cord there is a dense ventromedial plexus that contains both 5-HT and DA, and in which network neurons project their dendritic arborizations (Zhang et al., 1996). Both 5-HT and DA are released during NMDA-induced locomotor activity. Increasing the concentration of 5-HT by inhibiting its reuptake mechanisms with citalopram decreases the locomotor burst frequency. This effect is mimicked by exogenous application of 5-HT or specific agonists for 5-HT_{1A} or $5\text{-HT}_{1B/1D}$ receptors (Harris-Warrick and

Cohen, 1985; Schotland et al., 1995; Zhang et al., 1996; Zhang and Grillner, 2000; Schwartz et al., 2005). Blockade of DA reuptake has a dual action, which consists of an initial increase in the locomotor frequency followed by a gradual decrease that seems to depend on the concentration of DA (Svensson et al., 2003). Indeed, application of a low DA concentration increases the locomotor frequency, while a higher concentration slows down the rhythm. Thus, the two colocalized monoamines in the lamprey spinal cord are released during activity of the locomotor rhythm and contribute to modulating the activity of the locomotor network. In the intact lamprey, injection of 5-HT or DA receptor agonists also affects the speed of locomotion in freely swimming animals in a manner similar to what is seen in the isolated spinal cord in vitro (Kemnitz et al., 1995).

5-HT mediates its effect on the locomotor network activity by regulating the intrinsic firing properties of the constituent neurons and their synaptic interactions. First, it affects the K_{Ca}-dependent slow afterhyperpolarization (sAHP) and the spike frequency adaptation of neurons, making them fire for a longer period than in control conditions and thereby delaying the burst termination and as consequence the burst frequency. The decrease in the sAHP is the result of a combined inhibition of K_{Ca} channels as well as high voltage–activated Ca^{2+} channels (Wallen et al., 1989; Hill et al., 2003). Finally, 5-HT also acts presynaptically to modulate inhibitory and excitatory synaptic transmission (Takahashi et al., 2001; Biro et al., 2006; Schwartz et al., 2007). Thus, 5-HT combines pre- and postsynaptic actions to regulate the activity of the locomotor network.

PRESYNAPTIC MODULATION AS AN INTEGRATIVE MECHANISM OF CIRCUIT OPERATION

The locomotor circuitry is not only a "passive" entity that is under constant modulatory influences, but it is also continuously providing phasic presynaptic modulation of synaptic transmission from network interneurons and sensory afferents (Fig. 34.2). In the lamprey, synaptic transmission from cutaneous sensory neurons (dorsal cells) is phasically modulated during locomotion (El Manira et al., 1997). This phase-dependent presynaptic inhibition allows the locomotor network to filter out activity of interneurons and sensory afferents that would interfere with the ongoing activity of the locomotor network. Presynaptic inhibition of sensory afferents has been shown to exist in many locomotor networks in vertebrates and invertebrates (Clarac and Cattaert, 1996; El Manira et al., 1997; Buschges and Manira, 1998; Rossignol et al., 2006; Rudomin, 2009). The existence of presynaptic inhibition in network interneurons has been infrequently demonstrated because of the difficulty in recording from presynaptic axons of identified network interneurons. In the lamprey, intra-axonal recordings from inhibitory and excitatory

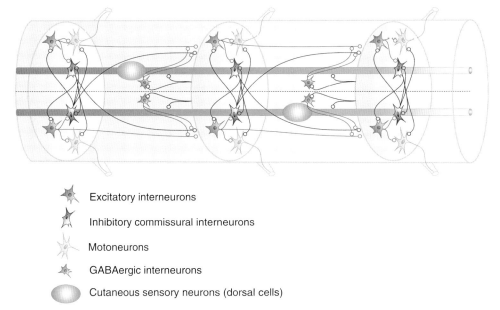

Excitatory interneurons

Inhibitory commissural interneurons

Motoneurons

GABAergic interneurons

Cutaneous sensory neurons (dorsal cells)

FIGURE 34–2. Locomotor-related presynaptic inhibition of synaptic transmission from interneurons and sensory afferents. Locomotor activity is generated by segmental networks consisting of ipsilateral glutamatergic interneurons (red) driving the activity of motoneurons (green), which project their axons through ventral roots. The left-right alternation is ensured by commissural glycinergic inhibitory interneurons (blue). The activity is propagated from rostral to caudal segments with a constant phase-lag to produce the waveform underlying swimming in the intact animal. Both excitatory and inhibitory interneurons project to caudal segments as well as rostral segments (not depicted). Cutaneous sensory feedback is conveyed to the spinal locomotor network by intraspinal sensory neurons called dorsal cells that project their axons in the dorsal column over several segments. During locomotion, the axons of excitatory and inhibitory interneurons as well as sensory dorsal cells receive phasic membrane potential depolarizations arising from the active spinal network and mediated presynaptic inhibition transmitter release from these neurons. The presynaptic inhibition in interneurons is mediated via GABAergic interneurons that activate $GABA_A$ and $GABA_B$ receptors on presynaptic terminals and depress transmitter release. Thus, the locomotor network can locally modulate synaptic transmission from interneurons and sensory afferents to filter out inputs that may interfere with the ongoing activity.

network interneurons revealed a phasic depolarization of their membrane that is correlated with the locomotor activity (Alford et al., 1991). These presynaptic depolarizations are mediated by activation of $GABA_A$ and $GABA_B$ receptors and mediate a depression of synaptic transmission from these axons (Alford et al., 1991; Alford and Grillner, 1991). By recruiting GABAergic interneurons, which mediate presynaptic inhibition, the locomotor network can locally set the strength of synaptic transmission in order to optimize the execution of the ongoing locomotor movement.

Concluding Remarks

It is becoming clear that the activity of networks underlying motor behavior is not only the result of fast synaptic transmission mediated by ionotropic receptors. Different G protein receptors also contribute to the generation of the network's activity. This review highlights how some modulatory systems act individually and in combination to determine the baseline activity level of the spinal locomotor network. These mechanisms are not only used for normal function of the locomotor circuitry but could also be used to change the excitability in the injured spinal cord and thus help recover motor function.

Acknowledgments

I would like to thank Drs. J. Ausborn, R. Hill, and E. Nanou for comments on the manuscript.

References

Alford S, Grillner S (1991) The involvement of GABAB receptors and coupled G-proteins in spinal GABAergic presynaptic inhibition. *J Neurosci* 11:3718–3726.

Alford S, Christenson J, Grillner S (1991) Presynaptic GABAA and GABAB Receptor-mediated phasic modulation in axons of spinal motor interneurons. *Eur J Neurosci* 3:107–117.

Alford S, Schwartz E, Viana di Prisco G (2003) The pharmacology of vertebrate spinal central pattern generators. *Neuroscientist* 9:217–228.

Auclair F, Lund JP, Dubuc R (2004) Immunohistochemical distribution of tachykinins in the CNS of the lamprey Petromyzon marinus. *J Comp Neurol* 479:328–346.

Biro Z, Hill RH, Grillner S (2006) 5-HT Modulation of identified segmental premotor interneurons in the lamprey spinal cord. *J Neurophysiol* 96:931–935.

Buchanan JT (1999) The roles of spinal interneurons and motoneurons in the lamprey locomotor network. *Prog Brain Res* 123:311–321.

Buchanan JT (2001) Contributions of identifiable neurons and neuron classes to lamprey vertebrate neurobiology. *Prog Neurobiol* 63:441–466.

Buschges A, Manira AE (1998) Sensory pathways and their modulation in the control of locomotion. *Curr Opin Neurobiol* 8:733–739.

Cangiano L, Grillner S (2005) Mechanisms of rhythm generation in a spinal locomotor network deprived of crossed connections: the lamprey hemicord. *J Neurosci* 25:923–935.

Christenson J, Cullheim S, Grillner S, Hokfelt T (1990) 5-hydroxytryptamine immuno-reactive varicosities in the lamprey spinal cord have no synaptic specializations—an ultrastructural study. *Brain Res* 512:201–209.

Clarac F, Cattaert D (1996) Invertebrate presynaptic inhibition and motor control. *Exp Brain Res* 112:163–180.

Dubuc R, Brocard F, Antri M, Fenelon K, Gariepy JF, Smetana R, Menard A, Le Ray D, Viana Di Prisco G, Pearlstein E, Sirota MG, Derjean D, St-Pierre M, Zielinski B, Auclair F, Veilleux D (2008) Initiation of locomotion in lampreys. *Brain Res Rev* 57:172–182.

El Manira A, Wallen P (2000) Mechanisms of modulation of a neural network. *News Physiol Sci* 15:186–191.

El Manira A, Tegner J, Grillner S (1997) Locomotor-related presynaptic modulation of primary afferents in the lamprey. *Eur J Neurosci* 9:696–705.

El Manira A, Kettunen P, Hess D, Krieger P (2002) Metabotropic glutamate receptors provide intrinsic modulation of the lamprey locomotor network. *Brain Res Rev* 40:9–18.

El Manira A, Kyriakatos A, Nanou E, Mahmood R (2008) Endocannabinoid signaling in the spinal locomotor circuitry. *Brain Res Rev* 57:29–36.

Grillner S (2003) The motor infrastructure: from ion channels to neuronal networks. *Nat Rev* Neurosci 4:573–586.

Grillner S, Wallen P, Saitoh K, Kozlov A, Robertson B (2008) Neural bases of goal-directed locomotion in vertebrates—an overview. *Brain Res Rev* 57:2–12.

Harris-Warrick RM, Cohen AH (1985) Serotonin modulates the central pattern generator for locomotion in the isolated lamprey spinal cord. *J Exp Biol* 116:27–46.

Hill RH, Svensson E, Dewael Y, Grillner S (2003) 5-HT inhibits N-type but not L-type Ca(2+) channels via 5-HT1A receptors in lamprey spinal neurons. *Eur J Neurosci* 18:2919–2924.

Jordan LM, Liu J, Hedlund PB, Akay T, Pearson KG (2008) Descending command systems for the initiation of locomotion in mammals. *Brain Res Rev* 57:183–191.

Katz PS, Frost WN (1996) Intrinsic neuromodulation: altering neuronal circuits from within. *Trends Neurosci* 19:54–61.

Kemnitz CP, Strauss TR, Hosford DM, Buchanan JT (1995) Modulation of swimming in the lamprey, Petromyzon marinus, by serotonergic and dopaminergic drugs. *Neurosci Lett* 201:115–118.

Kettunen P, Hess D, El Manira A (2003) mGluR1, but not mGluR5, mediates depolarization of spinal cord neurons by blocking a leak current. *J Neurophysiol* 90:2341–2348.

Kettunen P, Kyriakatos A, Hallen K, El Manira A (2005) Neuromodulation via conditional release of endocannabinoids in the spinal locomotor network. *Neuron* 45:95–104.

Krieger P, El Manira A, Grillner S (1996) Activation of pharmacologically distinct metabotropic glutamate receptors depresses reticulospinal-evoked monosynaptic EPSPs in the lamprey spinal cord. *J Neurophysiol* 76:3834–3841.

Krieger P, Grillner S, El Manira A (1998) Endogenous activation of metabotropic glutamate receptors contributes to burst frequency regulation in the lamprey locomotor network. *Eur J Neurosci* 10:3333–3342.

Kyriakatos A, El Manira A (2007) Long-term plasticity of the spinal locomotor circuitry mediated by endocannabinoid and nitric oxide signaling. *J Neurosci* 27:12664–12674.

Kyriakatos A, Molinari M, Grillner S, Sillar KT, El Manira A (2008) Source and modulatory actions of nitric oxide in the lamprey spinal locomotor network. *Soc Neurosci Abst* 575.5.

Nanou E, Kyriakatos A, Kettunen P, El Manira A (2009) Separate signalling mechanisms underlie mGluR1 modulation of leak channels and NMDA receptors in the network underlying locomotion. *J Physiol* 587:3001–3008.

Parker D, Grillner S (1998) Cellular and synaptic modulation underlying substance P-mediated plasticity of the lamprey locomotor network. *J Neurosci* 18:8095–8110.

Parker D, Zhang W, Grillner S (1998) Substance P modulates NMDA responses and causes long-term protein synthesis-dependent modulation of the lamprey locomotor network. *J Neurosci* 18:4800–4813.

Pelletier G, Steinbusch HW, Verhofstad AA (1981) Immunoreactive substance P and serotonin present in the same dense-core vesicles. *Nature* 293:71–72.

Rossignol S (1996) Visuomotor regulation of locomotion. *Can J Physiol Pharmacol* 74:418–425.

Rossignol S, Dubuc R, Gossard JP (2006) Dynamic sensorimotor interactions in locomotion. *Physiol Rev* 86:89–154.

Rudomin P (2009) In search of lost presynaptic inhibition. *Exp Brain Res* 196:139–151.

Schotland J, Shupliakov O, Wikstrom M, Brodin L, Srinivasan M, You ZB, Herrera-Marschitz M, Zhang W, Hokfelt T, Grillner S (1995) Control of lamprey locomotor neurons by colocalized monoamine transmitters. *Nature* 374:266–268.

Schwartz EJ, Gerachshenko T, Alford S (2005) 5-HT prolongs ventral root bursting via presynaptic inhibition of synaptic activity during fictive locomotion in lamprey. *J Neurophysiol* 93:980–988.

Schwartz EJ, Blackmer T, Gerachshenko T, Alford S (2007) Presynaptic G-protein-coupled receptors regulate synaptic cleft glutamate via transient vesicle fusion. *J Neurosci* 27:5857–5868.

Sillar KT, Kyriakatos A, Molinari M, Grillner S, El Manira A (2008) Nitric oxide potentiation of lamprey locomotor network activity. *FENS Abst* 4:021–16.

Svensson E, Woolley J, Wikstrom M, Grillner S (2003) Endogenous dopaminergic modulation of the lamprey spinal locomotor network. *Brain Res* 970:1–8.

Takahashi M, Freed R, Blackmer T, Alford S (2001) Calcium influx-independent depression of transmitter release by 5-HT at lamprey spinal cord synapses. *J Physiol* 532: 323–336.

Thörn Perez C, Hill RH, Grillner S (2007) Endogenous tachykinin release contributes to the locomotor activity in lamprey. *J Neurophysiol* 97:3331–3339.

Thörn Pérez C, Hill RH, El Manira A, Grillner S (2009) Endocannabinoids mediate tachykinin induced effects in the lamprey locomotor network. *J Neurophysiol* 102:1358–1365.

Van Dongen PA, Hokfelt T, Grillner S, Verhofstad AA, Steinbusch HW (1985a) Possible target neurons of 5-hydroxytryptamine fibers in the lamprey spinal cord: immunohistochemistry combined with intracellular staining with Lucifer yellow. *J Comp Neurol* 234:523–535.

Van Dongen PA, Hokfelt T, Grillner S, Verhofstad AA, Steinbusch HW, Cuello AC, Terenius L (1985b) Immunohistochemical demonstration of some putative neurotransmitters in the lamprey spinal cord and spinal ganglia: 5-hydroxytryptamine-, tachykinin-, and neuropeptide-Y-immunoreactive neurons and fibers. *J Comp Neurol* 234:501–522.

Wallen P, Buchanan JT, Grillner S, Hill RH, Christenson J, Hokfelt T (1989) Effects of 5-hydroxytryptamine on the afterhyperpolarization, spike frequency regulation, and oscillatory membrane properties in lamprey spinal cord neurons. *J Neurophysiol* 61:759–768.

Zhang W, Grillner S (2000) The spinal 5-HT system contributes to the generation of fictive locomotion in lamprey. *Brain Res* 879:188–192.

Zhang W, Pombal MA, El Manira A, Grillner S (1996) Rostrocaudal distribution of 5-HT innervation in the lamprey spinal cord and differential effects of 5-HT on fictive locomotion. *J Comp Neurol* 374:278–290.

35

Tadpole Swimming Network

Keith T. Sillar and Wenchang Li

Xenopus laevis frog tadpoles near the time of hatching have proven to be an excellent model system in which to explore the neural mechanisms responsible for the initiation, maintenance, sensory adaptation, and termination of rhythmic locomotor activity in vertebrates. The underlying neural network (Fig. 35.1) is one of the most completely understood in any vertebrate. Detailed knowledge has accrued over a period of nearly 40 years, spanning most of the career of the founding father and major contributor of research in the area, Professor Alan Roberts at the University of Bristol, England. This information has not only highlighted operational features of rhythm generators that have been shown to be conserved in more complex vertebrates, it has also served as an invaluable platform from which to investigate associated issues that are of fundamental importance in neuroscience such as motor programme switching, transmitter corelease, network development, neuromodulation, and metamodulation of network operation. There are many advantages of this simple model system, including the presence of a well-defined network output (reviewed in Roberts et al., 1998) that relates directly to the behavior of the animal under study, namely swimming locomotion.

Network Activation

Network organization and swimming mechanisms are best understood in stage 37/38 *Xenopus* tadpoles. These newly hatched animals spend much of their time motionless, hanging by a thread of mucus secreted from the rostral cement gland. Cement gland afferents, with cell bodies in the trigeminal ganglion (TG), excite mid-hindbrain reticulospinal (MHR; Fig. 35.1) GABA neurons, which provide descending inhibition to the spinal cord via $GABA_A$

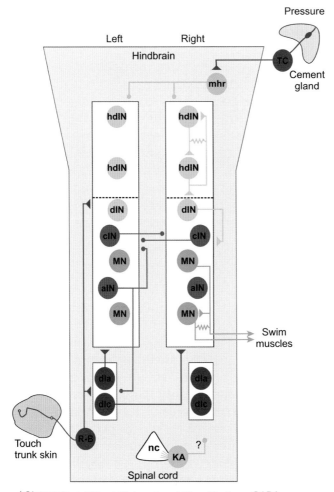

◀ Glutamate   ◁ ACh   ◁ Glutamate + ACh   • Glycine   ∘ GABA

Figure 35–1. Tadpole swimming circuit. Tadpole skin sensory pathway: free peripheral nerve endings of R-B sense mechanical stimuli in the skin; sensory impulses then propagate along R-B central axons in the spinal cord, directly exciting sensory interneurons strongly and central pattern generator (CPG) neurons weakly via glutamatergic synapses; the sensory information is amplified and relayed to CPG neurons by dorsolateral ascending interneurons (dlas) ipsilaterally and dorsolateral commissural interneurons (dlcs) contralaterally. Swimming CPG comprises four types of neurons: Some MNs have short descending central axons that form cholinergic/electrical synapses onto other synergic MNs; cINs produce glycinergic reciprocal inhibition suppressing the activity of neurons on the opposite side; aINs feedback glycinergic inhibition onto ipsilateral CPG neurons to limit multiple firing and also their corollary discharges produce sensory-gating inhibition in the sensory interneurons; dINs use both glutamate and ACh at their synapses and they are the neurons producing phasic excitation that drives all CPG neuron firing during swimming; the rostral populations of dINs in the hindbrain and rostral spinal cord (hdINs) form feedback excitation loops among themselves and are critical for swimming maintenance. Dotted lines are drawn to divide hdINs and dINs, but they seem to form one continuum of neurons with similar properties. Swimming stopping pathway: depression in cement gland activates trigeminal ganglion cells (TC); TCs then excite GABA-ergic mhrs, which inhibit CPG neurons on both sides of the spinal cord. KAs have GABA immunoreactivity, but they are not active in fictive swimming. nc, Neural canal; KA, Kolmer-Agdhur cell.

receptors (Roberts et al., 1998). When stimulated, however, they respond by performing a series of rhythmic swimming movements produced by alternating contractions of the left and right myotomal muscles which propagate rostrocaudally (reviewed in Roberts et al., 1998). They will swim until the rostral cement gland contacts an object in the environment at which point the ensuing release of GABA terminates swimming. There are three main sensory pathways that can activate swimming: *(1) skin cell pathway*, in which the skin is electrically excitable, generating a cardiac-like action potential when touched that propagates across the entire skin surface, gains access to the CNS via the trigeminal nerve, and through unknown mechanisms triggers swimming; *(2) Rohon-Beard pathway*, in which R-B neurons are SP immunoreactive mechanosensory primary afferents with free nerve endings in the skin that discharge when their receptive field is stimulated and contact spinal dorsolateral sensory interneurons, which amplify the input and trigger swimming (Li et al., 2003); and *(3) pineal gland*, in which changes in illumination can be detected by pineal afferents, which activate a descending pathway and trigger swimming via excitatory output in the hindbrain and spinal cord (Roberts et al., 1998).

Network Architecture

Aside from the sensory pathway interneurons, which are part of the R-B pathway (Li et al., 2003), rhythmic swimming is produced by four main classes of spinal interneuron with a prominent role also for excitatory interneurons located in the caudal hindbrain (Fig. 35.1). *(1)* Myotomal motoneurons (MNs) form a continuous row along the spinal cord and exit via ventral roots to excite segmented muscles by activating nicotinic ACh receptors. They make central connections with homonymous MNs via cholinergic and electrical connections (Roberts et al., 1998). *(2)* Commissural interneurons (cINs) are glycinergic with crossing axons, which contact contralateral network neurons and mediate reciprocal inhibition during swimming. *(3)* Descending interneurons (dINs) are glutamatergic but also release ACh (Li et al., 2004) to excite ipsilateral network neurons via descending axons. *(4)* Ascending interneurons (aINs) contain GABA but predominantly release glycine. They function as a sensory gate to integrate skin sensory pathway inflow during swimming in a phase-dependent manner. Hindbrain dINs (hdINs) are similar to dINs but are located more rostrally and play a key role in generating and maintaining swimming (Li et al., 2006). About half of hdINs also possess ascending axons. Functionally, hdINs and dINs are similar and may form a continuum of excitatory interneurons (Fig. 35.1) whose density is tapered from the hindbrain to more caudal regions of the spinal cord. Kolmer-Aghdur (KA) cells have cilia projecting into the neural canal (nc) and are GABA immunoreactive (Roberts et al., 1998).

NETWORK OPERATION

During fictive swimming in α-bungarotoxin immobilized tadpoles, MNs generally fire only one spike per cycle. The rhythm frequency ranges approximately between 10 and 20 Hz with the upper frequency normally occurring at the start of an episode which then slowly decelerates before stopping. All network neurons on the same side fire synchronously, and their activity alternates between left and right sides. MNs receive mid-cycle, strychnine-sensitive inhibitory postsynaptic potentials (IPSPs), which presumably help ensure alternation. However, rhythmic activity persists in strychnine, so inhibition appears not to be critical for rhythm generation. The strength of the IPSP may be an important regulator of swimming frequency. Myotomal motoneurons receive excitation during swimming (Roberts et al., 2008) from three sources: *(1)* glutamate and ACh released from dINs and hdINs, which produces fast rise, slow fall excitatory postsynaptic potentials (EPSPs) via coactivation of NMDA, AMPA, and nicotinic ACh receptors. The activation of NMDA receptors generates long EPSPs whose duration exceeds the cycle period and so can sum from one cycle to the next to produce tonic excitation; *(2)* a cholinergic input from other MNs; and *(3)* electrical coupling via contacts made from other MNs. The summation of these three types of EPSP ensures that most MNs fire on most cycles of an episode.

Once initiated, swim episodes can continue for several minutes and several thousand cycles. The eventual termination may depend upon the gradual accumulation of inhibitory adenosine (Dale and Gilday, 1996) and the gradual dropout of interneuron firing as the episode progresses. The spinal cord possesses a limited capacity to generate self-sustaining activity (although continuous activity can be produced by application of exogenous excitants like NMDA). Episode durations diminish with progressively more caudal spinalization, indicating that normally the rhythm is maintained by neurons located in the rostral spinal cord and caudal hindbrain where positive feedback among excitatory interneurons is important (Li et al., 2006).

NETWORK MATURATION

The single-spike-per-cycle pattern recorded at the hatching stage (37/38) is replaced during early larval development by an intrinsically more flexible swimming pattern in stage 42 larvae in which longer ventral bursts occur and MNs discharge variably and multiply in each cycle (reviewed in Sillar et al., 1998). The basic coordination of swimming does not change (left-right alternation, rostrocaudal delay), but cycle-by-cycle changes in swim frequency and intensity are now common and the rostrocaudal delay scales with swim frequency, unlike at hatching, but in common with swimming in other vertebrates like the lamprey. The acquisition of ventral root bursts follows a rostrocaudal path with bursts appearing first in rostral segments of stage 40 larvae, then more caudally by stage 42, consistent with a descending influence on the network during the first day or two of larval development. Evidence supports the conclusion that the in-growth of serotonergic projections emanating from the raphe nucleus is causal to the maturation of swimming (McLean et al., 2001).

There is no evidence for the appearance of new spinal neuron types over the same period of development, but the properties of the constituent neurons change. As well as MNs firing multiply during swimming (perhaps due in part to a reduction of electrical coupling between MNs), they respond to activation of NMDA receptors by producing slow (ca. 0.5 Hz) intrinsic oscillations in membrane potential contingent upon coactivation of 5-HT receptors. These oscillations affect network output by triggering increases in swim frequency and intensity over many consecutive cycles (Sillar et al., 1998; McLean et al., 2001).

NETWORK NEUROMODULATION

Tadpole swimming activity is sensitive to neuromodulation from a variety of intrinsic and extrinsic sources (Fig. 35.2A). *Intrinsic* modulation involving purine transmission has been described (Dale and Gilday, 1996); ATP released during swimming causes initial excitation by blocking K^+ channels but with time is broken down to adenosine by ectonucleotidases, which block Ca^{2+} channels to eventually terminate swimming. *Extrinsic* modulation (Fig. 35.2A) provides a suite of mechanisms for controlling swim frequency and intensity. By stage 42 the network can be differentially modulated by aminergic systems extrinsic to the spinal cord, which innervate the network over the late-embryonic to early larval period (McDearmid et al., 1997). 5-HT and noradrenaline (NA) produce fast intense and slow weak forms of network output, respectively. This rheostatic control of a single anatomically defined network relies upon the amines' effects on reciprocal glycinergic inhibition. 5-HT decreases and NA increases the amplitude of mid-cycle IPSPs via a direct effect on the release machinery (McDearmid et al., 1997).

Again by stage 42, particular nuclei of the brainstem can synthesize nitric oxide (NO) as a cotransmitter that inhibits swimming by enhancement of both GABA transmission to shorten swim episodes and glycine transmission to slow the rhythm frequency. The effects on GABA transmission are direct and likely occur in the brainstem by NO potentiation of release from MHR neurons, which are retained at larval stages (McLean et al., 2001). However the effects of NO on glycinergic synapses are indirect; NO effects are occluded by NA inhibition (but not vice versa), providing evidence that NO effects involve *metamodulation* (reviewed in Sillar et al., 2007).

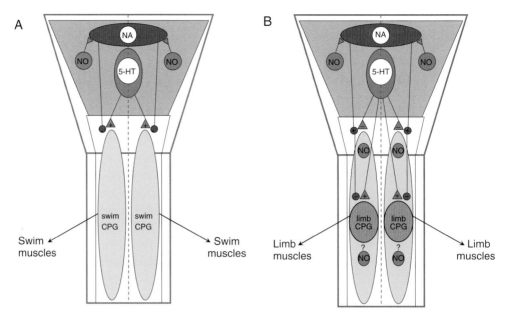

FIGURE 35-2. Development and modulation of tadpole network. (A) By stage 42 the axial swimming network is modulated by both intrinsic spinal influences and extrinsic brainstem modulatory pathways. 5-HT leads to fast, intense swimming, but NA produces the opposite effect leading to slower, weaker swimming. NO is produced by three neuron clusters in the brainstem (only one symbolized). One of NO's main actions is to act as a metamodulator by enhancing the effects of NA, presumably in the brainstem; for additional effects, see text. (B) At prometamorphic stage 61, the limb extensor/flexor system has appeared; the limb nerves are capable of independent bursting or participating in a conjoint rhythm with the axial system. This flexible coupling is modulated by 5-HT and NA but effects are, in effect, the reverse of earlier stages (see text for details). NO is now expressed by spinal neuron clusters but the function of NO during metamorphosis is unknown.

NETWORK PLASTICITY DURING METAMORPHOSIS

Little is known about the changes that occur in the anatomy of the rhythm generator for axial-based swimming after stage 37/38, but it seems likely that the basic network, like its function, is similar throughout the tadpole phase. However, a new network to control the limbs is added during the metamorphic period. Initially, around stage 56 limb MNs are coactive with the axial network with the latter system acting as a slave oscillator—limb extensor and flexor MNs on the same side are coactive and burst in alternation with their contralateral homologs (reviewed in Sillar et al., 2007). By stage 61 the limbs are more developed and now limb MNs adopt a different coordination suitable to drive the kicking movements that propel the froglet after metamorphosis—extensor and flexor bursts alternate within each side, but homologous

MNs are coactive across the spinal cord. Coupling between the limb and axial networks is flexible with both systems contributing to thrust. At these pro-metamorphic stages the limb and axial rhythms can be coexpressed either in a single coordinated output or as two rhythms with independent cadences (Rauscent et al., 2009). At this stage there has been an important switch in the influence of aminergic systems such that 5-HT couples independent limb and axial rhythms into a conjoint pattern by decelerating the axial rhythm and accelerating the limb rhythm, in effect the opposite influence to earlier stages (Fig. 35.2A). Noradrenaline, in contrast, uncouples the conjoint rhythm by exerting an opposing effect of slowing the limb and speeding up the axial rhythm (Fig. 35.2B). Nitric oxide–generating neurons, restricted to the brainstem at earlier stages, are now present in the spinal cord (Sillar et al., 2007).

REFERENCES

Dale N, Gilday D (1996) Regulation of rhythmic movements by purinergic neurotransmitters in frog embryos. *Nature* 383:259–263.

Li WC, Soffe SR, Roberts A (2003) The spinal interneurons and properties of glutamatergic synapses in a primitive vertebrate cutaneous flexion reflex. *J Neurosci* 23:9068–9077.

Li WC, Soffe SR, Roberts A (2004) Glutamate and acetylcholine corelease at developing synapses. *PNAS* 101:15488–15493.

Li WC, Soffe SR, Wolf E, Roberts A (2006) Persistent responses to brief stimuli: feedback excitation among brainstem neurons. *J Neurosci* 26:4026–4035.

McDearmid J, Scrymgeour-Wedderburn JFS, Sillar KT (1997) Aminergic modulation of glycine release in a spinal network controlling swimming. *J Physiol* 503(1):1473–1482.

McLean DL, Merrywest SM, Sillar KT (2001) The development of neuromodulatory systems and the maturation of motor patterns in amphibian tadpoles. *Brain Res Bull* 53:595–603.

Rauscent A, Einum J, Le Ray D, Simmers J, Combes D (2009) Opposing aminergic modulation of distinct spinal locomotor networks and their functional coupling during amphibian metamorphosis. *J Neurosci* 28:1163–1174.

Roberts A, Soffe SR, Wolf ES, Yoshida M, Zhao FY (1998) Central circuits controlling locomotion in young frog tadpoles. *Ann NY Acad Sci* 860:19–34.

Roberts A, Li WC, Soffe SR, Wolf E (2008) Origin of excitatory drive to a spinal locomotor network. *Brain Res Rev* 57:22–28.

Sillar KT, Reith CA, McDearmid J (1998) The development and aminergic modulation of a spinal network controlling swimming in *Xenopus* larvae. *Ann NY Acad Sci* 860:318–332.

Sillar KT, Combes D, Ramanathan S, Molinari M, Simmers AJ (2007) Neuromodulation and developmental plasticity in the locomotor system of anuran amphibians during metamorphosis. *Brain Res Rev* 57:94–102.

36

Spinal Circuit for Escape in Goldfish and Zebrafish

Joseph R. Fetcho

Escape or startle responses are vital to organisms: they can mean the difference between life and death when attacked by a predator. In fishes, the escape behavior is a rapid bend of the body and tail away from a potential threat. The response occurs within milliseconds after a stimulus and lasts only tens of milliseconds (Eaton et al., 1991). When properly executed, it is a fast, powerful body bend to only one side that takes precedence over any other ongoing movements of the body and tail. The behavior is initiated by the firing of one of a bilateral pair of hindbrain reticulospinal neurons called the Mauthner cells. These neurons receive many sensory inputs, and the output of each cell occurs via an axon that crosses in the brain and extends the length of the spinal cord on the opposite side of the body. The activation of the Mauthner cell on one side by a strong sensory input thus triggers an escape bend to the opposite side, turning the fish away from a threat. The circuit of the Mauthner cell in spinal cord based upon studies of goldfish and zebrafish is shown in Figure 36.1 (Fetcho and Faber, 1988; Faber et al., 1989; Fetcho 1991). This circuit, repeated all along spinal cord, has several features that are well matched to the behavioral demands of the movements of the escape.

Mauthner Axonal Size

The Mauthner axon is the largest axon in the fish, as are the axons of escape neurons in other species. This has historically been attributed to minimizing the latency of the motor response by activating muscle as quickly as possible. The additional reduction in latency afforded by the size (about 0.5 msec) is, however, small relative to the overall duration of the behavior, raising doubts about this explanation. Another, not mutually exclusive, possibility is that

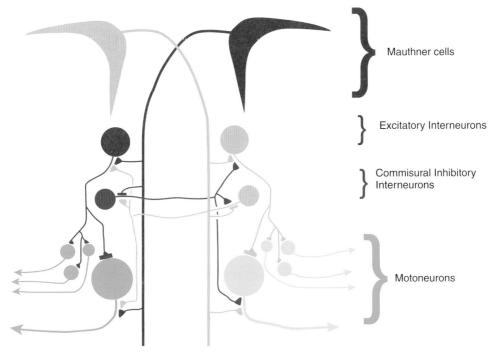

FIGURE 36–1. The circuit of the Mauthner cell in spinal cord based upon studies of goldfish and zebrafish.

the significance of the axonal size is related to beating other axons to the spinal cord to activate inhibitory networks that block those other neurons from producing conflicting outputs (Eaton et al., 1995). Vetoing other behaviors to establish precedence could be essential in such a do-or-die response.

DIRECT EXCITATION OF MOTONEURONS

The speed and forcefulness of the escape are accomplished by a massive activation of axial muscle. This activation is produced, in part, by two components of the spinal escape network. An initial massive spike of muscle activity is produced by a monosynaptic, glutamatergic excitatory connection between the Mauthner axon and all of one class of axial motoneurons called primary motoneurons (Fetcho and Faber, 1988). These motoneurons are the largest of the motoneurons and, as a group, they innervate every fast, large muscle fiber in the body of the fish. The bulk of the axial muscle fibers in goldfish and zebra fish are fast and the result is as massive and synchronous a force production as is likely possible for the fish. This serves two roles. First, it overcomes the inertia of the water to get the body moving. In addition, an escape response must sometimes occur in the midst of swimming, when muscle fibers on the side opposite the escape bend might have just been activated. Because these fibers relax slowly relative to the timing of the escape response, the massive pulse of muscle activation by the direct connections of the Mauthner cell with motoneurons might serve to simply overwhelm any residual muscle force on the opposite side. Following the initial pulse of activation of the muscle is a longer latency activation of motoneurons that sustains the muscle activation to varying extents depending on the extent of the escape turn. This is mediated in part via the excitation by the Mauthner axon of an ipsilaterally projecting excitatory interneuron with a descending axon that branches extensively to contact both primary and secondary motoneurons (Fetcho and Faber, 1988). This pathway may allow for variation in the overall extent of the escape turn through control of the motoneuronal activation at the level of the descending interneuron.

COMMISSURAL INHIBITION

While the initial massive pulse of motor output may overwhelm any residual muscle force on the opposite side of the escape, powerful central networks block any further activation of motoneurons on the side opposite the escape bend. The Mauthner axons electrotonically excite a glycinergic commissural interneuron that monosynaptically inhibits both the primary motoneurons on the opposite side and the descending interneurons that excite motoneurons on that side (Fetcho and Faber, 1988). This blocks contralateral motor activity from interfering with the escape bend. The inhibitory interneuron also inhibits its inhibitory counterpart on the opposite side, thus preventing it from interfering with the escape bend.

In a rare instance of a test of interneuronal function in a vertebrate, the critical role of these interneurons has been tested in a zebrafish enhancer trap line in which these neurons were selectively fluorescently labeled (Satou et al., 2009). Removing them with a laser in the transparent larval fish led to an impaired escape response that resulted from an aberrant bilateral muscle activation due to occasional firing of Mauthner neurons on both sides in response to a stimulus.

SYNAPSE LOCATION

One of the striking features of the connections of the Mauthner axon is that its output connections are located at the initial segment of the axon of the primary motoneurons and the descending interneurons and at the first node of Ranvier of the commissural inhibitory cell (Fetcho and Faber 1988; Yasargil and Sandri, 1990). This location makes the connections very powerful drivers of the neurons and places the synapses in the network in a position where they can override other inputs located on the somata or dendrites of the cells. Like the Mauthner connections, the synapses of the commissural inhibitory cells on contralateral neurons are also located at the initial segments or first nodes, where they can powerfully veto the activation of those cells by other inputs, including the other Mauthner axon. These design features point to the high priority of the circuits for escape behavior, as they are positioned to take precedence over other ongoing behaviors.

Relationship to Other Spinal Networks

The escape network has many parallels to swimming networks, which also involve ipsilateral descending excitatory neurons and commissural inhibitory cells. The inhibitory neurons in the escape networks appear dedicated to the escape behavior because they are not activated in swimming. The descending interneurons in escapes, however, are shared with swimming at high swimming speeds.

The design of the escape network for taking priority over other networks in spinal cord is driven home by experiments looking at the interaction between the firing of a single spike in the Mauthner cell and ongoing rhythmic swimming (Svoboda and Fetcho, 1996). In this situation, the Mauthner cell not only overrides the ongoing swimming to produce a motor output, it also resets the rhythm of the swimming networks so that subsequent swimming occurs in phase with the escape motor activity to allow a smooth transition back into swimming after an escape turn.

In summary, the spinal escape network is designed to produce a very fast and powerful unilateral motor output that takes precedence over other motor responses. The design features of the network, including the direct motor connections, the synapse locations, the inhibition of conflicting motor responses, and the fast conducting axons, are features of such high-priority motor circuits broadly.

References

Eaton RC, DiDomenico R, Nissanov J (1991) Role of the Mauthner cell in sensorimotor integration by the brain stem escape network. *Brain Behav Evol* 37:272–285.

Eaton RC, Hofve JC, Fetcho JR (1995) Beating the competition: the reliability hypothesis for Mauthner axon size. *Brain Behav Evol* 45:183–194.

Faber DS, Fetcho JR, Korn H (1989) Neuronal networks underlying the escape response in goldfish. General implications for motor control. *Ann NY Acad Sci* 563:11–33.

Fetcho JR (1991) Spinal network of the Mauthner cell. *Brain Behav Evol* 37:298–316.

Fetcho JR, Faber DS (1988) Identification of motoneurons and interneurons in the spinal network for escapes initiated by the mauthner cell in goldfish. *J Neurosci* 8:4192–4213.

Satou C, Kimura Y, Kohashi T, Horikawa K, Takeda H, Oda Y, Higashijima S (2009) Functional role of a specialized class of spinal commissural inhibitory neurons during fast escapes in zebrafish. *J Neurosci* 29:6780–6793.

Svoboda KR, Fetcho JR (1996) Interactions between the neural networks for escape and swimming in goldfish. *J Neurosci* 16:843–852.

Yasargil GM, Sandri C (1990) Topography and ultrastructure of commissural interneurons that may establish reciprocal inhibitory connections of the Mauthner axons in the spinal cord of the tench, Tinca tinca L. *J Neurocytol* 19:111–126.

37

Locomotor Circuits in the Developing Rodent Spinal Cord

Ole Kiehn and Kimberly J. Dougherty

Limbed locomotion in mammals involves recurring activation of flexor and extensor muscles within a limb, and coordinated activity between the left and right legs and between forelimbs and hindlimbs in four-legged animals. The phasing and timing of this complex motor act is to a large degree generated by neuronal circuitries or central pattern generators (CPGs) located in the spinal cord. The isolated spinal cord preparation from newborn rodents has become a model system for studying the neuronal basis underlying limbed locomotion in mammals. This preparation is a robust in vitro preparation for combined electrophysiological recording and experimental manipulation of rhythmic locomotor activity that closely resembles the pattern of motor loco-motor activity in the intact animal. Moreover, the mouse is amenable to molecular genetic experiments. By taking advantage of this excellent model system, some of the basic organizational principles of the mammalian loco-motor CPG are now beginning to be unraveled.

Locomotor Initiation

In intact animals, locomotor initiating signals, presumably originating in the forebrain, are funneled through the basal ganglia and conveyed to locomotor regions in diencephalon and mesencephalon and then to excitatory neurons in the mid-hindbrain that project to the spinal cord. In rodents, the mid-hindbrain excitatory neurons may include both glutamatergic reticulospinal neurons (Atsuta et al., 1990; Zaporozhets et al., 2006) and descending sero-tonergic neurons in the parapyramidal region of the medulla (Liu and Jordan, 2005) that alone or in a parallel fashion provide the external excitatory drive needed to initiate and maintain the activity of the locomotor CPG. Increased drive increases the frequency of locomotion. Activation of sensory afferents originating in the limb and tail can also activate hindlimb locomotion. The sensory modalities of these inputs include proprioception and pain (Iizuka et al., 1997; Blivis et al., 2007).

LOCALIZATION OF LOCOMOTOR NETWORK

The cervical enlargement contains the locomotor network involved in fore-limb movements (Juvin et al., 2005), while the lower thoracic and lumbar spinal cord contain the hindlimb locomotor network (Kiehn and Kjaerulff, 1998). These networks are coordinated through long propriospinal neurons. The localization of the hindlimb locomotor network has been studied in detail (Kiehn and Kjaerulff, 1998). It has a ventral location in the cord and a rhyth-mogenic capacity that is distributed in a rostral to caudal direction. This orga-nizational principle indicates that the hindlimb locomotor network is distributed in a modular fashion along the cord.

BASIC NETWORK ORGANIZATION AND OPERATION

The key features of the locomotor network in rodents are as follows: (1) rhythm generation, (2) flexor-extensor alternation, and (3) left-right coor-dination. Both non-NMDA (AMPA and kainate) and NMDA receptors con-tribute to intrinsic rhythm generation, which is not dependent on fast inhibitory transmission and can persist in the hemicord (Kiehn and Butt, 2003; Kiehn, 2006; Kiehn et al., 2008). These observations suggest that rhythm generation is made by a core of ipsilaterally projecting glutamatergic neurons possibly distributed along the cord. Flexor-extensor alternation depends on ipsilateral ionotropic glycinergic and GABAergic transmission. Motor neu-rons are driven into rhythmicity by alternating synaptic glutamatergic excita-tion and glycinergic/GABAergic inhibition, implying that rhythm-generating networks are organized in reciprocal fashion. Left-right coordination is medi-ated by commissural neurons with axons crossing ventrally in the cord, giving rise to either alternation or synchrony.

Information about the organization of the rodent CPG also comes from genetic manipulation and tracing studies of classes of spinal interneurons that have been classified according to their developmental expression of cer-tain transcription factors. These include five classes of interneurons desig-nated V0, V1, V2a, V2b, and V3 and marked by the expression of transcription factors Evx-1/2, En1, Chx10, Gata2/3, and Sim1, respectively (Jessell, 2000; Goulding and Pfaff, 2005; Kiehn, 2006). V0 interneurons constitute a group of mixed excitatory and inhibitory neurons that have strictly contralateral connections, V1 neurons are inhibitory and ipsilaterally projecting, V2a neurons are excitatory ipsilaterally projecting, V2b are inhibitory ipsilaterally projecting, and V3 interneurons are predominantly contralaterally projecting and excitatory. Two other groups of molecularly defined groups of interneurons in the ventral cord, cells that express the transcription factor Hb9 (Wilson et al., 2005; Hinckley et al., 2005) and cells that express the axon guidance molecule EphA4 (Butt et al., 2005), have also been characterized.

Figure 37.1 shows a proposed diagram of the rodent segmental locomotor CPG driving flexor and extensor motor neurons. The network organization reflects that both flexor and extensor motor neurons receive an asymmetric excitatory and inhibitory drive with inhibition dominating mostly in extensor motor neurons (Endo and Kiehn, 2008). In this model, populations of excitatory and inhibitory last-order interneurons are driven by common upstream excitatory interneurons, which themselves also directly project onto motor neurons.

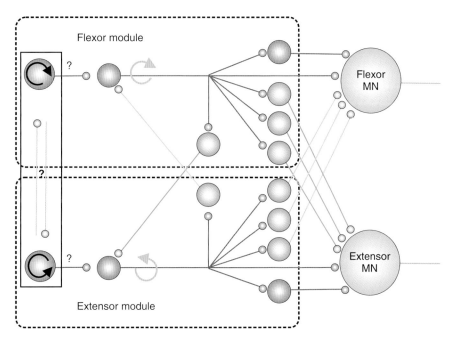

FIGURE 37–1. Segmental organization of flexor-extensor modules in rodent locomotor central pattern generator (CPG). Rhythmicity in flexor and extensor motor neurons (MNs) is driven by alternating synaptic excitation and inhibition. This drive is asymmetric, with inhibitory dominance indicated as a difference in the number of last-order excitatory (red) and inhibitory (blue) interneurons. The last-order interneurons are under common control of flexor or extensor-related excitatory neurons that may or may not have rhythm-generating properties (indicated by curved arrows). This layer is reciprocally connected to ensure flexor-extensor alternation. The diagram also indicates a further upstream rhythm-generating layer for which there is no direct experimental evidence but may be included to more easily explain the so-called non-resetting deletions reported in previous locomotor studies. (Adapted from Endo and Kiehn, 2008)

The segmental network is organized into flexor and extensor modules, which are reciprocally connected via inhibitory interneurons. The excitatory neurons may have rhythm-generating capability by themselves and/or be driven by further upstream rhythm-generating neurons. Of known candidate neurons for the last-order inhibitory neurons in the diagram are V1-related interneurons, including Ia interneurons and Renshaw cells (Gosgnach et al., 2006; Kiehn, 2006) and V2b interneurons (Goulding and Pfaff, 2005). The last-order excitatory interneurons may include Hb9 neurons (Hinckley et al., 2005; Wilson et al., 2005; Hinckley and Ziskind-Conhaim, 2006; Wilson et al., 2007) and EphA4 neurons (Butt et al., 2005) as well as other not-yet-defined populations of excitatory neurons. This network will account for flexor-extensor alternation segmentally. Repeated flexor/extensor modules along the cord may then be connected through propriospinal neurons. Motor neurons may boost their activity through excitatory collaterals (Zhu et al., 2007) and gap junction coupling (Tresch and Kiehn, 2000).

Figure 37.2 shows how the ipsilateral network is connected to network structures involved in left-right coordination. This diagram only shows connectivity at the segmental level and only accounts for activity in one locomotor phase (e.g., flexor phase). During normal locomotion, flexors and extensors are activated in a reciprocal pattern. The reciprocal pattern of flexor and extensors seen ipsilaterally and contralaterally are, however, easily incorporated in the summary diagram by duplicating the ipsilaterally and contralaterally projecting neuronal populations.

The left-right alternation is subserved by a dual inhibitory system acting on contralateral motor neurons in the same segment (Quinlan and Kiehn, 2007). The direct pathway is mediated via glycinergic/GABAergic commissural interneurons, while the indirect pathway mediates contralateral inhibition via glutamatergic commissural interneuron acting on local inhibitory interneurons (Renshaw cells and unknown inhibitory neurons). In parallel to the inhibitory pathways are crossed excitatory pathways mediated by glutamatergic commissural interneurons. The dual inhibitory system assists in providing alternating activity, while the direct excitatory pathway might be active during conditions of synchronous activity. To obtain this effect, crossed connections should also connect to the rhythm-generating core on the other side. These connections have not been confirmed and are indicated by dotted lines in the figure. The direct excitatory pathway is driven directly from the rhythm-generating core (Crone et al., 2008; Crone et al., 2009), while the dual inhibitory system is driven indirectly via a population of V2a excitatory interneurons that are characterized by their expression of the transcription factor Chx10 (Crone et al., 2008). The V2a interneurons also seem to convey sensory activation of the locomotor network (Crone et al., 2008). The commissural interneurons involved in the dual inhibitory pathway may be neurons originating from both the V0 and V3 interneuron populations (Lanuza et al., 2004; Crone et al., 2008; Zhang et al., 2008). During slow locomotion the

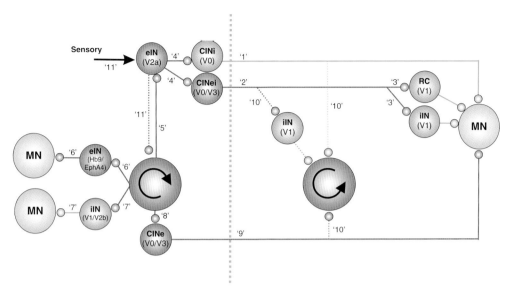

FIGURE 37–2. Organization of flexor-extensor and left-right segmental spinal locomotor networks in rodent. Diagram that shows the segmental ipsilateral flexor-extensor network connected to left-right coordinating networks. This diagram only shows connectivity at the segmental level and only accounts for activity in one locomotor phase (e.g., flexor motor neurons). Left-right alternating networks are mediated by a dual inhibitory commissural system that comprises two commissural interneurons (CINs) populations: (*a*) a set of glycinergic/GABAergic commissural interneurons (CINi) that inhibit contralateral motor neurons (MNs) directly ("1"), and (*b*) a set of excitatory commissural interneurons (CINei) that inhibit motor neurons indirectly ("2") by acting on contralateral inhibitory interneurons (IN) that include Renshaw cells (RCs) and other inhibitory neurons ("3"). A group of excitatory neurons expressing the transcript factor Chx10 called V2a inteneurons provide inputs to the dual inhibitory pathway ("4") and are driven by the rhythm-generating core ("5") (indicated as one neuronal population). The rhythm-generating core also provides excitation (directly or indirectly) to MNs ("6") and drives inhibitory neurons projecting to MNs ("7") (only one module is include in this drawing). The rhythm-generating core drives ("8") the commissural interneuron systems (CINe) responsible for left-right synchrony ("9"). Possible connections of CINs to rhythm-generating neurons on the contralateral side are indicated by dotted lines ("10"). V2a interneurons also seem be a gate for sensory signal that evoke locomotion ("11"). Vertical dotted line indicates midline (adapted from Crone et al., 2008). The diagram includes the proposed position in the segmental network structure of the seven known groups of molecularly defined interneuron populations found in the ventral spinal cord: V0, V1, V2a, V2b, V3, Hb9, and EphA4 neurons.

rhythm-generating core may bypass the V2a pathway and directly activate the left-right alternating circuits (Crone et al., 2009) (connections not shown in Figure 37.2).

The site for action of the descending initiating systems is not known except that reticulospinal neurons project directly to motor neurons (Szokol et al., 2008).

Cellular properties that may contribute to rhythm and pattern generation have been found in rodent spinal interneurons with no clear preferences to specific populations of interneurons. These properties include postinhibitory rebound firing that helps escape inhibition generated by h-channels and T-type calcium channels, delayed activation generated by activation of potassium channels with slow kinetics, spike-frequency adaptation properties, and properties that amplify and prolong synaptic inputs and promote bursting generated by persistent inward conductances (Kiehn et al., 1996; Wilson et al., 2005; Zhong et al., 2007; Zhong et al. 2010; Tazerart et al., 2008; Dougherty and Kiehn, 2010).

Neuromodulation

Extrinsic neuromodulation of network activity may change the frequency and amplitude of locomotion. This neuromodulation includes monoaminergic (Schmidt and Jordan, 2000; Sqalli-Houssaini and Cazalets, 2000; Clarac et al., 2004) and GABAergic modulation (Cazalets et al., 1998). The exact site for the actions of these neuromodulators in the network is not known, although it has been shown that some of the neuronal excitability of commissural interneurons and motor neurons can increase by depolarizing neurons, reducing afterhyperpolarization amplitude, decreasing the action potential threshold, and increasing h-channel conductance at rest (Berger and Takahashi, 1990; Kjaerulff and Kiehn, 2001; Zhong et al., 2006).

References

Atsuta Y, Garcia-Rill E, Skinner RD (1990) Characteristics of electrically induced locomotion in rat in vitro brain stem-spinal cord preparation. *J Neurophysiol* 64:727–735.

Berger AJ, Takahashi T (1990) Serotonin enhances a low-voltage-activated calcium current in rat spinal motoneurons. *J Neurosci* 10:1922–1928.

Blivis D, Mentis GZ, O'Donovan M J, Lev-Tov A (2007) Differential effects of opioids on sacrocaudal afferent pathways and central pattern generators in the neonatal rat spinal cord. *J Neurophysiol* 97:2875–2886.

Butt SJ, Lundfald L, Kiehn O (2005) EphA4 defines a class of excitatory locomotor-related interneurons. *Proc Natl Acad Sci USA* 102:14098–14103.

Cazalets JR, Bertrand S, Sqalli-Houssaini Y, Clarac F (1998) GABAergic control of spinal locomotor networks in the neonatal rat. *Ann NY Acad Sci* 860:168–180.

Clarac F, Pearlstein E, Pflieger JF, Vinay L (2004) The in vitro neonatal rat spinal cord preparation: a new insight into mammalian locomotor mechanisms. *J Comp Physiol A Neuroethol Sens Neural Behav Physiol* 190:343–357.

Crone SA, Quinlan KA, Zagoraiou L, Droho S, Restrepo CE, Lundfald L, Endo T, Setlak J, Jessell TM, Kiehn O, Sharma K (2008) Genetic ablation of V2a ipsilateral interneurons disrupts left-right locomotor coordination in mammalian spinal cord. *Neuron* 60: 70–83.

Crone SA, Zhong G, Harris-Warrick R, Sharma K (2009) In mice lacking V2a interneurons, gait depends on speed of locomotion. *J Neurosci* 29:7098–7109.

Dougherty K, Kiehn O (2010) Firing and cellular properties of V2a interneurons in the rodent spinal cord. *J Neurosci* 30(1):24–37.

Endo T, Kiehn O (2008) Asymmetric operation of the locomotor central pattern generator in the neonatal mouse spinal cord. *J Neurophysiol* 100:3043–3054.

Gosgnach S, Lanuza GM, Butt SJ, Saueressig H, Zhang Y, Velasquez T, Riethmacher D, Callaway EM, Kiehn O, Goulding M (2006) V1 spinal neurons regulate the speed of vertebrate locomotor outputs. *Nature* 440:215–219.

Goulding M, Pfaff SL (2005) Development of circuits that generate simple rhythmic behaviors in vertebrates. *Curr Opin Neurobiol* 15:14–20.

Hinckley CA, Ziskind-Conhaim L (2006) Electrical coupling between locomotor-related excitatory interneurons in the mammalian spinal cord. *J Neurosci* 26:8477–8483.

Hinckley CA, Hartley R, Wu L, Todd A, Ziskind-Conhaim L (2005) Locomotor-like rhythms in a genetically distinct cluster of interneurons in the mammalian spinal cord. *J Neurophysiol* 93:1439–1449.

Iizuka M, Kiehn O, Kudo N (1997) Development in neonatal rats of the sensory resetting of the locomotor rhythm induced by NMDA and 5-HT. *Exp Brain Res* 114:193–204.

Jessell TM (2000) Neuronal specification in the spinal cord: inductive signals and transcriptional codes. *Nat Rev Genet* 1:20–29.

Juvin L, Simmers J, Morin D (2005) Propriospinal circuitry underlying interlimb coordination in mammalian quadrupedal locomotion. *J Neurosci* 25:6025–6035.

Kiehn O (2006) Locomotor circuits in the mammalian spinal cord. *Annu Rev Neurosci* 29:279–306.

Kiehn O, Kjaerulff O (1998) Distribution of central pattern generators for rhythmic motor outputs in the spinal cord of limbed vertebrates. *Ann NY Acad Sci* 860:110–129.

Kiehn O, Butt SJ (2003) Physiological, anatomical and genetic identification of CPG neurons in the developing mammalian spinal cord. *Prog Neurobiol* 70:347–361.

Kiehn O, Johnson BR, Raastad M (1996) Plateau properties in mammalian spinal interneurons during transmitter-induced locomotor activity. *Neuroscience* 75:263–273.

Kiehn O, Quinlan KA, Restrepo CE, Lundfald L, Borgius L, Talpalar AE, Endo T (2008) Excitatory components of the mammalian locomotor CPG. *Brain Res Rev* 57:56–63.

Kjaerulff O, Kiehn O (2001) 5-HT modulation of multiple inward rectifiers in motoneurons in intact preparations of the neonatal rat spinal cord. *J Neurophysiol* 85:580–593.

Lanuza GM, Gosgnach S, Pierani A, Jessell TM, Goulding M (2004) Genetic identification of spinal interneurons that coordinate left-right locomotor activity necessary for walking movements. *Neuron* 42:375–386.

Liu J, Jordan LM (2005) Stimulation of the parapyramidal region of the neonatal rat brain stem produces locomotor-like activity involving spinal 5-HT7 and 5-HT2A receptors. *J Neurophysiol* 94:1392–1404.

Quinlan KA, Kiehn O (2007) Segmental, synaptic actions of commissural interneurons in the mouse spinal cord. *J Neurosci* 27:6521–6530.

Schmidt BJ, Jordan LM (2000) The role of serotonin in reflex modulation and locomotor rhythm production in the mammalian spinal cord. *Brain Res Bull* 53:689–710.

Sqalli-Houssaini Y, Cazalets JR (2000) Noradrenergic control of locomotor networks in the in vitro spinal cord of the neonatal rat. *Brain Res* 852:100–109.

Szokol K, Glover JC, Perreault MC (2008) Differential Origin of reticulospinal drive to motor neurons innervating trunk and hind limb muscles in the mouse revealed by optical recording. *J Physiol* 586(21):5259–5276.

Tazerart S, Vinay L, Brocard F (2008) The persistent sodium current generates pacemaker activities in the central pattern generator for locomotion and regulates the locomotor rhythm. *J Neurosci* 28:8577–8589.

Tresch MC, Kiehn O (2000) Motor coordination without action potentials in the mammalian spinal cord. *Nat Neurosci* 3:593–599.

Wilson JM, Hartley R, Maxwell DJ, Todd AJ, Lieberam I, Kaltschmidt JA, Yoshida Y, Jessell TM, Brownstone RM (2005) Conditional rhythmicity of ventral spinal interneurons defined by expression of the Hb9 homeodomain protein. *J Neurosci* 25: 5710–5719.

Wilson JM, Cowan AI, Brownstone RM (2007) Heterogeneous electrotonic coupling and synchronization of rhythmic bursting activity in mouse Hb9 interneurons. *J Neurophysiol* 98:2370–2381.

Zaporozhets E, Cowley KC, Schmidt BJ (2006) Propriospinal neurons contribute to bulbospinal transmission of the locomotor command signal in the neonatal rat spinal cord. *J Physiol* 572:443–458.

Zhang Y, Narayan S, Geiman E, Lanuza GM, Velasquez T, Shanks B, Akay T, Dyck J, Pearson K, Gosgnach S, Fan CM, Goulding M (2008) V3 spinal neurons establish a robust and balanced locomotor rhythm during walking. *Neuron* 60:84–96.

Zhong G, Diaz-Rios M, Harris-Warrick RM (2006) Serotonin modulates the properties of ascending commissural interneurons in the neonatal mouse spinal cord. *J Neurophysiol* 95:1545–1555.

Zhong G, Masino MA, Harris-Warrick RM (2007) Persistent sodium currents participate in fictive locomotion generation in neonatal mouse spinal cord. *J Neurosci* 27: 4507–4518.

Zhong G, Droho S, Crone SA, Dietz S, Kwan AC, Webb WW, Sharma K, Harris-Warrick RM (2010) Electrophysiological characterization of V2a interneurons and their locomotor-related activity in the neonatal mouse spinal cord. *J Neurosci* 30(1):172–182.

Zhu H, Clemens S, Sawchuk M, Hochman S (2007) Expression and distribution of all dopamine receptor subtypes (D(1)-D(5)) in the mouse lumbar spinal cord: a real-time polymerase chain reaction and non-autoradiographic in situ hybridization study. *Neuroscience* 149:885–897.

38

The Lamprey Postural Circuit

Tatiana G. Deliagina, Pavel V. Zelenin, and
Grigori N. Orlovsky

The lamprey swims due to the lateral body undulations. During swimming, orientation of the lamprey in the sagittal (pitch) and transversal (roll) planes is stabilized in relation to the gravity vector by means of the postural control systems driven by vestibular input (Deliagina et al., 1992a; Deliagina et al., 1992b; Deliagina and Fagerstedt, 2000; Pavlova and Deliagina, 2002; Ullén et al., 1995). Vestibular-driven mechanisms also contribute to stabilization of the direction of swimming in the horizontal (yaw) plane (Karayannidou et al., 2006). Any deviations from the stabilized orientation are reflected in vestibular signals (Deliagina et al., 1992b), which cause corrective motor responses. In the pitch and yaw planes, the corrections occur due to the body bending in the corresponding plane (Fig. 38.1A, Pitch and Yaw). In the roll plane, the corrections occur due to a change of the direction of locomotor body undulations, from lateral to oblique (red arrows in Fig. 38.1A, Roll) (Zelenin et al., 2003).

The principal elements of the postural network in the lamprey are shown in Figure 38.1B. Vestibular afferents through the neurons of vestibular nuclei affect reticulospinal (RS) neurons. The RS neurons form the main descending pathway that transmits all commands from the brainstem to the spinal cord, including commands for postural corrections. The spinal network transforms RS commands into the motor pattern of postural corrections. This network includes segmental interneurons, as well as four motoneuron pools in each segment that innervate the dorsal and ventral parts of a myotome on the right and left sides.

Responses of individual vestibular afferents and RS neurons to natural stimulations of vestibular organs (i.e., rotation of the whole lamprey or isolated brainstem-vestibular organ preparation in different planes) have

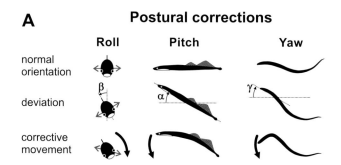

A **Postural corrections**

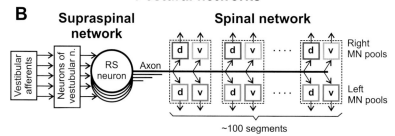

Postural networks

B

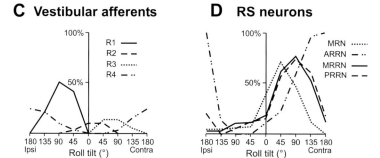

Vestibular responses in roll plane

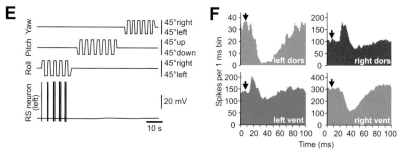

RS neuron of roll control system

FIGURE 38–1. (*A*) During regular swimming, the lamprey stabilizes its orientation in the sagittal (pitch) plane, in the transversal (roll) plane, and in the horizontal (yaw) plane. Deviations from the stabilized orientation in these planes (angles α, β, and γ, respectively) evoke corrective

FIGURE 38–1. Continued
motor responses (large arrows) aimed at restoration of the initial orientation. (*B*) Commands for correcting the orientation are formed on the basis of vestibular information and are transmitted from the brainstem to the spinal cord by reticulospinal (RS) neurons. Motor output of each segment is generated by four motoneuron (MN) pools controlling the dorsal and ventral parts of a myotome on the two sides (*d* and *v* pools). (*C* and *D*) Angular zones of activity of vestibular (otolith) afferents (*C*) and RS neurons (*D*) when the preparation was rotated in the roll plane (abscissa, % of active neurons; the data for different groups [R1–R4)]of otolith afferents in *C*, and for different reticular nuclei, MRN, ARRN, MRRN, and PRRN in *D* are presented separately). (*E* and *F*) An RS neuron that contributed to stabilization of the roll angle. (*E*) The neuron fired spikes in response to right (contralateral) roll tilts. (*F*) The neuron evoked excitation in the left (ipsilateral) ventral and right (contralateral) dorsal branches of the ventral roots, and inhibition in the right ventral and left dorsal branches. Arrows indicate the time of arrival of the RS spike to the segment 30, in which the motor output was monitored. (*A, B, E,* and *F* were adapted from Zelenin et al., 2007; *D* was adapted from Deliagina et al., 1992a)

been studied. Deviation of the lamprey from the stabilized orientation in any plane causes activation of a specific group of canal and otolith afferents (Deliagina et al., 1992b). In each of the main planes (pitch, roll, yaw), two groups of canal afferents responding to rotation in opposite directions were found. Rotation in the roll and pitch planes revealed specific groups of otolith afferents with different zones of spatial sensitivity. The groups with different angular zones of activity in the roll plane are shown in Figure 38.1C. Due to these vestibular inputs (Fig. 38.1B), RS neurons respond to rotation in different planes (Deliagina et al., 1992a; Deliagina and Fagerstedt, 2000; Pavlova and Deliagina, 2002; Karayannidou et al., 2006). Figure 38.1D shows the angular zones of activity of RS neurons from different reticular nuclei when the preparation was rotated in the roll plane. The overwhelming majority of RS neurons are maximally active at 45°–90° of the contralateral tilt (Fig. 38.1D). They are driven by the vestibular afferents responding to the ipsilateral roll tilt (Fig. 38.1C) and originating from the contralateral labyrinth (Fig. 38.2A; Deliagina et al., 1992a; Deliagina et al., 1992b; Deliagina and Pavlova, 2002). In each of the main planes (pitch, roll, yaw), two antagonistic groups of RS neurons responding to rotation in opposite directions were revealed (Deliagina et al., 1992a; Deliagina and Fagerstedt, 2000; Pavlova and Deliagina, 2002; Karayannidou et al., 2006). These groups are shown in Figures 38.2A and D for the roll and pitch control systems, and their tilt-related activity is presented in Figures 38.2B and E, respectively. This activity can be considered as the command for postural correction addressed from the supraspinal postural network to the spinal postural network.

The spinal network transforms RS commands into the motor pattern of postural corrections. It was proposed that two groups of RS neurons (activated by rotation in a particular plane but in opposite directions), through the spinal network, cause rotation of the animal in the direction opposite to the

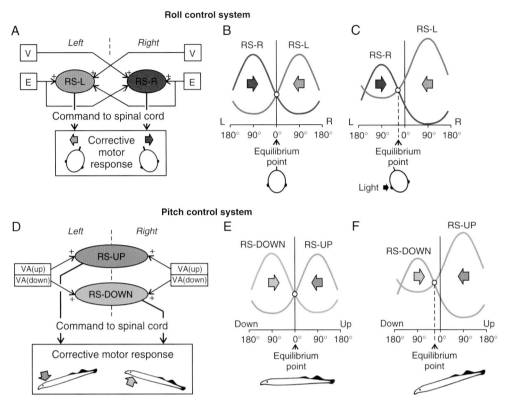

FIGURE 38–2. (*A–C*) The roll control system. (*A*) The left and right groups of RS neurons (RS-L and RS-R) are driven mainly by vestibular afferents from the contralateral vestibular organ (V); they cause ipsilateral rotation of the lamprey (red and blue arrows in *A–C*). (*B*) Due to vestibular inputs, these two groups are activated with the contralateral roll tilt and rotate the lamprey in opposite directions. The system stabilizes the orientation with equal activities of the two groups (equilibrium point). (*C*) Unilateral eye illumination (input E in *A*) causes an asymmetrical bias in the RS-L and RS-R activities and a shift of the equilibrium point, which results in a roll tilt of the lamprey toward the source of light. (*D–F*) The pitch control system. (*D*) Two groups of RS neurons (RS-UP and RS-DOWN) are driven by vestibular afferents responding to nose-up (VA-up) and nose-down (VA-down) tilts of the animal, respectively. The RS-UP and RS-DOWN groups cause downward and upward turning of the lamprey, respectively (orange and green arrows in *D–F*). (*E*) Due to vestibular inputs, RS-UP and RS-DOWN groups are activated with upward and downward tilts, respectively, and cause turning the lamprey in opposite directions. The system stabilizes the orientation with equal activities of the two groups (equilibrium point). (*F*) Raising the water temperature causes an asymmetrical bias in the RS-UP and RS-DOWN activities and a shift of the equilibrium point, which results in downward tilt of the lamprey. (*A–C* were adapted from Deliagina and Fagerstedt, 2000; *D–F* were adapted from Pavlova and Deliagina, 2002)

initial turn (which activates the neurons), and the system will thus stabilize the orientation with equal activities of the two antagonistic groups (equilibrium point in Fig. 38.2B and E) (Deliagina et al., 1992a; Deliagina and Fagerstedt, 2000; Pavlova and Deliagina, 2002; Karayannidou et al., 2006). Normally, this occurs at the dorsal-side-up and horizontal orientation of the body in the roll and pitch planes, correspondingly, as well as during swimming along a rectilinear trajectory in the yaw plane. The stabilized orientation can be gradually changed under the effect of some environmental factors. Asymmetrical eye illumination causes asymmetry in the activities of the two antagonistic groups of RS neurons of the roll control system, and a shift of the equilibrium point, which results in a change of the stabilized orientation (Fig. 38.2C) (Deliagina and Fagerstedt, 2000; Deliagina and Pavlova, 2002). This postural reaction (the dorsal light response) is a protective reflex, turning the dark dorsal side of the body toward the light and thereby decreasing the risk of detection by predators. In the pitch control system, the stabilized orientation can be changed by raising the water temperature, which affects the two groups of RS neurons differently and thus shifts the equilibrium point toward the nose-down orientation (Fig. 38.2F) (Pavlova and Deliagina, 2002), which allows the animal to reach deeper and colder water layers.

The relationships between vestibular inputs to individual RS neurons and their motor effects were recently studied (Zelenin et al., 2007). In these experiments, *first*, the vestibular-driven activity of an RS neuron was determined by rotating the preparation in different planes. *Second*, the same RS neuron was stimulated, and its effects on the motor output of the spinal cord (the activity in dorsal and ventral branches of the ventral roots innervating dorsal and ventral myotomes, respectively) were detected by means of the RS spike-triggered averaging technique. These effects (functional spinal projections of the RS neuron) were found to be similar along the whole extent of the axon, and they could thus be characterized by a combination of influences on the four motoneuron pools in any segment.

It was shown that the majority of RS neurons responded to rotation in only one of the three main planes (Pavlova and Deliagina, 2002; Zelenin et al., 2007), as illustrated in Figure 38.1E. This particular neuron fired spikes in response to contralateral roll tilts and did not respond to rotation in the pitch and yaw planes. Motor effects of this neuron are shown in Figure 38.1F. They included activation of the motoneuron pools projecting to the ipsiventral and contradorsal myotomes, and inhibition of those projecting to the ipsidorsal and contraventral myotomes. In the swimming lamprey, this pattern would lead to a change of the direction of locomotor body undulations, from lateral to oblique, and to a roll torque rotating the body in the direction opposite to the tilt that activated the neuron (Fig. 38.1A, Roll) (Zelenin et al., 2007).

In the majority of RS neurons, a strong correlation between their vestibular inputs and motor effects was found. Usually, a neuron produced a motor pattern (or a part of the pattern) causing a torque, which would oppose the initial rotation that activated the neuron. Such closed-loop microcircuits, formed by individual RS neurons responding to rotation in a given plane, operate in parallel to generate the resulting motor responses. A small proportion of RS neurons responded to rotation in more than one plane. Most of these neurons produced the motor pattern that represented the common part of the patterns of postural corrections caused by rotation in the corresponding planes.

These data on the sensory-motor transformation performed by individual RS neurons support the conceptual models of the postural control systems formulated earlier (Fig. 38.2) (Deliagina et al., 1992a; Deliagina and Fagerstedt, 2000; Deliagina and Orlovsky, 2002; Pavlova and Deliagina, 2002; Karayannidou et al., 2006).

To conclude, the operation of postural networks in the lamprey is based on interaction between two antagonistic vestibular reflexes. The lamprey stabilizes the body orientation in a particular plane at which the antagonistic reflexes are equal to each other. The closed-loop microcircuits, formed by individual RS neurons responding to deviation of the body orientation in a given plane from the stabilized orientation and affecting the motor output, operate in parallel to generate the resulting corrective motor response. The gradual change of the stabilized orientation is performed through the change of the gain in the reflex chains. These principles of operation of postural networks are similar to those revealed in the evolutionary remote species of marine mollusk *Clione limacina* (Deliagina and Orlovsky, 2002).

REFERENCES

Deliagina TG, Fagerstedt P (2000) Responses of reticulospinal neurons in intact lamprey to vestibular and visual inputs. *J Neurophysiol* 83:864–878.
Deliagina TG, Orlovsky GN (2002) Comparative neurobiology of postural control. *Curr Opin Neurobiol* 12:652–657.
Deliagina TG, Pavlova EL (2002) Modifications of vestibular responses of individual reticulospinal neurons in the lamprey caused by a unilateral labyrinthectomy. *J Neurophysiol* 87:1–14.
Deliagina TG, Orlovsky GN, Grillner S, Wallén P (1992a) Vestibular control of swimming in lamprey. II. Characteristics of spatial sensitivity of reticulospinal neurons. *Exp Brain Res* 90:489–498.
Deliagina TG, Orlovsky GN, Grillner S, Wallen P (1992b) Vestibular control of swimming in lamprey. 3. Activity of vestibular afferents. Convergence of vestibular inputs on reticulospinal neurons. *Exp Brain Res* 90:499–507.
Karayannidou A, Zelenin PV, Orlovsky GN, Deliagina TG (2006) Responses of reticulospinal neurons in the lamprey to lateral turns. *J Neurophysiol* 97:512–521.
Pavlova EL, Deliagina TG (2002) Responses of reticulospinal neurons in intact lamprey to pitch tilt. *J Neurophysiol* 88:1136–1146.
Ullén F, Deliagina TG, Orlovsky GN, Grillner S (1995) Spatial orientation of lamprey. 1. Control of pitch and roll. *J Exp Biol* 198:665–673.
Zelenin PV, Grillner S, Orlovsky GN, Deliagina TG (2003) The pattern of motor coordination underlying the roll in the lamprey. *J Exp Biol* 206:2557–2566.
Zelenin PV, Orlovsky GN, Deliagina TG (2007) Sensory-motor transformation by individual command neurons. *J Neurosci* 27:1024–1032.

39

Respiratory Central Pattern Generator

Jack L. Feldman

The importance of breathing to life is obvious. We must breathe continuously and reliably from birth, during wakefulness and sleep. Breathing is a behavior that is precisely modulated by metabolic demand that ranges over an order of magnitude, and it is integrated well with a broad range of volitional (e.g., speech) and emotional (e.g., sighing, crying, and laughing) behaviors. Disorders of breathing are legion, and their consequences significant. Genetic disorders such as central congenital hypoventilation syndrome, Rett syndrome, and Prader-Willi disease result from mutations of single genes, yet how the particular breathing phenotypes result is unknown. At the late stages of various neurodegenerative diseases, such as Parkinson disease, multiple systems atrophy, and amyotrophic lateral sclerosis, patients develop serious central respiratory problems during sleep that we hypothesize is the cause of death in many cases (McKay et al., 2005). Four to six percent of the adult population suffers from some form of sleep apnea that has substantial impact on individual health and longevity and represents a public health problem of substantial proportion. Of experimental interest, breathing is the only mammalian behavior that can be studied in preparations ranging from slices (Fig. 39.1) to awake behaving animals.

Here we focus on the mammalian respiratory central pattern generator (CPG) capable of regulating blood O_2, CO_2, and pH that lies within the brainstem and spinal cord. A substantial component of the CPG plays no significant role in rhythm generation but is devoted to transforming the rhythm into an appropriate pattern of motor activity to pump air and regulate airway resistance. While breathing may be considered a simple motor act whose goal is to inflate and deflate the lung, the energetics of these movements appears to be highly optimized. This optimization, reflected in circuitry in the caudal brainstem and spinal cord devoted to pattern formation, should not be

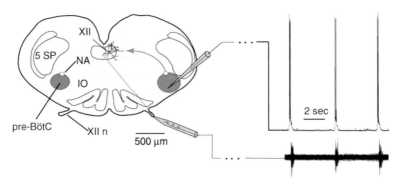

FIGURE 39–1. A thin slice (300–750 μm) of neonatal rodent brainstem generates respiratory rhythm in the twelfth nerve (XII n) that originates in the preBötzinger Complex (pre-BötC).

surprising given that breathing is the only continuous skeletal muscle behavior and accounts for ~5% of total body metabolism at rest.

The respiratory CPG has at least a three-layer structure: *(1) sensory*—chemoreception for O_2, CO_2/pH, mechanoreception for energetics and reflexes; *(2) rhythm generation*—primarily for determination of timing; and *(3) pattern formation*—transformation of the rhythm into an energetically appropriate output that integrates with other behaviors (e.g., locomotion, swallowing, and defecation). Furthermore, neurons involved in pattern formation can be divided into at least two subgroups, *respiratory premotoneurons* receiving presumptive direct input from the source(s) of rhythm generation and projecting to *respiratory muscle motoneurons*, either directly, as in the case of the bulbospinal projection to phrenic motoneurons that innervate the diaphragm, or via another layer of interneurons, as in the case of intercostal motoneurons, where there is an intermediary projection to segmental interneurons.

BASIC CONNECTIONS WITHIN THE BRAINSTEM

Several nuclei form the core of the brainstem respiratory CPG (Fig. 39.2).

Ventral Respiratory Column

Spanning the ventrolateral brainstem from the spinomedullary junction to the seventh nerve, the ventral respiratory column (VRC) (Feldman and McCrimmon, 2007) can be divided into at least four parts.

PreBötzinger Complex

This is an essential bilateral structure for the generation of normal respiratory rhythm (Tan et al., 2008) that lies just rostral to the rostral ventral respiratory

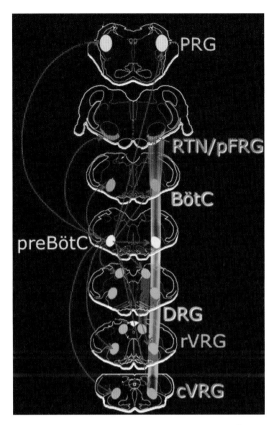

Figure 39–2. Brainstem respiratory central pattern generator. Representative transverse sections of rat brainstem showing three clusters of neurons widely regarded as essential elements of the central pattern generator (based on Fig. 37.3 in Feldman and McCrimmon, 2007). Long ventrolateral column, extending from the cVRG to the RTN/pFRG defines the ventral respiratory column. Dorso medially is a population in the ventrolateral portion of the solitary tract, referred to as the dorsal respiratory group (DRG). In the dorsolateral rostral pons in the regions of the Kölliker-Fuse and parabrachial nucleus is another population referred to as the pontine respiratory group (PRG). Red arrows represent widespread projections of excitatory preBötzinger Complex (preBötC) neurons (Tan et al., 2010).

group (rVRG) and caudal to the Bötzinger Complex (BötC). It is ~1 mm in diameter with several thousand neurons in adult rat. Several lines of evidence demonstrate its critical role. First of all, lesions selectively targeting preBötzinger Complex (preBötC) neurons that express the neurokinin 1 receptor (NK1R, ~300/side in rat), which are glutamatergic, in intact *unanesthetized* adult rats profoundly affect breathing. Breathing in all sleep/wake states is relatively unaffected until ~50% of these neurons are destroyed. With 50%–80% destruction, breathing during wakefulness appears normal, but during sleep there are apneas during REM sleep, and when >80% are killed, breathing during wakefulness is ataxic and absent during sleep (Gray et al., 2001;

McKay et al., 2005). Secondly, rapid silencing of preBötC neurons expressing somatostatin (Sst; ~500/side in rat with some overlap with NK1R neurons), which are glutamatergic, produces profound and persistent apnea in anesthetized or awake adult rats (Fig. 39.B1; Tan et al., 2008). Third, juvenile rats with brainstem transection just rostral to the preBötC continue to generate a rhythmic inspiratory-dominated breathing pattern but do not appear able to generate expiratory muscle activity (Janczewski and Feldman, 2006). Fourth, medullary slices containing the preBötC generate a respiratory-related rhythm (Smith et al., 1991).

At rest, mammals are inspiratory breathers, and expiration is passive. The preBötC is postulated to be the site for generation of this inspiratory rhythm (Feldman and Del Negro, 2006; Janczewski and Feldman, 2006), as well as the site generating gasps and sighs (Lieske et al., 2000). During exercise or other stimuli that significantly increase ventilation, active expiration is initiated and maintained. The source may not be the preBötC, but rather a second oscillator that may be in the RTN/pFRG (see below).

The intrinsic connectome for the preBötC remains to be determined, but a few details are known. The preBötC contains an interconnected network of both glutamatergic and glycinergic neurons. Excitatory but not inhibitory interactions are essential for generation of rhythm. Excitatory preBötC Sst neurons project to all core brainstem regions (Fig. 39.2), including VRC, but not to the spinal cord, and provide a connectivity substrate for its essential role in breathing (Tan et al., 2010).

In in vitro slices, the connectivity of excitatory preBötC neurons, based on a limited sample of dual recordings (23 pairs), appears sparse (Rekling et al., 2000; Fig. 39.3). Thus, ~16% of inspiratory neuron pairs have unidirectional excitatory connections and a nonoverlapping ~16% are electrotonically coupled. The group pacemaker hypothesis posits that the key to rhythm generation is in the dynamics of the microcircuit network interactions among preBötC neurons (Feldman and Del Negro, 2006).

The preBötC microcircuit in vitro contains three well-characterized phenotypes of inspiratory neurons, Types 1–3 (Rekling et al., 1996), distinguished by (1) inspiratory burst afterhyperpolarization or afterdepolarization pattern; (2) discharge pattern; (3) response to substance P (SP; ligand for the neurokinin 1 receptor-NK1R) and µ-opiates ligand for the µ-opiate receptor-µOR; and (4) pacemaker-like activity (in a subset of Type 1 neurons). Type 1 neurons have a postinspiratory burst hyperpolarization that can last the entire interval between inspiratory bursts, begin to depolarize and fire action potentials up to several hundred milliseconds prior to inspiratory motor onset, have SP and µORs, and some have pacemaker properties. Type 2 neurons have a short postinspiratory burst hyperpolarization, begin firing around the time of inspiratory motor onset, have limited response to SP and may not have µORs, and do not appear to have pacemaker properties. Type 3 neurons have a prominent postinspiratory burst afterdepolarization. We postulate that Type 1 neurons play an essential role in rhythmogenesis (irrespective of their

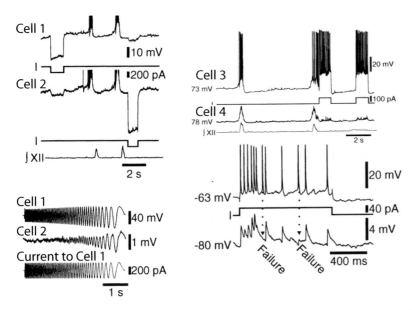

FIGURE 39–3. preBötC inspiratory neurons are bidirectionally electrically (*left*) and unidirectionally synaptically (*right*) coupled. (*Left top*) Simultaneous recordings from two type-1 neurons. Hyperpolarizing pulse in either neuron elicits an attenuated hyperpolarizing response in the other electrically coupled neuron. (*Left bottom*) ZAP current injection into cell 1 elicits an attenuated, filtered response in cell 2. (*Right top*) Simultaneous recordings from two type-1 neurons, one with no injected current (*top trace*) and the other at a hyperpolarized membrane potential. Transient depolarizing current injection into cell 3 elicits repetitive firing in that cell and trains of excitatory postsynaptic potentials in cell 4. (*Right bottom*) Expanded traces from shaded area at top showing failures in the coupled neurons. (From Rekling et al., 2000)

potential pacemaker properties (Feldman and Del Negro, 2006). Evidence of the *necessity* of pacemaker neurons, once widely accepted, is no longer compelling and is a source of some controversy (Feldman and Del Negro, 2006). While establishing whether pacemaker neurons are essential for rhythmogenesis would be useful, such an advance would still not delineate (other) key mechanisms because there are many other preBötC neuronal properties that must play a role. We know little about the connectivity among preBötC neurons, their morphology, the receptor and channel distributions on the somatodendritic membranes, and so on, and the relationship of these properties to physiological phenotype (e.g., Types 1–3).

Caudal Ventral Respiratory Group

An important role of this population is to function as a premotor relay to expiratory motoneurons in the spinal cord. From the spinomedullary junction to the level of the obex, the caudal ventral respiratory group (cVRG) is characterized by bulbospinal premotoneurons with expiratory-modulated firing patterns that project via spinal interneurons to motoneurons innervating abdominal and internal intercostal muscles. Their excitatory expiratory drive appears to derive from the retrotrapezoid nucleus/parafacial respiratory group (Janczewski et al., 2002). They likely receive inspiratory-modulated inhibitory inputs from the BötC and/or the preBötC.

Rostral Ventral Respiratory Group

An important role of this population is to function as a premotor relay to inspiratory motoneurons in the spinal cord. Overlapping caudally with the cVRG and spanning rostrally to past the obex, the rostral ventral respiratory group (rVRG) is characterized by bulbospinal premotoneurons with inspiratory-modulated firing patterns that project monsynaptically to phrenic motoneurons innervating the diaphragm and via spinal interneurons to motoneurons innervating external intercostal muscles. They likely receive expiratory-modulated inhibitory inputs from the BötC and inspiratory-modulated inputs from the preBötC.

Bötzinger Complex

The precise role of this structure is not known. Lying just rostral to the preBötC, this region contains inhibitory neurons that may provide reciprocal inhibition to premoto- and motoneurons. For example, expiratory BötC neurons appear to inhibit inspiratory premotoneurons during their normally silent phase. They may play a role in generating postinspiratory activity (Burke et al., 2010).

As one marches further rostral, there are two additional regions of great interest that may represent a continuation of the VRC (Feldman and McCrimmon, 2007).

Retrotrapezoid Nucleus/Parafacial Respiratory Group

This region plays an important role in central chemoreception (for CO_2) and is postulated to be critical for generation of active expiratory movements acting as an "expiratory oscillator"; it lies mostly ventral to the seventh nucleus and projects to the VRC, including the preBötC. A critical subset of retrotrapezoid nucleus (RTN) neurons expresses the transcription factor PHOX-2B, which is essential for its development and whose absence leads to neonatal death in mice (Dubreuil et al., 2008) and serious breathing pathologies in humans (central congenital hypoventilation syndrome; Gronli et al., 2008). These PHOX-2B neurons, which are glutamatergic, appear to play an important role in central chemoreception for CO_2 (Guyenet, 2008). Another population of neurons in this region that may be coextensive with RTN PHOX-2B neurons is referred to as the parafacial respiratory group (pFRG) (Onimaru and Homma, 2003). The pFRG was initially identified in fetal rodents as a rhythmogenic site postulated to play a critical role in the development of the respiratory CPG and sustain breathing during the early postnatal period (Thoby-Brisson et al., 2009); subsequently, it was identified in brainstem preparations from neonatal rodents, and shown to contain many neurons with respiratory rhythmic activity (Onimaru and Homma, 2003). Their rhythmic activity during fetal development precedes that of the preBötC and appears to be critical for proper development of respiratory rhythm at birth (Thoby-Brisson et al., 2009). Some of these neurons with expiratory discharge patterns project to the caudal VRG to provide expiratory drive to abdominal muscle motoneurons (Janczewski et al., 2002). Transection of the brainstem removing the RTN/pFRG but leaving the preBötC intact eliminates active expiration but has only a modest effect on inspiration. This underlies the two-oscillator hypothesis: the preBötC generates inspiratory rhythm and the RTN/pFRG (in addition to whatever role it plays in central chemoreception) generates active expiratory output (Feldman and Del Negro, 2006; Janczewski and Feldman, 2006).

Pontine Respiratory Group

The role of this region is likely to be in modulating the respiratory pattern and integrating ascending and descending information related to breathing and various behaviors. This most rostral portion of the brainstem respiratory CPG is in the dorsolateral rostral pons, including the Kölliker-Fuse and the parabrachial nuclei; and may be the rostral-most brainstem extension of the VRC. This region was once postulated to be essential for normal breathing, and under certain conditions in anesthetized or decerebrate mammals, lesions in the pontine respiratory group (PRG) can produce significant disturbances in breathing, particularly apneusis (long inspiratory breathholds followed by pauses of varying duration). However, the current view is that it is more likely playing a key modulatory role in generation of respiratory pattern. This region is likely a key integration site for ascending visceral and regulatory information projecting to more rostral regulatory regions such as the hypothalamus. The PRG may also integrate descending commands to modulate breathing.

In addition, at least two other brainstem structures need to be considered: *dorsal respiratory group (DRG)*, located in the dorsomedial medulla near the solitary tract, that may mediate mechanosensory information arriving via the vagus nerve, for example, pulmonary afferents, and *the raphe nuclei*, which project to various of the aforementioned structures to modulate their function particularly related to the postulated role of these neurons as chemoreceptors for CO_2 (Richerson, 2004) via the release of serotonin and colocalized peptides, such as substance P.

There are dozens of papers describing various aspects of the connectivity among these regions, using a wide range of techniques from antidromic mapping, correlations of spiking activity, and tract-tracing neuroanatomy. The most comprehensive compilation of this work is by Lindsay and colleagues (Segers et al., 2008), who have used correlation analysis to infer an extensive array of connections among the various populations.

What Is Missing?

There are many models for respiratory rhythmogenesis, and by and large they share a common problem: they are not testable because they make no substantive predictions that could falsify them. The principal reason is that there are so many unknown values for key parameters, such as strength of synaptic connection between various phenotypes of neurons (e.g., Types 1–3, glutamatergic, glycinergic), that one could always find a set of parameters, all with reasonable values within the "physiological range," to fit the data. Another deficiency in contemporary models for generation of respiratory rhythm is that they are based on assumptions or simplifications that are not physiological. For example, in many models of respiratory rhythmogenesis, neurons are points, ignoring data on the importance of dendrites of preBötC neurons (Morgado-Valle et al., 2008) and are uniformly connected, ignoring data that their interconnections are most likely sparse (Fig. 39.3) (Rekling et al., 2000).

So what is missing? Sufficient data are needed so that models are highly constrained; the fewer free parameters, the more useful a model can be. What do we need to know? For single preBötC neurons, we need to establish their functional phenotype in context of behavior (firing before/during/after rhythmic inspiratory bursting in slices) and how that relates to their morphology and the somatodendritic distribution of receptors and channels. We need to know their synaptic organization (location, transmitters/receptors, gap junctions) and their intracellular signaling mechanisms. Finally, we need to know their interneuronal connectivity (local and distant) and projections (recurrent, reciprocal, chain, and so on).

Summary

Many of the key nuclei comprising the respiratory CPG, especially the pre-BötC, are identified and localized, and we know a great deal about many of their functional roles. Yet we cannot understand how the neurons comprising the respiratory CPG function so well until we delineate the circuitry, particularly the microcircuitry characterizing each structure. The longstanding and recent advances in mapping such circuits in other regions of the brain, particularly the olfactory bulb and retina, promise that we will achieve a major advancement in understanding breathing in the near future.

ACKNOWLEDGMENT

The work in our laboratory has been generously supported by grants from the National Institutes of Health.

REFERENCES

Burke PGR, Abbott SBG, McMKullan S, Goodchild AK, Pilowsky PM (2010) Somatostatin selectively ablates post-inspiratory activity after injection into the Bötzinger Complex. *Neuroscience* 167: 528–539.

Dubreuil V, Ramanantsoa N, Trochet D, Vaubourg V, Amiel J, Gallego J, Brunet JF, Goridis C (2008) A human mutation in Phox2b causes lack of CO_2 chemosensitivity, fatal central apnea, and specific loss of parafacial neurons. *Proc Natl Acad Sci USA* 105(3):1067–1072.

Feldman JL, Del Negro CA (2006) Looking for inspiration: new perspectives on respiratory rhythm. *Nat Rev Neurosci* 7:232–242.Feldman JL, McCrimmon DR (2007) Neural control of breathing. In: Squire LR, Berg D, Bloom F, du Lac S, Ghosh A, and Spitzer N, eds. *Fundamental Neuroscience*, 3rd ed., pp. 855–872. New York: Academic Press.

Gray PA, Janczewski WA, Mellen N, McCrimmon DR, Feldman JL (2001) Normal breathing requires preBötzinger complex neurokinin-1 receptor-expressing neurons. *Nat Neurosci* 4:927–930.

Gronli JO, Santucci BA, Leurgans SE, Berry-Kravis EM, Weese-Mayer DE (2008) Congenital central hypoventilation syndrome: PHOX2B genotype determines risk for sudden death. *Pediatr Pulmonol* 43:77–86.

Guyenet PG (2008) The 2008 Carl Ludwig Lecture: retrotrapezoid nucleus, CO2 homeostasis, and breathing automaticity. *J Appl Physiol* 105:404–416.

Janczewski WA, Feldman JL (2006) Distinct rhythm generators for inspiration and expiration in the juvenile rat. *J Physiol* 570:407–420.

Janczewski WA, Onimaru H, Homma I, Feldman JL (2002) Opioid-resistant respiratory pathway from the preinspiratory neurones to abdominal muscles: in vivo and in vitro study in the newborn rat. *J Physiol* 545:1017–1026.

Lieske SP, Thoby-Brisson M, Telgkamp P, Ramirez JM (2000) Reconfiguration of the neural network controlling multiple breathing patterns: eupnea, sighs and gasps. *Nature Neuroscience* 6:600–607.

McKay LC, Janczewski WA, Feldman JL (2005) Sleep-disordered breathing after targeted ablation of preBötzinger complex neurons. *Nat Neurosci* 8:1142–1144.

Morgado-Valle C, Beltran-Parrazal L, Difranco M, Vergara JL, Feldman JL (2008) Somatic Ca^{2+} transients do not contribute to inspiratory drive in preBotzinger Complex neurons. *J Physiol* 586:4531–4540.

Onimaru H, Homma I (2003) A novel functional neuron group for respiratory rhythm generation in the ventral medulla. *J Neurosci* 23:1478–1486.

Rekling JC, Champagnat J, Denavit-Saubie M (1996) Electroresponsive properties and membrane potential trajectories of three types of inspiratory neurons in the newborn mouse brain stem in vitro. *J Neurophysiol* 75:795–810.

Rekling JC, Shao XM, Feldman JL (2000) Electrical coupling and excitatory synaptic transmission between rhythmogenic respiratory neurons in the preBötzinger complex. *J Neurosci* 20:RC113.

Richerson GB (2004) Serotonergic neurons as carbon dioxide sensors that maintain pH homeostasis. *Nat Rev Neurosci* 5:449–461.

Segers LS, Nuding SC, Dick TE, Shannon R, Baekey DM, Solomon IC, Morris KF, Lindsey BG (2008) Functional connectivity in the pontomedullary respiratory network. *J Neurophysiol* 100:1749–1769.

Smith JC, Ellenberger HH, Ballanyi K, Richter DW, Feldman JL (1991) Pre-Bötzinger complex: a brainstem region that may generate respiratory rhythm in mammals. *Science* 254:726–729.

Tan W, Janczewski WA, Yang P, Shao XM, Callaway EM, Feldman JL (2008) Silencing preBötzinger complex somatostatin-expressing neurons induces persistent apnea in awake rat. *Nat Neurosci* 11:538–540.

Tan W, Pagliardini S, Yang P, Janczewski WA, Feldman JL (2010) Projections of pre-Bötzinger Complex neurons in adult rats. *J Comp Neurol* 518:1862–1878.

Thoby-Brisson M, Karlen M, Wu N, Charnay P, Champagnat J, Fortin G (2009) Genetic identification of an embryonic parafacial oscillator coupling to the preBotzinger complex. *Nat Neurosci* 12:1028–1035.

Part II
Invertebrates

Section 1

Visual System

40

Neurons and Circuits Contributing to the Detection of Directional Motion across the Fly's Retina

Nicholas J. Strausfeld

Studies of the insect compound eye, and its underlying circuitry, have a long and venerable tradition. Particularly the eyes of cyclorrhaphan flies—calliphorids, muscids, and drosophilids—have all offered insights into the physiological optics of the retina and the relationship of photoreceptors with neurons that comprise the first synaptic levels of the optic lobes: the lamina and medulla. What is striking about these levels is their many similarities with the external and internal plexiform layers of vertebrate retinae.

Already in the early 1900s it was recognized that in flies groups of six photoreceptors, distributed under as many separate lenslets, each provide converging axons to a single target in the underlying neuropil: relay neurons called lamina monopolar cells (Vigier, 1907; Vigier, 1908; Vigier, 1909). These observations were taken up by Cajal (1909) and culminated in Cajal and Sanchez's stunning monograph on the insect optic lobes (1915). Half a century later, elegant optical experiments demonstrated that six photoreceptors that "look" at the same point in space indeed send their axons to a single retinotopic column in the lamina (Braitenberg, 1967; Kirschfeld, 1967). Each set of six receptors that have identical optical alignment has been termed a *visual sampling unit* or *VSU* (Franceschini, 1975).

Studies of the synaptic relationships of these photoreceptors with postsynaptic neurons were initiated in the 1960s by Trujillo-Cenóz (1966; see also, Trujillo-Cenóz and Melamed, 1970), with further work by Boschek (1971). These investigations employed serial electron microscopy sectioning and then reconstructing sequences of profiles. A volume of work on the lamina of *Drosophila* has used similar serial section strategies with the added refinement of computer-assisted alignment and three-dimensional rendering to provide a comprehensive reconstruction of the circuitry relating to a single retinotopic columns ("optic cartridge") in that species (Meinertzhagen and O'Neil, 1991; Meinertzhagen and Sorra,2001).

The following description departs from serial section studies in that it is based on combining selective impregnation of single neurons by the Golgi method with electron microscopy. Neurons were first identified in their entirety in thick plastic sections that were then re-embedded and thin-sectioned for electron microcopy. In the electron beam, an impregnated neuron is seen as an electron opaque precipitate within certain profiles. But synapses onto these profiles, and among other unimpregnated profiles, can be resolved. In this manner, a series of studies with José Campos-Ortega provided a comprehensive data set on which the following descriptions are based (Campos-Ortega and Strausfeld, 1972; Campos-Ortega and Strausfeld, 1973; Strausfeld and Campos-Ortega, 1973a; Strausfeld and Campos-Ortega, 1973b; Strausfeld and Campos-Ortega, 1977). In addition, immunocytological studies suggesting peptide, transmitter, or receptor affinities of certain neurons underpin that data set (Sinakevitch and Strausfeld, 2004). Golgi impregnations further demonstrate subsets of neurons in the medulla, and at deeper levels, that are common across many species of dipterous insects as well as those of other groups such as the Hymenoptera (Buschbeck and Strausfeld, 1996) and indeed malacostracan crustaceans (N. Strausfeld, unpublished data). Further, intracellular recordings have been obtained from small retinotopic neurons that are disposed to relay information from lamina outputs to levels of the optic lobes, particularly those in which reside wide-field and directionally selective motion-sensitive neurons (Douglass and Strausfeld, 1995–2005). All these approaches conspire to provide a neuromorphic circuit that explains the results of most studies of motion perception by flying insects (Higgins et al.,2004).

The present account summarizes, first, the synaptic organization of the house fly lamina (Fig. 40.1). Presynaptic sites are table-like structures (in optimal cross-section, appearing somewhat like a fattened Greek letter π) apposing a flatten postsynaptic density. In this figure, various reds indicate assumed excitatory channels, and blue indicates inhibitory channels. The T1 cell is enigmatic and is distinguished by light blue; its polarity is unknown, as its transmitter. Likewise, the L4 and L5 monopolar cells, which do not receive inputs from photoreceptors, have as yet unknown functions and their transmitters are unresolved. They are shown as green profiles. Figure 40.2 summarizes the chemical identities, as far as evidence suggests, of the major lamina neurons and their target neuropil, the medulla, with its links to wide-field neurons of a deep tectum-like neuropil called the lobula plate (Fig. 40.2). As in Figure 40.1, the color coding of neurons in these diagrams reflects their probable excitatory (various reds) and inhibitory (blue) roles and chemical identities. Again, the exception is the T1 neuron, the polarity and transmitter of which are unknown, but which may be centripetal. Likewise, a second amacrine is given as a green profile indicating its unknown role and chemical identity. Figure 40.3 maintains the color code. The figure suggests functional relationships among the described neurons that compute motion in two successive steps. First is a peripheral and radially symmetrical circuit that is reminiscent of a Hassenstein-Reichardt (1956) circuit, but it provides no information about directional or orientation—just motion. Second is an organization that is similar to the Barlow-Levick (1965) circuit for directional motion. The final figure (Fig. 40.4) is a functional circuit derived from the collective data: synaptology, transmitter identity, and intracellular recordings that demonstrate directional properties of neurons that target the lobula plate. Again, excitatory and inhibitory elements are depicted in red and blue, respectively.

Synaptic Organization in the First Visual Neuropil (the Lamina) of the House Fly Visual System

The first level of the insect visual system, called the lamina, is best known from two species: the house fly *Musca domestica* and the fruit fly *Drosophila melanogaster* (Strausfeld, 1976; Fischbach and Dittrich, 1989). Both reveal the same principal types of neurons, as revealed by Golgi impregnation. However, already at this level of the system there are subtle morphological differences between these species and, presumably therefore, others too.

Data from *Musca domestica* are based on electron micrographic reconstructions of preidentified Golgi-impregnated neurons (Strausfeld and Campos-Ortega, 1977). Published data from *Drosophila* are somewhat at variance with that from *Musca*, however, and were obtained by serial sectioning down the length of a single optic cartridge. Within that constraint, the strategy allows exact counts of synapses between profiles, the neuronal identities of which rely on partial serial reconstruction (Meinertzhagen and O'Neil, 1991).

Explanation of Figure 40.1

Groups of six photoreceptor endings (two shown only at each retinotopic column, labeled R1–6) define a regular system of subunits across the lamina. The outer segments of each group sample the same point in space although distributed beneath six different lenses of the compound eye.

The same set of neural elements occurs at each retinotopic column (also called "optic cartridge") of the lamina. Three relay neurons, called L1, L2, and L3 monopolar cells, receive inputs from photoreceptors, L3 receiving about a third of the synaptic input sites compared with either L1 or L2. The L2 neuron differs from L1 in that L2 provides feedback synapses onto L1 and onto some, if not all, the R1–R6 terminals. These feedback pathways are restricted to the inner third of the lamina (at open arrow, upper left panel of Fig. 40.1). A system of interneurons, called L4 neurons (shown in green), provides a

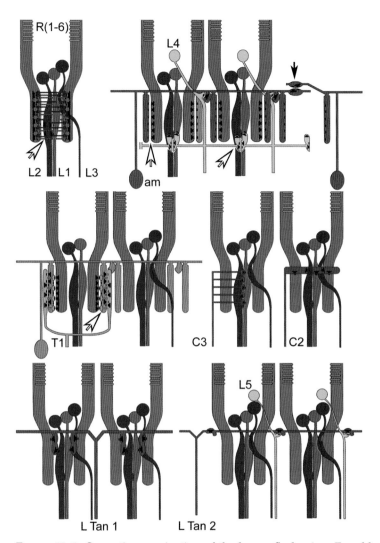

Figure 40–1. Synaptic organization of the house fly lamina. For abbreviations, arrows, color code, and full explanation, see text.

regular network of axon collaterals beneath the entire lamina. In *Musca*, L4 receives its inputs (open arrow) from the type 1 lamina amacrine ("am," in top right panel). L4 axon collaterals are pre- and postsynaptic to each other and are presynaptic onto the axons of L1 and L2. Studies of *Drosophila* have not identified amacrine-to-L4 connections, however. Also, genetic studies of the fruit fly lamina, using targeted cell knockouts, reveal functional asymmetry of L1 and L2, which may be due to a systematic asymmetry of synaptic connections by the L4 network onto these neurons (Rister et al., 2007).

A fourth centripetal cell, called T1 (shown light blue), receives inputs both from photoreceptor terminals and from amacrine processes (indicated at open arrow, left middle panel). Amacrines are also pre- and postsynaptic to each other (solid arrow, top right panel).

The lamina receives substantial inhibitory centrifugal input. Two GABA immunoreactive neurons, called C2 and C3 (shown in blue), terminate at single columns. C2 synapses onto the three monopolar cells L1–L3 at their outer necks, just distal to input synapses onto these three centripetal neurons. C3 climbs the retinotopic column, providing inputs to L1 and L2, but not L3—at least in *Musca*. Two types of wide-field centrifugal cells (shown in blue) terminate in the lamina. The type 1 tangential cell (L Tan 1, lower left) provides inputs to L1, L2, and L3. The type 2 (L Tan 2) lamina tangential (lower right) provides input to the enigmatic midget monopolar cells, L5 (shown in green), which appears to derive its inputs only from this species of centrifugal cell, at least in *Musca*. It is not known how strongly inhibitory centrifugal GABAergic inputs to the lamina are or the degree to which they contribute to feedback control of the L1, L2 outputs. Notably, the centripetal L3 neuron receives least feedback from GABAergic centrifugal neurons. Nor does it appear to be involved in local feedback pathways that have been suggested (in *Drosophila*'s lamina) to contribute to gain control of the photoreceptor output (Zheng et al., 2006). It has been proposed that L3 provides a third color channel to the medulla (Anderson and Laughlin, 2000), in parallel with the outputs of two photopic receptors in each ommatidia, called R7 and R8. These do not participate in lamina circuitry and are omitted here.

"Common Denominator" Neurons of Insect Optic Lobes

An advance has been the identification of a subset of optic lobe neurons that are ubiquitous to many different species of Diptera (flies) and other types of neopteran insects (Buschbeck and Strausfeld, 1996). These neurons also can be labeled by antisera raised against different transmitters or their precursors, peptides, and receptor proteins (Sinakevitch and Strausfeld, 2004). Their various immunoreactive affinities are color coded in Figure 40.2. The same colors have been used, where applicable for Figure 40.1 neurons. The layer relationships of these "common denominator" neurons are summarized in Figure 40.2. Crucially, their deepest elements, the T5 cells, supply inputs to large output neurons (shown gray) in the optic lobe's lobula plate—a tectum-like neuropil that integrates information about visual motion across the retina and which is essential for visually stabilized flight. Motion-induced responses have been recorded from some of these small retinotopic neurons at levels 1–3 of the system (e.g., nondirectionally selective responses by Tm neurons, strongly directionally responses by T5 neurons; see Strausfeld et al., 2006).

Explanation of Figure 40.2

Level 1 represents the lamina. Level 2 summarizes some of the assumed relationships in the outer layers of the second optic neuropil, the medulla. Level 3 indicates relationships with T5 cell dendrites within a superficial layer of the third optic neuropil, the lobula. T5 cells supply level 4, the tectum-like lobula plate, and its layers of large tangential neurons. Only one of these is represented here.

Receptor terminals are known to be histaminergic (Hardie, 1989). The L1, L2, and L3 monopolar cells show glutamate-like immunoreactivity, as do the large wide-field tangential neurons in the lobula plate at level 4 (Sinakevitch and Strausfeld, 2004). Comparisons across different species of Diptera (flies) reveal "common denominator" neurons associated with such lobula plate outputs. In the lamina these are the L2 monopolar cells, the T1 efferent neurons, and the type 1 lamina amacrine (synaptic connections of type 2 amacrine are not yet known but may provide amacrine-to-amacrine pathways).

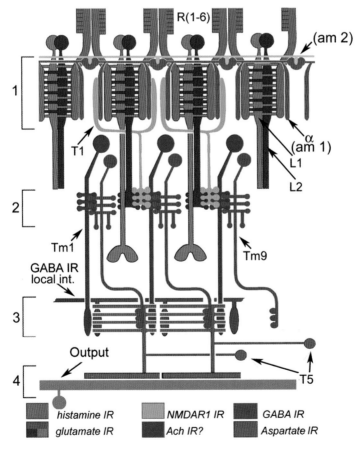

FIGURE 40–2. "Common denominator" neurons of insect optic lobes. For abbreviations and details, see full explanation in text.

The ending of each L2 neuron interdigitates into, and is tightly apposed with, the ending of the T1 neuron from the same lamina column. This, in turn, is tightly apposed to the dendrites of a pair of relay cells in the second neuropil of the optic lobes, called the medulla. These are the Tm 1 transmedullary cells, one of which, Tm 9 (shown blue), is revealed by antibodies raised against the inhibitory transmitter gamma amino butyric acid (GABA). These transmedullary cells terminate at level 3 on a system of small interneurons, called T5 cells. T5 cells are unusual in that they occur as quartets in each retinotopic column. Their dendrites extend across two to three projected columns. T5 axons terminate on output neurons of the lobula plate.

Unpublished observations have identified L2 terminals presynaptic onto the lamella-like profiles of the T1 ending. This is tightly opposed to the dendrites of Tm 1. Curiously, although densities in Tm1 profiles suggest postsynaptic sites, no synapses have been observed in the T1 terminal itself. Possible it acts to impede, and thus impose a delay of, the passage of transmitter molecules from L2 onto Tm1, whereas there may be direct connections between L2 and the inhibitory relay neuron Tm9. While synaptic relationships of these cells in the medulla are hardly known, their precise layer relationships, which are constant across species of *Diptera*, suggest likely connections. Their putative chemical identities are indicated, with GABA immunoreactive neurons possibly providing crucial elements in the computation of directional motion.

A Proposed Circuit for Computing Directional Motion

Synaptic relationships at the level of the lamina and topological (layer) relationships among "common denominator" neurons provide one possible explanation of how nondirectional motion might be computed in the lamina. Information about the direction and orientation of motion would be obtained at levels 2 and 3. The first step in this sequence of events is summarized in the upper part A of Figure 40.3. The color coding is as for Figure 40.2.

Explanation of Figure 40.3

Each L2 monopolar cell is postsynaptic to six photoreceptor endings that view the same point in space. These are termed a visual sampling unit (VSU). This set is denoted by the VSU labeled "C" that supplies one L2 monopolar neuron. The VSUs "1–6", which sample six surrounding points in space supply, via amacrine ("am") connections, the dendrites of the T1 neuron represented as a ring encircling L2. Amacrine-to-amacrine relays (at "am" boxes) en route to T1 provide delays in information transfer from a VSU to T1. Other than its monosynaptic delay there is no other delay between the direct connection between VSU "C" and L2, however.

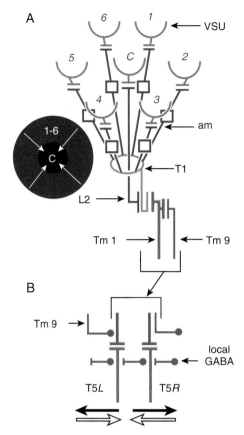

FIGURE 40.3. Proposed circuit for computing directional motion in the lamina. For abbreviations and details, see full explanation in text.

Axons of L2 and T1 from the same lamina column converge at the dendrites of Tm neurons (Tm1, Tm1b) in the medulla (level 2), as diagrammed in Figure 40.2. The ending of T1 interposes between the L2 ending and Tm1 dendrites. A change in the membrane potential of L2 denoting a change of luminance at VSU "C" would be integrated with a delayed membrane potential change in T1 denoting a change of luminance at any of the VSUs "1–6." Thus, at this level, motion is computed by Tm1 neurons as nondirectional (as indicated by the concentric fields and arrows, left). This model, which is based on identified neurons, immunocytology, and rare intracellular recordings from T1, Tm1, and amacrine cells (see Douglass and Strausfeld, 1995, 1996, 2005), proposes that cholinergic Tm1 cells also provide inputs to an aminobutyric acid (GABA) immunoreactive Tm9 neuron, or that Tm9 is itself also postsynaptic to L2, via T1.

The lower part B of Figure 40.3 summarizes the proposed circuit for directional motion selectivity by T5 neurons supplying large tangential neurons in the lobula plate. Four T5 neurons represent each lamina column, and thus each VSU. But for simplicity only one pair of T5 channels of a retinotopic column will be considered here. Each medulla column provides the outputs Tm1 and Tm9. These terminate on T5 neurons. One T5 neuron (here denoted by T5*L*) receives an input from the cholinergic Tm1, the ending of which is postsynaptic to the inhibitory GABA-immunoreactive neuron Tm9 that originates from an adjacent retinotopic column. This Tm 9 cell also encodes non-directional motion. This combination results in one T5 neuron (T5*L*) responding preferentially to leftward motion. Its partner T5 neuron (T5*R*) receives its excitatory input from the cholinergic Tm1. Its inhibitory input, however, is proposed to derive from a local GABA-immunoreactive element that extends from T5*L* to T5*R*. Such local inhibitory elements, like the Tm9 neurons, have been identified (Sinakevitch et al., 2003). They are provided by a wide-field intrinsic neuron that spreads among all the T5 cell dendrites in layer 3 of the system. As a consequence, T5*R* is preferentially inhibited by leftward motion but excited by rightward motion.

A Neuromorphic Circuit

A computational model of directional motion units based on data summarized earlier (Higgins et al., 2004) identifies each L2 monopolar neuron represented by a negative first-order high-pass filter (HPF). Amacrine cells are shown as receiving inputs from photoreceptors and have noninverting responses identical to those of photoreceptors (see Douglass and Strausfeld, 2005). In the model, the amacrine to T1 synapse includes a relaxed high-pass filter and sign inversion, as suggested by recordings from T1 neurons (Douglass and Strausfeld, 1995). A first-order low-pass filter (LPF) interposes in the path from the amacrine cell to T1. T1 is modeled as a summation. The synapse of L2 onto Tm1 with an intervening ending of T1 (as in Fig. 40.2) synapse is modeled as a synapse directly between L2 to Tm1, the latter modeled as summing L2 and T1 inputs. Tm9 is modeled to receive the same inputs as Tm1, but it is delayed with respect to Tm1 by a first-order low-pass filter. The time constants and descriptions for the various types of filters are given in Higgins et al. (2004).

Neuroanatomical observations show that the dendrites of T5 neurons are visited by several neighboring Tm1 neuron endings. The T5 neurons are thus assumed to integrate Tm inputs from several neighboring retinotopic columns in the lamina to produce a motion output. The orientation of neighboring retinotopic columns determines the preferred-null axis of the resulting T5 cells. In the model, Tm1 and Tm9 cells converge onto a T5 cell, with Tm9 crossing to a neighboring optic cartridge as shown by GABA-immunolabeling of Tm9 neurons (Sinakevitch et al., 2003). Tm1 excites T5 cells. Tm9 synapses onto T5 with a shunting inhibitory connection. An inhibitory interneuron, also identified as extending across all T5 dendrites, receives in the model the same inputs as both T5 cells combined and inhibits both T5 cells such that its responses are subtracted from both T5 cells.

Explanation of Figure 40.4

This figure provides a neuronally based one-dimensional computational EMD model that provides direction-selective T5 units. Rectification inherent in shunting inhibition expression makes T5 sensitive to transiently decreasing intensity levels. In this two-dimensional model, each T1 unit carries low-pass filtered inputs from all six surrounding VSUs, rather than just the left and right neighbors as shown here. The local inhibitory interneuron (LIN) is the wide-field inhibitory local interneuron at the level of T5 dendritic trees.

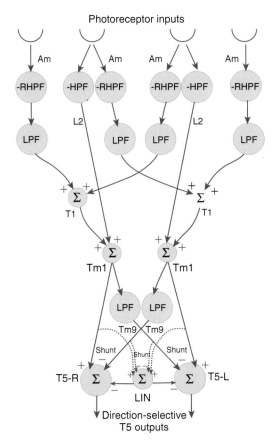

Figure 40.4. Proposed circuit for elementary motion detection (EMD). RHPF: relaxed high-pass filter; VSU: visual sampling unit. For other abbreviations and details, see full explanation in text.

THE 540-MILLION-YEAR EXPERIMENT

Finally, a word of caution: the cellular organization of the fly lamina is the result of about 540 million years of evolution. It began approximately in the Lower Mid-Cambrian when there were simple crustacean-like antecedents equipped with compound eyes but probably a simple visual system like those in extant "living fossils," such as the branchiopod crustacean *Triops*. This evolutionary basal crustacean has just two optic neuropils, similar to layers 1 and 4 of flies. Its lamina has L2-like neurons as its outputs and is equipped with amacrine-like connections. The L2 type neurons terminate in a tectum-like neuropil that is equipped with prominent tangential processes. These provide outputs to premotor neurons in the brain. A similar system equips scutigeromorph centipedes, also endowed with compound eyes. Such simple systems suggest that some form of motion computation must have been provided by a basic system of amacrines and L2-like neurons, and it is likely that these essential elements form the core of the sophisticated motion-computing circuits present in modern insects and crustaceans.

However, it is also just as probable that additional motion-computing circuits have evolved or that additional components of the present system exist. There is evidence from studies of *Phaenicia* that some T5 neurons may contain inhibitory transmitter and thus may directly interact at the lobula plate to amplify or even provide the directional selectivity of its large tangential neurons (Strausfeld et al., 1995). Studies of *Drosophila* have identified GABA receptors on the dendrites of lobula plate giant tangential neurons (Raghu et al., 2007). Finally, many outputs from the optic lobes diverge to other central regions of the brain. Recordings from neurons that extend from the medulla to the lobula (Douglass and Strausfeld, 2003), and centrally from this neuropil (Okamura and Strausfeld, 2007), demonstrate their directional motion sensitivity. This implies the existence of other motion-computing circuits in addition to those suggested here.

REFERENCES

Anderson JC, Laughlin SB (2000) Photoreceptor performance and the co-ordination of achromatic and chromatic inputs in the fly visual system. *Vision Res* 40:13–31.
Barlow HB, Levick WR (1965) The mechanism of directionally selective units in the rabbit's retina. *J Physiol* 178:477–504.
Boschek CB (1971) On the fine structure of the peripheral retina and lamina ganglionaris of the fly, *Musca domestica*. *Z Zellforsch Mikrosk Anat* 118:369–409.
Braitenberg V (1967) Patterns of projections in the visual system of the fly. I. Retina lamina projections. *Exp Brain Res* 3:271–298.
Buschbeck EK, Strausfeld NJ (1996) Visual motion-detection circuits in flies: small field retinotopic elements responding to motion are evolutionarily conserved across taxa. *J Neurosci* 16:4563–4578.

Cajal SR (1909) Nota sobre la structural de la retina de la Mosca. *Trab Lab Invest Biol Univ Madr* 7:217–227.

Cajal SR, Sánchez SD (1915) Contribución al conocimiento de los centros nerviosos de los insectos. Parte I Retina y centros opticos. *Trab Lab Invest Biol Univ Madrid* 13:1–168.

Campos-Ortega JA, Strausfeld NJ (1972) The columnar organization of the second synaptic region of the visual system of *Musca domestica* L. I. Receptor terminals in the medulla. *Z Zellforsch* 124:561–582.

Campos-Ortega JA, Strausfeld NJ (1973) Synaptic connections of intrinsic cells and basket arborizations in the external plexiform layer of the fly's eye. *Brain Res* 59:119–136.

Douglass JK, Strausfeld NJ (1995) Visual motion detection circuits in flies: peripheral motion computation by identified small-field retinotopic neurons. *J Neurosci* 15:5596–5611.

Douglass JK, Strausfeld NJ (1996) Visual motion-detection circuits in flies: parallel direction- and non-direction-sensitive pathways between the medulla and lobula plate. *J Neurosci* 16:4551–4562.

Douglass JK, Strausfeld NJ (1998) Functionally and anatomically segregated visual pathways in the lobula complex of a calliphorid fly. *J Comp Neurol* 396:84–104.

Douglass JK, Strausfeld NJ (2000) Optic flow representation in the optic lobes of Diptera: modeling the role of T5 directional tuning properties. *J Comp Physiol A* 186:783–797.

Douglass JK, Strausfeld NJ (2003) Retinotopic pathways providing motions elective information to the lobula from peripheral elementary motion detecting circuits. *J Comp Neurol* 457:326–344.

Douglass JK, Strausfeld NJ (2005) Sign-conserving amacrine neurons in the fly's external plexiform layer. *Visual Neurosci* 22:345–358.

Fischbach KF, Dittrich APM (1989) The optic lobe of *Drosophila melanogaster*.1. A Golgi analysis of wild-type structure. *Cell Tissue Res* 258:441–475

Franceschini N (1975) Sampling of the visual environment by the compound eye of the fly: fundamentals and applications. In: Snyder AW and Menzel R, eds. *Photoreceptor Optics*, pp. 98–125. Berlin: Springer.

Hardie RC (1989) A histamine-activated chloride channel involved in neurotransmission at a photoreceptor synapse. *Nature* 339:704–706

Hassenstein B, Reichardt W (1956) Systemtheoretische Analyse der Zeit-, Reihenfolgen- und Vorzeichenauswertung bei der Bewegungsperzeption des Rüsselkäfers *Chlorophanus*. *Z Naturforsch* 11:513–524.

Higgins CM, Douglass JK, Strausfeld NJ (2004) The computational basis of an identified neuronal circuit for elementary motion detection in dipterous insects. *Visual Neurosci* 21:567–586.

Kirschfeld K (1967) Die Projektion der optischen Umwelt auf das Raster der Rhabdomere im Komplexauge von *Musca*. *Exp Brain Res* 3:248–270.

Meinertzhagen IA, O'Neil SD (1991) Synaptic organization of columnar elements in the lamina of the wild type in *Drosophila melanogaster*. *J Comp Neurol* 305:232–263.

Meinertzhagen IA, Sorra KE (2001) Synaptic organization in the fly's optic lamina: few cells, many synapses and divergent microcircuits. *Prog Brain Res* 131:53–69.

Okamura JY, Strausfeld NJ (2007) Visual system of calliphorid flies: Motion- and orientation-sensitive visual interneurons supplying dorsal optic glomeruli. *J Comp Neurol* 500:189–208.

Raghu SV, Joesch M, Borst A, Reiff DF (2007) Synaptic organization of lobula plate tangential cells in Drosophila: gamma-aminobutyric acid receptors and chemical release sites. *J Comp Neurol* 502:598–610.

Rister J, Pauls D, Schnell B, Ting CY, Lee CH, Sinakevitch I, Morante J, Strausfeld NJ, Ito K, Heisenberg M (2007) Dissection of the peripheral motion channel in the visual system of *Drosophila melanogaster*. *Neuron* 56:155–170.

Sinakevitch I, Strausfeld NJ (2004) Chemical neuroanatomy of the fly's movement detection pathway. *J Comp Neurol* 468:6–23.

Sinakevitch I, Douglass JK, Scholtz G, Loesel R, Strausfeld NJ (2003) Conserved and convergent organization in the optic lobes of insects and isopods, with reference to other crustacean taxa. *J Comp Neurol* 467:150–172.

Strausfeld NJ (1976) *Atlas of an Insect Brain*. Berlin: Springer.

Strausfeld NJ, Campos-Ortega JA (1973a) L3, the 3rd 2nd order neuron of the 1st visual ganglion of the "neural superposition" eye of *Muscadomestica*. *Z Zellforsch* 139: 397–403.

Strausfeld NJ, Campos-Ortega JA (1973b) The L4 monopolar neuron: a substrate for lateral interaction in the visual system of the fly, *Musca domestica*. *Brain Res* 59:97–117.

Strausfeld NJ, Campos-Ortega JA (1977) Vision in insects: pathways possibly underlying neural adaptation and lateral inhibition. *Science* 195:894–897.

Strausfeld NJ, Kong A, Milde JJ, Gilbert C, Ramaiah L (1995) Oculomotor control in calliphorid flies: GABAergic organization in heterolateral inhibitory pathways. *J Comp Neurol* 361:298–320.

Strausfeld N, Douglass J, Campbell H, Higgins C (2006) Parallel processing in the optic lobes of flies and the occurrence of motion computing circuits. In: Warrant E and Nilsson D-E, eds. *Invertebrate Vision*, pp. 349–398. Cambridge, England: Cambridge University Press.

Trujillo-Cenóz O (1966) Compound eye of Dipterans—anatomical basis for integration. *J Ultrstruct Res* 16:395–398.

Trujillo-Cenóz O, Melamed J (1970) Light and electron microscopical study of one of the systems of centrifugal fibres found in the lamina of muscoid flies. *Z Zellforsch* 110: 336–349.

Vigier P (1907) Sur la réception de l'excitant lumineux dans les yeux composés des insectes, en particulier chez les Muscides. *CR Acad Sci (Paris)* 145:633–636.

Vigier P (1908) Sur l'existence réelle et le rôle des appendices piriformes des neurones. Le neurone périoptique des Diptères. *CR Soc Biol (Paris)* 64:959–961.

Vigier P (1909) Mécanisme de la synthèse des impressions lumineuses recueillies par les yeux composés des Diptères. *CR Acad Sci (Paris)* 148:1221–1223.

Zheng L, de Polavieja GG, Wolfram V, Asyali MH, Hardie RC, Juusola M (2006) Feedback network controls photoreceptor output at the layer of first visual synapses in Drosophila. *J Gen Physiol* 127:495–510.

41

The Optic Lamina of Fast Flying Insects as a Guide to Neural Circuit Design

Simon B. Laughlin

Lamina circuits are designed to transfer information effectively and efficiently from photoreceptors to interneurons in the face of two neural constraints: a limited dynamic range and synaptic noise. The similarities between lamina circuits and their vertebrate equivalents in the outer plexiform layer of the retina suggest that both systems have adopted common solutions to shared problems (Laughlin and Hardie, 1978), and the lamina's explicitly modular structure and its facility for high-quality in vivo intracellular recordings have enabled us to work some of these principles out (Järvilehto and Zettler, 1973; Shaw, 1984; van Hateren, 1992; Laughlin, 1994; Juusola et al., 1996). Thus, we can appreciate how a neural circuit is designed to perform a well-defined task both effectively and efficiently.

The lamina circuits of two fast flying diurnal insects, blowflies (Diptera) and dragonflies (Odonata), support high-quality recordings and respond similarly (Laughlin and Hardie, 1978), but we switched our work to blowflies for two good reasons. First, we have remarkably comprehensive wiring diagrams (Meinertzhagen and Sorra, 2001; Chapter 40, this volume), and, second, there is an impressive body of quantitative work on fly visual behavior (Buchner, 1981)—a veritable fly psychophysics. The field is turning increasingly to the fly *Drosophila* for genetic techniques (Zheng et al., 2006; Rister et al., 2007).

The lamina offers electrophysiologists a number of technical advantages. One can record from identified lamina neurons in intact, unanesthetized flies and test their performance over the full operating range of fly vision, from 4 photons/photoreceptor s^{-1} at the threshold for motion detection to 10^7 photons/photoreceptor s^{-1} in sunlight. High-quality recordings support a variety of systems identification techniques, starting with step and impulse responses (Järvilehto and Zettler, 1973; Laughlin and Hardie, 1978; Laughlin et al., 1987), and progressing to white noise analysis (French and Järvilehto, 1978; Juusola et al., 1995; de Ruyter van Steveninck and Laughlin, 1996) and natural stimuli (van Hateren, 1997; Juusola and de Polavieja, 2003). The resulting measurements of signals, noise, and transfer functions have been combined with powerful modeling and analysis techniques from systems and communications engineering to demonstrate that the lamina is designed to code natural signals efficiently (Laughlin, 1981; Srinivasan et al., 1982; van Hateren, 1992). We will start our brief account of lamina function with an overview of structure and function, and then we will move on to the design strategies that improve circuit performance.

The lamina is a retinotopic array of neural modules. Each of these cartridges collects the axons of photoreceptors with the same field of view to map, one to one, the optical image sampled by the ommatidia. Golgi-EM studies and serial EM reconstructions have identified every lamina neuron, mapped their connections and their synapses, and shown that the numbers of neurons, connections, and synapses are consistent from cartridge to cartridge (Meinertzhagen and Sorra, 2001; Chapter 40, this volume). A cartridge contains 19 neuronal elements and is enveloped in a sheath of epithelial glial cells, which creates a resistance barrier by restricting the flow of ions and other solutes in extracellular space (Shaw, 1984).

The majority of the lamina's synapses are involved in a relatively simple set of circuits (Fig. 41.1). The achromatic photoreceptors R1–6 drive four glutamatergic interneurons, L1, L2, L3, and the amacrine cell. L1, L2, and L3 are the large monopolar cells (LMCs), and they project to different levels in the same medulla cartridge. An amacrine cell sends processes into several cartridges and is in a position to pool signals. L2 feeds back onto the photoreceptor terminals and amacrines feed back onto the R1–6 terminals and sideways to L3. The basket fiber, T1, is anatomically centrifugal, with its cell body and some dendrites in the medulla, and is strongly driven in the lamina by amacrines. The amacrines dominate the relatively small number of lateral connections between cartridges made within the lamina. Each amacrine cell sends branches over the distal surface of the lamina to participate in the circuits of approximately 40 cartridges, but the branches are so fine that they may not transmit much information.

The electrical responses of the circuits involving R1–R6 and the three LMCs (L1, L2, and L3) have been exceptionally well characterized using intracellular recordings. The photoreceptors R1–R6 drive the LMCs at fast sign-inverting synapses where histamine is a fast transmitter, binding and directly opening postsynaptic chloride channels (Hardie, 1989; Stuart et al., 2007). Photoreceptors and LMCs produce graded responses (Figs. 41.1 and 41.2), and during transfer these analog signals are inverted, amplified, and high-pass filtered to produce pronounced LMC transients (Laughlin, 1994; Juusola et al., 1996). Signal transfer changes with increasing light level; the gain remains constant but transients are heightened and sharpened (Fig. 41.2).

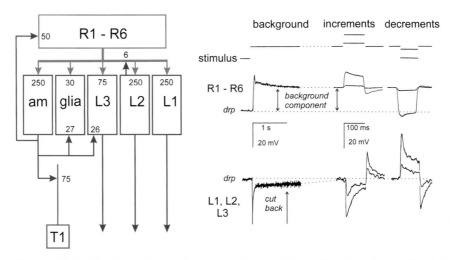

FIGURE **41–1.** Circuits and signals in a lamina cartridge. The six achromatic photoreceptors R1–R6 are presynaptic to four lamina interneurons and to epithelial glial cells. Three major interneurons—L1, L2, and L3—project their axons retinotopically to a single medulla cartridge. The intrinsic amacrine cell (am) drives the centrifugal neuron T1, feeds back onto R1–R6, and feeds sideways to L3 and glia. L2 feeds back onto R1–R6. The numbers show the numbers of synapses in *Drosophila* (rounded off from Meinertzhagen and Sola, 2001), and the numbers are about four times higher in *Musca*. For a more complete, authoritative account of circuitry, see Chapter 40 (this volume). Intracellular recordings of responses show that the photoreceptors code both the background light intensity and increments and decrements about the background. The interneuron responses cut back sharply to the dark resting potential (*drp*) to eliminate the background component, and this enables them to amplify responses to increments and decrements.

Note that the majority of R1–R6 recordings have been made in the retina, before R1–R6 axons enter the lamina. Interactions within the photoreceptor terminals undoubtedly play a role in amplification and high-pass filtering (Weckström et al., 1992), but the mechanisms have not been established.

The sharp cutback in the LMC response eliminates the response to background light, freeing up the neuron's response range to code increments and decrements in light level with high gain (Fig. 41.1). By amplifying these contrast signals (Figs. 41.1 and 41.2) the circuits produce a robust neural image of objects in visual scenes (van Hateren, 1997). Recordings from vertebrate bipolar cells (Fig. 41.3) suggest that they also eliminate background components and amplify (Laughlin and Hardie, 1978), whereas individual bipolar cells are selective for either ON or OFF, and each LMC codes both ON and OFF (Figs. 41.1 and 41.3).

Lamina circuits and codes are designed to improve performance (Laughlin, 1994), according to principles that apply in many neural circuits. To achieve a good signal to noise ratio (SNR), the six photoreceptor terminals drive lamina circuits with hundreds of specialized synapses. Each of these synapses

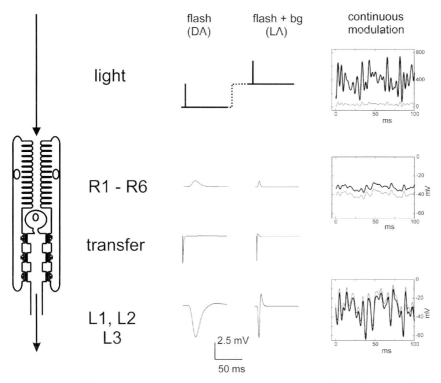

Figure 41–2. The transformation and enhancement of signals as they pass from photoreceptors to neurons L1, L2, and L3 via a large array of parallel synapses. Impulse responses elicited
by flashes of light delivered in darkness (DA) and in the presence of steady background light
(LA) show that light adaptation changes the impulse response for signal transfer. The sharpening of the large monopolar cell response optimizes information transfer (van Hateren, 1992).
A complicated moving pattern will continuously modulate the light signal, and the top panel
shows how increasing illumination 10-fold produces a proportionate increase in the mean and
depth of modulation. The photoreceptor–L1, L2, L3 circuit extracts the modulation, enhances
it, and renders it independent of background light level. This processing achieves an important constancy: reflecting objects produce the same signals at different levels of illumination.
Note that these are diagrammatic traces constructed from measured transfer functions. Noise
is not included.

is a tetrad, with a tabular presynaptic density facing postsynaptic processes
from L1, L2, an amacrine cell, and either L3 or a glial cell. In the housefly
Musca there are 200 tetrads per photoreceptor terminal and in *Drosophila*
about 40 (Meinertzhagen and Sorra, 2001).

To maintain sensitivity and SNR, each tetrad continuously releases vesicles at high rates, both in the dark and in light (Juusola et al., 1996). Noise
analysis indicates release rates of 240 vesicles s^{-1} per tetrad, totaling 250,000
vesicles per LMC per second (Laughlin, 1994). Despite this high rate, intrinsic
noise still limits LMC signal quality, contributing 50% of LMC noise in bright
light when the SNR is at its best (Laughlin et al., 1987). Thus, the high release

rates engineered into lamina circuits are pushing down a strong noise constraint. Vertebrate photoreceptors and hair cells use ribbon synapses to release vesicles at high rates, suggesting that depressing synaptic noise is an important first step (Sterling and Matthews, 2005). Rod and cone output synapses also have two or three postsynaptic elements, suggesting that the projection of vesicles from one release site to multiple postsynaptic elements is advantageous (Meinertzhagen and Sorra, 2001). Perhaps dyads, triads, and tetrads are designed to use vesicles efficiently, to balance inputs, and equalize vesicle noise in postsynaptic elements.

Lamina circuits use graded electrical signals to transfer information at high rates: 1000 bits s^{-1} in R1–R6, and by summing six photoreceptor signals LMCs reach 1650 bits s^{-1} (de Ruyter van Steveninck and Laughlin, 1996). Analog transmission and signal processing avoid the information bottleneck of spike coding (single spiking neurons transmit less than 500 bits s^{-1}) and circuits in vertebrate outer retina also use graded responses (Fig. 41.3).

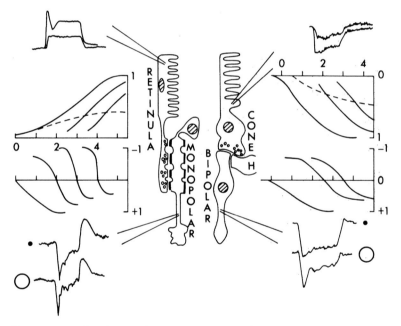

FIGURE 41–3. Similarities between coding in fly retina and vertebrate retina. Photoreceptors code light intensity by amplitude of graded response, albeit of opposite polarities. Plots of response amplitude versus log intensity show that photoreceptors code the background light level (dashed line) and shift their broad dynamic ranges moderately to avoid saturation. Like the monopolar cells L1, L2, and L3, the vertebrate bipolar cells have a much narrower dynamic range. Both cell types adapt rapidly to shift their response/log intensity curves in step with the background, and this removes most of the background signal. A comparison between responses to point light sources (filled circle) and extended sources (open circle) shows that both monopolars and bipolars are subject to lateral inhibition. Response from ON and OFF bipolar have been combined to produce single response/log intensity curves. (Reproduced from Laughlin and Hardie, 1978, with permission)

Measurements of signal and noise power spectra and counts of tetrads give estimates of the information capacity of a single synapse. This capacity, 55 bits s^{-1} is the maximum rate the synapse could achieve, were it driven by an ideal noise-free input. Fifty-five bits s^{-1} is good for a synapse with a contact area of 0.1 μm^2 (de Ruyter van Steveninck and Laughlin, 1996), but without comparable data or a detailed theoretical analysis of biophysical limits, we cannot say how good. Have lamina tetrads overcome a fearsome constraint that boutons have failed to master, or are the low vesicle release probabilities of boutons engineered into these synapses to allow for plasticity and homeostasis? The information capacity of the parallel array of all 1200 tetrads (the rate calculated for an ideal noise free input signal) is 2100 bit s^{-1}, just 30% higher than the maximum bit rate achieved with visual stimuli. This narrow margin confirms that synaptic noise is a constraint and similar tight margins are found in mammalian retina (Sterling and Freed, 2007). The fact that these primary sensory circuits are designed to avoid overcapacity suggests that pushing down synaptic noise is an expensive business.

Measurements and basic theory show that using parallel synaptic arrays to reduce the effects of synaptic noise is indeed costly (Laughlin et al., 1998). Twelve hundred photoreceptor output synapses, each transmitting at 55 bits s^{-1}, could deliver 74,250 bits s^{-1}, but only when each synapse is given a different set of bits to transmit. When all synapses work in parallel to transmit the same set of bits in a parallel array, the array's capacity drops by 97%, to 2210 bits s^{-1}. If the circuit included a complicated set of mechanisms that fed every lamina tetrad with a different set of bits, then just 50 of these statistically independent tetrads could do the job of transmitting at 2210 bits s^{-1}. But this ideal solution seems to be impractical. The circuit sticks to the simpler solution of attempting to send all of the bits through each and every synapse. In this case, the information rate is pushed up by increasing the SNR, and because SNR increases with the square root of the number of synapses, this is a costly solution. Increase in the information rate from 55 bits s^{-1} for 1 synapse to 2210 bits s^{-1} for 1200 synapses raises the energy cost of synaptic transfer from around 10^4 Adenosine-5'-triphosphate (ATP) molecules per bit to 10^6 ATP molecules per bit (Laughlin et al., 1998). But this expensive investment in transmission is worthwhile for the visual system as whole (van Hateren and Laughlin, 1990). The six photoreceptors driving LMCs consume 4.5 × 10^{10} ATP molecules per second to capture the information, almost 20 times the cost of transmitting their information in an LMC.

Given the high cost of high performance in these lamina circuits, it is not surprising that the coding performed by these circuits is also well designed (van Hateren, 1992; Laughlin, 1994). Signal amplification is matched to input signal statistics to increase the information capacity of LMCs. Like other graded synapses, the relationship between presynaptic (R1–R6) and postsynaptic (LMC) response amplitude is sigmoidal. The gain in the midregion of this instantaneous nonlinearity is adjusted so that the relationship between stimulus contrast and LMC response coincides with the cumulative probability distribution of the contrasts experienced by LMCs, when they are coding natural scenes. At any given bright background light level, this match reduces redundancy by ensuring that all LMC response levels are used equally often (Laughlin, 1981; Laughlin et al., 1987). The slope in the mid-region of the characteristic curve is quite high, 6.5, and two factors might contribute: the sensitivity of neurotransmitter release to presynaptic voltage (Laughlin et al., 1987) and a Hill coefficient of approximately 3 for histamine binding to postsynaptic chloride channels (Hardie, 1989). Increasing the transmitter release sensitivity improves the synaptic SNR by recruiting more vesicles, and differences among fly species suggest that the Hill coefficient is used to adjust synaptic gain to visual ecology (Skingsley et al., 1995).

High-gain circuits are prone to saturation, and lamina circuits avoid this by adjusting the characteristic curve for synaptic transfer to maintain operations within the prevailing range of photoreceptor input (Laughlin et al., 1987; van Hateren, 1997). As in vertebrate bipolar cells, this matches a narrow response range to the background light level (Fig. 41.3). The LMC's transient response to a step suggests that the circuit is continuously updating its estimate of the mean photoreceptor input and then using this estimate to reposition the characteristic curve. In fact, the circuits are implementing predictive coding, using weighted means that are computed over time and across space (via lateral antagonism) to predict the current signal (Srinivasan et al., 1982). By definition, predictable components contain no information; they are redundant. Consequently, uninformative components are eliminated, leaving the circuit free to code the informative.

To make good predictions, the lamina circuits take account of input statistics. The dominant factor turns out to be photon noise (Srinivasan et al., 1982), which becomes increasingly severe as light levels fall. Lamina circuits adapt progressively to falling light levels by increasing the intervals of space and time over which they compute predictions, and these changes are precisely regulated to generate best estimates. Thus, predictive coding helps to explain why in many visual systems the spatial and temporal filtering of neural images is intensity dependent, with lateral inhibition and temporal inhibition increasing progressively with light level. These applications of information theory to measurements of natural image statistics and lamina transfer functions (Laughlin, 1981; Srinivasan et al., 1982) vindicated Horace Barlow's farsighted proposal that neural coding eliminates redundancy to pack information neatly into neurons (Barlow, 1961).

Indeed, LMCs are brilliant packagers. These neurons adapt their spatiotemporal filtering remarkably precisely to the statistics of photon noise and natural images, so as to maximize the amount of information they transmit (van Hateren, 1992). As light levels increase, LMCs progressively strengthen lateral inhibition and convert their flash (impulse) response from an integrator of noisy inputs at low light levels to a short sharp differentiator of reliable inputs at high light levels (Fig. 41.2). Given that many visual systems have to solve similar packaging problems, it should come as no surprise that van Hateren's optimization explains how and why spatial, temporal, and chromatic coding adapt to light level in mammalian vision (van Hateren, 1992).

The mechanisms used by lamina circuits to adapt coding to light level have not been fully worked out, but the evidence suggests more principles of circuit design. The increasing attenuation of low frequencies at higher light levels is implemented as signals are transferred from photoreceptors to LMCs. As light levels increase, the impulse response for synaptic transfer develops a pronounced off-transient of opposite polarity (Fig. 41.2), and this attenuates the more slowly changing signals and implements predictive coding. This antagonistic component is remarkably fast at high light levels (time constant = 2 ms), suggesting that fast feedback on the presynaptic terminals of photoreceptors and histamine receptor desensitization are involved. This suggestion is supported by the fact that LMC response transients and lateral inhibition are produced by reducing the postsynaptic chloride conductance. Regulating the input before opening postsynaptic channels is energy efficient, because producing a transient by activating a slower opponent postsynaptic conductance is like stamping on the brake without taking your foot off the gas. Presynaptic regulation also helps synapses to make better use of their limited supply of vesicles and range of release rates (Laughlin, 1994). The need to use energy and vesicles efficiently and economically could help to explain why the outputs of many sensory receptors, for example, cones, olfactory neurons, and mechanosensors, are regulated presynaptically (Nawroth et al., 2007; Sterling and Freed, 2007).

Both the lamina circuitry and the electrophysiological data allow for considerable complexity in the mechanisms that regulate and tune photoreceptor outputs and LMC responses (Laughlin, 1994). For example, the large transients that LMC produce when they are driven hard contain contributions from several LMC conductances, acting nonlinearly on different time scales. In exciting new studies of *Drosophila* lamina, blocking synapses with the temperature-sensitive *Shibire* mutation shows that lamina circuits depolarize the photoreceptor terminals (Zheng et al., 2006). Whether this feedback promotes synaptic gain and, via the involvement of L2 and am, provides a negative feedback that contributes to the production of transient responses is an interesting question.

The photoreceptor output synapses are probably regulated ephaptically by prominent changes in extracellular potential within the cartridge (Shaw, 1984; Laughlin, 1994). The potentials are produced by current flowing across the resistance barrier formed around the cartridge by epithelial glial cells. The extracellular potential is a slower, attenuated version of the photoreceptor input and is, therefore, well placed to produce transient responses by backing off the presynaptic membrane potential at photoreceptor output synapses. This extracellular mechanism must be less noisy than direct synaptic feedback because, like neurotransmitter spillover, it is produced by a large number of sources, and the mechanism is certainly economical because it uses existing currents. There is evidence that ephaptic transmission mediates the feedback of horizontal cells onto cone output synapses (Kamermans and Fahrenfort, 2004), and this raises the possibility that we have another design principle: use extracellular routes (field potentials, spillover) to compute and apply a reliable and economical estimate of the mean level of activity in a population of neurons. Could the large numbers of synapses made onto epithelial glial cells (Fig. 41.1) be modulating ephaptic feedback to help tune LMC responses to light level?

Neural circuit design in the lamina does not miss a trick. Even the passive transmission of signals along the LMC axon, from input synapses in the lamina to output synapses in the medulla, is tuned to improve performance (van Hateren and Laughlin, 1990). The frequency response for transmission is matched to the signal spectrum to selectively attenuate higher frequency synaptic noise. Because the high conductance synaptic array in the lamina drives a passive axon with a very high membrane resistance, transmission is remarkably good. Very low frequencies travel 500 µm down the 3 µm diameter axon without significant loss, and a 200 Hz signal is attenuated by approximately 50%.

But is everything in the lamina sweetness and light, a Panglossian dream? It seems odd that circuits that eliminate redundancy to code information with near optimum efficiency (van Hateren, 1992) are replicated three times, in L1, L2, and L3 (Fig. 41.1). Needless to say, there is a sensible rationale for this profligacy. The three LMCs deliver information to different circuits in the medulla for different purposes (see Chapter 40, this volume). L3 provides an achromatic input to the chromatic pathway established by the direct projection of photoreceptors R7 and R8 to the medulla. L1 and L2 project to different medulla layers, and genetic dissection suggests that each neuron has a distinct role in driving motion-sensitive behavior (Rister et al., 2007).

To wind up this quick tour, observe that work on the optic lamina of insects continues the successful comparative approach to understanding vision, established by Hartline's studies of *Limulus* lateral eye (Hartline, 1969). Producing data that are good enough to suggest and test new hypotheses and theories, especially tractable preparations, reveals design principles. These principles illuminate studies of more complicated systems because it is easier to find a needle in a haystack when you know that you are looking for a needle.

References

Barlow HB (1961) Possible principles underlying the transformation of sensory messages. In: Rosenblith WA, ed. *Sensory Communication*, pp. 217–234. Cambridge, MA: MIT Press.

Buchner E (1981) Behavioural analysis of spatial vision in insects. In: Ali MA, ed. Photoreception and vision in invertebrates, pp. 561–621. New York: Plenum.

de Ruyter van Steveninck RR, Laughlin SB (1996) The rate of information-transfer at graded-potential synapses. *Nature* 379:642–645.

French AS, Järvilehto M (1978) Transmission of information by 1st and 2nd order neurons in fly visual-system. *J Comp Physiol* 126:87–96.

Hardie RC (1989) A histamine-activated chloride channel involved in neurotransmission at a photoreceptor synapse. *Nature* 339:704–706.

Hartline HK (1969) Visual receptors and retinal interaction. *Science* 164:270–278.

Järvilehto M, Zettler F (1973) Electrophysiological-histological studies of some functional properties of visual cells and second order neurons of an insect retina. *Z Zellforsch* 136:291–306.

Juusola M, de Polavieja GG (2003) The rate of information transfer of naturalistic stimulation by graded potentials. *J Gen Physiol* 122:191–206.

Juusola M, Uusitalo RO, Weckström M (1995) Transfer of graded potentials at the photoreceptor interneuron synapse. *J Gen Physiol* 105:117–148.

Juusola M, French AS, Uusitalo RO, Weckström M (1996) Information-processing by graded-potential transmission through tonically active synapses. *Trends Neurosci* 19:292–297.

Kamermans M, Fahrenfort I (2004) Ephaptic interactions within a chemical synapse: hemichannel-mediated ephaptic inhibition in the retina. *Curr Opin Neurobiol* 14:531–541.

Laughlin SB (1981) A simple coding procedure enhances a neuron's information capacity. *Z Naturforsch* 36c:910–912.

Laughlin SB (1994) Matching coding, circuits, cells, and molecules to signals—general principles of retinal design in the fly's eye. *Prog Retinal Eye Res* 13:165–196.

Laughlin SB, Hardie RC (1978) Common strategies for light adaptation in the peripheral visual systems of fly and dragonfly. *J Comp Physiol* 128:319–340.

Laughlin SB, Howard J, Blakeslee B (1987) Synaptic limitations to contrast coding in the retina of the blowfly *Calliphora*. *Proc R Soc Lond B* 231:437–467.

Laughlin SB, de Ruyter van Steveninck RR, Anderson JC (1998) The metabolic cost of neural information. *Nat Neurosci* 1:36–41.

Meinertzhagen IA, Sorra KE (2001) Synaptic organization in the fly's optic lamina: few cells, many synapses and divergent circuits. *Prog Brain Res* 131:53–69.

Nawroth JC, Greer CA, Chen WR, Laughlin SB, Shepherd GM (2007) An energy budget for the olfactory glomerulus. *J Neurosci* 27:9790–9800.

Rister J, Pauls D, Schnell B, Ting CY, Lee CH, Sinakevitch I, Morante J, Strausfeld NJ, Ito K, Heisenberg M (2007) Dissection of the peripheral motion channel in the visual system of Drosophila melanogaster. *Neuron* 56:155–170.

Shaw SR (1984) Early visual processing in insects. *J Exp Biol* 112:225–251.

Skingsley DR, Laughlin SB, Hardie RC (1995) Properties of histamine-activated chloride channels in the large monopolar cells of the dipteran compound eye—a comparative study. *J Comp Physiol A* 176:611–623.

Srinivasan MV, Laughlin SB, Dubs A (1982) Predictive coding—a fresh view of inhibition in the retina. *Proc R Soc Lond B* 216:427–459.

Sterling P, Matthews G (2005) Structure and function of ribbon synapses. *Trends Neurosci* 28:20–29.

Sterling P, Freed M (2007) How robust is a neural circuit? *Visual Neurosci* 24:563–571.

Stuart AE, Borycz J, Meinertzhagen IA (2007) The dynamics of signaling at the histaminergic photoreceptor synapse of arthropods. *Prog Neurobiol* 82:202–227.

van Hateren JH (1992) Theoretical predictions of spatiotemporal receptive-fields of fly LMCs, and experimental validation. *J Comp Physiol A* 171:157–170.

van Hateren JH (1997) Processing of natural time series of intensities by the visual system of the blowfly. *Vision Res* 37:3407–3416.

van Hateren JH, Laughlin SB (1990) Membrane parameters, signal transmission, and the design of a graded potential neuron. *J Comp Physiol A* 166:437–448.

Weckström M, Juusola M, Laughlin SB (1992) Presynaptic enhancement of signal transients in photoreceptor terminals in the compound eye. *Proc R Soc Lond B* 250:83–89.

Zheng L, de Polavieja GG, Wolfram V, Asyali MH, Hardie RC, Juusola M (2006) Feedback network controls photoreceptor output at the layer of first visual synapses in *Drosophila*. *J Gen Physiol* 127:495–510.

Section 2

Olfactory System

42

Microcircuits for Olfactory Information Processing in the Antennal Lobe of *Manduca sexta*

*Hong Lei, Lynne A. Oland, Jeffery A. Riffell,
Aaron Beyerlein, and John G. Hildebrand*

Olfactory circuits of all animals, from flies to mice, face a common challenge of extracting meaningful odor cues from background odors. In the following paragraphs we summarize what we have learned from our ongoing voyage toward the goal of fully understanding how the neural circuits the antennal lobe (AL) of a moth determine diverse physiological responses that ultimately mediate the animal's natural behavior. We start with a description of the different types of cellular elements that participate in the glomerular circuitry, then focus on the functional organization of these elements, and finally attempt to explain the observed physiological responses in the context of behavior using the understood operating principles of the AL circuits. For the convenience of narration, we describe the connections from the perspective of output neurons of the circuits—uniglomerular projection neurons (uPNs).

CELLULAR ELEMENTS CONSTITUTING GLOMERULAR CIRCUITRY IN A MOTH'S ANTENNAL LOBE

The AL of the hawkmoth, *Manduca sexta*, shares the glomerular organization and numerical simplicity—ca. 1200 cell bodies in the AL—present in most insect primary olfactory centers. Located in the ventrolateral deutocerebrum of the moth's brain, the AL is an easily distinguishable ovoid structure. Its predominant inputs are the axons of olfactory receptor cells (ORCs) entering AL via the antennal nerve, although CO_2-sensitive olfactory receptors on the mouthparts also send direct inputs to the AL (Guerenstein et al., 2004). An array of condensed neuropilar knots—glomeruli—populate a layer that surrounds a central core of coarse neuropil. No neuronal cell bodies are ever found in the neuropil; they reside instead in several clusters outside the neuropil.

Synaptic contacts between AL input, output, and intrinsic cells are restricted to the interior of the 63 glomeruli (three of which are sexually dimorphic; Rospars and Hildebrand, 2000). The neural circuitry of a glomerulus comprises four main types of neurons and one nonneuronal element, and its basic architecture is consistent throughout the entire array of glomeruli (Boeckh and Tolbert, 1993).

The neuropil of each glomerulus in the adult AL is almost completely surrounded by a multilayered envelope of glial cells, which comprises the cell bodies and processes of 75–100 "simple" glial cells (SGCs; Fig. 42.1A, circular magenta bodies) and up to 10 "complex" glial cells (CGCs; Fig. 42.1A, CGC, green processes; Oland et al., 1999; Oland et al., 2010). Simple glial cells extend multiple, mostly unbranched processes extensively connected by scalariform-like junctions (Tolbert and Hildebrand, 1981). Modeling suggests this glial envelope greatly limits passage of small molecules/ions (e.g., K$^+$) between

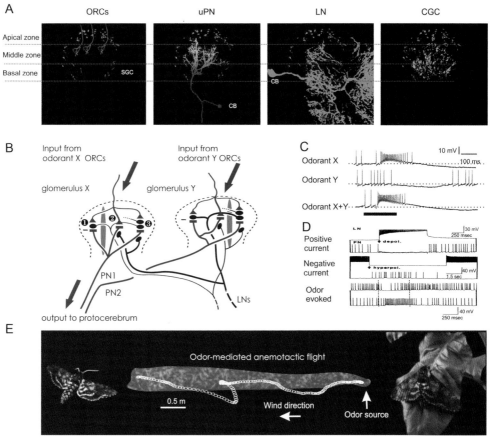

FIGURE 42–1. Functional organization of neural circuits in the glomeruli of *M. sexta* antennal lobe. (*A*) Semirealistic representation of olfactory receptor cells (ORCs; orange), a uniglomerular projection neuron (uPN; yellow-pink-red gradient), an LN (light blue), and complex

FIGURE **42–1.** Continued

glial cells (CGC) (green) in a glomerulus surrounded by simple glial cells (SGCs) (magenta), demonstrating the relative spatial relationship among these cellular elements. The cell bodies and processes of SGCs form multilayered envelope bordering glomerular neuropil. Images are not scaled, and the same glial border is used in all four panels. The dotted lines denote the approximate locations of three zones of a glomerulus based on intraglomerular synaptic organization. The apical zone is occupied by the ORC axons entering a glomerulus, free of synapses. The middle zone is where the uPNs receive inputs from cholingergic ORCs (distal branches, yellow) and GABAergic LNs as well as the serotonergic centrifugal neuron (proximal neurites, orange). Output synapses are also present on the proximal neurites of uPNs in the middle zone. The basal zone is primarily where the uPN output synapses are distributed (pink). Interglomerular interactions are likely initiated in the basal zone. The uPN cell bodies (CB) are located outside the glomerular neuropil in the medial group of AL neuronal somata. Multiglomerular LNs (likely GABAergic, light blue) are the major source of glomerular inhibition. CGCs (green, stained with an antibody against MasGAT, a GABA transporter) heavily branch in the basal zone of a glomerulus with a few branches extending into the middle zone and may regulate extracellular GABA concentration. (B) A diagram showing possible synaptic interactions within and between glomeruli. Two glomeruli are represented, each receiving a specific olfactory input (orange arrow for odorant X or purple arrow for odorant Y). The uPN innervating each glomerulus has three types of synaptic connections: *(1)* inhibitory through an inverting LN synapse (ORC→LN→uPN); *(2)* excitatory (ORC→uPN); and *(3)* excitatory through disinhibition involving two LNs in series (ORC→LN→LN→uPN). The elliptical green bars represent modulatory elements of the circuitry such as the CGCs and centrifugal neurons. (C) Lateral inhibition is often evident from uPN's inverted response profile: excitatory response to odorants that are specific ligands for the glomerulus (e.g., glomerulus X) the uPN innervates (upper panel), but inhibitory response to odorants that specifically activate neighboring glomeruli (e.g., glomerulus Y) (*middle*). The mixture of X and Y excites the uPN (*bottom*). (D) Paired intracellular recordings show that injection of positive currents into LNs results in suppression of firing on the connected PNs (*upper*); injection of negative currents into LNs causes PNs to increase firing rate (*middle*); odor stimuli produce responses of opposite polarity on LNs and PNs (*bottom*). (E) Circuits depicted in (B) may be involved in controlling the odor-mediated anemotactic flight behavior of moths. The elongated clouds between the two moths represent a sex-pheromone plume emitted from the female moth sitting on leaf. The flight track, which is digitized from a wind-tunnel experiment, consists of a series of circles that denote the video frames captured at 0.016 sec intervals. When a mixture of the two key pheromone components is used as an odor source in the wind-tunnel, both the cumulus and the first toroid of the MGC are activated, as indicated in (B) (glomerulus X for the cumulus; Y for the first toroid). The moth can follow the odor plume in most of the time, thus displaying a relatively straight flight track.

glomeruli (Goriely et al., 2002). Only rarely do the SGCs extend processes into glomerular neuropil. Complex glial cells have comparatively larger cell bodies and extend multiple usually robust branches around the inside edge of the glial envelope and through the middle of the glomerulus. These end in multiple fine, often vellate branches, especially at the base of the glomerulus.

Uniglomerular projection neurons (uPNs) provide the predominant output from the AL to higher brain areas. Typically the primary neurite of a uPN extends from its soma into the coarse neuropil of AL, where it bifurcates with one branch leaving the AL and the other preceding toward a glomerulus

(Fig. 42.1A, uPN). Near the base of a glomerulus the primary neurite gives rise to the second- and third-order branches that further ramify into distal dendrites. The axons of uPNs are connected to the protocerebrum via one major and four comparatively minor output tracts (Homberg et al., 1988).

Along with ORCs and uPNs, the spiking local interneurons of the AL (LNs) (Fig. 42.1A, LN) are the third principal cell type of glomerular circuitry. An LN soma extends a single neurite into the coarse neuropil, which then ramifies into numerous processes that arborize in a few (Christensen et al., 1993), many, or all glomeruli (Matsumoto and Hildebrand, 1981). These amacrine-like cells form synaptic connections with uPNs, ORCs, and themselves (Anton and Homberg, 1999) and have been shown to be the major source of inhibition in glomerular circuitry (Waldrop et al., 1987; Christensen et al., 1993).

Neurons that project from the protocerebrum to the glomerular circuitry are classified as centrifugal neurons (CNs), the fourth circuit element in AL. Centrifugal neurons are heterogenous, including protocerebral cells that extend projections into both ALs (Homberg et al., 1988), a protocerebral cell that sends processes to only the ipsi- and contralateral labial-palp-organ glomeruli (Guerenstein et al., 2004), and the extensively studied serotonin-immunoreactive (SI) neuron (Kent et al., 1987; Dacks et al., 2008). These cellular elements (ORCs, uPNs, LNs, CNs, and CGCs) participate in the circuitry within glomeruli, organizing glomerular output to higher centers.

INTRAGLOMERULAR SYNAPTIC ORGANIZATION

Based on its synaptic organization a glomerulus can be divided roughly into three zones (Fig. 42.1A). The apical zone is occupied by the ORC axons entering the glomerulus, free of synapses. Interactions between different types of neurons take place in the middle and basal zones.

Double labeling using neurobiotin for sensory neurons and Lucifer Yellow staining for uPNs, in conjunction with confocal microscopy, demonstrates that the afferent terminals overlap with the distal dendrites of uPNs at the upper portion of the middle zone (Sun et al., 1997). Although confirmation at the electronmicroscopic (EM) level is lacking, this result suggests that ORCs directly synapse on PN dendrites, as shown in cockroach preparations (Distler and Boeckh, 1996). In *M. sexta,* EM examination of the distal dendrites of uPNs reveals only input synapses (Sun et al., 1997), leading to designation of this region as the "input region" (Fig. 42.1A, PN, fine yellow branches). Likely cholinergic (Homberg et al., 1995; Torkkeli et al., 2005), ORCs are candidates for supplying excitatory input to these synapses (Fig. 42.1A, ORCs; Fig. 42.1B; Stengl et al., 1990; Sun et al., 1997; Fig. 42.1B, circled 2). Other evidence from *M. sexta* also suggests that ORCs may synapse on LNs (Fig. 42.1B, circled 1); electrical shock of the antennal nerve elicits monosynaptic spiking response in LNs (Christensen et al., 1993) and inhibitory potentials in uPNs (Hansson et al., 1991). Serial synapses that may constitute disinhibitory pathways (Fig. 42.1B, circled 3) also appear in the glomerular neuropil of *M. sexta* (Sun et al., 1997).

Both input and output synapses are present on uPN branches located at the lower portion of the middle zone (Fig. 42.1A, PN, orange; Sun et al., 1997)—thus the term "mixed input/output region." Possible synaptic partners at this region include GABAergic LNs (Fig. 42.1A, LN), as shown in cockroaches (Malun, 1991). The site of modulation from centrifugal neurons may also be at the mixed input/output region. For example, each AL receives innervation from a single serotonin-immunoreactive (SI) neuron (Sun et al., 1993). Although the soma of the SI neuron is located in the lateral group of AL neuronal cell bodies, it does not have arborizations in the ipsilateral AL. Its primary process extends to several regions in the ipsi- and contralateral protocerebra, the putative input regions, and then turns to enter the contralateral AL, where the cell ramifies extensively. Within a glomerulus the distal branches of the SI neuron do not overlap with the ORC terminals. Instead, these fine branches form a large number of output synapses within the mixed input/output region of uPNs.

In the basal zone of glomeruli, mainly output (presynatic) synapses have been identified on uPNs (Fig. 42.1A, uPN, magenta; Sun et al., 1997). The most likely postsynaptic cells in the basal zone are LNs (Fig. 42.1A, LN). The PN→LN connections are confirmed by observations of reciprocal lateral inhibition between glomeruli of the macroglomerular complex (MGC)—a group of three large, male-specific glomeruli devoted to processing information about sex pheromones. The uPNs that innervate two of the MGC glomeruli, called the cumulus and the first toroid, respectively, can be inhibited laterally by the stimulus (a pheromone component) that activates the adjacent MGC glomerulus (Fig. 42.1C; Lei et al., 2002). Because ORC input to the cumulus and the first toroid is segregated by responsiveness to individual components of the pheromone blend (Christensen et al., 1995) and because pheromone-responsive ORCs are excitatory (Kaissling et al., 1989; Kalinova et al., 2001), the best explanation for the lateral inhibition resides in PN→LN→PN connection, where the GABAergic LN inverts the effect of the ligand on the two uPNs that innervate cumulus and first toroid. This circuit is further confirmed with dual intracellular recordings from connected LN-PN pairs. When depolarizing current is injected into an LN, the firing of PNs is suppressed; in contrast, injection of hyperpolarizing current into LNs increases PN firing rate (Fig. 42.1D; Christensen et al., 1993).

In addition to transmitting information across glomeruli, the LN→PN circuit also participates in intraglomerular signal processing. One piece of supporting evidence comes from the early inhibitory postsynaptic potential (or I_1), which is typically recorded in uPNs innervating the cumulus or first toroid. This potential is shown to be mediated by GABAergic LNs (Christensen et al., 1998). When using a pheromone component to activate either the cumulus or the first toroid, I_1 can be elicited in the uPNs innervating that glomerulus. This suggests that LNs can directly modulate the activity of uPNs without lateral input from other glomeruli (Heinbockel et al., 1999).

GABA release, triggered either by intraglomerular input or by interglomerular input, has a profound effect on the response dynamics of uPNs. In addition, and not usually considered, complex glial cells (Fig. 42.1A, CGC) may have a nonsynaptic role in regulating extracellular GABA levels, particularly in the middle/basal portion of each glomerulus, where they branch extensively and enwrap large dendritic processes of PNs and LNs. These glial cells express a GABA transporter in the adult (as well as during metamorphic development) and are capable of taking up GABA (Oland et al., 1999).

INHOMOGENEOUS INTERGLOMERULAR CONNECTIONS

Functional connections between glomeruli are not restricted to adjacent glomeruli; rather, they appear to be inhomogeneously distributed in the AL. This is often reflected in the stimulus-evoked activity in spatially distinct and separate glomeruli, where different stimuli evoke unique patterns of glomerular activity. For instance, the activity pattern of glomeruli in the AL of *M. sexta* is odor specific, but activated glomeruli can be located both adjacent to and distant from one another (Collmann et al., 2004; Lei et al., 2004; Reisenman et al., 2008; Riffell et al., 2009). The LN network in the AL mediates such interactions. Recent work examining inhibitory interactions among uPNs in *D. melanogaster* and *M. sexta* has shown that odor-evoked inhibition is not limited to adjacent glomeruli, and hence it does not reflect the neighboring dominance of lateral inhibitory interactions (Silbering and Galizia, 2007; Reisenman et al., 2008). Instead, spatially segregated inhibitory activity appears to dominate among uPNs that are distant from one another. It is currently unknown whether, as in fruit flies (Shang et al., 2007), a subset of LNs are cholinergic.

Both intra- and interglomerular circuits affect the response profiles of uPNs. To understand how the circuits generate an accurate internal representation of external stimuli, it is important to know the natural features of an odor stimulus to which the circuits are tuned.

PHYSIOLOGICAL FUNCTIONS OF GLOMERULAR MICROCIRCUITS

GABAergic Inputs Shape Uniglomerular Projection Neuron Outputs

In nature, air turbulence quickly converts clouds of volatile molecules emitted from an odor source into countless odor filaments that are carried by wind to a distance far from the source. One of the challenges the moth's olfactory circuits must meet is to resolve the spatiotemporal dynamics of such natural odor plumes. To do this, glomerular circuits must enable uPNs to respond quickly to brief stimulation elicited by contact with an odor filament, and quickly recover in order to process successive odor strands. In a moth's AL, LNs play a major role in enabling uPNs to resolve odor pulses. The early IPSP (I_1) is positively correlated with the ability of a uPN to follow repeatedly encountered odor pulses (Heinbockel et al., 2004), suggesting that GABAergic LNs may determine how well a uPN can encode odor dynamics, perhaps through the LN→PN connection concentrated in the mixed input/output region of glomeruli. If true, blocking this connection should result in a decreased ability to track odor pulses in natural plumes. This hypothesis has gained support from a recent study in which a known GABA$_A$ receptor antagonist, bicuculline methiodide, was injected into the ALs of *M. sexta*. Treated moths subsequently showed disturbed odor-tracking behavior and location of an odor source in wind-tunnel assays (Lei et al., 2009).

Lateral Inhibition Promotes Intraglomerular Spiking Synchrony

LN→PN connections can be activated by odor stimulation of a neighboring glomerulus, resulting in inhibition of the glomerulus in which the monitored PN arborizes (Fig. 42.1B). This circuit effectively organizes an ensemble of glomeruli to enhance the contrast of representation for a specific input. For instance, to signal the presence of bombykal (one of the key sex-pheromone components of *M. sexta*), the LN→PN circuit suppresses the activity level of PNs in the cumulus while strongly activating those in the first toroid. A reciprocal relationship is observed upon stimulation with the second key pheromone component. Importantly, when the mixture of these two components is presented to a male moth, the cumulus and the first toroid inhibit each other, but uPNs associated with either of these glomeruli still show excitatory responses. This result indicates that lateral inhibition mediated by LN→PN circuit does not simply shunt PN responses. Instead, it serves to coordinate intraglomerular PN activity. The stronger the inhibition transmitted from the neighboring glomerulus, the greater the intraglomerular synchrony of spiking in PNs (Lei et al., 2002). When only one pheromone component is presented, the corresponding glomerulus (either cumulus or first toroid) is activated through the ORC→PN pathway, but the neighboring glomerulus is inhibited through the LN→PN connection. On the other hand, when the mixture of both key components is present, uPNs in both the cumulus and the first toroid are activated and the reciprocal lateral inhibition increases the synchrony of firing of PNs in both glomeruli. This enhanced synchrony may serve to encode the fact that each pheromone component is received in the presence of the other.

Interglomerular Synchrony Underlies Mixture Coding

The often sparse and specific lateral connections between glomeruli have important implications for olfactory coding mechanisms. The LN network, instead of acting to increase the contrast between neighboring glomeruli, may act to impose a temporal patterning on uPN responses (Lei et al., 2004; Riffell et al., 2009), thereby binding the uPN outputs. These interactions may be particularly important for the coding of complex odor mixtures and discrimination of behaviorally relevant olfactory stimuli (Lei et al., 2002; Riffell et al., 2009). Two kinds of neural codes, one derived from the spatial distribution of odor-evoked glomerular activity and the other based on the synchrony between neurons recorded across a large area of the moth's AL, have been compared for their efficiency to encode odor mixtures of varying concentrations. Behavioral findings show that the moths are equally attracted by a suite of same odor mixtures but diluted 10- to 1000-fold, a result that is consistent with the synchrony code but not supported by the spatial code. This feature of signal processing may be important in allowing a moth to fly toward an odor source (flower or mating partner) over a long distance, where the odor concentration varies considerably.

Centrifugal Modulation of Antennal Lobe Circuits

Among the reported centrifugal neurons influencing the AL (Homberg et al., 1988; Dacks et al., 2005), only the unique serotonin-immunoreactive (SI) neuron has been studied in detail in *M. sexta* (Kent et al., 1987; Kloppenburg et al., 1999; Kloppenburg and Heinbockel, 2000; Dacks et al., 2008). Exogenously applied serotonin increases the depolarization and also the response duration level of pheromone-responsive uPNs. The increased excitability can be attributed to the downregulation of both transient (I_A) and sustained $(I_{(K)v})$ potassium conductances (Kloppenburg et al., 1999). Serotonin appears to affect the entire population of uPNs associated with a glomerulus, based on the observation that the amine increases the duration and magnitude of local field-potential oscillations (Kloppenburg and Heinbockel, 2000). Furthermore, an expanded examination at the AL level demonstrates that serotonin not only increases the gain of AL circuitry but also enhances its discrimination of different odors (Dacks et al., 2008).

Our exploration of microcircuits in a microbrain is far from complete. For instance, little attention has been paid to multiglomerular projection neurons that perhaps are important in encoding odor mixtures (Lei and Vickers, 2008) or to the physiological roles of different subtypes of LNs and centrifugal neurons. How glial cells contribute to or modulate the circuit's activity is also unknown. The effort continues.

References

Anton S, Homberg U (1999) Antennal lobe structure. In: Hansson BS, ed. *Insect Olfaction*, pp. 97–124. Berlin: Springer.

Boeckh J, Tolbert LP (1993) Synaptic organization and development of the antennal lobe in insects. *Microsc Res Tech* 24:260–280.

Christensen TA, Waldrop BR, Harrow ID, Hildebrand JG (1993) Local interneurons and information-processing in the olfactory glomeruli of the moth *Manduca sexta*. *J Comp Physiol A* 173:385–399.

Christensen TA, Harrow ID, Cuzzocrea C, Randolph PW, Hildebrand JG (1995) Distinct projections of two populations of olfactory receptor axons in the antennal lobe of the sphinx moth *Manduca sexta*. *Chem Senses* 20:313–323.

Christensen TA, Waldrop BR, Hildebrand JG (1998) Multitasking in the olfactory system: context-dependent responses to odors reveal dual GABA-regulated coding mechanisms in single olfactory projection neurons. *J Neurosci* 18:5999–6008.

Collmann C, Carlsson MA, Hansson BS, Nighorn A (2004) Odorant-evoked nitric oxide signals in the antennal lobe of *Manduca sexta*. *J Neurosci* 24:6070–6077.

Dacks AM, Christensen TA, Agricola H-J, Wollweber L, Hildebrand JG (2005) Octopamine-immunoreactive neurons in the brain and subesophageal ganglion of the hawkmoth *Manduca sexta*. *J Comp Neurol* 488:255–268.

Dacks AM, Christensen TA, Hildebrand JG (2008) Modulation of olfactory information processing in the antennal lobe of *Manduca sexta* by serotonin. *J Neurophysiol* 99:2077–2085.

Distler PG, Boeckh J (1996) Synaptic connection between olfactory receptor cells and uniglomerular projection neurons in the antennal lobe of the american cockroach, *Periplaneta americana*. *J Comp Neurol* 370:35–46.

Goriely AR, Secomb TW, Tolbert LP (2002) Effect of the glial envelope on extracellular K($^+$) diffusion in olfactory glomeruli. *J Neurophysiol* 87:1712–1722.

Guerenstein PG, Christensen TA, Hildebrand JG (2004) Sensory processing of ambient CO_2 information in the brain of the moth *Manduca sexta*. *J Comp Physiol A* 190: 707–725.

Hansson BS, Christensen TA, Hildebrand JG (1991) Functionally distinct subdivisions of the macroglomerular complex in the antennal lobe of the male sphinx moth *Manduca sexta*. *J Comp Neurol* 312:264–278.

Heinbockel T, Christensen TA, Hildebrand JG (1999) Temporal tuning of odor responses in pheromone-responsive projection neurons in the brain of the sphinx moth *Manduca sexta*. *J Comp Neurol* 409:1–12.

Heinbockel T, Christensen TA, Hildebrand JG (2004) Representation of binary pheromone blends by glomerulus-specific olfactory projection neurons. *J Comp Physiol A* 190:1023–1037.

Homberg U, Montague RA, Hildebrand JG (1988) Anatomy of antenno-cerebral pathways in the brain of the sphinx moth *Manduca sexta*. *Cell Tissue Res* 254:255–281.

Homberg U, Hoskins SG, Hildebrand JG (1995) Distribution of acetylcholinesterase activity in the deutocerebrum of the sphinx moth *manduca-sexta*. *Cell Tissue Res* 279:249–259.

Kaissling KE, Hildebrand JG, Tumlinson JH (1989) Pheromone receptor cells in the male moth *Manduca sexta*. *Arch Insect Biochem Physiol* 10:273–279.

Kalinova B, Hoskovec M, Liblikas I, Unelius CR, Hansson BS (2001) Detection of sex pheromone components in *Manduca sexta* (L.). *Chem Senses* 26:1175–1186.

Kent KS, Hoskins SG, Hildebrand JG (1987) A novel serotonin-immunorective neuron in the antennal lobe of the sphinx moth *Manduca sexta* persists throughout postembrionic life. *J Neurobiol* 18:451–465.

Kloppenburg P, Ferns D, Mercer AR (1999) Serotonin enhances central olfactory neuron responses to female sex pheromone in the male sphinx moth *Manduca sexta*. *J Neurosci* 19:8172–8181.

Kloppenburg P, Heinbockel T (2000) 5-hydroxytryptamine modulates pheromone-evoked local field potentials in the macroglomerular complex of the sphinx moth Manduca sexta. *J Exp Biol* 203:1701–1709.

Lei H, Vickers N (2008) Central processing of natural odor mixtures in insects. *J Chem Ecol* 34:915–927.

Lei H, Christensen TA, Hildebrand JG (2002) Local inhibition modulates odor-evoked synchronization of glomerulus-specific output neurons. *Nat Neurosci* 5:557–565.

Lei H, Christensen TA, Hildebrand JG (2004) Spatial and temporal organization of ensemble representations for different odor classes in the moth antennal lobe. *J Neurosci* 24:11108–11119.

Lei H, Riffell JA, Gage SL, Hildebrand JG (2009) Contrast enhancement of stimulus intermittency in a primary olfactory network and its behavioral significance. *J Biol* 8:21.

Malun D (1991) Synaptic relationships between GABA-immunoreactive neurons and an identified uniglomerular projection neuron in the antennal lobe of *Periplaneta americana*: A double-labeling electron microscopic study. *Histochemistry* 96:197–207.

Matsumoto SG, Hildebrand JG (1981) Olfactory mechanisms in the moth *Manduca sexta*: Response characteristics and morphology of central neurons in the antennal lobes. *Proc Roy Soc Lond B* 213:249–277.

Oland LA, Marrero HG, Burger I (1999) Glial cells in the developing and adult olfactory lobe of the moth *Manduca sexta*. *Cell Tissue Res* 297:527–545.

Oland LA, Gibson NJ, Tolbert LP (2010) Localization of a GABA transporter to glial cells in the developing and adult olfactory pathway of the moth *Manduca sexta*. *J Comp Neurol* 518:815–838.

Reisenman CE, Heinbockel T, Hildebrand JG (2008) Inhibitory interactions among olfactory glomeruli do not necessarily reflect spatial proximity. *J Neurophysiol* 100:554–564.

Riffell JA, Lei H, Christensen TA, Hildebrand JG (2009) Characterization and coding of behaviorally significant odor mixtures. *Curr Biol* 19:335–340.

Rospars JP, Hildebrand JG (2000) Sexually dimorphic and isomorphic glomeruli in the antennal lobes of the sphinx moth *Manduca sexta*. *Chem Senses* 25:119–129.

Shang Y, Claridge-Chang A, Sjulson L, Pypaert M, Miesenbock G (2007) Excitatory local circuits and their implications for olfactory processing in the fly antennal lobe. *Cell* 128:601–612.

Silbering AF, Galizia CG (2007) Processing of odor mixtures in the *Drosophila* antennal lobe reveals both global inhibition and glomerulus-specific interactions. *J Neurosci* 27:11966–11977.

Stengl M, Homberg U, Hildebrand JG (1990) Acetylcholinesterase activity in antennal receptor neurons of the sphinx moth *Manduca sexta*. *Cell Tissue Res* 262:245–252.

Sun XJ, Tolbert LP, Hildebrand JG (1993) Ramification pattern and ultrastructural characteristics of the serotonin immunoreactive neuron in the antennal lobe of the moth *Manduca sexta*: A laser scanning confocal and electron microscopic study. *J Comp Neurol* 338:5–16.

Sun XJ, Tolbert LP, Hildebrand JG (1997) Synaptic organization of the uniglomerular projection neurons of the antennal lobe of the moth *Manduca sexta*: A laser scanning confocal and electron microscopic study. *J Comp Neurol* 379:2–20.

Tolbert LP, Hildebrand JG (1981) Organization and synaptic ultrastructure of glomeruli in the antennal lobes of the moth *Manduca sexta*: A study using thin sections and freeze-fracture. *Proc Roy Soc Lond B* 213:279–301.

Torkkeli PH, Widmer A, Meisner S (2005) Expression of muscarinic acetylcholine receptors and choline acetyltransferase enzyme in cultured antennal sensory neurons and non-neural cells of the developing moth *Manduca sexta*. *J Neurobiol* 62:316–329.

Waldrop B, Christensen TA, Hildebrand JG (1987) GABA-mediated synaptic inhibition of projection neurons in the antennal lobes of the sphinx moth, *Manduca sexta*. *J Comp Physiol A* 161:23–32.

43

Antennal Lobe of the Honeybee

Randolf Menzel and Jürgen Rybak

The antennal lobe (AL) of an insect is the functional analog of the olfactory bulb in mammals. The first-level synaptic interaction between large numbers of multiple-type olfactory receptor neurons (ORNs) serves the function of reliable coding of a vast range of general odors, the fast identification of an odor, the selective coding of mixtures of compounds, and the separation between odor identity and odor concentration. Honeybees learn and discriminate a seemingly unlimited number of odors (natural and artificial), categorize odor mixtures as unique stimuli, identify odors within 250 ms, and generalize odors according to the respective combinatorial glomerulus activity patterns in the AL (Galizia and Menzel, 2000). Therefore, the AL is the first-order neuropil serving basic functions of odor discrimination, categorization, generalization, and learning. The social life of the honeybee is guided by a range of well-described pheromones. No difference between general odor and pheromone coding was found so far with the exception of the queen pheromone coding in males (drone bees) (Sandoz et al., 2007).

We first describe the internal organization of the AL, its inputs, and its outputs. Then we analyze the local circuit of a prototypical glomerulus.

INPUT, CROSS-INTERNAL ORGANIZATION, AND
OUTPUT OF THE ANTENNAL LOBE

The AL receives about 60,000 ORNs predominantly originating in olfactory receptor cells of pore plates on the antenna, but CO_2 receptors, thermoreceptors, and hygroreceptors are known to be distributed along the flagellum of the antenna, too. The antennal nerve is split into four tracts, T1 to T4, innervating different groups of glomeruli of the AL (Fig. 43.1), the dorsoanterior (da) group with 70 glomeruli, the medial (m) group with seven glomeruli, the ventroposterior (vp) group with 70 glomeruli, and the posterior (p) group with seven glomeruli (Fig. 43.1A). Glomeruli are the anatomical and functional units of the AL and constitute sites of synaptic interaction between different neuron types. The output of the glomeruli comprise several tracts of projection neurons (PNs), which receive either input only from one glomerulus (uniglomerular PNs belonging to the lateral and median PN, l-PN, and m-PN; Fig. 43.1A) or from many glomeruli (multiglomerular PNs which form several tracts including the mediolateral ml-PN tracts; Fig. 43.1B). L- and m-PNs project to the lip and basal ring regions of the mushroom body and to the lateral horn (LH). The m-PN arborize more in the core of the LH. As a consequence of the different pathways that the two uniglomerular PNs take, the l-PNs reach the LH first and send collaterals to the mushroom body, whereas the m-PNs project first to the mushroom body and send collaterals to the LH. L-PNs receive input only from glomeruli of ORN tract 1 (dorsoanterior group), and m-PNs from the remaining three groups of glomeruli (Fig. 43.1A).

Multiglomerular ml-PNs innervate predominantly the T1 and T3 glomeruli, and they project ipsilaterally to subcompartments of the protocerebral lobe: the LH, the lateral protocerebral lobe (LPL), and a ring-like neuropil around the alpha lobe of the mushroom body. A subpopulation of ml-PNs (mlPN [2]) are immunoreactive (ir) to an antibody against GABA and may represent an efferent pathway (dotted triangles in Fig. 43.1B). Two other types of multiglomerular PNs have been described, both of which project also to the contralateral side of the brain (Abel et al., 2001). Those PNs that arborize in glomeruli of T4 respond to contact chemoreceptive stimuli like the wax surface of the comb. Glomeruli of the ORN T4 appear to collect input from CO_2 receptors, thermoreceptors, and hygroreceptors. A possible feedback neuron, the ALF-1, connects the mushroom bodies with many glomeruli in the antennal lobe (not shown in Fig. 43.1; Kirschner et al., 2006).

The AL houses a relatively large number (4000) of local interneurons (LIs). Many of them are GABA ir, and some are putatively glutamatergic and/or histaminergic. The existence of excitatory LIs has been demonstrated in the AL of the fly, *Drosophila melanogaster*, but it is also likely in the honeybee AL.

A multiple set of modulatory neurons innervate large proportions (possibly all) glomeruli with broadly arborizing axodendrons, for example, dopamine ir neurons, a 5HT ir neuron, histamine ir neurons, and octopamine ir neurons. The latter are particularly interesting because it was shown that one of these octopamine ir neurons, the VUMmx1, serves the function of an appetitive reinforcer during reward odor learning in bees (Hammer, 1993). Since pharmacological stimulation of octopamine receptors in the AL as a substitute for the reward leads to odor learning, it has been concluded and confirmed by additional experiments that an associative memory trace is

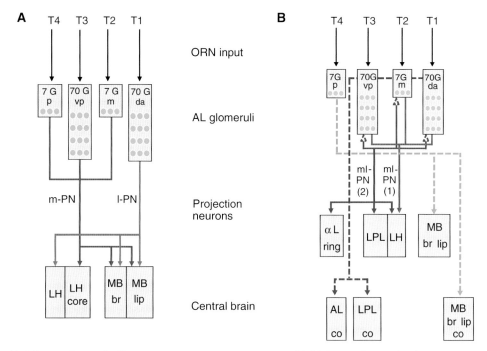

Uniglomerular projection neurons   **Multiglomerular projection neurons**

FIGURE 43–1. Basic organization of the olfactory receptor neuron (ORN) input to the antennal lobe (AL) glomeruli, and their output via projection neurons (PN) to neuropils in the central brain: the mushroom body (MB), with its calycal subdivisions, lip, and basal ring (br), the lateral horn (LH), the lateral protocerebral lobe (LPL), and the ring neuropil around the α lobe of the MB (α L ring). The approximately 160 glomeruli of the AL are segregated into four groups according to their input from four sensory tracts (T1 to T4) of the antennal nerve. T1–T3 contain the axons of olfactory receptor neurons (ORN) predominantly at the tip of the flagellum in pore plates. T4 contains the axons from contact chemoreceptors, CO_2, temperature, and hygroreceptors. The numbers in the boxes give the number of glomeruli respectively, and the notations indicate the position of the respective group of glomeruli (da, dorsoanterior with 70 G; m, medial with 7 G; vp, ventroposterior with 70 G; p, posterior with 7 G). (*A*) Organization of outputs of uniglomerular projection neurons (lateral and medial PNs, l-PN, m-PN). The l-PNs receive input from the T1 glomeruli, m-PN from the T2–T4 glomeruli. L-PNs project first to the lateral horn (LH), and then to the lip and basal ring of the mushroom body calyx (MB lip, br), M-PNs project first to the lip of the mushroom body and then to the core of the LH. (*B*) Organization of AL inputs and outputs of multiglomerular PNs. Mediolateral PNs (ml-PN) receive input predominantly from T1 and T3 glomeruli, and they project to the LPL and the alpha lobe ring like neuropil, ml-PN(1) solely to the lateral horn (LH). Some of the ml-PNs (ml-PN [2]) are immunoreactive (ir) to an antibody against GABA, and they may represent an efferent pathway (dotted triangles). Other multiglomerular PNs project also to the other side of the brain. Those branching in T1–T3 glomeruli arborize in the contra lateral AL and LPL (AL co and LPL co, shown in hatched violet line); those branching in T4 glomeruli arborize in both the ipsi and the contralateral mushroom body calyces (MB br lip, CO) (hatched green line).

formed in the AL (Menzel and Giurfa, 2001). Several neuropeptides were found by immunocytochemistry, but their functions are unknown.

LOCAL CIRCUITRY WITHIN THE ANTENNAL LOBE AND WITHIN A PROTOTYPICAL GLOMERULUS

Other than in *Drosophila* but similar to lobsters and moths, the circuitry connecting glomeruli in the honeybee AL involves two morphological distinct types of spiking inhibitory LIs that are distinct by their heterogeneous (different branching patterns in different glomeruli) and homogenous (similar branching patterns in different glomeruli) branching patterns. Excitatory LI has been predicted for the honeybee for various reasons, but clear experimental evidence and relation to a morphological type is lacking. Electron microscopical studies of the synaptic contacts within the glomerulus of insects exist for the moth *Manduca* (see Chapter 42) and the cockroach *Periplaneta* (Boeckh and Tolbert, 1993), but unfortunately not for the honeybee. Figure 43.2A gives the documented synapses between the four major neural elements of the *Periplaneta* glomerulus, the ORNs, the PN, and two kinds of LIs, identified GABA ir LIs, and unidentified LI possibly excitatory LIs. It is believed that the observed synaptic connections are representative for an insect glomerulus. Other than in the mammalian olfactory bulb all synapses in the insect AL are confined to glomeruli besides very few potential synapses of PN at the exit of PN from the AL. Direct synapses from ORN to PN exist but are rare. Both GABA ir and unidentified LIs receive input from ORNs and synapse back on ORNs. Recurrent pathways also exist between LIs of both types and PNs. (Fig. 43.2A).

On average, each of the 160 or so glomeruli in the honeybee AL receives input from about 375 ORNs and is innervated by about 25 LIs. It has not been documented yet whether the ORNs from the antennae terminating in one glomerulus express the same receptor protein as it is known from *Drosophila*.

ENSEMBLE ENCODING OF GENERAL ODORS

Imaging the Ca^{2+} activity upon odor stimulation indicates a spatial and combinatorial olfactory code (Galizia and Menzel, 2000; ftp://www.neurobiologie.fu-berlin.de/honeybeeALatlas/). Although such studies have been possible so far only for glomeruli of the T1 ORN tract, accumulating evidence from intracellular recordings of both l- and m-PNs proves that such a coding scheme applies also to the other glomeruli (Krofczik et al., 2008), possibly with the exception of T4 glomeruli. Blocking GABA receptors in the AL proves that *(1)* odor responses of PN dendrites increase or decrease (possibly by a loss of disinhibition) depending on the particular odor, *(2)* the time

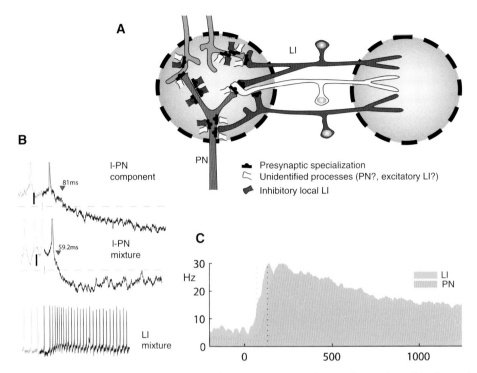

Figure 43–2. (*A*) Synaptic contacts within a prototypic glomerulus as found in the cockroach *Periplaneta americana* (modified from Boeckh and Tolbert, 1993). Two olfactory receptor neuron (ORNs) are shown in green providing excitatory input with inhibitory components. They terminate predominantly on local interneurons (LI, blue and yellow), but also on projection neurons (PN, red). Two types of LI are distinguished on the basis of immunoreactivity (ir) to an antibody against GABA, identified LI (GABA ir, blue) and unidentified LI (no GABA ir, yellow). (*B*) Membrane potential of a l-PN to an elemental odor (component, upper row) and to a tertiary mixture (mixture, second row) of odorants. Stimulus onset is indicated by the dash. The initial response is an inhibitory component as indicated by the red triangles. The numbers give the response latencies in milliseconds (ms). The third row depicts the membrane potential of an LI showing a fast excitatory response to a tertiary mixture. (*C*) The average temporal firing rate profiles of LIs (green) and PNs (gray) show that LIs respond faster than PNs and with a shorter phasic response. (Modified from Krofczik et al., 2008)

courses of the odor responses in PN dendrites change, *(3)* the synchrony between PN spikes and local field potential is weakened, and *(4)* odor discrimination of the animal is compromised. These results clearly document that the LIs contribute essentially to odor coding in the AL. More direct evidence for the role of both inhibitory and excitatory LI comes from intracellular recording of PNs. *(1)* On average PN response onset is delayed by about 60 ms with respect to the fast interneuron response onset (Fig. 43.2C), indicating that PN activation is at least to some extent mediated via excitatory interneurons. *(2)* Excitatory input from LNs also explains how fast responding inhibitory LIs effectively suppress responses in PNs before excitatory inputs drive PN firing. *(3)* L-PN and m-PNs have broad overlapping chemoprofiles. This broad tuning can be explained if excitation is mediated by LNs. In contrast to the mammalian olfactory bulb, ordered functional segregation (e.g., according to C-chain length or functional group) is generally not found in the insect AL, although there might be a tendency toward such chemotopy in the honeybee AL. *(4)* The two types of uniglomerular PNs differ with respect to coding of odor mixtures. L-PNs are inhibited in response to a mixtures but not to its individual components (odor mixture suppression is also called synthetic mixture coding; Figure 43.2 B), whereas m-PNs code mixtures in an elemental way by responding to the most effective component. Thus, odor complexity is likely to be decoded by those target neurons to which l-PNs terminate. *(5)* Since the onset of inhibition in l-PNs matches the excitatory response onset of LNs, which in turn is significantly shorter than in PNs, lateral inhibition is responsible for the observed mixture suppression effect. This interpretation is supported by results of Ca^{2+}-imaging studies of T1 glomeruli showing that increasing the number of components in the mixture enhances suppressive interglomerular interactions. Taken together, these results indicate that interglomerular connectivity is not merely a means to regulate and normalize the overall activation level of PNs. Rather, LNs are involved in rapidly processing the ORN input, altering and shaping the PN output depending on compounds and their mixtures.

ACKNOWLEDGMENT

We are grateful to Tilman Franke for providing us with the modified version of the scheme from Boeckh and Tolbert (1993) shown in Figure 43.2A.

REFERENCES

Abel R, Rybak J, Menzel R (2001) Structure and response patterns of olfactory interneurons in the honeybee, *Apis mellifera. J Comp Neurol* 437:363–383.
Boeckh J, Tolbert LP (1993) Synaptic organization and development of the antennal lobe in insects. *Micros Res Tech* 24:260–280.
Galizia CG, Menzel R (2000) Odour perception in honeybees: coding information in glomerular patterns. *Curr Op Neurobio* 10:504–510.
Hammer M (1993) An identified neuron mediates the unconditioned stimulus in associative olfactory learning in honeybees. *Nature* 366:59–63.
Kirschner S, Kleineidam CJ, Zube C, Rybak J, Grünewald B, Rössler W (2006) Dual olfactory pathway in the honeybee, *Apis mellifera. J Comp Neurol* 499:933–952.
Krofczik S, Menzel R, Nawrot MP (2008) Rapid odor processing in the honeybee antennal lobe network. *Front Comput Neurosci* 2:1–9.
Menzel R, Giurfa M (2001) Cognitive architecture of a mini-brain: the honeybee. *Trends Cog Sci* 5:62–71.
Sandoz JC, Deisig N, de Brito Sanchez MG, Giurfa M (2007) Understanding the logics of pheromone processing in the honeybee brain: from labeled-lines to across-fiber patterns. *Front Behav Neurosci* 1:5. doi: 10.3389/neuro.08.005.2007.

44

Mushroom Body of the Honeybee

Jürgen Rybak and Randolf Menzel

The mushroom body (MB) in the insect brain is composed of a large number of densely packed neurons called Kenyon cells (KCs) (Drosophila, 2500; honeybee, 170,000). In most insect species, the MB consists of two cap-like dorsal (developmentally: frontal) structures, the calyces, which contain the dendrites of KCs and two to four lobes formed by collaterals of branching KC axons. Although the MB receives input and provides output throughout its whole structure, the neuropil part of the calyx receives predominantly input from sensory projection neurons (PNs) of second or higher order, and the lobes send output neurons to many other parts of the brain, including recurrent neurons to the MB calyx (Fig. 44.1B). Widely branching supposedly modulatory neurons innervate the MB at all levels (calyx, peduncle, lobes), including the somata of KCs in the calyx (dopamine) (Fig. 44.1A).

Two major classes of KCs, KI and KII, can be distinguished with different dendritic morphologies and axonal projections with respect to their location in the peduncle and lobes. KII cells exhibit claw-like specializations in narrow dendritic trees, whereas wide-field branching KI cells span larger areas in the calyx neuropil. The different subtypes of both cell types are found in all calycal neuropils. KII cell somata are located outside the calycal wall, whereas KI cell somata are clustered in the central (small somata) and the adjacent zones (large-diameter somata) within the calyces. KII axons project through the peduncle into the ventral α-lobe and anterior β-lobe, whereas KI cell axons occupy horizontally stratified areas of the dorsal α-lobe and posterior β-lobe, representing the calycal zone's lip, collar, and basal ring (Rybak and Menzel, 1993; Strausfeld, 2002).

Projection neurons originating in the sensory neuropils innervate the calyx in an orderly fashion. Olfactory PNs branch predominantly in the lip, visual

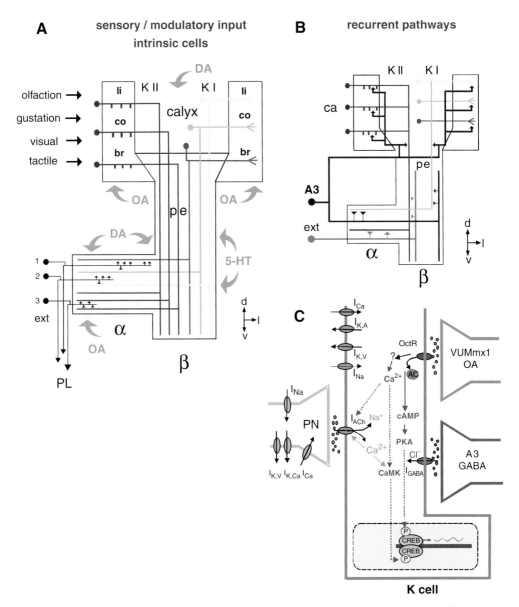

Figure 44–1. Diagram of the mushroom body and its components. (*A*) Sensory fiber tracts (dark arrows from the left) invade the calyx (ca) in an orderly fashion as indicated. Intrinsic Kenyon cells type KI and type KII exhibit broad (KI) or narrow (KII) dendritic trees within the calycal subcompartments (li, lip; co, collar; br, basal ring). The modality specific inputs are maintained by KI cells whose axons project through the peduncle (pe) to discrete strata of the α- and β-lobes (α,β) indicated by different colors. KII cells project from all calycal zones to the ventral α-lobe and anterior β-lobe. The innervations of modulatory neurons (5-HT: serotonine; OA: octopamine; DA: dopamine) are indicated by the orange arrows. Extrinsic neurons (ext, 1, 2, 3) exhibit either single- or double-stranded layers in the lobes (only shown for the α-lobe here) and project to the protocerebral lobes (PL). d, dorsal; v, ventral; l, lateral. (*B*) Recurrent (feedback) pathway. An extrinsic group of the MB neurons (type A3) form fine dendritic bands

in the α-lobe and send collaterals within the MB to the β-lobe and peduncle. Outside the MB these cells project to the calyces (ca) and exhibit axonal terminals within subregions of the calyx synapsing onto KC (red and yellow-green). Ext, extrinsic GABA-ir cell with no recurrent loop to the calyces innervates the ventral α-lobe and its corresponding zones of the α-lobe and peduncle. (C) Schematic representation of the ion channels, receptors, and signaling cascades in a Kenyon cell (K cell) as they are relevant for associative plasticity. PN (green), olfactory projection neurons from the antennal lobe; VUMmx1, ventral unpaired median neuron from the maxillary neuromere number 1. VUMmx1 was found to serve the reinforcing function in olfactory conditioning. It is immunoreactive to octopamine (OA). A3 (recurrent) neurons are immunoreactive to GABA. (Modified from Grünewald et al., 2004)

PNs in the collar, mechanosensory PNs in the basal ring, and gustatory PNs in a small region between lip and collar. The intrinsic structure imposed by the PNs onto the KCs is kept throughout the MB with respect to the course of the KI cells leading to an orderly subdivision of the lobes with the KCs of the lip projecting to the ventral zone of the dorsal α-lobe, those of the collar in the layer above, and those of the basal ring in the most dorsal part. A corresponding layering is found in the KC collaterals forming the β-lobe (Fig. 44.1A). In summary, due to the subdivision of KCs into two classes, the ventral part of the α-lobe is composed of KII-type KCs homogenously distributed throughout all calycal zones, and the dorsal α-lobe is layered by the orderly projection of the KI cells from distinct calycal subcompartments (Fig. 44.1A).

The branching pattern of extrinsic neurons of the α-lobe (ext in Fig. 44.1A) and the density of synaptic contacts between KCs and extrinsic neurons lead to a stratification (Fig. 44.1A and B). One class of extrinsic neurons (A3 neurons) is at least partly immunoreactive (ir) to a GABA antibody and is considered to provide inhibitory recurrent information from the α-lobe, β-lobe, and peduncle to the calyx besides local inhibitory connections between KCs and extrinsic neurons in both α- and β-lobe (Grünewald, 1999; Ganeshina and Menzel, 2001) (Figs. 44.1B and 44.2B). A3 neurons forming dendritic strata within the α-lobe also project parallel to the KC type KI into the peduncle and the β-lobe. Outside the MB they run to the calyces and terminate in all subcompartments close to KC type KI and KII. Other GABA-ir cells (ext in Fig. 44.1B) invade the ventral α-lobe and peduncle and run parallel to axons from KII cells (Fig. 44.1B)

FUNCTIONAL ORGANIZATION OF THE CALYX

Projection neurons diverge and converge on KCs in the calyx, leading to a matrix of connectivity (Fig. 44.2A). On average, one KC receives input from 10–15 olfactory PNs, and each of these PNs provides input to 50–100 KCs. Each PN presynaptic bouton comprises a microcircuit composed of

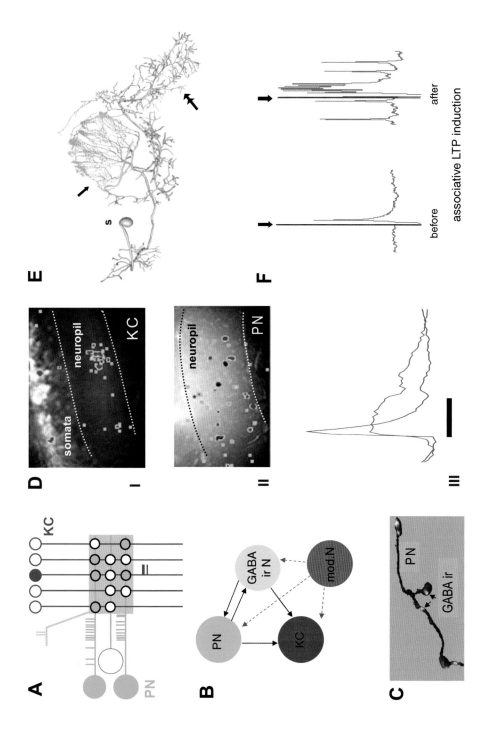

A

KC

II

PN

B

GABA
ir N

mod.N

PN

KC

C

PN

GABA ir

D

I

somata

neuropil

KC

II

neuropil

PN

III

E

s

F

before   after

associative LTP induction

FIGURE 44–2. (*A*) Schematic wiring diagram of olfactory projection neurons (PN) and Kenyon cells (KC) in MB calyx. Action potentials in the PNs (dark green) lead to phasic responses in the KC (gray) that receive coincident input (filled red KC). (*B*) Pre- and postsynaptic inhibition, in the local microcircuit of a microglomerulus of the MB calyx. Reciprocal synaptic connections exist between GABA-ir recurrent neurons (GABA ir N) and projection neurons (PN). Additionally, GABA-ir N synapse onto KC. mod.N, modulatory neurons (e.g., VUMmx1). (*C*) Light m·cro-scopic analysis allows the reconstruction of axonal terminals of a projection neuron (PN) collateral and its close attachment of putative synaptic contacts of GABA-ir profiles (GABA ir) (light spots, arrows). (*D*) Odor-induced activity in KC (in D I) and PN boutons (in D II) in the lip neuropil of the MB calyx (neuropil). D III: Ca²⁺ responses of KC (red, phasic excitation) and PN (green) to an odor stimulus (black bar). (*E*) Digital 3-D reconstruction of the PE1 neuron, an extrinsic neuron of the mushroom bodies (MB) α-lobe. The PE1 receives input via extensive arborizations in the MB (arrow). It projects with elaborate arborizations to the lateral protocerebral lobe (double-arrow). (*F*) Pairing titanic (100 Hz, 1 sec) stimulation of KC with depolarization of PE1 leads to associative LTP indicated by a prolonged (>30 min) increase of responses to single test stimuli of the KC. The traces show the responses to a single test stimulus before and after pairing of the titanic stimulation with depolarization. (Modified from Menzel and Manz, 2005)

multiple outputs to KCs, input and output from and to GABA ir profiles representing the recurrent A3 neurons, output from these profiles onto KC spines, and tentative modulatory neurons with en passant synapses containing dense core vesicles (Fig. 44.2B). Putative inhibitory input from A3 neurons also reaches PNs on axodendrons (Fig. 44.2C). Olfactory PN boutons respond to odor stimulation with excitation and/or inhibition in an odor identity–specific way. Olfactory, gustatory, and tactile stimuli excite KCs (type KII) transiently and lead to Off rebound excitation, indicating delayed release from inhibition (Fig. 44.2D). Odors are coded in KCs (type KII) in a sparse way both in the temporal and the population domain, indicating that inhibition via recurrent A3 neurons is a prominent feature of the circuit (Szyszka et al., 2005).

Repeated stimulation leads to stimulus-specific decrease of KC responses. Pairing an odor as conditioned stimulus with sucrose reward in an associative learning situation induces prolonged KC responses. After conditioning, KC responses to the rewarded odor recover from repetition-induced decrease, while the responses to a nonrewarded odor decrease further. The spatiotemporal pattern of activated KCs changes both for the learned and the specifically not learned odor (Szyszka et al., 2008). These results document that KC responses are subject to nonassociative plasticity during odor repetition and undergo associative plasticity after appetitive odor learning.

NMDA-like receptors are localized throughout the bee brain, including the MB. Silencing the expression of the NR1 subunit of the NMDA receptor (NMDAR) in the MB by RNAi impairs selectively the acquisition phase and the formation of middle-term memory leaving long-term memory intact. It is concluded that NMDARs are not coincidence detectors in the MB but are rather involved in the formation of particular memories.

A putative octopaminergic neuron, the VUMmx1, represents the reinforcing function during olfactory conditioning of the bee. It receives input from sucrose receptors in the suboesophageal ganglion. The axon arborizes bilaterally in both antennal lobes, the lateral horns, and the lip and basal ring regions of the calyces (Hammer, 1993).

FUNCTIONAL ORGANIZATION OF THE α-LOBE

One of the α-lobe extrinsic neurons, the PE1, is an identified single neuron projecting to the lateral protocerebral lobe of the bee brain (Fig. 44.2E). It responds to a large range of stimuli (olfactory, gustatory, tactile, and visual) with prolonged excitation. Pairing electrical stimulation of KCs with depolarization of PE1 leads to associative long-term potentiation (LTP) (Menzel and Manz, 2005) (Fig. 44.2F). PE1 also changes its response properties during olfactory reward learning. In such a paradigm PE1 reduces its response specifically to the learned odor. PE1 receives putative inhibitory input from

A3 neurons (Okada et al., 2007). Since A3 neurons also change their odor responses during associative odor learning, it is possible that the reduced PE1 responses to the learned odor result from an increase of inhibition induced by the learned odor via A3 neurons. However, it is also possible that associative plasticity is an intrinsic property of PE1 itself. In such a case, behavioral learning would have to induce long-term depression (LTD) rather than LTP. The transition between LTD and LTP may be controlled by local Ca^{2+} activities in spines because it is known to occur in mammalian cortical neurons. The main limitation in delineating the processes of long-term synaptic plasticity in MB extrinsic neurons and its relation to behavioral learning lies in the fact that the transmitter(s) of KCs and the respective receptors in the extrinsic neurons are not known.

GLOBAL PROPERTIES OF THE MUSHROOM BODY

The high divergence of sensory input to KCs, their sparse coding properties, and the necessity of coincident activity via PNs makes it likely that KCs code sensory modalities in a highly specific way. The MB extrinsic neurons, however, integrate across sensory modalities and change their properties during learning in multiple ways. It is, therefore, concluded that the MB recodes the highly dimensional sensory space at it input sites (calyx) into a low-dimensional space of meaning and value with a special emphasis on acquired forms of recoding at its output sites (lobes). This view is supported by the finding that the MB in insects is a necessary structure for olfactory learning and memory consolidation. In honeybees, reversible blocking neural activity in the MB by local cooling leads to retrograde amnesia within a few minutes after a single learning trial. Furthermore, several cellular pathways in MB neurons are known to be related to learning and memory formation, and manipulation of the structural integrity of the MB during ontogeny leads to compromised olfactory learning in adult bees.

REFERENCES

Ganeshina OT, Menzel R (2001) GABA-immunoreactive neurons in the mushroom bodies of the honeybee: An electron microscopic study. *J Comp Neurol* 437:335–349.

Grünewald B (1999) Morphology of feedback neurons in the mushroom body of the honeybee, *Apis mellifera*. *J Comp Neurol* 404:114–126.

Grünewald B, Wersing A, Wüstenberg DG (2004) Learning channels. Cellular physiology of odor processing neurons within the honeybee brain. *Acta Biol Hungarica* 55:53–63.

Hammer M (1993) An identified neuron mediates the unconditioned stimulus in associative olfactory learning in honeybees. *Nature* 366:59–63.

Menzel R, Manz G (2005) Neural plasticity of mushroom body-extrinsic neurons in the honeybee brain. *J Exp Bio* 208(22):4317–4332.

Okada R, Rybak J, Manz G, Menzel R (2007) Learning-related plasticity in PE1 and other mushroom body-extrinsic neurons in the honeybee brain. *J Neurosci* 27(43): 11736–11747.

Rybak J, Menzel R (1993) Anatomy of the mushroom bodies in the honey bee brain: the neuronal connections of the alpha-lobe. *J Comp Neurol* 334:444–465.

Strausfeld NJ (2002) Organization of the honey bee mushroom body: representation of the calyx within the vertical and gamma lobes. *J Comp Neurol* 450:4–33.

Szyszka P, Ditzen M, Galkin A, Galizia CG, Menzel R (2005) Sparsening and temporal sharpening of olfactory representations in the honeybee mushroom bodies. *J Neurophysiol* 94:3303–3313.

Szyszka P, Galkin A, Menzel R (2008) Associative and non-associative plasticity in Kenyon cells of the honeybee mushroom body. *Front Syst Neurosci* 2:1–10.

Section 3

Motor Systems

45

The Tritonia Swim Central Pattern Generator

Paul S. Katz

Tritonia diomedea is a sea slug that escapes from predatory starfish by rhythmically flexing its entire body in the dorsal and ventral directions (Katz, 2007). This escape swim behavior is produced by a central pattern generator (CPG) without the need for sensory feedback (Dorsett et al., 1969). There are several features of the neural basis for this response that make it of particular interest for neuroscientists. One is that the CPG is a *network oscillator*; bursting arises as an emergent property of the neurons and their connectivity (Getting, 1989a). Another interesting feature is that the CPG contains state-dependent, *intrinsic neuromodulation*: one of the CPG neurons uses the neurotransmitter serotonin (5-HT) to modulate the strength of synapses made by the other CPG neurons under certain conditions (Katz and Frost, 1996). Finally, this CPG seems to have evolved from a nonoscillatory network (Katz et al., 2001). A number of older reviews are available about this system (Getting and Dekin, 1985; Getting, 1989a). A recent review is available at http://scholarpedia.org/article/Tritonia_swim_network.

There are four levels to the swim network: afferent neurons, gating interneurons, CPG neurons, and efferent neurons (Fig. 45.1A). All of the published identified neurons have been catalogued in NeuronBank (http://NeuronBank.org). The accession IDs of neurons included in this chapter provide unique URLs to access further information about those neurons, including citations.

INPUT TO THE CENTRAL PATTERN GENERATOR

The afferent neurons, the S cells (Tri0002367), have receptive fields on the body surface. They respond tonically to chemosensory stimuli such as extracts

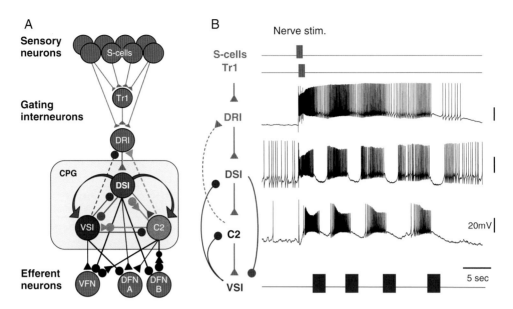

FIGURE 45–1. The *Tritonia* swim network and motor pattern. (*A*) Schematic diagram of the neurons in the swim network. The S cells are primary sensory neurons that innervate the body surface. Tr1 is a trigger neuron. DRI is a gating command neuron. The central pattern generator (CPG) consists of dorsal swim interneurons (DSI), C2, and ventral swim interneurons (VSI). The efferent flexion neurons, VFN, DFN-A, and DFN-B, convey the pattern of activity to the muscles. Combinations of triangles and circles indicate multicomponent synapses. The arrows indicate serotonergic neuromodulatory actions. (*B*) Simultaneous intracellular microelectrode recordings from DRI, DSI, and C2 showing the motor pattern. Juxtaposed are the relative timing of bursts in the S cells, Tr1, and VSI-B. On the left is another rendition of the neuronal circuitry showing the positive feedback loop and the inhibition from VSI.

from starfish tube feet, but phasically to mechanosensory stimuli (Getting, 1976).

The S cells provide glutamatergic, excitatory synaptic input to the trigger neuron, Tr1 (Tri0002513), activating it briefly at the onset of a swim motor pattern. Tr1 is silent for the remainder of the swim motor pattern (Frost et al., 2001). Brief stimulation of Tr1 is sometimes sufficient to trigger a swim episode.

Both Tr1 and the S cells synapse on DRI (Tri0002471), which acts as a gating command neuron (Frost and Katz, 1996). DRI fulfills the classic definition of a command neuron (Kupfermann and Weiss, 1978): activity in DRI is both necessary and sufficient for the swim motor pattern production. Stimulation of DRI at physiological spike frequencies reliably activates the swim motor pattern, whereas hyperpolarization of DRI prevents sensory stimulation from initiating the motor pattern. Furthermore, hyperpolarization of DRI after the initiation of the motor pattern causes it to prematurely terminate (Frost and Katz, 1996; Frost et al., 2001).

THE CENTRAL PATTERN GENERATOR

The CPG consists of three cell types: the dorsal swim interneurons (DSIs; Tri0001043), the ventral swim interneurons (VSI-A, Tri0002406; VSI-B, Tri0002436), and interneuron C2 (Tri0002380). There are three DSIs on each side of the brain: DSI-A, B, and C. For the purposes of this review, the DSIs will be considered equivalent (but see http://scholarpedia.org/article/Tritonia_swim_network).

The DSIs receive strong monosynaptic excitatory input from DRI, which causes them to fire at more than 40 Hz at the onset of a swim motor pattern (Fig. 45.1B). The DSIs then excite C2 to fire a burst of action potentials. C2 provides polysynaptic excitatory input back to DRI, initiating a positive feedback loop: DRI to DSI to C2 and back to DRI.

C2 also provides excitation to VSI-B. Following a C2 spike train, there is a delay before VSI-B starts firing. This delay has been attributed to a strong A-current in VSI-B and the slow nature of C2-evoked excitatory synaptic potentials (Getting, 1983).

VSI-B inhibits both DSI and C2, momentarily interrupting its own excitatory input and allowing the positive feedback loop to reinitiate. This inhibition from VSI-B is critical for the system to oscillate.

As the motor pattern continues, the spike rate in the DSIs gradually declines, causing a longer delay before C2 is excited. This results in a progressive increase in cycle duration until the motor pattern finally ceases.

After the end of the rhythmic motor pattern, the DSIs continue to fire at 1–5 Hz for well over 1 hour. This tonic firing is thought to play a role in the increased tendency of *Tritonia* to crawl after a swim because the DSIs excite neurons that activate cilia involved in crawling (Popescu and Frost, 2000).

STATE-DEPENDENT INTRINSIC NEUROMODULATION

The DSIs are serotonergic (Katz et al., 1994; McClellan et al., 1994) and use serotonin for both synaptic transmission and neuromodulation (Katz and Frost, 1995). The synaptic transmission consists of both fast and slow synaptic potentials (Getting, 1981). The fast excitatory postsynaptic potentials (EPSPs) are mediated by ligand-gated ion channels, whereas the slow EPSPs are mediated by G protein–coupled receptors (Clemens and Katz, 2001).

The DSIs also use serotonin to modulate the synaptic strength and membrane properties of C2 and VSI-B. At least some of these actions are mediated by G protein–coupled receptors (Clemens and Katz, 2003). The ability of one neuron within a neuronal circuit to evoke neuromodulatory actions on other neurons in the same circuit has been termed *intrinsic neuromodulation* to distinguish it from *extrinsic neuromodulation*, where the neuromodulatory inputs arise from other parts of the nervous system (Katz and Frost, 1996).

   The effect of DSI on VSI-B is of particular interest because it is state- and
timing-dependent (Fig. 45.2; Sakurai and Katz, 2003; Sakurai and Katz, 2009).
If VSI-B is stimulated to fire an action potential within 15 sec of a brief DSI
spike train, then VSI-B synaptic strength is enhanced (Figs. 45.2A, B). The
strength of this synapse is also affected by homosynaptic plasticity; if VSI is
stimulated to fire a train of action potentials, then the synapse is greatly
potentiated for over 10 minutes. When the synapse is in the potentiated state,
DSI stimulation depotentiates the synapse, resetting it to a basal amplitude
(Figs. 45.2B, D). This synaptic reset occurs even when DSI is stimulated
more than 15 s before the next VSI spike (Fig. 45.2D). In the unpotentiated
state, stimulating DSI more than 15 s before VSI has no effect on the synapse
(Fig. 45.2C). Thus, the effect of serotonin in this circuit is dependent upon the
timing of when it is released with respect to activity in other CPG neurons,
and it is dependent upon the state of those other neurons.
   Although the role played by intrinsic neuromodulation is not fully known,
it is clear that serotonergic signaling is necessary for production of the swim
motor pattern. Methysergide, a serotonin receptor antagonist that specifically
blocks the neuromodulatory actions of the DSIs, also blocks production of the
swim motor pattern (McClellan et al., 1994). Furthermore, blocking G protein
signaling or adenylyl cyclase in C2 blocks production of the swim motor
pattern (Clemens and Katz, 2003). Thus, state-dependent, intrinsic neuromodulation might be important for adjusting the strength of synapses during

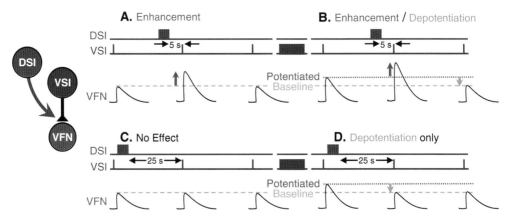

Figure 45–2. State- and timing-dependent neuromodulation. This schematic shows that the
effect of dorsal swim interneurons (DSI) on ventral swim interneuron (VSI) synaptic strength
is a function of the state of the VSI synapse and on the timing between DSI and VSI spikes. In
the top traces (A and B), DSI is stimulated to fire a burst of action potentials 5 sec before the
next VSI spike. In the bottom traces (C and D), the DSI spike train precedes the next VSI spike
by 25 sec. In A and C, the VSI synapse is in the basal state. A VSI spike train (5 Hz for 15 sec)
causes the synapse to enter a potentiated state (B and D). The effect of DSI on VSI output to
VFN for each of these four conditions is shown. The arrows indicate the change in synaptic
strength. (Based on Sakurai and Katz, 2009)

a swim motor pattern and then resetting them afterward (Sakurai and Katz, 2009).

A Model of the Central Pattern Generator

The CPG was modeled using an integrate and fire approach (Getting, 1989b). This model is available at ModelDB (http://senselab.med.yale.edu/ModelDB/), accession number 93326. This early model showed that the basic synaptic organization and firing properties of the neurons were sufficient to account for the periodicity and pattern of bursting in the neurons. However, this model was developed prior to the discovery of DRI, so the input to the circuit was approximated with a declining ramp of depolarization. The model also predated the discovery of intrinsic neuromodulation by the DSIs. A more recent model suggests that the known neuromodulatory actions are important for establishing the oscillatory state (Calin-Jageman et al., 2007). Furthermore, evaluation of parameter space suggests that the oscillatory state is very stable. Thus, dynamic processes must come into play for shifting the network from its resting, nonoscillatory state into an oscillator.

Evolution of the Swim Central Pattern Generator

Tritonia is somewhat unusual in its mode of swimming. Most nudibranch mollusc species do not swim at all and several species swim with side-to-side or lateral flexion movements rather than dorsoventral flexions. Despite these differences in the behaviors of the animals, homologous neurons can be identified across species (Newcomb and Katz, 2007). Homologs of the DSIs were identified in the lateral flexion swimmer, *Melibe leonina*, where they were found not to be rhythmically active during that animal's swim motor pattern (Newcomb and Katz, 2008). Furthermore, serotonin was not necessary for the swim motor pattern. Instead, the DSI homologs could initiate the swim motor pattern, thereby acting as extrinsic modulators of the swim CPG. Thus, the role of serotonin and the DSIs as an intrinsic neuromodulator in *Tritonia* differs in another species with a different form of swimming behavior.

The sea slug, *Pleurobranchaea californica*, does swim like *Tritonia* with dorsoventral body flexions. Its brain contains homologs of DSI and C2, which function as part of the CPG (Jing and Gillette, 1999). *Pleurobranchea* is not a nudibranch, but belongs to a sister taxon and is thus more distantly related to *Tritonia* than all of the nonswimming and lateral swimming nudibranchs. This has led to the hypothesis that the appearance of similar neuronal circuitry in *Tritonia* and *Pleurobranchaea* is an example of parallel evolution, where homologous features independently have come to have similar function (Katz and Newcomb, 2007).

This work indicates that the neuronal circuits can exhibit phylogenetic plasticity. In one case homologous neurons serve different functions to produce different behaviors; in the other case, homologous neurons independently evolved to serve the same function. There may be favored states among neurons to produce particular types of circuits from the common organization of the nervous system.

REFERENCES

Calin-Jageman RJ, Tunstall MJ, Mensh BD, Katz PS, Frost WN (2007) Parameter space analysis suggests multi-site plasticity contributes to motor pattern initiation in Tritonia. *J Neurophysiol* 98:2382–2398.
Clemens S, Katz PS (2001) Identified serotonergic neurons in the *Tritonia* swim CPG activate both ionotropic and metabotropic receptors. *J Neurophysiol* 85:476–479.
Clemens S, Katz PS (2003) G Protein signaling in a neuronal network is necessary for rhythmic motor pattern production. *J Neurophysiol* 89:762–772.
Dorsett DA, Willows AOD, Hoyle G (1969) Centrally generated nerve impulse sequences determining swimming behavior in *Tritonia. Nature* 224:711–712.
Frost WN, Katz PS (1996) Single neuron control over a complex motor program. *Proc Natl Acad Sci USA* 93:422–426.
Frost WN, Hoppe TA, Wang J, Tian LM (2001) Swim initiation neurons in *Tritonia diomedea. Am Zool* 41:952–961.
Getting PA (1976) Afferent neurons mediating escape swimming of the marine mollusc, *Tritonia. J Comp Physiol* 110:271–286.
Getting PA (1981) Mechanisms of pattern generation underlying swimming in *Tritonia.* I. Neuronal network formed by monosynaptic connections. *J Neurophysiol* 46:65–79.
Getting PA (1983) Mechanisms of pattern generation underlying swimming in *Tritonia.* III. Intrinsic and synaptic mechanisms for delayed excitation. *J Neurophysiol* 49: 1036–1050.
Getting PA (1989a) A network oscillator underlying swimming in *Tritonia.* In: Jacklet JW, ed. *Neuronal and Cellular Oscillators,* pp. 215–236. New York: Marcel Dekker, Inc.
Getting PA (1989b) Reconstruction of small neural networks. In: Koch C and Segev I, eds. *Methods in Neuronal Modeling: From Synapses to Networks,* pp. 171–194. Cambridge, MA: MIT Press.
Getting PA, Dekin MS (1985) *Tritonia* swimming: a model system for integration within rhythmic motor systems. In: Selverston AI, ed. *Model Neural Networks and Behavior,* pp. 3–20. New York: Plenum Press.
Jing J, Gillette R (1999) Central pattern generator for escape swimming in the notaspid sea slug *Pleurobranchaea californica. J Neurophysiol* 81:654–667.
Katz PS (2007) Tritonia. *Scholarpedia* 2:3504.
Katz PS, Frost WN (1995) Intrinsic neuromodulation in the *Tritonia* swim CPG: serotonin mediates both neuromodulation and neurotransmission by the dorsal swim interneurons. *J Neurophysiol* 74:2281–2294.
Katz PS, Frost WN (1996) Intrinsic neuromodulation: altering neuronal circuits from within. *Trends Neurosci* 19:54–61.
Katz PS, Newcomb JM (2007) A tale of two CPGs: phylogenetically polymorphic networks. In: Kaas JH, ed. *Evolution of Nervous Systems,* pp. 367–374. Oxford, England: Academic Press.

Katz PS, Getting PA, Frost WN (1994) Dynamic neuromodulation of synaptic strength intrinsic to a central pattern generator circuit. *Nature* 367:729–731.

Katz PS, Fickbohm DJ, Lynn-Bullock CP (2001) Evidence that the swim central pattern generator of *Tritonia* arose from a non-rhythmic neuromodulatory arousal system: implications for the evolution of specialized behavior. *Am Zool* 41:962–975.

Kupfermann I, Weiss KR (1978) The command neuron concept. *Behav Brain Sci* 1:3–39.

McClellan AD, Brown GD, Getting PA (1994) Modulation of swimming in *Tritonia*: Excitatory and inhibitory effects of serotonin. *J Comp Physiol A* 174:257–266.

Newcomb JM, Katz PS (2007) Homologues of serotonergic central pattern generator neurons in related nudibranch molluscs with divergent behaviors. *J Comp Physiol A Neuroethol Sens Neural Behav Physiol* 193:425–443.

Newcomb JM, Katz PS (2009) Different functions for homologous serotonergic interneurons and serotonin in species-specific rhythmic behaviours. *Proc Biol Sci* 276(1654): 99–108.

Popescu IR, Frost WN (2000) One circuit for two types of locomotion in *Tritonia diomedea*. *Soc Neurosci Abstr* 26:163.12.

Sakurai A, Katz PS (2003) Spike timing-dependent serotonergic neuromodulation of synaptic strength intrinsic to a central pattern generator circuit. *J Neurosci* 23: 10745–10755.

Sakurai A, Katz PS (2009) State-, timing-, and pattern-dependent neuromodulation of synaptic strength by a serotonergic interneuron. *J Neurosci* 29:268–279.

46

The Heartbeat Neural Control System of the Leech

Ronald L. Calabrese

In medicinal leeches (*Hirudo sp.*), heartbeat is an automatic function that is continuous (Wenning et al., 2004a; Wenning et al., 2004b; Kristan et al., 2005). Rhythmic constrictions of the muscular lateral vessels that run the length of the animal (the hearts) move blood through the closed circulatory system. The hearts are coordinated so that one beats in a rear-to-front progression (peristaltically), whereas the other one beats synchronously along most of its length. Every 20–40 beats, the two hearts switch coordination states. The hearts are innervated in each segment by heart excitatory (HE) motor neurons, a bilateral pair of neurons found in the 3rd through 18th segmental ganglion (HE[3] through HE[18]) (Fig. 46.1). The heart motor neurons are rhythmically active and the coordinated activity pattern of the segmental heart motor neurons determines the constriction pattern of the hearts. The same coordination modes—peristaltic and synchronous—observed in the hearts occur in the heart motor neurons: on one side they are active in a rear-to-front progression, while on the other they are active nearly synchronously (Fig. 46.1), and the coordination of the motor neurons along the two sides switches approximately every 20–40 heartbeat cycles. The rhythmic activity pattern of the heart motor neurons derives from the cyclic inhibition that they receive from the heartbeat central pattern generator (CPG). The CPG comprises nine bilateral pairs of identified heart (HN) interneurons that occur in the first seven ganglia (HN[1]–HN[7]; Kristan et al., 2005) and ganglia 15 and 16 (HN[15] and HN[16]; Wenning et al., 2008; Fig. 46.1). Heart interneurons make inhibitory synapses with heart motor neurons and among themselves; in addition, certain heart interneurons are electrically coupled (Fig. 46.2B).

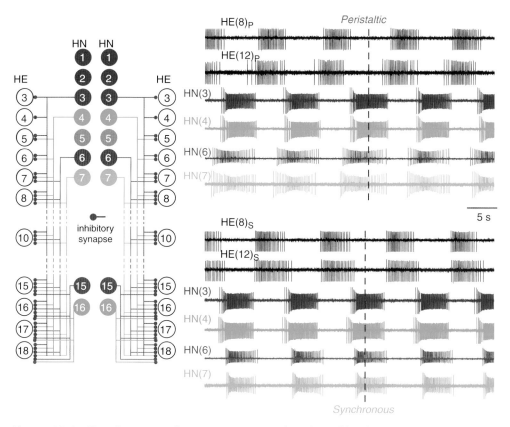

FIGURE 46–1. Heartbeat central pattern generator of medicinal leeches. Heart motor neurons and interneurons and their respective activity patterns: peristaltic and synchronous. (*Left*) Identified heart interneurons and motor neurons and their synaptic connections. All neuron symbols denote single identified neurons. All synaptic connections are inhibitory and are shown with the standard blue filled circles. (*Right*) Peristaltic and synchronous activity patterns of the major premotor HN interneurons and two of the segmental motor neurons that they all innervate. All recordings are extracellular (loose patch) and ipsilateral. The top (peristaltic) and bottom (synchronous) panels were from the same side (segmental ganglion number of each neuron is indicated) prior to and after a switch in coordination mode from peristaltic to synchronous. The dashed vertical lines mark the middle spike of the HN(4) burst to show the ipsilateral phase relations of the neuronal activity in each panel. Interneurons are color coded to indicate their ganglion of origin.

THE HALF-CENTER OSCILLATORS

The HN(1)–HN(4) heart interneurons constitute a core network that sets beat timing throughout the heartbeat CPG. The other five pairs of heart interneurons are followers of these front pairs. There are two foci of oscillation in the beat timing network organized by strong reciprocal inhibitory synapses between the bilateral pairs of HN(3) and HN(4) interneurons in these ganglia

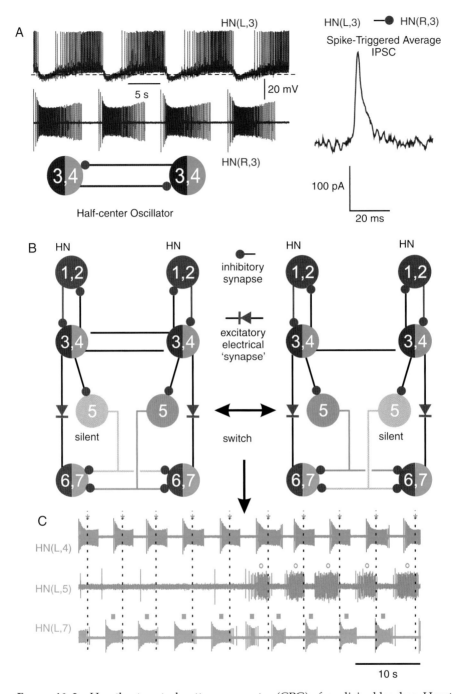

FIGURE 46–2. Heartbeat central pattern generator (CPG) of medicinal leeches. Heart interneurons: Synaptic connections and activity patterns. (A) A half-center oscillator is formed by the HN(3) and HN(4) bilateral (*Right* and *Left*) interneurons pairs and is based on strong reciprocal inhibition. Extracellular recording from the HN(R,3) interneuron and intracellular recording of the HN(L,3) interneuron, showing alternating burst activity in the half-center oscillator.

FIGURE 46–2. Continued
(*Right*) Spike-triggered-averaged IPSC of the HN(R,3) to HN(L,3) inhibitory synapse recorded in voltage clamp. (*B*) Synaptic connections among the first seven pairs of heart interneurons and the configuration of the CPG in the two different coordination modes corresponding to Left peristaltic (left diagram) and Right peristaltic (right diagram). Inhibitory synaptic connections are shown with the standard blue filled circles, and excitatory electrical "synapses" are shown with red rectifier symbols. Interneurons are color coded to indicate their ganglion of origin and those with similar properties and connections are lumped together for ease of illustration. (*C*) Activity (extracellular recording) of two ipsilateral premotor and an ipsilateral switch interneurons during a switch in coordination mode from left peristaltic (HN[5] switch interneuron silent) to left synchronous (HN[5] switch interneuron rhythmically bursting). The dashed vertical lines mark the middle spike of the HN(4) burst (colored symbols mark the middle spikes of all bursts) to show the ipsilateral phase relations of the neuronal activity.

(Fig. 46.2A), which combined with the intrinsic membrane properties of these neurons pace the oscillation. Thus, each of these two reciprocally inhibitory heart interneuronal pairs is a half-center oscillator (Fig. 46.2A). Synaptic and several intrinsic currents contribute to the oscillatory activity of oscillator interneurons (Angstadt and Calabrese, 1989, 1991; Lu et al., 1997; Kristan et al., 2005). These include a fast Na current that mediates spikes; two low-threshold Ca currents (one rapidly inactivating [I_{CaF}] and one slowly inactivating [I_{CaS}]); three outward currents (a fast transient K current [I_A] and two delayed rectifier-like K currents, one inactivating [I_{K1}] and one persistent [I_{K2}]); a hyperpolarization-activated inward current (I_h, a mixed Na/K current with a reversal potential of –20 mV); and a low-threshold persistent Na current (I_p). The inhibition between oscillator interneurons consists of both spike-mediated and graded components. Oscillation in an HN half-center oscillator is a subtle mix of escape and release (Hill et al., 2001). Escape from inhibition is due to the slow activation of I_h in the inhibited oscillator interneuron. Release from inhibition results from a waning of the depolarization in the active oscillator interneuron due to the slow inactivation of its I_{CaS}, which slows its spike rate and thereby reduces its spike-mediated inhibition of the contralateral oscillator interneuron.

COORDINATION IN THE BEAT TIMING NETWORK

The HN(1) and HN(2) heart interneurons act as coordinating interneurons, serving to couple the two half-center oscillators (Hill et al., 2002; Masino and Calabrese, 2002a; Masino and Calabrese, 2002b; Jezzini et al., 2004). The HN(1) and HN(2) interneurons do not initiate spikes in their own ganglion; instead they have two spike initiating sites, one in G3 and the other in G4.

Normally, the majority (>85%) of spikes in the coordinating neurons are initiated in G4. The coupling between the G3 and G4 segmental oscillators causes the HN(3) and HN(4) oscillator interneurons on the same side to be active roughly in phase (Fig. 46.1), although a subtle phase lead by the HN(4) oscillator interneurons is an aspect of CPG output that may be important for proper heart motor neuron coordination. The mechanisms of coordination within the timing networks are consistent with interaction between two independent half-center oscillators that mutually entrain one another and assume the period of the faster oscillator, which then leads in phase.

HEARTBEAT MOTOR PATTERN SWITCHING BY SWITCH INTERNEURONS

Switching between the peristaltic and synchronous modes is accomplished by the pair of HN(5) switch interneurons. The HN(3) and HN(4) oscillator interneurons on one side inhibit the switch heart interneuron on the same side (Fig. 46.2B; Kristan et al., 2005). The HN(5) switch interneurons bilaterally inhibit the HN(6) and HN(7) heart interneurons. Only one of the switch interneurons produces impulse bursts during any given heartbeat cycle; the other switch interneuron is silent, although it receives rhythmic inhibition from the beat timing oscillator (Fig. 46.2C; Gramoll et al., 1994). With a period approximately 20–40 times longer than the period (6–10 sec) of the heartbeat cycle, the silent switch interneuron is activated and the previously active one is silenced (Fig. 46.2). The switch interneurons determine which side is in the peristaltic versus the synchronous activity mode by linking the timing oscillator to the HN(6) and HN(7) heart interneurons. Because only one switch interneuron is active at any given time, there is an asymmetry in the coordination of the heart interneurons on the two sides: the HN(3), HN(4), HN(6), and HN(7) heart interneurons are active roughly in phase on the side of the active switch interneuron, whereas the HN(6) and HN(7) interneurons lead the HN(3) and HN(4) interneurons in phase on the side of the silent switch interneuron (Figs. 46.1 and 46.2C). Because HN(3), HN(4), HN(6), and HN(7) are all premotor inhibitory interneurons, the heart motor neurons are coordinated synchronously on the side of the rhythmically active switch interneuron, whereas the motor neurons are coordinated peristaltically (rear-to-front progression) on the side of the silent switch interneuron. The observed switches in the coordination state of the heart motor neurons, therefore, reflect switches in the activity state of the switch interneurons (Fig. 46.2C). The recently discovered HN(15) and HN (16) interneurons are clearly premotor and provide input to the rearmost heart motor neurons (Fig. 46.1) (Wenning et al., 2008). Less is known about how they integrate within the CPG. They appear to receive electrical (excitatory) input from the HN(6) and HN(7) interneurons and are switched in phase with these inputs when they in turn are switched by the switch interneurons (Seaman and Calabrese, 2008).

There are no synaptic connections between the switch interneurons, even though spontaneous switches in the activity state are always reciprocal. In the silent state, switch neurons have a persistent outward current that is not voltage sensitive and reverses around –60 mV (Gramoll et al., 1994). This current turns off in a switch to the active state. Thus, in its silent state, a switch interneuron is inhibited by a persistent leak current. Switching appears to be controlled by an unidentified independent timing network extrinsic to the switch neurons that alternately imposes a tonic inhibitory leak alternately on one of the two switch interneurons.

The heartbeat CPG can be conceptualized as two timing networks: a beat timing network comprising the first four pairs of heart interneurons (two oscillator pairs and two coordinating pairs) and an unidentified switch timing network that governs the activity of the switch interneurons. The two timing networks converge on the switch interneurons, and together with the HN(6), HN(7), HN(15), and HN(16) heart interneurons they make up the heartbeat CPG. The output of the CPG is configured into two coordination states of the heart motor neurons by the alternating activity states of the two switch interneurons.

References

Angstadt JD, Calabrese RL (1989) A hyperpolarization-activated inward current in heart interneurons of the medicinal leech. *J Neurosci* 9(8):2846–2857.

Angstadt JD, Calabrese RL (1991) Calcium currents and graded synaptic transmission between heart interneurons of the leech. *J Neurosci* 11(3):746–759.

Gramoll S, Schmidt J, Calabrese RL (1994) Switching in the activity state of an interneuron that controls coordination of the hearts in the medicinal leech (Hirudo medicinalis). *J Exp Biol* 186:157–171.

Hill AA, Lu J, Masino MA, Olsen OH, Calabrese RL (2001) A model of a segmental oscillator in the leech heartbeat neuronal network. *J Comput Neurosci* 10(3):281–302.

Hill AA, Masino MA, Calabrese RL (2002) Model of intersegmental coordination in the leech heartbeat neuronal network. *J Neurophysiol* 87(3):1586–1602.

Jezzini SH, Hill AA, Kuzyk P, Calabrese RL (2004) Detailed model of intersegmental coordination in the timing network of the leech heartbeat central pattern generator. *J Neurophysiol* 91(2):958–977.

Kristan WB, Jr., Calabrese RL, Friesen WO (2005) Neuronal control of leech behavior. *Prog Neurobiol* 76(5):279–327.

Lu J, Dalton JF, 4th, Stokes DR, Calabrese RL (1997) Functional role of Ca^{2+} currents in graded and spike-mediated synaptic transmission between leech heart interneurons. *J Neurophysiol* 77(4):1779–1794.

Lu J, Gramoll S, Schmidt J, Calabrese RL (1999) Motor pattern switching in the heartbeat pattern generator of the medicinal leech: membrane properties and lack of synaptic interaction in switch interneurons. *J Comp Physiol A* 184(3):311–324.

Masino MA, Calabrese RL (2002a) Phase relationships between segmentally organized oscillators in the leech heartbeat pattern generating network. *J Neurophysiol* 87(3):1572–1585.

Masino MA, Calabrese RL (2002b) Period differences between segmental oscillators pro-
    duce intersegmental phase differences in the leech heartbeat timing network. *J Neuro-
    physiol* 87(3):1603–1615.
Seaman RC, Calabrese RL (2008) Synaptic variability and stereotypy in the leech heart-
    beat CPG. *Soc Neurosci Abstr* 371 7.
Wenning A, Cymbalyuk GS, Calabrese RL (2004a) Heartbeat control in leeches.
    I. Constriction pattern and neural modulation of blood pressure in intact animals.
    J Neurophysiol 91(1):382–396.
Wenning A, Hill AA, Calabrese RL (2004b) Heartbeat control in leeches. II. Fictive motor
    pattern. *J Neurophysiol* 91(1):397–409.
Wenning A, Norris BJ, Seaman RC, Calabrese RL (2008) Two additional pairs of premo-
    tor heart interneurons in the leech heartbeat CPG: the more the merrier. *Soc Neurosci
    Abstr* 371 6.

47

The Leech Local Bending Circuit

William B. Kristan, Jr.

When a medicinal leech is touched lightly (enough to indent the skin but not enough to cause pain) in the middle of its body, that part of the body bends away from the site of the touch. If the touch site is moved around the circumference of a segment, the location of the peak of the bend varies continuously, rather than having a limited number of movement directions that might, for instance, correspond to sensory receptive fields or motor movement fields (Lewis and Kristan, 1998b). The leech's body is essentially a long tube, consisting of a body wall that encloses the viscera. The major muscles in the body wall are longitudinal and circular, which contract when the appropriate motor neurons (MNs) fire. Although both sets of muscles are activated in local bending (Zoccolan and Torre, 2002), the circuitry for local bending has been established only for the longitudinal component (Kristan et al., 2005).

The leech is an annelid, a segmented worm, which has a fixed number of segments (32), of which 21 are a series of roughly identical midbody segments. The local bend circuitry is repeated in each of the segmental ganglia associated with these segments. The basic circuit consists of three layers of neurons: pressure-sensitive mechanoreceptive neurons (P cells), which excite a layer of local bend interneurons (LBIs), which in turn excite MNs (Fig. 47.1). There are four P cells per ganglion, 17 identified LBIs (there may be a small number yet to be found), and 22 longitudinal MNs. The system is completely feedforward except for the inhibitory lateral connections among MNs: the inhibitory motor neurons (iMNs) release GABA that hyperpolarizes both the muscle fibers and the excitatory motor neurons (eMNs) to the same muscles.

The receptive field of each P cell around the circumference extends about halfway around the body, with a central peak of responsiveness that

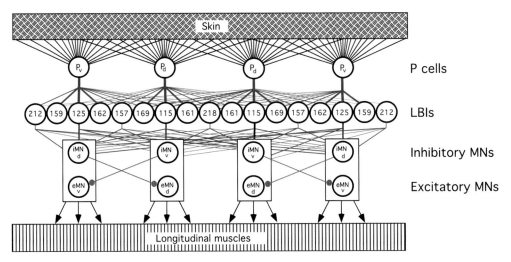

FIGURE 47–1. The leech local bending circuit. Each of the four P cells innervates nearly half the skin around the circumference of each segment, as indicated by the black lines emanating from each P cell that terminates in the skin. Each P cell's sensitivity to touch decreases away from the center of the innervation field, indicated by decreasing density of branches at both edges of its receptive field. The receptive field of each neuron overlaps with half the receptive field of the two P cells on either side. Each P cell makes excitatory synaptic connections to all 17 local bend interneurons (LBIs), which make excitatory connections to excitatory (eMN) and inhibitory (iMN) motor neurons of the segmental longitudinal muscles. (The numbers identify individual LBIs; Lockery and Kristan, 1990b.) The subscripts indicate whether the sensory and motor neurons innervate dorsal (d) or ventral (v) territories. Red lines connecting neurons indicate excitatory chemical synaptic contacts (the upper cells are presynaptic to the cells below), and the blue lines with dots at the end indicate inhibitory connections. The skin and longitudinal muscles are represented as though they were cut longitudinally along the ventral axis and flattened out. The ends, therefore, are the ventral midline and the middle is the dorsal midline.

diminishes monotonically in both directions. (The responses to stimuli at any location around the circumference are the same whether the stimulus is applied more anteriorly or more posteriorly within the same segment.) A plot of its response as a function of distance around the circumference is fit nicely by a cosine function (Lewis and Kristan, 1998c; Lewis, 1999; Baca et al., 2005). The P cells are named by the location of the center of their receptive fields: dorsal (d) and ventral (v) on the left and the right. This arrangement ensures that any moderate touch to the skin usually activates two P cells. The location of the touch is encoded by the relative firing rates in the two P cells activated (Lewis and Kristan, 1998a; Lewis, 1999; Thomson and Kristan, 2006) and by the difference in latency of their first spikes (Thomson and Kristan, 2006). The encoding of relative latencies provides a more precise localization of the stimulus location, but by stimulating pairs of adjacent P cells electrically, the decoding of relative latency proved to be very coarsely decoded.

Each P cell makes excitatory connections onto every LBI (Lockery and Kristan, 1990b), so that the receptive field of each LBI is the entire surface of a segment. Once again, the plot of response sensitivity at different circumferential positions is fit nicely by a cosine function (Kristan et al., 1995), but in this case, the width of the receptive field is twice that of the P cells. This receptive field shape is produced by an orderly progression in the strength of the synaptic contacts between P cells and LBIs: typically, one P cell has a strong synaptic connection and the P cells with adjacent fields have weaker connections, and the fourth P cell has an even weaker connection (Lockery and Kristan, 1990b). The relative strengths of the synaptic connections are represented in Figure 47.1 by the thicknesses of the lines between P cells and LBIs.

The strengths of the connections from the LBIs onto MNs were inferred from the effects of stimulating individual and pairs of P cells, which defines their receptive fields (Lockery and Kristan, 1990a). Again, the receptive fields of the MNs could be fitted nicely by a cosine function that included the entire circumference of the animal, but in this case, half of the receptive field was inhibitory (Kristan et al., 1995), reflecting the fact that touching a given location causes longitudinal contraction at that location and inhibition at the opposite location (Kristan, 1982). By hyperpolarizing the iMNs (they are electrically coupled, so hyperpolarizing one strongly effectively hyperpolarizes all of them), the connections from the LBIs were found to be exclusively excitatory (Lockery and Kristan, 1990a). The conclusion from these experiments is that the connections from the LBIs onto MNs is all to all, with every LBI making a connection to every MN, again with decreasing strengths from LBIs with progressively more distant receptive fields (Fig. 47.1). All the lateral inhibition, therefore, is provided by the connections of iMNs onto eMNs. It should be noted that the iMNs receive inputs from LBIs as though they were eMNs to the opposite side: dorsal iMNs are most strongly excited by inputs from the opposite ventral P cell, and the ventral iMNs get most strongly excited by the dorsal P cell on the opposite side.

Until recently, the only inhibition identified in this circuit was the lateral inhibition among the MNs, as indicated in Figure 47.1. These connections are GABAergic and can be blocked by bicuculline (Baca et al., 2008). In applying bicuculline, however, there was a big surprise: not only did it block the lateral inhibition, it also caused a significant increase in the response at all locations, including at the site of stimulation, where there was thought to be only excitation. In addition, the response on the opposite side—thought to be exclusively inhibitory—became weakly excitatory. Using voltage-sensitive dyes and modeling, the circuitry in Figure 47.2 has been deduced: there is an inhibitory cell or group of cells (Inhib) that receive input from all P cells and inhibit all LBIs (and possibly all eMNs—the modeling could not rule this out). This means that every MN receives a balance of excitation and inhibition wherever P cells are activated. Whether the balance favors activation or inhibition depends on the relative strengths of the inputs. Why have such a seemingly

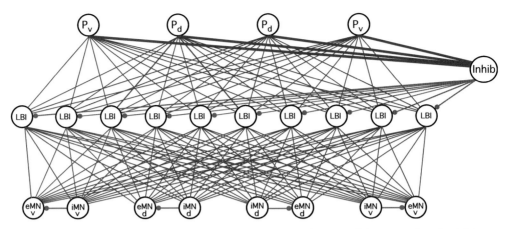

FIGURE 47–2. Simplified local bending circuit, with the addition of the generalized inhibitory connections. P cell activation is summed by an unidentified neuron or systems of neurons (Inhib) that provides a generalized inhibition to all local bend interneurons (a sampling of the 17 LBIs). The connections from inhibitory motor neurons onto excitatory motor neurons produces lateral inhibition. Labels for the P cells and motor neurons are as in Figure 47.1.

wasteful arrangement, with every excitation having to overcome a significant inhibition? Many suggestions have been made: it may linearize the input/output curve; it allows for vector averaging of the response to two or simultaneous stimuli; and changing the generalized inhibition can change the gain of the system. Whatever the function, it appears that many nervous systems use a balance between excitation and inhibition to control their output, and this relatively simple leech circuit should prove to be a good one to test the various functional suggestions.

REFERENCES

Baca SM, Thomson EE, Kristan WB (2005) Location and intensity discrimination in the leech local bending response quantified using optic flow and principal components analysis. *J Neurophysiol* 93:3560–7352.

Baca SM, Marin-Burgin A, Wagenaar DA, Kristan WB, Jr. (2008) Widespread inhibition proportional to excitation controls the gain of a leech behavioral circuit. *Neuron* 57:276–289.

Kristan WB, Jr. (1982) Sensory and motor neurones responsible for the local bending response in leeches. *J Exp Biol* 96:161–180.

Kristan WB, Jr., Lockery SR, Lewis JE (1995) Using reflexive behaviors of the medicinal leech to study information processing. *J Neurobiol* 27:380–389.

Kristan WB Jr., Calabrese RL, Friesen WO (2005) Neuronal basis of leech behaviors. *Prog Neurobiol* 76:279–327.

Lewis JE (1999) Sensory processing and the network mechanisms for reading neuronal population codes. *J Comp Physiol A* 185:373–378.

Lewis JE, Kristan WB, Jr. (1998a) A neuronal network for computing population vectors in the leech. *Nature* 391:76–79.

Lewis JE, Kristan WB, Jr. (1998b) Quantitative analysis of a directed behavior in the medicinal leech: implications for organizing motor output. *J Neurosci* 18:1571–1582.

Lewis JE, Kristan WB, Jr. (1998c) Representation of touch localization by a population of leech touch sensitive neurons. *J Neurophysiol* 80:2584–2592.

Lockery SR, Kristan WB, Jr. (1990a) Distributed processing of sensory information in the leech. II. Identification of interneurons contributing to the local bending reflex. *J Neurosci* 10:1816–1829.

Lockery SR, Kristan WB, Jr. (1990b) Distributed processing of sensory information in the leech. II. Identification of interneurons contributing to the local bending reflex. *J Neurosci* 10:1816–1829.

Thomson EE, Kristan WB (2006) Encoding and decoding touch location in the leech CNS. *J Neurosci* 26:8009–8016.

Zoccolan D, Torre V (2002) Using optic flow to characterize sensory-motor interactions in a segment of the medicinal leech. *J Neurophysiol* 86:2475–2488.

48

Neuronal Circuits That Generate Swimming Movements in Leeches

Wolfgang Otto Friesen

The central neuronal system that generates swimming undulations in medicinal leeches comprises an iterated, concatenated set of segmental oscillators coupled by extensive intersegmental synaptic interactions. Segmental circuits of the central oscillator include intersegmental interneurons (INs) identified by cell morphology, physiological properties, and synaptic interactions. Somata of INs have diameters of about 15 µm; most are inhibitory and bilaterally paired (cells 27, 28, 33, 60, 115, and 123), whereas the exceptional cell 208 is unpaired and excitatory (Fig. 48.1A; Stent et al., 1978; Kristan et al., 2005). These INs control the activity of motoneurons (MNs)—inhibitory (DI, VI) and excitatory (DE, VE)—that are the output elements of the central oscillator system. The inhibitory MNs synapse onto the central oscillatory INs and thereby participate in generating the oscillatory pattern itself. Within segmental ganglia, the interactions are almost exclusively via non-spike-mediated chemical or electrical synapses. Axons of the INs project either rostrally or caudally in the lateral intersegmental connectives of the nerve cord, with a projection span of about six segments (Fig. 48.1B). The exceptional cell 208 projects considerably further. Intersegmental connections are mediated by impulses, which require about 15 ms to traverse the span between adjacent ganglia.

Excitation and Control

Sensory stimulation initiates swimming activity through a cascade of interactions beginning with sensory receptors and ending with the rhythmic antiphasic activation and inhibition of dorsal and ventral longitudinal muscle

A   Intrasegmental phase groups

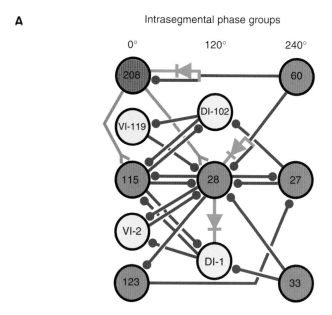

B   Anterior ganglion   Posterior ganglion

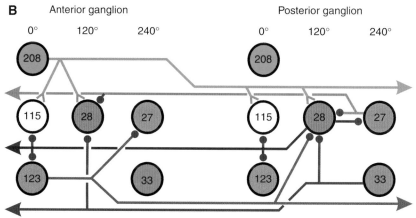

Figure 48–1. Rhythm-generating and coordinating circuits. (*A*) Intrasegmental synaptic connections between swim oscillator interneurons (INs) and inhibitory motoneurons (MNs). Inhibitory neurons occur as bilateral pairs; the sole excitatory IN, cell 208, is unpaired. Some of the synapses shown occur between contralateral cells, and others between ipsilateral cells. (*B*) Intersegmental interactions between the INs (color-coded lines denote intersegmental projections). Only INs with identified intersegmental targets are shown; however, all oscillator INs make intersegmental synapses with neurons up to five or six segments away. Both intra- and intersegmental connections are depicted for the INs. Arrows indicate the direction of projections to more distant ganglia. Neurons are arranged in columns that reflect the phase of their membrane potential oscillations during swimming. Cell designations: 208—unpaired excitatory oscillator IN; 27, 28, 33, 60, 115, and 123—paired inhibitory oscillator INs; DI-1 and DI-102—dorsal inhibitor MNs; VI-2 and VI-119—ventral inhibitor MNs; dark blue terminals (filled circles) designate inhibitory synapses; "Y" terminals denote excitatory synapses; diode symbols denote rectifying electrical synapses. Most bilateral homologous are coupled by nonrectifying electrical connections (not shown). (See review by Kristan et al., 2005)

(Fig. 48.2). One continuous strand in this cascade begins with pressure (P) and nociceptive (N) sensory neurons, which drive trigger neurons (Tr1 and Tr2) via monosynaptic connections (Fig. 48.2A; Brodfuehrer and Friesen, 1986). Trigger neuron somata are located in the subesophageal ganglion; their axons project rearward via intersegmental connectives, with broadly distributed input and output sites. Brief stimulation of trigger neurons elicits bouts of swimming activity, with no correlation between durations of trigger neuron activity and evoked swim episodes. Stimulation of Tr2 can both elicit swimming activity and terminate ongoing swim episodes; hence, this cell acts, perhaps via segmental IN cell 256, as a toggle switch (O'Gara and Friesen, 1995; Taylor et al., 2003). Multiple, currently unknown factors determine whether any particular stimulus initiates swimming.

In addition to trigger neurons, the highest level of control includes a pair of swim inhibitory neurons (SINs) with somata also found in the subesophageal ganglion. These cells may be part of a swim-inactivating system (Brodfuehrer and Thorogood, 2001), whereas another set of neurons, the swim excitatory (SE1) cells, may function as gain-control elements (Fig. 48.2A). Subesophageal neuron, cell SRN1 (not shown), may contribute to rhythm generation because this IN exhibits oscillations that are phase-locked to the swimming rhythm.

The swim-gating neurons, segmental cells 204 and 205, receive monosynaptic excitatory input from trigger neurons and occupy a third level in the swim-initiation cascade, below sensory and trigger neuron (Fig. 48.2B; Weeks, 1982). Somata of these unpaired excitatory INs are limited to ganglia in the posterior half of nerve cord, with intersegmental axons that project to most ganglia (cell 205 projects only rostrally). Depolarization of cell 204 drives the expression of swimming activity in intact nerve cords or in a nearly isolated ganglion (Fig. 48.2B, inset; Weeks, 1981; Debski and Friesen, 1986). Swim initiation—by any stimulus—depolarizes these gating cells, which remain depolarized throughout swim episodes. The sources of this persistent depolarization remain unidentified. Glutamate, acting via non-NMDA receptors, appears to be the primary transmitter in the swim initiation cascade (Brodfuehrer and Thorogood, 2001).

FUNCTIONAL ASPECTS OF THE CENTRAL OSCILLATOR

Individual ganglia are capable of generating the rudiments of the swimming rhythm when swim-gating neurons are stimulated or serotonin is bath-applied. Perhaps because strong excitation is lacking, robustness of swim-like activity observed in isolated ganglia is low; such restricted preparations approximate only weakly the crisp bursting observed in longer chains of leech nerve cord ganglia. Studies with analog neuromimes and computer

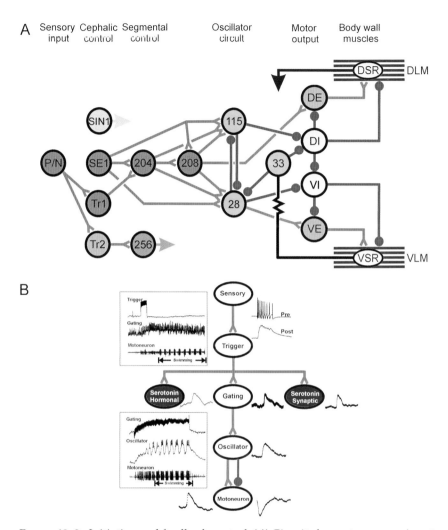

FIGURE 48–2. Initiation and feedback control. (*A*) Circuit elements: sensory input, cephalic control, central oscillator, sensory feedback, and motoneuron (MN) output. Arrows indicate that postsynaptic targets have not been identified. (*B*) Information flow for swim initiation and system modulation by serotonin. Traces beside the main axis show typical profiles of synaptic interactions along the cascade. Insets (*left*) illustrate the physiology and function of trigger and gating neurons. DE and DI, dorsal excitor and inhibitor MNs; DLM and VLM, dorsal and ventral longitudinal muscle; DSR and VSR, dorsal and ventral stretch receptors; P/N, pressure/nociceptive cells; SIN1, swim inhibitor neuron; SE1, swim excitor neuron; Tr1 and Tr2, trigger neurons; 204, swim-gating neuron; 256, cell postsynaptic to Tr2; VE and VI, ventral excitor and inhibitor MNs. (Thanks to Professor John Hackett and Dr. Ruey-Jane Fan for assistance in the construction of *B*)

simulations demonstrate that the identified intrasegmental circuits can generate these rudimentary oscillations (Fig. 48.1A; Taylor et al., 2000; Wolpert and Friesen, 2000). Although intersegmental interactions have been viewed as being relatively strong, recent modeling studies support another viewpoint: namely, that intersegmental coordinating interactions, although very extensive, are individually much weaker than synaptic connections between oscillator INs within ganglia (Kristan et al., 2005; Zheng et al., 2007). These recent analytical and simulation studies confirmed earlier theoretical work by demonstrating that the anterior-to-posterior intersegmental phase lags are quantitatively predictable from (1) the asymmetry in intersegmental synaptic interactions among oscillatory swim INs (Fig. 48.1B) and (2) a deduced "U"-shaped gradient in cycle periods expressed in short changes of nerve cord ganglia (Pearce and Friesen, 1988; Hocker et al., 2000; Cang and Friesen, 2002). Thus, the identified intersegmental topology of the swim circuit can account quantitatively for intersegmental coordination of segmental oscillatory circuits, although the 10° per segment phase lags are considerably smaller than those observed in swimming leeches.

SENSORY FEEDBACK

Shaping neuronal activity patterns to generate effective movements is fundamentally important for animal survival. Sensory inputs from neurons embedded in the leech body wall provide the feedback needed to convert the basic activity pattern generated by the central oscillator into the effective swim undulations of intact leeches. This feedback decreases cycle period by about 50% (0.7 sec, isolated CNS; 0.4 sec, intact animal) and approximately doubles intersegmental phase lags (Pearce and Friesen, 1984). Peripheral stretch receptors (tension transducers) associated with dorsal and ventral longitudinal body wall muscles are the likely source of this feedback. These sensory neurons have peripheral somata, with dendrites that insert into longitudinal muscle fibers (Blackshaw, 1993; Huang et al., 1998). Their giant, nonspiking axons project into segmental ganglia with broad, individually identifiable terminal arbors (Fan and Friesen, 2006). Within ganglia, tension information is conveyed to an oscillator IN, cell 33, by strong electrical connections (Fig. 48.2A; Cang et al., 2001). The functional role of ventral stretch receptors (VSRs) can be deduced from experiments that showed: (1) the VSRs undergo rhythmic membrane potential oscillations phase-locked to muscle tension changes, (2) injection of repeated current pulses into a VSR alters intersegmental phase lags and entrains ongoing swim oscillations, and (3) transient depolarization of the VSR shifts the swim phase (Cang and Friesen, 2000; Yu and Friesen, 2004). Given these effects, tension information transduced by stretch receptors and conveyed electrotonically to oscillator INs can be expected to shape intersegmental phase lags and, perhaps, decrease cycle period.

NEUROMODULATION

The propensity of leeches to locomote by swimming is regulated by neuroactive substances, most notably by serotonin. Leeches with high levels of serotonin in the blood swim more often and isolated nerve cord preparations exhibit swimming activity spontaneously when serotonin is bath applied (Willard, 1981; Hashemzadeh-Gargari and Friesen, 1989). The effect occurs slowly, with swimming activity achieving half-maximal rates about 15 min after treatment with 50 µM serotonin; return to control levels requires about 30 min of washout. The effects of serotonin are restricted to swim propensity; there are few, if any, alterations in cycle period, impulse frequency, or duration of episodes. The slow modulatory effects of bath-applied serotonin are mimicked by prolonged stimulation of the serotonergic Retzius cells (Fig. 48.2B, "serotonin hormonal"). Swimming is more rapidly activated by depolarization of cells 21/61, which appear to release serotonin more locally (Fig. 48.2B, "serotonin synaptic"; Kristan and Nusbaum, 1983).

The swim-gating neuron, cell 204, provides one potential locus for the action of serotonin. Bath application of serotonin modulates the cellular properties of this neuron slightly; functionally, serotonin converts this cell from a gating control neuron into a trigger cell. Although the specific mechanisms underlying this conversion remain unclear, in the presence of elevated serotonin brief intracellular depolarization of cell 204 triggers full-length swim episodes (Angstadt and Friesen, 1993).

The circuit diagrams provided here clearly show that not every neuron and synapse underlying the initiation, generation, feedback control, and modulation of leech swimming movements has been discovered. Nevertheless, these circuits can account for qualitative and quantitative aspects of rhythm generation, intersegmental coordination, muscle feedback control, and hormonal modulation of swimming locomotion in medicinal leeches. Modeling results demonstrate that circuit topology is critical for rhythm generation and control, with cellular properties, such as postinhibitory rebound and impulse adaptation also playing important roles.

ACKNOWLEDGMENTS

This work was supported by a grant from the National Science Foundation (IOS-0615631). Thanks to Olivia Mullins for her helpful comments on a draft of the manuscript.

REFERENCES

Angstadt JD, Friesen WO (1993) Modulation of swimming behavior in the medicinal leech. I. Effects of serotonin on the electrical properties of swim-gating cell 204. *J Comp Physiol A* 59:223–234.

Blackshaw SE (1993) Stretch receptors and body wall muscle in leeches. *Comp Biochem Physiol* 105A:643–652.

Brodfuehrer PD, Friesen WO (1986) From stimulation to undulation: an identified pathway for the control of swimming activity in the leech. *Science* 234:1002–1004.

Brodfuehrer PD, Thorogood MSE (2001) Identified neurons and leech swimming behavior. *Prog Neurobiol* 63:371–381.

Cang J, Friesen WO (2000) Sensory modification of leech swimming: rhythmic activity of ventral stretch receptors can change intersegmental phase relationships. *J Neurosci* 20:7822–7829.

Cang J, Friesen WO (2002) Model for intersegmental coordination of leech swimming, central and sensory mechanisms. *J Neurophysiol* 87:2760–2769.

Cang J, Yu X, Friesen WO (2001) Sensory modification of leech swimming: Interactions between ventral stretch receptors and swim-related neurons. *J Comp Physiol A* 187:569–579.

Debski EA, Friesen WO (1986) The role of central interneurons in the habituation of swimming activity in the medicinal leech. *J Neurophysiol* 55:977–994.

Fan RJ, Friesen WO (2006) Morphological and physiological characterization of stretch receptors in leeches. *J Comp Neurol* 494:290–302.

Hashemzadeh-Gargari H, Friesen WO (1989) Modulation of swimming activity in the medicinal leech by serotonin and octopamine. *Comp Biochem Physiol* 94C:295–302.

Hocker CG, Yu X, Friesen WO (2000) Heterogeneous circuits generate swimming movements in the medicinal leech. *J Comp Physiol* 186:871–883.

Huang Y, Jellies J, Johansen KM, Johansen J (1998) Development and pathway formation of peripheral neurons during leech embryogenesis. *J Comp Neurol* 397:394–402.

Kristan WB, Nusbaum MP (1983) The dual role of serotonin in leech swimming. *J Physiol* 78:743–747.

Kristan WB, Calabrese R, Friesen WO (2005) Neuronal control of leech behavior. *Prog Neurobiol* 76:279–327.

O'Gara BO, Friesen, WO (1995) Termination of leech swimming activity by a previously identified swim trigger neuron. *J Comp Physiol* 177:627–636.

Pearce RA, Friesen WO (1984) Intersegmental coordination of leech swimming, comparison of in situ and isolated nerve cord activity with body wall movement. *Brain Res* 299:363–366.

Pearce RA, Friesen WO (1988) A model for intersegmental coordination in the leech nerve cord. *Biol Cybernetics* 58:301–311.

Stent GS, Kristan WB, Friesen WO, Ort CA, Poon M, Calabrese RL (1978) Neuronal generation of the leech swimming movement. *Science* 200:1348–1356.

Taylor AL, Cottrell GW, Kristan WB (2000) A model of the leech segmental swim central pattern generator. *Neurocomputing* 32–33:573–584.

Taylor AL, Cottrell GW, Kleinfeld D, Kristan WB (2003) Imaging reveals synaptic targets of a swim-terminating neuron in the leech CNS. *J Neurosci* 23:11402–11410.

Weeks JC (1981) Neuronal basis of leech swimming: separation of swim initiation, pattern generation and intersegmental coordination by selective lesions. *J Neurophysiol* 45:698–723.

Weeks JC (1982) Synaptic basis of swim initiation in the leech. I. Connections of a swim initiating neuron cell 204 with motor neurons and pattern generating "oscillator" neurons. *J Comp Physiol A* 148:253–263.

Willard AL (1981) Effects of serotonin on the generation of the motor program for swimming by the medicinal leech. *J Neurosci* 1:936–944.

Wolpert SX, Friesen WO (2000) On the parametric stability of a central pattern generator. *Neurocomputing* 32:603–608.

Yu X, Friesen WO (2004) Entrainment of leech swimming activity by the ventral stretch receptor. *J Comp Physiol A* 190:939–949.

Zheng M, Friesen WO, Iwasaki T (2007) Systems-level modeling of neuronal circuits for leech swimming. *J Comput Neurosci* 22:21–38.

49

The Crustacean Stomatogastric Nervous Systems

Eve Marder

The crustacean stomatogastric nervous system has become one of the premier preparations used for the study of the mechanisms underlying the generation of rhythmic motor patterns. Figure 49.1 shows the overall organization of the stomatogastric nervous system. The stomatogastric ganglion (STG) contains about 30 neurons, most of which are motor neurons that innervate more than 40 sets of striated muscles that move the animal's stomach (Maynard and Dando, 1974; Selverston and Moulins, 1987; Harris-Warrick et al., 1992). Descending projection neurons from the two commissural ganglia (CoGs) and the single esophageal ganglion (OG) are important for the generation of the motor patterns produced by the STG (Nusbaum and Beenhakker, 2002). Identified sensory neurons project either into the CoGs to activate descending modulatory neurons or directly into the STG (Nusbaum and Beenhakker, 2002; Beenhakker et al., 2007; Marder and Bucher, 2007).

Over the years a number of different crustacean species, including lobsters (notably *Panulirus interruptus* and *Homarus americanus*) and crabs (most commonly *Cancer borealis*), have been used to study the stomatogastric nervous system. Although there are minor species differences in firing patterns, connectivity, stomach anatomy, muscle innervation, and neuromodulator distribution, the principal conclusions hold for all crustaceans.

The stomatogastric nervous system generates four motor patterns: the pyloric rhythm, the gastric mill rhythm, the esophageal rhythm, and the cardiac sac rhythm (Maynard and Dando, 1974; Selverston and Moulins, 1987; Harris-Warrick et al., 1992; Marder and Bucher, 2007). The mechanisms that give rise to the pyloric and gastric mill rhythms have been much more extensively studied than those of the other two rhythms. An important feature of the STG that facilitated these studies of both the pyloric and gastric mill

470

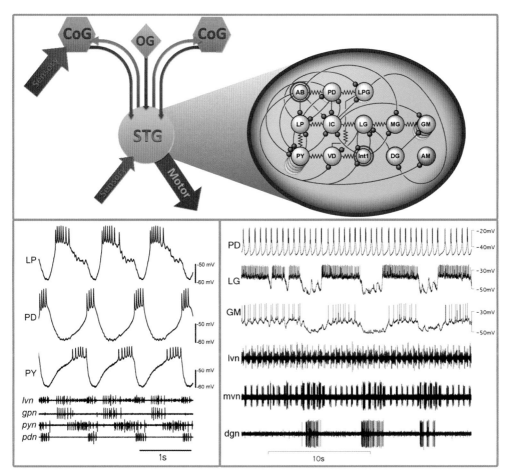

FIGURE 49–1. (*Top*) Overview of the stomatogastric nervous system. The stomatogastric ganglion (STG) receives descending modulatory inputs from the anterior commissural ganglia (CoGs) and the esophageal ganglion (OG). Sensory inputs project both to the STG and to the CoGs to influence the activity of the projection neurons. In the connectivity diagram, electrical synapses are shown as resistor symbols and chemical inhibitory synapses by circles. The pyloric rhythm is shown in the simultaneous intracellular recordings from the lateral pyloric (LP), pyloric (PY), and pyloric dilator (PD) neurons and from the motor nerves shown below in extracellular recordings. The pdn shows the activity of the two PD neurons, the pyn shows the activity of the ≈5 PY neurons, and the LP neuron is seen as the largest unit on the lvn and on the gpn. In the right-hand set of recordings, the simultaneous recordings show activity of the pyloric and gastric rhythms at the same time. The top trace is an intracellular recording from a PD neuron. The next two traces are intracellular recordings from two of the gastric circuit neurons, the lateral gastric (LG) and gastric mill (GM) neurons. The extracellular recordings show activity of other circuit elements, most notably the dorsal gastric nerve (dgn) shows rhythmic dorsal gastric (DG) activity. gpn, gastro-pyloric nerve; lvn, lateral ventricular nerve; mvn, median ventricular nerve; pdn, pyloric dilator nerve; pyn, pyloric nerve. (Figure composition is thanks to Gabrielle Gutierrez. Recordings of the pyloric rhythm on the bottom left are from Lamont Tang. Recordings on the bottom right from Gabrielle Gutierrez.)

circuits is that, unlike the case in most motor systems in which the central pattern generator consists of interneurons that eventually drive motor neurons, in the STG the motor neurons themselves are part of the central pattern generating circuits.

The pyloric rhythm (Fig. 49.1, bottom left) controls the movements of the pylorus, which filters and sorts food. The fast pyloric rhythm (period, ≈1 sec) is driven by a three-neuron pacemaker kernel consisting of a single anterior burster (AB) and the two pyloric dilator (PD) neurons. These neurons rhythmically inhibit the lateral pyloric (LP) and pyloric (PY) neurons that fire in a triphasic rhythm (LP, PY, PD) that is continuously active in the animal (Selverston and Moulins, 1987; Harris-Warrick et al., 1992; Marder and Bucher, 2007). The LP neuron fires before the PY neurons because of the strength and time courses of the synapses from the pacemaker kernel and the intrinsic properties of the LP and PY neurons (Marder and Bucher, 2007).

The gastric mill rhythm controls the movements of the single medial tooth and two lateral teeth within the stomach which grind food (Selverston and Moulins, 1987; Nusbaum and Beenhakker, 2002). The slower gastric mill rhythm (period, ≈ 5–15 sec) is intermittent, and it is activated by sensory and modulatory inputs (Nusbaum and Beenhakker, 2002). Unlike the pyloric rhythm, which maintains a characteristic sequence of activity, the gastric mill rhythm is found in numerous forms, depending on how it is activated (Selverston and Moulins, 1987; Nusbaum and Beenhakker, 2002; Beenhakker et al., 2007). In Figure 49.1 (bottom right) the gastric mill rhythm is seen in the intracellular recordings from the lateral gastric (LG) and gastric mill (GM) neurons and in the extracellular recording from the dorsal gastric nerve (dgn), which shows the activity of the dorsal gastric (DG) neuron. In these recordings, DG neuron activity alternates with those of the LG and GM neurons. Unlike the pyloric rhythm, which is driven by a pacemaker, the gastric mill rhythm is "assembled" by the action of interactions between one or several of the descending modulatory inputs to the STG and neurons of the gastric mill circuit (Nusbaum and Beenhakker, 2002).

The connectivity diagram shown in Figure 49.1 is for the STG of the crab *C. borealis*, from which the recordings in Figure 49.1 were taken. This connectivity diagram shows that the neurons conventionally considered part of the pyloric network (highlighted in red) are highly interconnected with those conventionally considered part of the gastric mill network (highlighted in blue). The interactions between the pyloric rhythm (as monitored by the intracellular recording from the PD neuron) and neurons of the gastric rhythm are seen in the pyloric timed depolarizations in the LG neuron and in the last DG neuron burst. The extensive connections among the neurons of the STG also provide the anatomical substrate for "switching" of some of the neurons of the STG, from firing in time with the pyloric rhythm to firing in time with the gastric mill rhythm (Weimann and Marder, 1994; Marder and Bucher, 2007).

Because of its small size and readily identifiable neurons, work on the STG, in addition to providing basic insights into the mechanisms underlying generation of rhythms in the nervous system, has contributed to our understanding of a number of problems in neuroscience (Marder and Bucher, 2007). These findings include the following: *(1)* Individual neurons and networks are modulated by many substances to alter intrinsic neuronal excitability and synaptic strength, thus reconfiguring neuronal networks. All of the neurons and synapses within a circuit are likely targets for neuromodulation. *(2)* There are a variety of long-term compensatory mechanisms that tend to maintain stable function after the neuromodulatory inputs are removed (Zhang et al., 2009). *(3)* Each identified neuron type in the STG may have a characteristic set of correlations in the expression of its ion channel genes (Schulz et al., 2007).

REFERENCES

Beenhakker MP, Kirby MS, Nusbaum MP (2007) Mechanosensory gating of propriocep-
tor input to modulatory projection neurons. *J Neurosci* 27:14308–14316.

Harris-Warrick RM, Marder E, Selverston AI, Moulins M (1992) *Dynamic Biological
Networks. The Stomatogastric Nervous System*. Cambridge, MA: MIT Press.

Marder E, Bucher D (2007) Understanding circuit dynamics using the stomatogastric
nervous system of lobsters and crabs. *Annu Rev Physiol* 69:291–316.

Maynard DM, Dando MR (1974) The structure of the stomatogastric neuromuscular
system in *Callinectes sapidus, Homarus americanus* and *Panulirus argus* (decapoda crus-
tacea). *Philos Trans R Soc Lond (Biol)* 268:161–220.

Nusbaum MP, Beenhakker MP (2002) A small-systems approach to motor pattern generation. *Nature* 417:343–350.

Schulz DJ, Goaillard JM, Marder EE (2007) Quantitative expression profiling of identified
neurons reveals cell-specific constraints on highly variable levels of gene expression.
Proc Natl Acad Sci USA 104:13187–13191.

Selverston AI, Moulins M, eds. (1987) *The Crustacean Stomatogastric System*. Berlin:
Springer-Verlag.

Weimann JM, Marder E (1994) Switching neurons are integral members of multiple
oscillatory networks. *Curr Biol* 4:896–902.

Zhang Y, Khorkova O, Rodriguez R, Golowasch J (2009) Activity and neuromodulatory
input contribute to the recovery of rhythmic output after decentralization in a central
pattern generator. *J Neurophysiol* 101:372–386.

50

The Swimming Circuit in the Pteropod Mollusc *Clione limacina*

Yuri I. Arshavsky, Tatiana G. Deliagina, and Grigori N. Orlovsky

The pelagic marine mollusc *Clione limacina* (class *Gastropoda*, subclass *Opisthobranchaea*, order *Pteropoda*), 3–5 cm in length, swims by rhythmically moving (1–2 Hz) two wing-like appendages (Arshavsky et al., 1985a; Satterlie et al., 1985; Fig. 50.1A). Each swim cycle consists of two phases: the dorsal (D) and ventral (V) wing flexions (Fig. 50.1B), which are produced by alternating contractions of antagonistic wing muscles. The nervous system of *Clione* consists of five pairs of ganglia. The wing movements are controlled by the pedal ganglia giving rise to the wing nerves. The neuronal circuit of the swim central pattern generator (CPG) is located in the pedal ganglia, which is able to generate the basic pattern of rhythmic activity after isolation from the organism (fictive swimming, Fig. 50.1C). Approximately 120 pedal neurons (out of 800 neurons in both ganglia) exhibit rhythmic activity during fictive swimming (for review, see Arshavsky et al., 1998; Arshavsky et al., 2009). According to their morphology, rhythmic neurons are divided into motoneurons (MNs), with axons exiting via the wing nerves to wing muscles, and interneurons (INs), with axons projecting to the contralateral ganglion.

Organization of Swim Central Pattern Generator

The main neuron groups constituting the swim CPG and their interconnections are shown in Figure 50.1E (Arshavsky et al., 1985b; Panchin et al., 1995; Sadreyev and Panchin, 2002). The motoneurons are not responsible for rhythm generation since the rhythm persisted after their photo-inactivation. The core of the swim CPG is formed by two antagonistic groups of INs: group 7 (glutamatergic neurons) and group 8 (cholinergic neurons), which are active

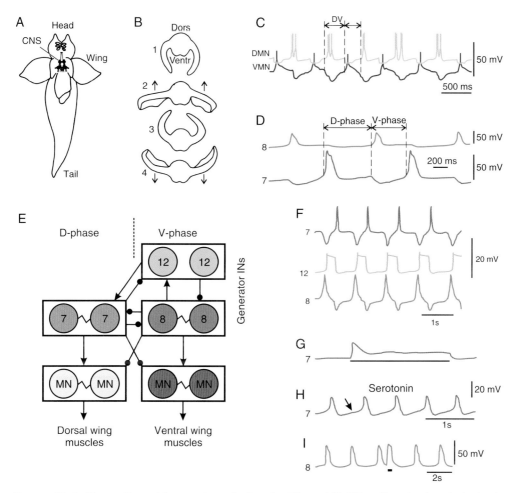

FIGURE 50–1. Generation of locomotory rhythm in *Clione*. (*A*) *Clione limacina* (ventral view). (*B*) Successive wing positions during a locomotor cycle (frontal view). (*C*) Activity of D-phase and V-phase motoneurons (DMN and VMN) during fictive swimming generated by the isolated pedal ganglia. (*D*) Activity of interneurons (INs) of groups 7 and 8 during fictive swimming. (*E*) Organization of swimming central pattern generator in *Clione*. Arrows, filled circles, and resistor symbols show chemical excitatory and inhibitory connections and electrical connections, respectively. (*F*) Activity of INs of groups 7, 8, and 12 during fictive swimming. (*G–I*) Activity of the group 7 and 8 INs after extraction from the ganglion. An IN extracted from the pedal ganglion does not generate rhythmic activity, spontaneous or in response to prolonged depolarization (horizontal line in *G*), but it is transferred in a rhythm-generating state by serotonin application (*H*). Extra excitation of the isolated IN by current injection shifts the phase of rhythmic activity (*I*). (*A, B,* and *D* were adapted from Arshavsky et al., 1985a. *C* was adapted from Arshavsky et al., 1985b. *E* was adapted from Arshavsky et al., 1985c. *F* is from Panchin et al., 1995. *G* and *H* were adapted from Panchin et al., 1996. *I* was adapted from Arshavsky et al., 1986.)

in the D- and V-phases of the swim cycle, respectively (Fig. 50.1D). In each swim cycle, the INs produce one prolonged (~100 ms), tetrodotoxin-resistant action potential. When excited, the INs produce the prolonged inhibitory postsynaptic potentials (IPSPs) in the antagonistic INs. These prolonged action potentials and IPSPs determine the duration of the cycle phases. In young specimens (2–6 mm long) that have a higher frequency of wing oscillations (5–6 Hz), the duration of IN action potentials is much shorter (~30 ms) than in adults.

Each group of INs contains about 20 cells, 10 cells per ganglion. All INs within both groups are electrically interconnected. This contributes to the synchronization of their activity in each ganglion and in both ganglia and, therefore, to synchronous rhythmic movements of the two wings. The INs of each group cause the mid-cycle IPSPs in the INs of the other group (Fig. 50.1D). This reciprocal inhibition determines the alternation of the D-phase and V-phase of the swim cycle. In addition to the reciprocal inhibition between the two groups of INs, there are two other factors contributing to the reliable transition from one phase of the cycle to another phase, and preventing each group of INs from repetitive firing before the antagonistic INs become excited. One of these factors is the postinhibitory rebound. By producing an IPSP in the antagonistic group of INs, a given group facilitates an excitation of the antagonistic INs in the opposite phase of the cycle, just after the termination of the IPSP.

The other factor contributing to the reliable transition from one phase of the cycle to the other phase is activation of an additional group of INs (group 12 in Fig. 50.1E). This occurs only during the more intense activity of the swim CPG, when the probability of disrupting the swim pattern is higher (Arshavsky et al., 1998). The group 8 INs (unlike the homogenous group 7 INs) includes two subgroups: the "early" subgroup (8e) and the "delayed" subgroup (8d). The 8d INs are excited with a slight delay following the 8e INs, due to their higher threshold. The 8e INs are always active during swim rhythm generation, whereas the 8d INs are excited only during more intensive activity of the swim CPG. The 8d INs initiate the rhythmic activity in the group 12 INs. These INs are nonspiking cells that generate long-lasting plateau potentials (Fig. 50.1F). The plateau potential is initiated in the V-phase of the cycle by the excitatory input from the 8d INs and terminated by the inhibitory input from the group 7 INs. When excited, group 12 INs produce a recurrent inhibitory effect on the group 8 INs and an excitatory effect on the group 7 INs. Therefore, group 12 INs promote the termination of the V-phase and initiation of the next D-phase of the swim cycle.

Thus, the neuronal network shown in Figure 50.1E includes several functional mechanisms that participate in forming the biphasic motor output. These mechanisms cooperate with each other to ensure a high reliability of network operation. For example, the swim CPG continues to generate a normal motor pattern (biphasic activity with a half-cycle shift between the

D- and V-phases) after the pharmacological elimination of inhibitory connections from the group 7 INs to the group 8 INs.

The INs of groups 7 and 8 control the activity of wing muscle MNs. They excite synergistic MNs and inhibit antagonistic ones (Fig. 50.1E). The synergistic MNs within each ganglion are electrically interconnected. Along with common inputs from INs, this contributes to synchronization of MN activity in the corresponding phase of the cycle.

ORIGIN OF RHYTHMIC ACTIVITY

Generation of the swim rhythm is based on the endogenous pacemaker activity of the INs of groups 7 and 8. When extracted from the ganglion, the INs continue the rhythmical generation of prolonged action potentials similar to those before extraction, or this rhythmic activity can be evoked by serotonin (Fig. 50.1G and H) (Arshavsky et al., 1986). Each neuron discharge is preceded by a pacemaker potential (arrow in Fig. 50.1H). An extra-excitation of the isolated IN causes a phase shift in its rhythmic activity (Fig. 50.1I), which is typical for pacemakers. In contrast to INs, the MNs do not generate swim-like rhythmic activity after isolation.

The other mechanism contributing to rhythm generation is the postinhibitory rebound. Although this mechanism is not sufficient for sustained rhythm generation, it facilitates the alternation between the two phases of the cycle. Thus, the endogenous pacemaker properties and the inhibitory interactions between the antagonistic INs reinforce one another in providing reliable rhythm generation.

COMMAND SYSTEM

Clione swims episodically, not continuously. Correspondingly, the rhythm-generating INs of groups 7 and 8 are not constitutive pacemakers (like the pacemakers that control heartbeats and respiratory movements); rather, they are conditional pacemakers that are transferred from the passive to the rhythm-generating state by command signals. The serotonergic command neurons, located in the cerebral ganglia and projecting to the pedal ganglia, play the most important role in the initiation of swimming in *Clione* and in the control of its intensity (Arshavsky et al., 1985a; Panchin et al., 1996; Satterlie and Norekian, 1996). The serotonergic command neurons produce activation of the swim INs; this results in initiation or acceleration of the swim rhythm. They also produce depolarization of the wing MNs; this results in an increase of their spike activity in the corresponding phase of the swim cycle (Fig. 50.2D).

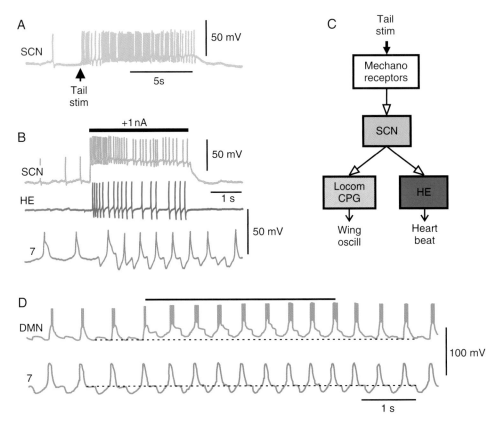

FIGURE 50–2. Serotonergic command neurons. (*A*) Response of the serotonergic command neuron (SCN) to tactile stimulation of the tail. (*B*) Excitation of the same SCN by current injection resulted in acceleration of the locomotory rhythm (monitored by the group 7 interneuron) and in activation of the heart excitatory neuron (HE). (*C*) Diagram of connections of SCN. It receives excitatory input from the tail mechanoreceptors and exerts an excitatory action on the locomotor central pattern generator (CPG) and on the heart exciter (HE). (*D*) Excitation of SCN by current injection (horizontal line) resulted in acceleration of the locomotory rhythm (monitored by the group 7 interneuron) and in increase of motoneuron discharges; dotted lines indicate initial levels of the membrane potential. (*A* and *B* were adapted from Arshavsky et al., 1992. *D* was adapted from Panchin et al., 1996.)

The serotonergic command neurons participate in the control of different forms of behavior. For example, they initiate and accelerate the swim rhythm during an escape reaction elicited by tail stimulation (Figs. 50.2A–C). The start and increase of locomotor activity, induced by the serotonergic command system, is not accompanied by the depolarization of pacemaker INs of groups 7 and 8 (Figs. 50.2B and D). This suggests that the effect of serotonin is realized through metabotropic receptors, whose activation leads to changes in membrane properties. One more effect of the serotonergic command neurons on the locomotor activity is an enhancement of the wing muscle contractility.

This command system also activates the circulatory system by exciting the heart excitatory neuron (Figs. 50.2B and C). Therefore, serotonergic mechanisms perform both command and integrative functions.

Conclusions

Even a neuronal network as simple as the swim CPG in *Clione* has a redundant organization. Every function that is crucial for generating the swim rhythm is determined not by one but by several complementary mechanisms that act in concert. This ensures a high reliability of the network operation.

References

Arshavsky YI, Beloozerova IN, Orlovsky GN, Panchin YV, Pavlova GA (1985a) Control of locomotion in marine mollusc *Clione limacina*. I. Efferent activity during actual and fictitious swimming. *Exp Brain Res* 58:255–262.

Arshavsky YI, Beloozerova IN, Orlovsky GN, Panchin YV, Pavlova GA (1985b) Control of locomotion in marine mollusc *Clione limacina*. II. Rhythmic neurons of pedal ganglia. *Exp Brain Res* 58:263–272.

Arshavsky YI, Beloozerova IN, Orlovsky GN, Panchin YV, Pavlova GA (1985c) Control of locomotion in marine mollusc *Clione limacina*. IV. Role of type 12 interneurons. *Exp Brain Res* 58:285–293.

Arshavsky YI, Deliagina TG, Orlovsky GN, Panchin YV, Pavlova GA, Popova LB (1986) Control of locomotion in marine mollusc *Clione limacina*. VI. Activity of isolated neurons of pedal ganglia. *Exp Brain Res* 63:106–112.

Arshavsky YI, Deliagina TG, Orlovsky GN, Panchin YV, Popova LB (1992) Interneurons mediating the escape reaction of the marine mollusc *Clione limacina*. *J Exp Biol* 164:307–314.

Arshavsky YI, Deliagina TG, Orlovsky GN, Panchin YV, Popova LB, Sadreyev RI (1998) Analysis of the central pattern generator for swimming in the mollusk *Clione*. *Ann NY Acad Sci* 860:51–69.

Arshavsky YI, Deliagina TG, Orlovsky GN (2009) Swimming: neural mechanisms. In: Squire LR, ed. *Encyclopedia of Neuroscience*, Vol. 9, pp. 651–661. Oxford, England: Academic Press.

Panchin YV, Sadreyev RI, Arshavsky YI (1995) Control of locomotion in marine mollusc *Clione limacina*. X. Effects of acetylcholine antagonists. *Exp Brain Res* 106:135–144.

Panchin YV, Arshavsky YI, Deliagina TG, Orlovsky GN, Popova LB, Selverston AI (1996) Control of locomotion in marine mollusc *Clione limacina*. XI. Effects of serotonin. *Exp Brain Res* 109:361–365.

Sadreyev RI, Panchin YV (2002) Effects of glutamate agonists on the isolated neurons from the locomotor network of the mollusc *Clione limacina*. *Neuroreport* 13:2235–2239.

Satterlie RA, Norekian TP (1996) Modulation of swimming speed in the pteropod mollusc, *Clione limacina*: role of a compartmental serotonergic system. *Invert Neurosci* 2:157–165.

Satterlie RA, LaBarbera M, Spencer AN (1985) Swimming in the pteropod mollusc *Clione limacina*. 1. Behavior and morphology. *J Exp Biol* 116:189–204.

51

The Circuit for Chemotaxis and Exploratory Behavior in *C. elegans*

Cornelia I. Bargmann

Over 20 years ago, a wiring diagram of the *C. elegans* nervous system was constructed from serial-section electron micrographs (White et al., 1986). The 302 neurons in the nervous system of the adult hermaphrodite consist of three overall classes: sensory neurons with specialized cilia or dendrites, motor neurons that form neuromuscular junctions, and interneurons that connect sensory neurons with motor neurons. Most sensory neurons and interneurons belong to bilaterally symmetric pairs with similar connections and morphologies, while many motor neurons belong to larger classes. The *C. elegans* nervous system presents an unusual situation in which neuroanatomical connections are extremely well defined, but the understanding of neuronal activity is fragmentary.

One *C. elegans* circuit that is relatively well characterized generates undirected search when animals are removed from food, and directed chemotaxis in odor gradients. Both of these behaviors are based upon temporally regulated turning behavior (Pierce-Shimomura et al., 1999; Hills et al., 2004; Wakabayashi et al., 2004; Gray et al., 2005). When animals are removed from food, a transient turning bout produces undirected local search (Tsalik and Hobert, 2003; Zhao et al., 2003; Hills et al., 2004; Wakabayashi et al., 2004; Gray et al., 2005). The turning bout lasts about 15 minutes and keeps animals in a restricted area; thereafter, suppression of turning results in dispersal. In the presence of an odor gradient, fine-grained temporally regulated turns produce a biased random walk for gradient climbing (Pierce-Shimomura et al., 1999). During chemotaxis, increases in odor levels transiently suppress turning, and decreases in odor levels increase turning.

The wiring diagram and quantitative behavioral analysis have been used to trace the circuit for local search behavior from sensory input to motor output (Gray et al., 2005) (Fig. 51.1). Based on cell ablations, some neurons

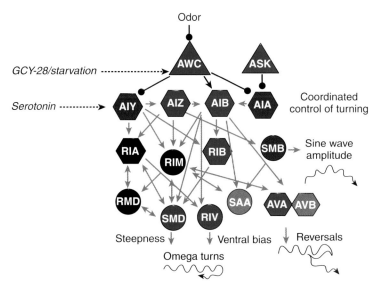

Figure 51–1. Turning circuit for chemotaxis and exploratory behavior. Sensory neurons are denoted as triangles, interneurons as hexagons, and motor neurons as circles. Blue neurons stimulate turning; red neurons inhibit turning; black neurons have mixed activity; gray neurons are untested (Gray et al., 2005). Black lines indicate excitatory (arrow) and inhibitory (stopped) synapses or synaptic inputs established by calcium imaging (Chalasani et al., 2007; Macosko et al., 2009). Dashed lines indicate sites of neuromodulation (Zhang et al., 2005; Tsunozaki et al., 2008). Gray arrows indicate anatomically predicted, but functionally unconfirmed connections from the wiring diagram (White et al., 1986).

stimulate turning—killing them leads to reduced turning during local search. The most important of these are AWC sensory neurons and AIB interneurons. Other neurons inhibit turning—killing them leads to increased turning during local search and/or dispersal. The most important of these are AIA and AIY interneurons. Several neurons such as RIM have mixed properties, stimulating some classes of turns while inhibiting others. AWC sensory neurons trigger both exploratory behavior and chemotaxis; the downstream interneurons have overlapping, but not identical roles in the two behaviors.

Calcium imaging experiments indicate that AWC and ASK sensory neurons detect changes in odor or food levels. AWC and ASK neurons are tonically active at rest and hyperpolarized by chemical stimuli: AWC is inhibited by specific odors or bacterially conditioned medium, and ASK is inhibited by pheromones or amino acids (Chalasani et al., 2007; Macosko et al., 2009; Wakabayashi et al., 2009). Both AWC and ASK are strongly activated if these chemical cues are removed, consistent with the evidence that they stimulate turning behavior following the removal of food.

AWC and ASK synapse onto a layer of interneurons that are also connected with each other, AIA, AIB, AIY, and AIZ. Cell ablations indicate that these interneurons coordinately control multiple classes of turning behaviors, including reversals and a sharp turn called an omega turn (Gray et al., 2005). AWC neurons are glutamatergic (Chalasani et al., 2007). AWC releases glutamate to activate AIB neurons through AMPA-type glutamate receptors, and it inhibits AIY neurons through glutamate-gated chloride channels. As a result, AIB is active upon odor removal, like AWC, whereas AIY becomes active upon odor addition. This circuitry can produce a coordinated switch in turning behavior. When food odors are present, AWC is inhibited, AIB is inactive, and AIY is tonically active; tonically active AIY suppresses turning. When food odors are removed, AWC becomes active, AIB becomes active and stimulates turning, and AIY is inhibited.

A second layer of interneurons and downstream motor neurons regulate specific classes of turns and features of turns (Gray et al., 2005). The AVA interneurons are backward command neurons; they are required for all kinds of reversals, including those triggered by the turning circuit, but they are not required for sharp omega turns. AVA synapses onto motor neurons in the body that drive backward locomotion. Conversely, the SMD and RIV motor neurons stimulate omega turns but are not required for reversals. SMD affects the steepness of the omega turn, while RIV biases omega turns toward the ventral side of the animal. Other neurons may affect gentler turns, including the SMB motor neurons that affect the amplitude of sinusoidal movement. The motor neurons in the turning circuit are largely cholinergic. The transmitters for most interneurons are unknown.

The same circuit that generates innate odor responses allows context and experience to modify odor preferences. Neuromodulatory inputs to interneurons or sensory neurons can reorganize the turning circuit to transform behavior. For example, *C. elegans* is susceptible to infection by common pathogenic bacteria, and it uses behavioral strategies as part of its antibacterial defense. Infection induces specific olfactory avoidance of pathogenic bacteria, a behavior that may be analogous to conditioned taste aversion (Zhang et al., 2005). Aversive odor learning requires serotonin, which promotes learning in many animals. Infection elevates serotonin, and exogenous serotonin accelerates learning, suggesting that serotonin provides an instructive learning signal. Serotonin converges on the turning circuit by activating the inhibitory serotonin receptor *mod-1* on AIY and AIB interneurons.

Another form of plasticity in the turning circuit affects sensory neurons. One of the two AWC olfactory neurons, AWC^{ON}, can direct either attraction or repulsion depending on the experience of the animal. In naïve animals, odors sensed by AWC^{ON} are attractive, but extended starvation in the presence of these odors switches AWC^{ON} to repulsion (Tsunozaki et al., 2008). Three signaling molecules that regulate the switch between attraction and repulsion—a receptor-like guanylate cyclase (GCY-28, Fig. 51.1), a diacylglycerol kinase, and a protein kinase C homolog—all act in AWC^{ON}, apparently to modulate presynaptic release. These results suggest that alternative modes of neurotransmission can couple one sensory neuron to opposite behavioral outputs.

The AIA, AIB, AIY, and AIZ interneurons in this circuit receive extensive synaptic input from other sensory neurons. Two of the best-characterized sensory inputs come from AFD and ASE neurons, which also regulate taxis behaviors (Fig. 51.2). AFD thermosensory neurons use a biased random walk to drive thermotaxis to a preferred temperature, a behavior that has a strong component of heat avoidance (Mori and Ohshima, 1995; Ryu and Samuel, 2002). AFD neurons are depolarized at warm temperatures to promote turning (Ramot et al., 2008). ASE neurons sense attractive salts and other water-soluble attractants. The left and right ASE neurons have different sensory properties: the ASER neuron is hyperpolarized by attractive salts, resembling AWC, and has the largest role in chemotaxis; the contralateral ASEL neuron is depolarized by attractive salts (Suzuki et al., 2008). Salt chemotaxis results from a combination of a biased random walk strategy and a directed turning strategy in salt gradients (Iino and Yoshida, 2009). The AIA/AIB/AIY/AIZ interneurons regulate both the biased random walk and directed turning to salts; AIZ has the largest role among the interneurons (Iino and Yoshida, 2009).

The patterns of synaptic connections made by ASE and AWC are almost identical, suggesting a common circuit mechanism for these two kinds of chemotaxis (Fig. 51.2). AFD synaptic partners overlap with those of ASE and AWC, but the exact patterns differ; for example, AFD forms gap junctions rather than chemical synapses with AIB. The circuit-level effects of these differences in connectivity are not known. ASE, AWC, and AFD are also linked by synapses to each other, but as yet there is no clear functional correlate of these anatomical connections.

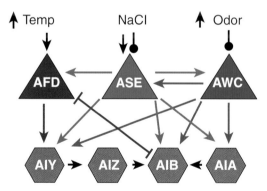

FIGURE 51–2. Convergent sensory inputs target a common set of interneurons. Sensory neurons are denoted as triangles and interneurons are denoted as hexagons. Sensory neurons and their connections are color-coded for ease of viewing. Arrows indicate chemical synapses defined by the wiring diagram, including excitatory and inhibitory synapses; H-bar indicates gap junctions (White et al., 1986). AFD neurons are depolarized by (repulsive) high temperatures; AWC neurons are hyperpolarized by attractive odors. ASE neurons are bilaterally asymmetric; attractive NaCl concentrations inhibit the ASER neuron and activate the ASEL neuron (Suzuki et al., 2008). The synaptic connections made by ASER and ASEL are similar to each other.

In addition to the synaptic inputs from sensory neurons described earlier, extrasynaptic inputs from mechanosensory dopaminergic neurons affect local search behavior (Hills et al., 2004). It is likely that the turning circuit is a common substrate for a variety of exploratory and taxis behaviors that are regulated by sensory inputs.

REFERENCES

Chalasani SH, Chronis N, Tsunozaki M, Gray JM, Ramot D, Goodman MB, Bargmann CI (2007) Dissecting a circuit for olfactory behaviour in Caenorhabditis elegans. *Nature* 450:63–70.
Gray JM, Hill JJ, Bargmann CI (2005) A circuit for navigation in Caenorhabditis elegans. *Proc Natl Acad Sci USA* 102:3184–3191.
Hills T, Brockie PJ, Maricq AV (2004) Dopamine and glutamate control area-restricted search behavior in Caenorhabditis elegans. *J Neurosci* 24:1217–1225.
Iino Y, Yoshida K (2009) Parallel use of two behavioral mechanisms for chemotaxis in Caenorhabditis *elegans*. *J Neurosci* 29:5370–5380.
Macosko EZ, Pokala N, Feinberg EH, Chalasani SH, Butcher RA, Clardy J, Bargmann CI (2009) A hub-and-spoke circuit drives pheromone attraction and social behaviour in C. elegans. *Nature* 458:1171–1175.
Mori I, Ohshima Y (1995) Neural regulation of thermotaxis in Caenorhabditis elegans. *Nature* 376:344–348.
Pierce-Shimomura JT, Morse TM, Lockery SR (1999) The fundamental role of pirouettes in Caenorhabditis elegans chemotaxis. *J Neurosci* 19:9557–9569.
Ramot D, MacInnis BL, Goodman MB (2008) Bidirectional temperature-sensing by a single thermosensory neuron in C. elegans. *Nat Neurosci* 11:908–915.
Ryu WS, Samuel AD (2002) Thermotaxis in Caenorhabditis elegans analyzed by measuring responses to defined thermal stimuli. *J Neurosci* 22:5727–5733.
Suzuki H, Thiele TR, Faumont S, Ezcurra M, Lockery SR, Schafer WR (2008) Functional asymmetry in Caenorhabditis *elegans* taste neurons and its computational role in chemotaxis. *Nature* 454:114–117.
Tsalik EL, Hobert O (2003) Functional mapping of neurons that control locomotory behavior in Caenorhabditis elegans. *J Neurobiol* 56:178–197.
Tsunozaki M, Chalasani SH, Bargmann CI (2008) A behavioral switch: cGMP and PKC signaling in olfactory neurons reverses odor preference in C. elegans. *Neuron* 59: 959–971.
Wakabayashi T, Kitagawa I, Shingai R (2004) Neurons regulating the duration of forward locomotion in Caenorhabditis elegans. *Neurosci Res* 50:103–111.
Wakabayashi T, Kimura Y, Ohba Y, Adachi R, Satoh YI, Shingai R (2009) In vivo calcium imaging of OFF-responding ASK chemosensory neurons in C. elegans. *Biochim Biophys Acta* 1790:765–769.
White JG, Southgate E, Thomson JN, Brenner S (1986) The structure of the nervous system of the nematode *Caenorhabditis elegans*. *Phil Transact R Soc Lond B* 314:1–340.
Zhang Y, Lu H, Bargmann C (2005) Pathogenic bacteria induce aversive olfactory learning in Caenorhabditis elegans. *Nature* 438:179–184.
Zhao B, Khare P, Feldman L, Dent JA (2003) Reversal frequency in Caenorhabditis elegans represents an integrated response to the state of the animal and its environment. *J Neurosci* 23:5319–5328.

52

The Neuronal Circuit for Simple Forms of Learning in *Aplysia*

Robert D. Hawkins, Craig H. Bailey, and Eric R. Kandel

The gill- and siphon-withdrawal reflex of *Aplysia* is a simple defensive behavior that is partly monosynaptic: sensory neurons synapse directly on motor neurons of the reflex. Despite this remarkable simplicity, the reflex can be modified by a variety of different forms of learning, including habituation, dishabituation, sensitization, and classical and operant conditioning (Hawkins, 1989; Hawkins et al., 2006a). Moreover, these forms of learning give rise to both short- and long-term memory depending on repetition. Because habituation, dishabituation, sensitization, and classical conditioning are reflected importantly in the monosynaptic connections between the sensory and motor neurons, it has proven possible to study the modifications of this connection in the intact animal, in various simplified behavioral preparations, and in isolated cell culture in which a single sensory neuron is cultured with a single motor neuron.

The different forms of learning that have been found here exhibit many of the behavioral properties of learning in mammals, suggesting that they may involve similar neuronal mechanisms (for review, see Hawkins, 1989). These behavioral properties include higher order features of classical conditioning that are thought to have a cognitive flavor, and thus may form a bridge between basic conditioning and more advanced forms of learning. To explore that idea further, we proposed in 1984 cellular mechanisms for several of the higher order features of conditioning based on *Aplysia* circuitry and molecular mechanisms known at the time (Hawkins and Kandel, 1984), and we incorporated those ideas in a quantitative model that was able to simulate a broad range of behavioral properties of habituation, sensitization, basic classical conditioning, and higher order features of conditioning (Hawkins, 1989). We also identified several properties that the model was not able to simulate.

In this chapter, we first summarize our current knowledge about the behavior, circuitry, and cellular and molecular mechanisms of learning of the reflex (for more extensive reviews and references, see Kandel, 2001; Bailey et al., 2004; Hawkins et al., 2006b). We then review the original model and suggest how recent advances might be able to explain some of the behavioral properties that the original model could not.

CELLULAR AND MOLECULAR MECHANISMS OF LEARNING OF THE GILL- AND SIPHON-WITHDRAWAL REFLEX

Short-Term Forms of Learning

A tactile stimulus to the siphon elicits defensive withdrawal of the gill and siphon in *Aplysia*. The neural circuit mediating the reflex consists partly of monosynaptic connections from siphon sensory neurons to gill and siphon motor neurons, as well as polysynaptic connections involving several identified interneurons (Fig. 52.1). Short-term habituation is largely due to homosynaptic depression of the sensory-motor neuron excitatory postsynaptic

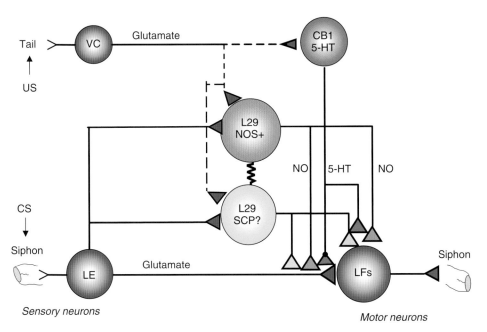

FIGURE 52–1. Partial circuit diagram for simple forms of learning of the siphon-withdrawal reflex. Dotted lines represent polysynaptic connections. CS, conditioned stimulus; US, unconditioned stimulus. Red, excitatory; green, purple, and yellow, facilitatory. Several additional excitatory and inhibitory interneurons have been omitted for simplicity (see Frost and Kandel, 1995).

potentials (EPSPs). Dishabituation and sensitization involve heterosynaptic facilitation of those EPSPs by modulatory transmitters including serotonin (5-HT), which contributes importantly to the facilitation and behavioral enhancement of the reflex.

Serotonin binds to presynaptic receptors linked to the production of cAMP. During short-term facilitation at rested synapses, cAMP activates cAMP-dependent protein kinase (PKA), which phosphorylates and closes K^+ channels, leading to broadening of subsequent action potentials and increased Ca^{2+} influx and transmitter release (Fig. 52.2). By contrast, short-term facilitation at depressed synapses involves activation of serotonin receptors coupled to PKC, which can act through a mechanism that is independent of spike broadening and is thought to involve vesicle mobilization. The spike broadening-independent component of facilitation may also involve Ca^{2+}/calmodulin-dependent protein kinase (CamKII).

Classical conditioning involves activity-dependent enhancement of the heterosynaptic facilitation by the occurrence of action potentials in the sensory neuron just before binding of the modulatory transmitter. The action potentials produce an increase in Ca^{2+}, which primes the adenylyl cyclase leading to greater production of cAMP, increased broadening, and enhanced transmitter release.

Long-Term Learning and the Growth of New Sensory Neuron Synapses

The reflex also undergoes long-term behavioral modifications, which involve a completely different type of mechanism: long-term habituation involves the retraction of preexisting synapses, and long-term sensitization involves growth of new synapses between the sensory and motor neurons. The molecular mechanisms contributing to long-term memory storage have been most extensively studied for sensitization. As with defensive behaviors in other species, the memory for sensitization of the gill-withdrawal reflex is graded, and retention is proportional to the number of training trials. A single tail shock produces short-term sensitization that lasts for minutes. Repeated tail shocks given at spaced intervals produce long-term sensitization that lasts for days or even weeks.

The simplicity of the neuronal circuit for the reflex has allowed the reduction of the analysis of the short- and long-term memory of sensitization to the cellular and molecular level. For example, the monosynaptic connections between sensory and motor neurons can be reconstituted in isolated cell culture, where 5-HT, a modulatory neurotransmitter normally released by sensitizing stimuli, can substitute for the shock to the neck or tail used during behavioral training in the intact animal. A single application of 5-HT produces short-term changes in synaptic effectiveness, whereas five spaced applications given over a period of 1.5 hr produce long-term changes lasting several days.

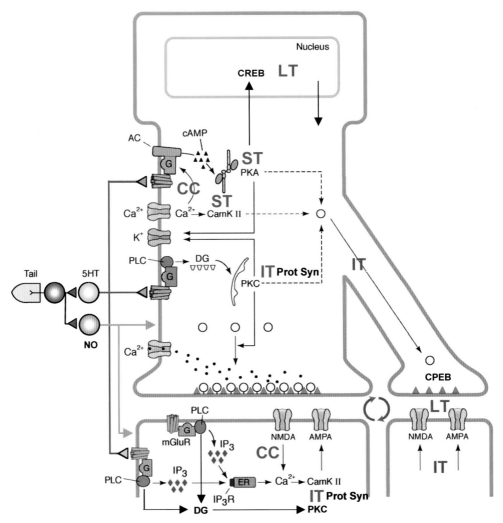

FIGURE 52–2. Cellular and molecular mechanisms of plasticity at sensory-motor neuron synapses that contribute to simple forms of learning. Short-term (ST) sensitization involves presynaptic PKA and CamKII. Intermediate-term (IT) sensitization involves presynaptic PKA and CamKII or PKC and protein synthesis, postsynaptic CamKII or PKC and protein synthesis, possible membrane insertion of AMPA receptors, and recruitment of pre- and postsynaptic proteins to new synaptic sites. Long-term (LT) sensitization involves gene regulation and growth of new synapses. Classical conditioning (CC) involves activity-dependent enhancement of presynaptic facilitation, Hebbian plasticity, and pre- and postsynaptic effects of nitric oxide (NO).

Studies of this monosynaptic connection between sensory and motor neurons both in the intact animal and in culture indicate that, phenotypically, the long-term changes are surprisingly similar to the short-term changes, consistent with the idea that long-term memory is a direct extension of short-term memory. A component of the increase in synaptic strength observed during both the short- and long-term changes is due, in each case, to enhanced release of transmitter by the sensory neuron, accompanied by a broadening of the action potential and an increase in excitability attributable to the depression of specific sets of potassium channels.

Despite this phenotypic similarity, the long-term cellular changes for sensitization differ fundamentally from the short-term changes in at least two important ways. First, the long-term change requires new transcription and new protein synthesis. Second, the long-term process involves a structural change. Long-term sensitization is associated with the growth of new synaptic connections by the sensory neurons onto their follower cells. The persistence of this structural change parallels the behavioral duration of the memory. This synaptic growth also can be reconstituted in sensory-motor neuron cocultures by repeated presentations of 5-HT. Whereas these learning-related anatomical changes are highly regulated and involve both pre- and postsynaptic changes, in this review we will focus primarily on the presynaptic component of the changes.

Initiation of Long-Term Facilitation: PKA and MAP Kinase Activate CREB-Related Transcription Factors

During the induction of long-term facilitation by repeated applications of 5-HT, PKA translocates to the nucleus and activates gene expression by phosphorylating the transcription factors that bind to the cAMP-responsive element (CRE), the CRE binding protein (CREB1). Microinjection of CRE containing oligonucleotides into sensory neurons inhibits the function of CREB1 and blocks long-term facilitation but has no effect on the short-term process. Injection of recombinant CREB1a phosphorylated in vivo by PKA leads to an increase in EPSP amplitude at 24 hr in the absence of any 5-HT stimulation. Not only is the CREB1 activator necessary for long-term facilitation, it is sufficient to induce long-term facilitation, albeit in reduced form and in a form that is not maintained beyond 24 hr.

The transcriptional switch in long-term facilitation requires not only activation of the CREB1 regulatory unit but also removal of an inhibitory constraint due to the CREB2 repressor. Injection of anti-ApCREB2 antibodies into *Aplysia* sensory neurons causes a single pulse of 5-HT, which normally induces only short-term facilitation lasting minutes, to evoke facilitation that lasts more than 1 day. This disinhibition requires both transcription and translation and is accompanied by the growth of new synaptic connections. Ap-CREB2 has both protein kinase C and MAP kinase phosphorylation sites, and MAP kinase is activated by 5-HT in *Aplysia* neurons. Like PKA, MAP kinase translocates to the nucleus with prolonged 5-HT treatment so as to activate the activators (CREB1) and relieve the repressors (CREB2).

The balance between CREB activator and repressor isoforms is also critically important in long-term behavioral memory, as shown in *Drosophila* and mice. Expression of an inhibitory form of CREB (dCREB-2b) blocks long-term olfactory memory but does not alter short-term memory in *Drosophila*. Overexpression of an activator form of CREB (dCREB-2a) increases the efficacy of massed training in long-term memory formation. Similarly, partial knockout of CREB-1 impairs long-term but not short-term hippocampal-dependent memory in mice. Conversely, reduced expression of ATF4, which is homologous to ApCREB-2, enhances long-term hippocampal-dependent memory formation (Chen et al., 2003).

The CREB-mediated response to extracellular stimuli can be modulated by a number of kinases (PKA, CaMKII, CaMKIV, RAK2, MAPK, and PKC) and phosphatases (PP1 and calcineurin). The CREB regulatory unit may therefore serve to integrate signals from various signal transduction pathways. This ability to integrate signaling as well as mediate activation or repression may explain why CREB is so central to memory storage in different contexts, implicit and explicit, invertebrate and vertebrate.

This question has been studied by Guan et al. (2002), who examined the role of CREB-mediated responses in long-term synaptic integration by studying the long-term interactions of two opposing modulatory transmitters important for behavioral sensitization in *Aplysia*. Toward that end they utilized a single bifurcated sensory neuron that contacts two spatially separated postsynaptic neurons (Martin et al., 1997). They found that when a neuron receives 5-HT and at the same time receives input from the inhibitory transmitter FMRFamide at another set of synapses, the synapse-specific long-term depression produced by FMRFamide dominates. These opposing inputs are integrated in the neuron's nucleus and are evident in the repression of the CCAAT-box-enhanced-binding-protein (C/EPB), a transcription regulator downstream from CREB that is critical for long-term facilitation. Whereas 5-HT induces C/EPB by activating CREB1 and recruiting the CREB-binding protein (CBP), a histone acetylase, to acetylate histones, FMRFamide displaces CREB1 with CREB2, which recruits a histone deacetylase to deacetylate histones. When 5-HT and FMRFamide are given together, FMRFamide overrides 5-HT by recruiting CREB2 and the deacetylase to displace CREB1 and CBP, thereby inducing histone deacetylation and repression of C/EBP. Thus, both the facilitatory and inhibitory modulatory transmitters that are important for long-term memory in *Aplysia* activate signal transduction pathways that alter nucleosome structure bidirectionally through acetylation and deacetylation of chromatin.

To follow further the sequence of steps whereby CREB leads to the stable, self-perpetuating long-term process, Alberini and colleagues (1994) focused on the CCAAT-box enhanced-binding protein (C/EBP) transcription factors, which they found were induced by exposure to 5-HT. Inhibition of ApC/EBP activity blocked long-term facilitation but had no effect on short-term facilitation. Thus, the induction of ApC/EBP seems to serve as an intermediate component of a molecular switch activated during the consolidation period.

A Molecular Model for the Stabilization of Synaptic Growth and Maintenance of Long-Term Facilitation: Self-Perpetuating Activation of Translational Regulators

As outlined earlier, the stability of long-term facilitation seems to result from the activation of a nuclear program and the persistence of structural changes at sensory neuron synapses, the decay of which parallels the decay of the behavioral memory. This raises two fundamental questions in the cell biology of memory storage. First, the activation of a nuclear program suggests that long-term memory could potentially be cell-wide. On the other hand, there might be a cellular mechanism to utilize a cell-wide process in a synapse-specific way. Second, if a change in synaptic strength and structure is indeed the underlying mechanism of long-term memory storage, then the experience-dependent molecular changes at the synapse must also be maintained for the duration of the memory. Since biological molecules have a relatively short half-life (hours to days) compared to the duration of memory (years), how is the altered molecular composition of a synapse maintained for such a long time?

To begin to address this question, Martin et al. (1997) focused on the role of local protein synthesis in the maintenance of synapse-specific, long-term plasticity. Toward that end they developed a culture system in *Aplysia* in which a single bifurcated sensory neuron of the gill-withdrawal reflex was plated in contact with two spatially separated gill motor neurons. In this culture system, repeated application of 5-HT to one synapse produces a CREB-mediated, synapse-specific long-term facilitation that is accompanied by the growth of new synaptic connections and persists for at least 72 hours. This long-term facilitation, as well as the long-lasting synaptic growth, can be captured by a single pulse of 5-HT applied at the opposite sensory-to-motor neuron synapse. In contrast to the synapse-specific forms, cell-wide long-term facilitation generated by repeated pulses of 5-HT at the cell body is not associated with growth and does not persist beyond 48 hours. However, this cell-wide facilitation also can be captured and growth can be induced in a synapse-specific manner by a single pulse of 5-HT applied to one of the peripheral synapses.

Thus, CREB-mediated transcription appears to be necessary for the establishment of all of these forms of long-term facilitation and for the initial maintenance of the synaptic plasticity at 24 hours. However, CREB-mediated transcription is not sufficient to maintain the changes beyond this time. To obtain persistent facilitation and specifically to obtain the growth of new synaptic connections, one needs, in addition to CREB-mediated transcription, a marking signal produced by a single pulse of 5-HT applied to the synapse. This single pulse of 5-HT has at least two marking functions. First, it produces a PKA-mediated covalent modification that marks the captured synapse for growth. Second, it stimulates rapamycin-sensitive local protein synthesis, which is required for the long-term maintenance of the plasticity and stabilization of the growth beyond 24 hours.

The finding of two distinct components for the marking signal first suggested that there is a mechanistic distinction between the initiation of long-term facilitation and synaptic growth (which require central transcription and translation but do not require local protein synthesis) and the stable maintenance of the long-term functional and structural changes that are dependent on local protein synthesis. How might this local protein synthesis at the synapse, which is necessary for stabilizing synaptic growth and long-term facilitation, be regulated? Since mRNAs are made in the cell body, the need for the local translation of some mRNAs suggests that these mRNAs may be dormant before they reach the activated synapse. If that were true, one way of activating protein synthesis at the synapse would be to recruit a regulator of translation that is capable of activating translationally dormant mRNAs.

Si et al. (2003a) began to search for such a molecule by focusing on the *Aplysia* homolog of cytoplasmic polyadenylation element-binding protein (CPEB), a protein capable of activating dormant mRNAs through the elongation of their polyA tail. CPEB was first identified in oocytes and subsequently in hippocampal neurons. In *Aplysia*, a novel neuron-specific isoform of CPEB is present in the processes of sensory neurons and stimulation with 5-HT increases the amount of CPEB protein at the synapse. The induction of CPEB is independent of transcription but requires new protein synthesis and is sensitive to rapamycin and to inhibitors of P13 kinase. Moreover, the induction of CPEB coincides with the polyadenylation of neuronal actin, and blocking CPEB locally at the activated synapse blocks the long-term maintenance of synaptic facilitation but not its early expression at 24 hours. These results suggest that CPEB has all the properties required of the local protein synthesis–dependent component of marking, and they support the idea that there are separate mechanisms for initiation of the long-term process and its stabilization. Moreover, these data suggest that the maintenance but not the initiation of long-term synaptic plasticity requires a new set of molecules in the synapse, and that some of these new molecules are made by CPEB-dependent translational activation. A similar neuronal isoform of CPEB, CPEB-3, has been found in mouse hippocampal neurons and is induced by the neurotransmitter dopamine. Interestingly, activation of a dopaminergic pathway is critical for the synaptic marking during mouse hippocampal long-term potentiation. This raises the possibility that dopamine-dependent regulation of mouse CPEB-3 might be similar to the serotonin-mediated regulation of *Aplysia* neuronal CPEB, and CPEB-3 can potentially act as a synaptic mark in mammalian synapses.

How might CPEB stabilize the late phase of long-term facilitation? The 5-HT-induced structural changes at the synapses between sensory and motor neurons include the remodeling of preexisting facilitated synapses, as well as the growth and establishment of new synaptic connections. The reorganization and growth of new synapses have two broad requirements: *(1)* structural (changes in shape, size, and number) and *(2)* regulatory (where and when to grow). The genes involved in both of these aspects of synaptic growth might be potential targets of apCPEB. The structural aspects of the synapses are dynamically controlled by reorganization of the cytoskeleton, which can be achieved either by redistribution of preexisting cytoskeletal components or by their local synthesis. The observation that N-actin and Tα1 tubulin (K. C. Martin et al., unpublished data) are present in the peripheral population of mRNAs at the synapse and can be polyadenylated in response to 5-HT suggests that at least some of the structural components for synaptic growth can be controlled through apCPEB-mediated local synthesis. In addition, recently CPEB has been found to be involved in the regulation of local synthesis of EphA2, a member of the family of receptor tyrosine kinases, which have been implicated in axonal path finding and the formation of excitatory synapses in the mammalian brain. Thus, CPEB might contribute to the stabilization of learning-related synaptic growth by controlling the synthesis of both the structural molecules such as tubulin and N-actin and the regulatory molecules such as members of the Ephrin family.

These findings in turn raise further questions: Is there a continuous need for the local synthesis of a set of molecules to maintain the learning-related synaptic changes over long periods of time? If so, how can these enduring changes be achieved by a translational regulator such as CPEB in the face of a continuous turnover of the protein? One possible answer to how a population of unstable molecules can produce a stable change in synaptic form and function comes from the subsequent finding by Si et al. (2003b) that the neuronal isoform of CPEB shares properties with prion-like proteins. Prions are proteins that can assume at least two stable conformational states. Usually one of these conformational states is active while the other is inactive. Furthermore, one of the conformational states, the prion state, is self-perpetuating, promoting the conformational conversion of other proteins of the same type. Work on yeast suggests the *Aplysia* neuronal CPEB exists in two stable, physical states that are functionally distinct. As with other prions, one of these states has the ability to self-perpetuate in a dominant epigenetic fashion. However, unlike the known prion proteins where the dominant state is the inactive form of the protein, surprisingly, in the case of *Aplysia* CPEB, the dominant form is the active form of the protein capable of activating translationally dormant mRNAs.

Postsynaptic Mechanisms Are Recruited during a Novel Intermediate Phase

The mechanisms just described all occur in the presynaptic sensory neurons. Synaptic growth requires pre- and postsynaptic changes coordinated by transsynaptic signaling (Sanes and Lichtman, 1999; McAllister, 2007), and several postsynaptic mechanisms of long-term facilitation have also been described including increased Ca^{2+}, protein synthesis, and the formation of new clusters of glutamate receptors (Trudeau and Castellucci, 1995; Zhu et al., 1997; Sherff and Carew, 2004; Cai et al., 2008; Li et al., 2009; Wang et al., 2009).

To try to analyze how these long-term processes are initiated, we and others identified an intermediate-term (hours) stage of facilitation that involves elements of the mechanisms of both short- and long-term facilitation, and might form a bridge between them. Thus, intermediate-term facilitation involves covalent modifications and protein but not RNA synthesis, as well as redistribution of pre- and postsynaptic proteins but not synaptic growth. Like long-term facilitation, intermediate-term facilitation also involves both pre- and postsynaptic molecular mechanisms and transsynaptic signaling. For example, whereas short-term facilitation by a 1 min exposure to 5-HT involves presynaptic PKA and CamKII, intermediate-term facilitation by a 10 min exposure to 5-HT involves both presynaptic (PKC and protein synthesis) and postsynaptic (Ca^{2+}, CamKII or PKC, and protein synthesis) mechanisms.

The postsynaptic mechanisms of intermediate-term facilitation might be induced in two different ways, which are not mutually exclusive. First, the pre- and postsynaptic mechanisms might be induced in parallel by activation of both pre- and postsynaptic receptors for 5-HT or other modulatory transmitters. Experiments on single motor neurons in cell culture support the involvement of postsynaptic 5-HT receptors linked to the production of IP3 (Villareal et al., 2007, 2009; Fulton et al., 2008). Alternatively, however, the postsynaptic mechanisms might be induced by enhanced spontaneous transmitter release from the presynaptic neuron. That idea could explain why induction of the postsynaptic mechanisms requires a longer exposure to 5-HT than induction of the presynaptic mechanisms of short-term facilitation, and also why some of the postsynaptic mechanisms of intermediate-term facilitation are similar to those involved in homosynaptic potentiation by weak tetanic stimulation of the presynaptic neuron. In support of that idea, experiments on sensory-motor neuron cocultures have shown that activation of presynaptic 5-HT receptors leads to an increase in spontaneous release of glutamate, which then activates postsynaptic metabotropic glutamate receptors, leading to increased production of IP3 and increased postsynaptic Ca^{2+} (Jin et al., 2007; Jin et al., 2008).

Spontaneous transmitter release acts through this signaling pathway to contribute to the induction of long-term as well as intermediate-term facilitation. Conversely, some of the presynaptic mechanisms of long-term facilitation are regulated by signaling from the postsynaptic neuron (Cai et al., 2008; Wang et al., 2009). Collectively, these results suggest that intermediate- and long-term plasticity involve both anterograde and retrograde signaling between the pre- and postsynaptic neurons, as occurs during synaptic growth.

Furthermore, this signaling is first engaged during intermediate-term plasticity, when it may recruit some of the early steps in a program that can lead to the growth of new synapses during long-term plasticity.

Plasticity of Identified Neurons and Synapses during Learning

In addition to performing behavioral experiments on learning in the intact animal and cellular and molecular experiments on analogs of learning at sensory-motor neuron synapses in vitro, one can bridge the behavioral and cellular levels by using a semi-intact preparation of the siphon-withdrawal reflex. With that preparation, it is possible to record the activity of identified neurons and the synaptic connections between them and to examine plasticity of those connections during behavioral learning.

The reflex in the simplified siphon-withdrawal preparation undergoes habituation, dishabituation, sensitization, and classical conditioning that are very similar parametrically to learning in the intact animal. The initial cellular experiments using that preparation found that monosynaptic EPSPs from LE siphon sensory neurons to LFS siphon motor neurons mediate approximately one-third of the reflex response, and they confirmed that each of these simple forms of learning involves plasticity of those EPSPs. Subsequent experiments have examined molecular mechanisms of synaptic plasticity during the learning.

In general, the results of those experiments have been similar to those in vitro (see earlier discussion), and they have also revealed some additional mechanisms. For example, whereas facilitation during short-term sensitization by a single tail shock involves presynaptic PKA and CamKII, facilitation during intermediate-term sensitization by four tail shocks involves both presynaptic (PKA, CamKII, and protein synthesis) and postsynaptic (Ca^{2+}, CamKII, and protein synthesis) mechanisms. Also similar to in vitro experiments, facilitation during behavioral dishabituation by four tail shocks differs from sensitization in that it involves presynaptic PKC rather than PKA. In addition, unlike sensitization with the same shock, dishabituation by four shocks does not involve protein synthesis or postsynaptic Ca^{2+} (Antonov et al., 2008). These results demonstrate that the site as well as the mechanisms of plasticity during learning depends on the stage of plasticity and also on the prior history of plasticity (metaplasticity).

A Computational Model for Higher Order Features of Learning

The Original Model

In addition to these relatively simple forms of learning, conditioning in *Aplysia* and other animals exhibits higher order features that may form a bridge to more advanced forms of learning. In 1984, we proposed cellular mechanisms for several of these higher order features of conditioning based on *Aplysia* circuitry and molecular mechanisms (Hawkins and Kandel, 1984) and incorporated those ideas in a computational model that was able to simulate a broad range of behavioral properties of habituation, sensitization, basic classical conditioning, and higher order features of conditioning (Hawkins, 1989). The model was based on the known molecular mechanisms of short-term learning and the properties of the L29 interneurons, which refers to a group of about five electrically coupled neurons in the abdominal ganglion. Intracellular stimulation of a single L29 neuron produces facilitation of the monosynaptic EPSP from a sensory neuron to a motor neuron and broadening of action potentials in the sensory neuron. In addition to being facilitatory interneurons, the L29 neurons are also major excitatory interneurons (SN-L29-MN) in the circuit for the siphon-withdrawal reflex. That dual role of the L29 neurons has a number of interesting implications, and it is an important feature of the computational model (Hawkins, 1989). A key feature of the model is that the L29 neurons produce facilitation and activity-dependent facilitation at all of the synapses of the sensory neurons, including those onto the L29 neurons themselves. In addition, the L29 neurons undergo spike accomodation during prolonged stimulation, due in part to inhibitory feedback from L30 interneurons.

A computational model incorporating these circuit, cellular, and molecular mechanisms was able to simulate most of the known behavioral properties of habituation, dishabituation, sensitization, and classical conditioning (Hawkins, 1989). These included both the stimulus and temporal specificity of conditioning and a higher order feature, contingency, all of which had already been demonstrated in *Aplysia* at the time. In addition, the computational model was able to simulate several other higher order features that had not yet been demonstrated in *Aplysia* but had been demonstrated in other invertebrates, including second order conditioning, blocking, overshadowing, and conditioned stimulus (CS) and unconditioned stimulus (US) preexposure effects. However, it was not able to simulate some other common features of conditioning, including response specificity (i.e., the conditioned response usually resembles the unconditioned response), posttraining US exposure effects, sensory preconditioning (which is generally thought to involve formation of a stimulus-stimulus association), and conditioned inhibition.

Possible Additions to the Model Based on Advances in Knowledge

Behavior

Since the original model was proposed, some additional features of conditioning in *Aplysia* have been demonstrated. First, conditioning of siphon withdrawal exhibits response specificity. The tail shock US used in most conditioning experiments produces backwards bending of the siphon, whereas a mantle shock US produces forward bending. After an initially neutral siphon touch CS is paired with a tail shock US, it produces backward bending as well. By contrast, after the same siphon CS is paired with a mantle shock US, it produces forward bending. Motor neurons that mediate the backward and forward bending responses have been identified, and possible cellular mechanisms of the response specificity of conditioning have been proposed, but as yet there has been no cellular analysis of that effect.

Gill withdrawal also exhibits second-order conditioning with two siphon CSs and a mantle shock US. In stage I, CS1 is paired with the US, and in stage II CS2 is paired with CS1. This training produces an increase in responding to CS2, compared to unpaired controls. Unlike first-order conditioning, which requires forward pairing of the CS and US, second-order conditioning occurs with either forward or simultaneous pairing of CS1 and CS2 in stage II. Furthermore, following simultaneous second-order conditioning, extinction of CS1 produces a decrease in responding to CS2. That result is formally similar to a posttraining US exposure effect, and it suggests that simultaneous second-order conditioning involves formation of a stimulus–stimulus (CS1–CS2) association.

Circuit

There has also been further characterization of facilitatory interneurons involved in enhancement of the reflex. A pair of identified serotonergic neurons in the cerebral ganglia (the CB1 neurons) are excited by noxious stimulation and produce facilitation of siphon sensory-motor neuron EPSPs. In addition, the L29 facilitatory neurons are excited by both the CS and US used in behavioral conditioning, consistent with the model of conditioning. However, different L29s respond to somewhat different USs. That result suggests that firing of an L29 can be thought of as the internal representation of a US, and facilitation of SN-L29 synapses could be the basis for learning an association between a CS and the internal representation of the US. That idea is the basis for second-order conditioning in the model, and it also suggests a possible mechanism for posttraining US exposure effects, in which exposure to the US after training alters the response to the CS. On the psychological level those effects are thought to be due to altering the internal representation of the US. On the cellular level they could be due to plasticity of either L29 excitability or L29–MN synapses following the US exposure, both of which occur under some circumstances.

The transmitter of the L29 neurons has not previously been known. Recently Antonov et al. (2007) found that nitric oxide (NO) contributes to facilitation and behavioral enhancement of the reflex, and that two to three of the five L29 neurons express an *Aplysia* neuronal-type isoform of nitric oxide synthase. An endogenous peptide (SCP) also contributes to facilitation and enhancement of the reflex, and mapping of SCP immunoreactive neurons suggests that the NOS-negative L29s might contain SCP.

Collectively, these results suggest that facilitation of sensory-motor neuron postsynaptic potentials and behavioral enhancement of the reflex involve three different types of facilitatory substances: a conventional modulatory transmitter (5-HT), a freely diffusible messenger molecule (NO), and a small peptide transmitter (SCP). Presumably each of these serves a different type of function. 5-HT is thought to mediate a general arousal/stress response to noxious stimulation. Strong tactile stimulation or shock leads to activation of a large population of serotonergic neurons (including CB1) and an increase in global 5-HT release over the CNS, resulting in greater expression of defensive behaviors in general and enhanced gill and siphon withdrawal in particular (Marinesco et al., 2004; Marinesco et al., 2006). By contrast, freely diffusible messenger molecules like NO have been hypothesized to affect neurons in the surrounding volume of tissue and thus to participate in local volume computation, which is thought to have different rules than traditional linear computation in neural circuits (Gally et al., 1990). The identification of individual serotonergic and nitrergic facilitatory neurons now makes it possible to test those ideas experimentally.

Postsynaptic Mechanisms of Intermediate- and Long-Term Facilitation

Our original model was based on presynaptic mechanisms of short-term (minutes) facilitation by 5-HT or tail shock. Long-term (days) facilitation involves a different type of mechanism, the growth of new synapses between the sensory and motor neurons. In principle, new synaptic connections might also form between neurons that did not previously have connections, thus fundamentally changing the neuronal circuit. Such rewiring via global overgrowth occurs during development, but it is not yet known whether it occurs during learning.

In addition, both intermediate- and long-term facilitation involve postsynaptic molecular mechanisms. It is also not yet known whether, like the presynaptic mechanisms of facilitation, the postsynaptic mechanisms are enhanced if activity in the motor neuron is paired with the modulatory transmitter. However, it seems plausible that spike activity might act synergistically with activation of postsynaptic 5-HT or metabotropic glutamate receptors to increase postsynaptic Ca^{2+}. If so, such activity-dependent enhancement of postsynaptic mechanisms of facilitation could contribute to the response specificity of behavioral sensitization and conditioning.

Hebbian Plasticity

Conditioning in the simplified siphon-withdrawal preparation is accompanied by pairing-specific facilitation of the EPSP from an LE sensory neuron to an LFS motor neuron, as well as increased evoked firing and membrane resistance of the LE neuron. These effects could all be due to activity-dependent enhancement of presynaptic facilitation caused by firing of sensory neurons in the CS pathway just before facilitatory neurons in the US pathway during conditioning. All three cellular effects are blocked by injecting a peptide inhibitor of PKA into the sensory neuron, consistent with the idea that facilitation of the EPSP during conditioning involves activity-dependent enhancement of the presynaptic PKA pathway.

However, pairing-specific facilitation of the EPSP might also be due to Hebbian plasticity caused by near coincident firing of the sensory and motor neurons during conditioning. The sensory-motor neuron EPSPs are glutamatergic, have AMPA- and NMDA-like components, and undergo Hebbian plasticity in vitro. Consistent with a role of that plasticity in behavioral conditioning, the conditioning and all three pairing-specific cellular effects are also blocked by the NMDA receptor antagonist APV or injection of the Ca^{2+} chelator BAPTA into the motor neuron. These results suggest that, unlike the facilitation during sensitization, the facilitation during conditioning involves both activity-dependent presynaptic facilitation and Hebbian plasticity. Furthermore, because postsynaptic BAPTA also blocks the changes in presynaptic membrane properties, the two mechanisms are evidently coordinated by retrograde signaling.

Why might conditioning involve these two different associative cellular mechanisms at the same synapses? One possibility is that a hybrid mechanism may have more desirable functional characteristics than either mechanism alone. Thus, although Hebbian plasticity can be highly synapse-specific and provides a natural mechanism for the response specificity of conditioning, it may not by itself support long-term plasticity (Bailey et al., 2000a). By contrast, activity-dependent presynaptic facilitation by widely projecting facilitatory neurons (such as the CB1 neurons) may not provide a high degree of synapse specificity, but it provides a natural mechanism for associating a stimulus with a strong reinforcing event that can promote long-term memory formation. Activity-dependent presynaptic facilitation also provides a natural mechanism for the stimulus specificity of conditioning, and it could contribute to response specificity as well if the facilitatory transmitter were targeted to specific synapses (Bailey et al., 2000b). In addition, whereas activity-dependent presynaptic facilitation requires the forward pairing typical of stimulus–response (S–R) learning, Hebbian plasticity generally requires the near simultaneous pairing typical of stimulus–stimulus (S–S) learning. Thus, the Hebbian component of facilitation might contribute more under conditions that favor S–S learning, such as simultaneous second-order conditioning.

Nitric Oxide

These results suggested that facilitation at sensory-motor neuron synapses during intermediate- or long-term sensitization and conditioning involves both pre- and postsynaptic mechanisms coordinated by extracellular signaling. Because the L29 neurons express NO synthase, Antonov et al. (2007) investigated the possible role of the extracellular signaling molecule NO in conditioning. Their results suggest that NO makes an important contribution to conditioning by acting directly in both the sensory and motor neurons to affect facilitation at the synapses between them. Furthermore, NO does not come from either the sensory or motor neurons but rather comes from another source, perhaps the L29 interneurons, consistent with the idea that it acts as a local "volume messenger" in the circuit for the reflex. Their results also suggest that NO plays different roles in several different pre- and postsynaptic mechanisms of facilitation, and that it is necessary but not sufficient for some of them.

One way that NO might be necessary but not sufficient is if it acts synergistically with another facilitatory transmitter such as 5-HT to enhance presynaptic and/or postsynaptic mechanisms of facilitation. Another way that NO might be necessary but not sufficient is if it acts synergistically with activity in the pre- or postsynaptic neurons. A key feature of the "volume computation" hypothesis is that a diffusible messenger like NO will produce different effects on different neurons in the surrounding volume depending on their spike activity at the time. Presynaptic activity is necessary for potentiation by NO in hippocampal neurons, and presynaptic activity enhances facilitation and changes in sensory neuron membrane properties in *Aplysia*. If NO paired with activity in either the sensory or motor neuron produces greater facilitation than NO alone, that effect could also contribute to the stimulus or response specificities of behavioral conditioning.

In summary, since the original model was proposed, there has been considerable progress in our knowledge of the behavior, circuitry, and cellular and molecular mechanisms of plasticity contributing to simple forms of learning of the gill- and siphon-withdrawal reflex. One emerging theme is that even in this simple reflex and circuit, multiple processes contribute and can interact in complex ways. For example, facilitation of transmitter release from the sensory neurons depends on the combined actions and interactions of 5-HT, SCP, NO, activity, and in some cases a retrograde signal from the motor neurons. This new knowledge has also suggested possible mechanisms for some behavioral properties of learning of the reflex that we could not previously explain. Again, these may involve multiple processes; for example, we have described several different mechanisms that could contribute to the response specificity of conditioning. It should now be possible to test these possibilities with additional behavioral, cellular, and modeling experiments in this system.

ACKNOWLEDGMENTS

Preparation of this manuscript was supported by the Howard Hughes Medical Institute and the Simon Foundation.

REFERENCES

Alberini CM, Ghirardi M, Metz R, Kandel ER (1994) C/EBP is an immediate-early gene required for the consolidation of long-term facilitation in Aplysia. *Cell* 76:1099–1114.

Antonov I, Ha T, Antonova I, Moroz LL, Hawkins RD (2007) Role of nitric oxide in classical conditioning of siphon withdrawal in *Aplysia. J Neurosci* 27:10993–11002.

Antonov I, Kandel ER, Hawkins RD (2008) Intermediate-term sensitization and disahbituation of the Aplysia siphon-withdrawal reflex involve different sites and mechanisms of facilitation. *Soc Neurosci Abstr* 880:12.

Bailey CH, Giustetto M, Huang YY, Hawkins RD, Kandel ER (2000a) Is heterosynaptic modulation essential for stabilizing Hebbian plasticity and memory? *Nat Rev Neurosci* 1:11–20.

Bailey CH, Giustetto M, Zhu H, Chen M, Kandel ER (2000b) A novel function for serotonin-mediated short-term facilitation in Aplysia: conversion of transient, cell-wide homosynaptic Hebbian plasticity into a persistent, protein synthesis-independent synapse-specific enhancement. *Proc Natl Acad Sci USA* 97:11581–11586.

Bailey CH, Kandel ER, Si K (2004) The persistence of long-term memory: a molecular approach to self-sustaining changes in learning-induced synaptic growth. *Neuron* 44:49–57.

Cai D, Chen S, Glanzman DL (2008) Postsynaptic regulation of long-term facilitation in Aplysia. *Curr Biol* 18:920–925.

Chen A, Muzzio IA, Malleret G, Bartsch D, Verbitsky M, Pavlidis P, Yonan AL, Vronskaya S, Grody MB, Cepeda I, Gilliam TC, Kandel ER (2003) Inducible enhancement of memory storage and synaptic plasticity in transgenic mice expressing an inhibitor of ATF4 (CREB-2) and C/EBP proteins. *Neuron* 39:655–669.

Frost WN, Kandel ER (1995) Structure of the network mediating siphon-elicited siphon withdrawal in *Aplysia. J Neurophysiol* 73:2413–2427.

Fulton D, Condro MC, Pearce K, Glanzman DL (2008) The potential role of postsynaptic phospholipase C activity in synaptic facilitation and behavioral sensitization in Aplysia. *J Neurophys* 100:108–116.

Gally JA, Montague PR, Reeke GN, Edelman GM (1990) The NO hypothesis: possible effects of a short-lived, rapidly diffusible signal in the development and function of the nervous system. *Proc Natl Acad Sci USA* 87:3547–3551.

Guan Z, Giustetto M, Lomvardas S, Kim JH, Miniaci MC, Schwartz JH, Thanos D, Kandel ER (2002) Integration of long-term-memory-related synaptic plasticity involves bidirectional regulation of gene expression and chromatin structure. *Cell* 111:483–493.

Hawkins RD (1989) A biologically based computational model for several simple forms of learning. *The Psychology of Learning and Motivation* 23:65–108.

Hawkins RD, Kandel ER (1984) Is there a cell biological alphabet for simple forms of learning? *Psych Rev* 91:375–391.

Hawkins RD, Clark GA, Kandel ER (2006a) Operant conditioning of gill withdrawal in Aplysia. *J Neurosci* 26:2443–2448.

Hawkins RD, Kandel ER, Bailey CH (2006b) Molecular mechanisms of memory storage in Aplysia. *Biol Bull* 210:174–191.

Jin I, Rayman JB, Puthanveettil S, Visvishrao H, Kandel ER, Hawkins RD (2007) Spontaneous transmitter release from the presynaptic sensory neuron recruits IP3 production in the postsynaptic motor neuron during the induction of intermediate-term facilitation in Aplysia. *Soc Neurosci Abstr* 429:13.

Jin I, Rayman JB, Puthanveettil S, Visvishrao H, Kandel ER, Hawkins RD (2008) Spontaneous transmitter release from the presynaptic neuron recruits postsynaptic mechanisms contributing to intermediate-term facilitation in Aplysia. *Soc. Neurosci Abstr* 880:23.

Kandel ER (2001) The molecular biology of memory storage: a dialogue between genes and synapses. *Science* 294:1030–1038.

Li HL, Huang BS, Vishwasrao H, Sutedja N, Chen W, Jin I, Hawkins RD, Bailey CH, Kandel ER (2009) Dscam mediates remodeling of glutamate receptors in *Aplysia* during de novo and learning-related synapse formation. *Neuron* 61:527–540.

Marinesco S, Wickremasinghe N, Kolkman KE, Carew TJ (2004) Serotonergic modulation in aplysia. II. Cellular and behavioral consequences of increased serotonergic tone. *J Neurophysiol* 92:2487–2496.

Marinesco S, Wickremasinghe N, Carew TJ (2006) Regulation of behavioral and synaptic plasticity by serotonin release within local modulatory fields in the CNS of Aplysia. *J Neurosci* 26:12682–12693.

Martin KC, Casadio A, Zhu H, E Y, Rose JC, Chen M, Bailey CH, Kandel ER (1997) Synapse-specific, long-term facilitation of Aplysia sensory to motor synapses: A function for local protein synthesis in memory storage. *Cell* 91:927–938.

McAllister AK (2007) Dynamic aspects of CNS synapse formation. *Annu Rev Neurosci* 30:425–450.

Sanes JR, Lichtman JW (1999) Development of the vertebrate neuromuscular junction. *Annu Rev Neurosci* 22:389–442.

Sherff CM, Carew TJ (2004) Parallel somatic and synaptic processing in the induction of intermediate-term and long-term synaptic facilitation in Aplysia. *Proc Natl Acad Sci USA* 101:7463–7468.

Si K, Giustetto M, Etkin A, Hsu R, Janisiewicz AM, Miniaci MC, Kim JH, Zhu H, Kandel ER (2003a) A neuronal isoform of CPEB regulates local protein synthesis and stabilizes synapse-specific long-term facilitation in Aplysia. *Cell* 115:893–904.

Si K, Lindquist S, Kandel ER (2003b) A neuronal isoform of the Aplysia CPEB has prion-like properties. *Cell* 115:879–891.

Trudeau LE, Castellucci VF (1995) Postsynaptic modifications in long-term facilitation in Aplysia: upregulation of excitatory amino acid receptors. *J Neurosci* 15:1275–1284.

Villareal G, Li Q, Cai D, Glanzman DL (2007) The role of rapid, local, postsynaptic protein synthesis in learning-related synaptic facilitation in Aplysia. *Curr Biol* 17:2073–2080.

Villareal G, Li Q, Cai D, Fink AE, Lim T, Bougie JK, Sossin WS, Glanzman DL (2009) Role of protein kinase C in the induction and maintenance of serotonin-dependent enhancement of the glutamate response in isolated siphon motor neurons of Aplysia californica. *J Neurosci* 29:5100–5107.

Wang DO, Kim SM, Zhao Y, Hwang H, Miura SK, Sossin WS, Martin KC (2009) Synapse- and stimulus-specific local translation during long-term neuronal plasticity. *Science* 324:1536–1540.

Zhu H, Wu F, Schacher S (1997) Site-specific and sensory neuron-dependent increases in postsynaptic glutamate sensitivity accompany serotonin-induced long-term facilitation at Aplysia sensorimotor synapses. *J Neurosci* 17:4976–4986.

Index

Note: Page numbers followed by "*f*" and "*t*" denote figures and tables, respectively.

Corrections/additions:
Stellate in index p. 512
Figure 1-2 p 9